95

There is currently a new "golden age" of dinosaur paleontology that has not been seen since the interval between the two World Wars. This revival stems from the hypothesis developed in the late 1960's and early 1970's that perhaps dinosaurs were active endotherms. In the last couple of decades the study of dinosaur eggs and babies, in particular, has proved to be one of the most exciting and profitable areas of dinosaur research.

Although it is over a hundred years since O.C. Marsh first recognized a baby dinosaur, that of the sauropod *Apatosaurus*, renewed interest stems from the discovery of a nesting ground in western Montana in 1978, and the realization that the study of baby dinosaur bones, eggs, and embryonic remains can tell us a great deal about dinosaur behavior, reproduction and physiology.

This is the first book solely devoted to this topic and reviews, in scientific detail, our present state of knowledge about this exciting area of paleontology. Chapters in the book discuss all aspects of the science, including the occurrence of eggs, nests and baby skeletons, descriptive osteology of juvenile skeletons, comparative histology of juvenile bone, analyses of eggs and egg shells, paleoenvironments of nesting sites, nesting behavior and developmental growth of baby dinosaurs.

The volume will be an invaluable addition to the book collections of vertebrate paleontologists and their graduate students.

Dinosaur Eggs and Babies

Dinosaur Eggs and Babies

Edited by

Kenneth Carpenter
Denver Museum of Natural History

Karl F. Hirsch
Denver Museum of Natural History

John R. Horner
Museum of the Rockies

CAMBRIDGE
UNIVERSITY PRESS

Published by the Press Syndicate of the University of Cambridge
The Pitt Building, Trumpington Street, Cambridge CB2 1RP
40 West 20th Street, New York, NY 10011-4211, USA
10 Stamford Road, Oakleigh, Melbourne 3166, Australia

First published 1994

Printed in the United States of America

Library of Congress Cataloging-in-Publication Data
Dinosaur eggs and babies / edited by Kenneth Carpenter, Karl F.
  Hirsch, John R. Horner.
        p.   cm.
    Includes index.
    ISBN 0-521-44342-3
    1. Dinosaurs – Infancy.   2. Dinosaurs – Eggs.   I. Carpenter,
Kenneth, 1949–   .   II. Hirsch, Karl F.   III. Horner, John R.
QE862.D5D4935   1994
567.9'1 – dc20   93-25263
                      CIP

A catalog record for this book is available from the British Library.

ISBN 0–521–44342–3 hardback

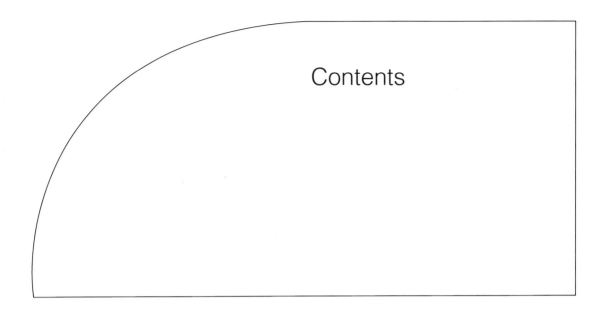

Contents

Contributors

Karen Alf
Department of Earth Sciences
Denver Museum of Natural History
2001 Colorado Blvd.
Denver, CO 80205
U.S.A.

Sunil Bajpai
Vertebrate Palaeontology Laboratory
Centre of Advanced Study in Geology
Panjab University, Chandigarh
India

A. Baltres
University of Bucharest
Faculty of Geology and Geography
Bvd. N. Balcescu Nr. 1
70111 Bucharest
Romania

Gérard Breton
Muséum d'Histoire Naturelle
Palais du Vieux-Marché
F-76600 Le Havre
France

Brooks B. Britt
Royal Tyrrell Museum of Palaeontology
Box 7500
Drumheller, Alberta TOJ OYO
Canada

Eric Buffetaut
URA 1433 du CNRS Laboratoire de Paléontologie des
        Vertebres
Université Paris VI
4 place Jussieu
75252 Paris Cedex 05
France

Olivier Buscail
Laboratoire de Paléontologie, URA-CNRS 327,
Institut des Sciences de l'Evolution,
Université de Montpellier II
Place Eugène Bataillon,
34095 Montpellier Cedex 05
France

Kenneth Carpenter
Department of Earth Sciences
Denver Museum of Natural History
2001 Colorado Blvd.
Denver, CO 80205
U.S.A.

Daniel Chure
Dinosaur National Monument
P.O. Box 128
Jensen, UT 84035
U.S.A.

Rémi Cousin
Muséum d'Histoire Naturelle
Palais du Vieux-Marché
F-76600 Le Havre
France

Philip J. Currie
Royal Tyrrell Museum of Palaeontology
Box 7500
Drumheller, Alberta TOJ OYO
Canada

Guillermo Faccio
Universidad de la Republica
Facultad de Agronomia
Catedra de Geologia
Garzon 780 Montevideo
Uruguay

Roger Fournier
Muséum d'Histoire Naturelle
Palais Longchamp
F-13004 Marseille
France

Dan Grigorescu
University of Bucharest
Faculty of Geology and Geography
Bvd. N. Balcescu Nr. 1
70111 Bucharest
Romania

Karl F. Hirsch
2950 South Perry
Denver, CO 80236
U.S.A.

John R. Horner
Museum of the Rockies
Montana State University
Bozeman, MT 59717
U.S.A.

Louis Jacobs
Department of Geological Sciences
Southern Methodist University
Dallas, TX 75275
U.S.A.

Sohn L. Jain
Geological Studies Unit
Indian Statistical Institute
Calcutta 700 035
India

Asit Jolly
Vertebrate Palaeontology Lab
Centre of Advanced Study in Geology
Panjab University, Chandigarh
India

James I. Kirkland
145 N. Main
Fruita, CO 81521
U.S.A.

Sergei Kurzanov
Paleontological Institute
Russian Academy of Sciences
Profsoyuznaya 123
Moscow 117321
Russia

Martin G. Lockley
Department of Geology
University of Colorado, Denver
1200 Larimer St.
Denver, CO 80217
U.S.A.

Jean Le Loeuff
URA 1433 du CNRS Laboratoire de Paléontologie des
    Vertebres
Université Paris VI
4 place Jussieu
75252 Paris Cedex 05
France

Pascale Mallan
Institut Géologique,
Palais de Rumine,
CH-1005 Lausanne
Switzerland

John M. Maurice
Department of Geological Sciences
Southern Methodist University
Dallas, TX 75275
U.S.A.

John McIntosh
278 Court St.
Middletown, CT 06457
U.S.A.

Konstantin E. Mikhailov
Paleontological Institute
Russian Academy of Sciences
Profsoyuznaya 123
Moscow 117321
Russia

Claudine Montgelard
Laboratoire de Paléontologie, URA-CNRS 327,
Institut des Sciences de l'Evolution,
Université de Montpellier II
Place Eugène Bataillon
34095 Montpellier Cedex 05
France

J. Moratalla
Unidad de Paleontología
Dto. Biología
Facultad de Ciencias
Universidad Autónoma
Cantoblanco
28049 Madrid
Spain

Phillip A. Murry
Department of Physical Sciences
Tarleton State University
Stephenville, TX 76402
U.S.A.

Bruce G. Naylor
Royal Tyrrell Museum of Palaeontology
Box 7500
Drumheller, Alberta TOJ OYO
Canada

David Norman
Sedgwick Museum
Cambridge University
Cambridge CB2 3EQ
England

Gregory S. Paul
3109½ North Calvert St.
Baltimore, MD 21218
U.S.A.

Fred Peterson
U.S. Geological Survey
Box 25046, MS-939
Denver, CO 80225
U.S.A.

J. E. Powell
Facultad de Ciencias Naturales
Universidad Nacional de Tucumán
Miguel Lillo 205
(4000) Tucumán
Argentina

M. Rusu
Faculty of Geology and Geography
University of Bucharest
Bvd. N. Balcescu Nr. 1
70111 Bucharest
Romania

Karol Sabath
Institute of Paleobiology
Polish Academy of Sciences
Al. Zwirki i Wigury 93
02–089 Warsaw
Poland

Ashok Sahni
Centre of Advanced Study in Geology
Panjab University
Chandigarh 160014
India

M. Seclamen
Faculty of Geology and Geography
University of Bucharest
Bvd. N. Balcescu Nr. 1
70111 Bucharest
Romania

Anil Sood
Department of Geology
University of Delhi
Delhi
India

S. Srinivasan
Vertebrate Palaeontology Laboratory
Centre of Advanced Study in Geology
Panjab University, Chandigarh
India

S. K. Tandon
Department of Geology
University of Delhi
Delhi
India

V. Teodorescu
Faculty of Geology and Geography
University of Bucharest
Bvd. N. Balcescu Nr. 1
70111 Bucharest
Romania

Christine Turner
U.S. Geological Survey
Box 25046, MS-939
Denver, CO 80225
U.S.A.

Monique Vianey-Liaud
Laboratoire de Paléontologie, URA-CNRS 327
Institut des Sciences de l'Evolution
Université de Montpellier II
Place Eugène Bataillon
34095 Montpellier Cedex 05
France

Jean-Pierre Watté
Muséum d'Histoire Naturelle
Palais du Vieux Marché
F-76600 Le Havre
France

David Weishampel
Department of Cell Biology and Anatomy
Johns Hopkins University
School of Medicine
Baltimore, MD 21205
U.S.A.

Dale A. Winkler
Department of Geological Sciences
Southern Methodist University
Dallas, TX 75275
U.S.A.

Zhao Zi-kui
Institute of Vertebrate Paleontology and
        Paleoanthropology
Academia Sinica
Beijing 100044
China

Foreword
A tribute to Robert Makela

JOHN R. HORNER

Vertebrate paleontology is a science with a diversity of professionals. There are the researchers, the collection managers, the preparators, the illustrators, the field crew chiefs, and the field crew. In some institutions there are many more. But it is surely a discipline where researchers cannot stand alone if they expect excellence in data collection and interpretation. If data collection is poor, the interpretation can be no better.

Data collection was the primary goal of Robert ("Bob") Makela. He was not so much interested in the interpretation, but instead wanted to be sure that if a question were later asked, the information could be acquired, even when the specimen was in a collection. Bob often made field jackets three or four times larger than they needed to be, just to be sure that the geological and taphonomic data were not lost. This was even after the maps with dips and strikes and geological samples had been made and collected. Bob was the ultimate Field Crew Chief, a teacher of science and paleontological method.

Bob's discoveries were vast; he planned and executed the excavation of Egg Mountain and discovered North America's first dinosaur embryo, numerous clutches of eggs, dinosaur babies, and nesting horizons. But it was the associated data collections that have made it possible to envision the environment and lives of these animals. Bob died in the field on June 26, 1987, a day after discovering the first pachycephalosaur from the Two Medicine Formation, and a few days after removing a clutch of hypacrosaur eggs with embryonic remains. A great deal of the field data incorporated in Chapters 8, 14, and 21 were derived from the notes and maps and specimens made and collected by Bob Makela.

Bob was born and raised in Great Falls, Montana, and throughout his childhood had an intense love for the outdoors and its inhabitants. Attending the University of Montana, he majored in biology and the teaching of science.

I first met Bob in a herpetology class at the university. He made his entrance into the class carrying a Gila monster in his arms. He pointed out that this highly poisonous animal would not bite if a person handled it gently. And Bob, even though a big, burly man who was once a Golden Gloves boxer, was ever so gentle with animals and people (at least those that he liked).

When Bob graduated from the university, he moved with his wife, Doris, and two sons, Jim and Jay, to the small town of Rudyard in north-central Montana. In Rudyard he took a job in the science department of the local high school, teaching earth science, chemistry, biology, paleontology, computer science, and several advanced courses. He took his earth science classes on field trips to study geology, and he set up a field meteorology lab. He took his paleontology classes into the field to collect, his advanced biology classes to Washington State to see and collect from the ocean, and had his computer classes make programs for other classes. He always seemed to most enjoy just teaching, as testified by the many teaching awards he won.

Bob and I first went into the field together in 1970, helping Dick Lund collect fish fossils from the Bear Gulch Limestone in central Montana. After a couple of years we decided that fish were nice, but that dinosaurs were better. In 1972 we began an intensive collection effort in the Judith River Formation north of Rudyard. Together with Larry French, another interested science teacher from a nearby town, we scoured the badlands. By 1973 we had found two partial skeletons, now known to be *Brachylophosaurus* and *Gryposaurus*. Bob built special packs for us to get field supplies, water, and beer into difficult sites. And at that time, none of us were quite sure how to extract information or specimens. None of us had ever seen or met a professional paleontologist other than Robert Fields, our geology advisor at the University of Montana. And Bob Fields was a mammal man, who, like so many others, simply picked up fossils already weathered out.

The only books that we had showed the old techniques used by E. D. Cope and Barnum Brown. So, Bob became the Field Crew Chief and reinvented or modernized the old paleontological field methods. Because we all had other jobs, we could only spend spring and fall weekends and partial summer weeks in the field. In 1975 I got a job at Princeton University where Don Baird allowed me to return to Montana in the summers to collect. So Bob and I began spending our entire summers searching the Judith River beds throughout Montana. In 1977 Bob and Larry discovered a rich microsite with mammals, north of Rudyard, which perked the interest of Bill Clemens at Berkeley. Bill brought out Mark Goodwin and a small field crew and began a mammal hunt. Bob thought mammals were all right, but that dinosaurs were better. So we continued our dinosaur hunt.

In 1978, with a lead from Bill Clemens, Bob and I drove to Bynum, Montana, where a woman named Marion Brandvold had found a dinosaur skeleton that she wanted identified. After explaining there wasn't enough of the skeleton to identify beyond hadrosaur, she showed us some small bones she had picked up west of Choteau. They were unquestionably the bones of baby hadrosaurs. A discovery made in a small rock shop in Bynum initiated the beginning of our paleontological careers. That year Bob and I extracted the first dinosaur nest with babies, and a skull of the hadrosaur we later named *Maiasaura peeblesorum*.

In 1979 I brought a crew of thirteen to Montana from New Jersey. Bob and I were the only part of the crew that had any experience in the field, so we had to improvise in every respect. Bob was officially the Crew Chief, and as such was in charge of creating a camp and making the decisions concerning fossil extraction. The camp had to be made grizzly bear-proof, as we were located in an area where twelve grizzlies had been seen that year. Bob took a 30 06 rifle to the field sites and always kept a sharp eye out while we dug. In 1979 Bob found two adult and two juvenile maiasaur skeletons, and was responsible for sending Fran Tannenbaum out to search for sites. In her search she discovered Egg Mountain, and the first intact dinosaur egg from North America.

From 1979 to 1984 the 1.5 sq-mile area called the Willow Creek Anticline was searched on hands and knees. Bob undertook the extraordinary excavation of Egg Mountain, which yielded fourteen clutches of *Orodromeus* eggs and three skeletons, several *Troodon* eggs (including one with embryonic remains) and skeletons of lizards and mammals.

In 1985 we left the Willow Creek Anticline and moved to other areas in the Two Medicine Formation. Bob always set up the camps, and took charge of all excavations. He created new mapping techniques and a variety of field gadgets intended to bring fossils from the field with ease. He created the ''Dino-Wheel,'' the ''Dino-Sled,'' and the ''Dino-Moon-Buggy.'' And he was always there for anyone that needed help or encouragement. Bob showed people of all ages how to work hard, have fun, and enjoy life.

This book is dedicated to Robert R. Makela, an extraordinary paleontologist with a vision of excellence in data collection and a discoverer of dinosaur eggs and babies.

Figure F.1. Robert Makela doing what he loved best, collecting dinosaurs.

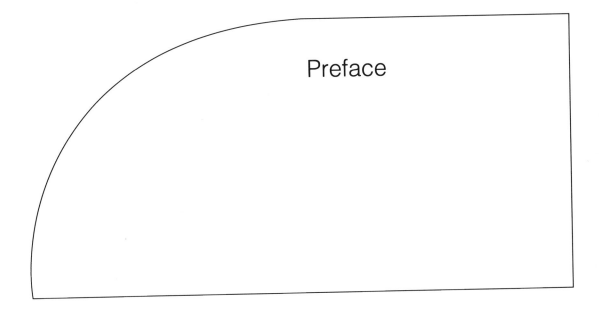

Preface

Bob Makela, to whom this book is dedicated, typifies the hard-working persons who gathered the specimens and data used by the various authors in this volume. Many of these support personnel are long dead, but their efforts have allowed us to advance the field of dinosaur paleontology one step further. This book is a tribute to all these hard working individuals, past and present. The editors also thank the authors for their contributions to this volume and hope that they will be pleased with the final result.

The editors would also like to extend their thanks to the following reviewers for the time and effort they put into the manuscripts, which greatly improved the quality of the papers: Walter Coombs, Philip Currie, Kyle Davis, Peter Dodson, Wann Langston, John McIntosh, William Sarjeant, Bryan Small, and David Weishampel.

Initial assistance for the project was provided by John Dolson and staff of the Amoco Production Co., Denver. The Denver Museum of Natural History and Museum of the Rockies provided financial and material support during the production of this volume.

Kenneth Carpenter, Karl F. Hirsch, and
John R. Horner
Denver, Colorado, 1993

Institutional abbreviations

AMNH — American Museum of Natural History, New York City, New York

BYUVP — Brigham Young University, Vertebrate Paleontology Collection, Provo, Utah

CMN — Canadian Museum of Nature, National Museums of Canada, Ottawa, Canada

CMNH — Carnegie Museum of Natural History, Pittsburgh, Pennsylvania

DMNH — Denver Museum of Natural History, Denver

DNM — Dinosaur National Monument, Jensen, Utah

GI SPS — Geological Institute, Section of Paleontology and Stratigraphy, Ulan Bator, Mongolia

IVPP — Institute of Vertebrate Paleontology and Paleanthropology, Beijing, China

LACM — Museum of Natural History of Los Angeles County, Los Angeles, California

MHNH — Muséum d'Histoire Náturelle, Le Havre, France

MOR — Museum of the Rockies, Bozeman, Montana

MPAB — Museo Paleontologico ''Alejandro Berro,'' Argentina

MWC — Museum of Western Colorado, Grand Junction, Colorado

OMNH — Oklahoma Museum of Natural History, Norman, Oklahoma

PIN — Paleontological Institute of the Russian Academy of Sciences, Moscow, Russia

RTMP — Royal Tyrrell Museum of Palaeontology, Drumheller, Canada

SMU — Southern Methodist University, Dallas, Texas

UCM — University of Colorado Museum, Boulder

USNM — U.S. National Museum (now National Museum of Natural History), Washington, D.C.

UU — University of Utah, Salt Lake City, Utah

UW — University of Wyoming, Laramie, Wyoming

YPM — Yale Peabody Museum, New Haven, Connecticut

YPM-PU — Yale Peabody Museum, Princeton University Collections, New Haven, Connecticut

ZPAL MgOv — Institute of Paleobiology of the Polish Academy of Sciences Oological Collection, Warsaw, Poland

Introduction

KENNETH CARPENTER,
KARL F. HIRSCH, AND
JOHN R. HORNER

We are currently enjoying a new Golden Age of Dinosaur Paleontology not seen since the interval between the two world wars. This revival stems from the hypothesis developed in the late 1960s and 1970s that perhaps dinosaurs were active endotherms. This suggestion came at a time when the Baby Boom generation was maturing and entering college. Many, including most of those contributing to this volume, were captivated by the idea that it was possible to determine something about dinosaur biology, physiology, and reproduction.

This volume, *Dinosaur Eggs and Babies*, was conceived while working on *Dinosaur Systematics: Approaches and Perspectives* (Carpenter & Currie, 1990). It was clear that a considerable amount of research had been done during the past two decades describing the microscopic structure of dinosaur eggs, structure of dinosaur egg clutches interpreted to be the remains of nests, and the skeletons of various very young hatchling dinosaurs. The editors of this volume felt that the time had come to organize and summarize this research before the sheer volume of material made this impossible. Hopefully, we could point the way for future research and possibly give it some cohesiveness.

Unlike other volumes in this series – *The Beginning of the Age of Dinosaurs* (Padian, 1986), *Dinosaur Tracks and Traces* (Gillette & Lockley, 1989) and *Dinosaur Systematics* (Carpenter & Currie, 1990) – this volume is not the result of a symposium. Instead, many of the contributors were solicited for specific overview chapters, while other chapters were voluntarily submitted for the volume.

The discovery of dinosaur eggs

Dinosaur eggs have often been considered novelties or museum curiosities not worthy of much scientific study. Consequently, they have been the subject of many misconceptions by the public and the film industry. Giant dinosaurs are assumed to have laid giant eggs; thus Ringo Starr in *Caveman* (1981) is shown staggering under the load of a dinosaur egg bigger than he is. In point of fact, the largest dinosaur eggs are a little smaller than a basketball or a soccer ball. Most eggs are considerably smaller, and vary in shape from spheres, to ovals, to tapering elongated bodies.

Another myth is that dinosaur eggs were unknown before the 1920s when the American Museum of Natural History announced their discovery in Mongolia. Actually, dinosaur eggs have been known for thousands of years, however, the discoverers did not make a media event out of it. Instead, they chose to shape the shell fragments, possibly for jewelry. At least that is the conclusion reached by archeologists accompanying the Central Asiatic Expeditions of the American Museum of Natural History (Andrews, 1932, p. 255). The shaped eggshells were found at a late paleolithic or early neolithic site (Perle, personal communication) near Shabarakh Usu (now called Bayn Dzak), in Mongolia.

In historical times, a dinosaur eggshell was first reported in 1859 from the egg-rich area of southern France (Introduction-Table 1). This discovery, discussed by Buffetaut and Le Loeff (Chapter 2), came not long after the naming of the Dinosauria by Owen in 1841. This little-known discovery of eggshell occurred about 64 years before the well-publicized discoveries in Mongolia by the American Museum of Natural History in 1923. Less known is that in 1859 a clutch (nest?) of fossilized eggs was discovered in the "Great Oolite" of England (Buckman, 1859). The eggs were small, about the size and shape of "*Gobiopteryx minuta*" eggs (see Mikhailov, Sabath, & Kurzanov in Chapter 7). These were the first fossil eggs to be named (*Oolithes bathonicae*, Buckman, 1859). The identity of the eggs remains uncertain, although they have been ascribed to the crocodiles by Buckman (1859) and to the pterosaurs by Carruthers (1871). Pending a restudy of the specimens, we cannot rule out the possibility that they are the eggs of a small dinosaur.

Since 1859, dinosaur eggs and eggshells, as well as baby dinosaurs, have been found on almost every continent, as documented by Carpenter and Alf in Chapter 1.

Studies of dinosaur eggshells

Not long after the discovery of dinosaur eggshells, Paul Gervais (1877) sought to identify the large eggshell fragments from the Provence region of France (Buffetaut & Le Loeff in Chapter 2). Gervais made the first detailed microstructural study of these eggs by comparing them with those of modern turtles, crocodiles, geckos, and birds. From this study, he concluded that the eggs from Provence were structurally most like those of turtles. However, because of the large size of the eggshells, he could not rule out the possibility that they were dinosaurian. Many years later, Sochava (1971) also concluded that the eggshells from France resembled those of turtles and assigned them to his testudoid group.

The technique of comparative microstructual analysis was continued by van Straelen and Denaeyer (1923) who used both normal and polarized light to study both radial and tangential sections of the Provence eggshells. van Straelen later conducted the first study of eggs collected by the American Museum of Natural History expeditions to Mongolia (van Straelen 1925, 1928). Originally, he recognized only a single egg type from Bayn Dzak and suggested that it was possibly the egg of *Protoceratops andrewsi*. A few years later, he concluded that a second egg type was present, based on shell thickness, surface ornamentation, and number of pore openings (van Straelen, 1928). This time he did not suggest who might be the egg-layer.

The use of polarized light microscopy (PLM) and normal light microscopy (LM) to study dinosaur eggshells continued with the studies of Lapparent (1947), Chow (1951, 1954), and Sochava (1969). A new technique, the scanning electron microscope (SEM), was introduced by Erben (1970) in his study of extant reptile and avian eggshells; also included were dinosaur eggshells from Provence, France. This technique, combined with PLM and LM, is becoming the standard for the analysis of fossilized eggshell (e.g., Hirsch & Quinn 1990).

Classification of dinosaur eggs

The early classification schemes for dinosaur eggshells apparently developed independently with little knowledge of other workers. Only recently, have attempts been made to utilize a single parataxonomical scheme as discussed in Chapters 7, 10, 11, and 12.

The earliest classification assigned names to the eggs on the basis of macrostructural characteristics. Size and shape were used by Buckman (1859) and Carruthers (1871). Their reasoning presupposed that such characters were as distinctive as the morphological features of bones used to separate dinosaur taxa. Shell thickness and surface ornamentation were also used by van Straelen (1928) for Mongolian eggs, but his conclusions were

Introduction Table 1. Chronological discovery of dinosaur eggs and babies

Date	Country	Material	Reference
1859	France	Eggshell	Buffetaut and Le Loewff; Chapter 2
1859	England	Nest?	Buckman, 1859
1883	U.S.A.	Baby	Marsh, 1883; Chapter 17
1908	Portugal	Egg	Lapparent and Zbyszewski, (1957)
1913	U.S.A.	Eggshell	Gilmore, field notes
1922	Mongolia	Egg	Andrews, 1932
1923	China	Eggs	Andrews, 1932
1923	Mongolia	Baby	Brown and Schlaikjer, 1940
1950	Tanzania	Eggs	Swinton, 1950
1951	Brazil	Eggs	Price, 1951
1954	Spain	Eggs	Lapparent and Aguirre, 1954
1961	Kazakstan	Eggshell	Bajanov, 1961
1964	Canada	Eggshell	Jepsen, 1964
1968	Peru	Eggshell	Sigé, 1968
1974	Argentina	Nest/baby	Bonaparte and Vince, 1974
1979	South Africa	Nest/baby	Kitching, 1979
1977	Uzbekistan	Eggs	Nesov and Kaznyshkin, 1986
1980	Uruguay	Eggs	Mones, 1980; Chapter 4
1982	India	Eggshell	Sahni and Gupta, 1982
1990	Romania	Nest	Chapter 6

contested by Brown and Schlaikjer (1940). Young (1954) used macrostructural characters to separate eggs from Laiyang, China, into two forms: a spherical egg (*Oolithes spheroides*) and an elongated egg (*O. elongatus*). Microscopic differences were not used in the original diagnosis, but appeared in a later paper (Chow, 1955). Length–diameter ratios were used by Young (1965) to separate *Oolites rugustus* into two subforms, one of which was shorter and broader than the other. The shortcoming of this macrostructural classification became apparent, however, with Chao and Chiang's (1974) discovery that *O. spheroides* could be further divided into five types, based on the thickness of the shell layer, thickness of the mammilary layer, and number of mammilae per square millimeter. As a result, they recognized five different species of *Oolithes* (most of which have since been recognized as distinct egg genera; see Chapter 12 by Zhao).

Since pore canals had long been noted as differing among eggs (e.g., van Straelen, 1925), Sochava (1969) used these differences as the foundation for his classification scheme. Sochava separated eggshells into angusticanalicuate, prolatocanaliculate, and multicanaliculate. Erben (1970) introduced a new pore canal type, tubocanaliculate, and Nesov and Kaznyshkin (1986) introduced foveocanaliculate and lagenocanaliculate.

Sochava (1971) later placed his prolatocanaliculate and angusticaniculate eggshells into a larger category he called the ornithoid group because of microstructural similarities with bird eggshells. The multicanaliculate eggshell (similar to the eggshells from Provence, France), were placed by Sochava into the testudoid group because of similarities with turtle eggshells. By doing so, Sochava formalized the distinction long noted that some dinosaur eggshells resembled those of birds, and others more closely resembled those of turtles. Dughi and Sirugue (1976) independently placed dinosaur eggs from southern France into a reptile and a bird group.

Other methods of classification that have been developed include designating the different egg types alphabetically (e.g., Jensen, 1966), numerically (e.g., Williams, Seymour, & Kérourio, 1984; Nesov & Kaznyshkin, 1986), and by shell-unit shape (Penner, 1985). The alphabetical and numerical systems of the various authors are usually not comparable; for example, one author's egg type A, is not necessarily the same as another author's type A (e.g., Jensen, 1966 versus Srivastava et al., 1986).

The most recent classifications (Sabath, 1991; Chapter 7 by Mikhailov, Sabath, & Kurzanov; Chapter 10 by Hirsch; Chapter 11 by Vianey-Liaud et al.; and Chapter 12 by Zhao) utilize the parataxonomic classification pioneered by Zhao (1975, 1979a). Zhao's classification scheme is now supplemented by the structural classification into basic groups, morphotypes, pore systems, and ornamentation (Mikhailov 1991, 1992). The

parataxonomical system, however, is not yet universally accepted. Sahni et al. (Chapter 13) prefer to use the identity of the egg-layer to identify the eggs from India, while Moratalla and Powell (Chapter 3) use the basic organizational groups of Hirsch and Quinn (1990).

Chemical analysis and organic material of dinosaur eggshell

The chemical analysis of dinosaur eggshell is a relatively new field of research, although the first study was conducted by van Straelen and Denaeyer (1923). They determined that the dinosaur eggshells from France were calcium carbonate ($CaCO_3$) and therefore little altered. Little additional analyses were conducted until sophisticated analytical equipment and techniques were developed in the 1960s. Voss-Foucart (1968) isolated two proteins by chromatography and verified their proteinlike macromolecular shape with SEM. She observed that the shell proteins were so similar to those of chicken eggs as to suggest a common ancestry.

The discovery of proteins in dinosaur eggshell is rather remarkable because they are not very stable chemically (Wyckoff, 1972). Usually, they degrade into amino acids which are more stable. But even these acids are soluble in water and so can be selectively removed from buried eggs by groundwater. Heat and pressure, such as found in the deep burial of sediments, can also affect amino acids. Nevertheless, Wyckoff (1972) was able to find various amino acids in a number of dinosaur eggshells. Kolesnikov and Sochava (1972) identified eighteen proteins from a dinosaur egg found in Mongolia, and noted a greater similarity between it and a chicken egg than between it and testudoid eggshell fragments from France. This suggests to us that amino acid studies of dinosaur eggshells might yet have some taxonomic value and perhaps might be an aid to evolutionary studies. Vianey-Liaud et al. discuss the results of their amino acid studies in Chapter 11.

Even more remarkable than the discovery of eggshell protein, is the report of organic material in eggshell. Van Straelen (1925) reported black organic material in the eggshell of *Protoceratops*. Chow (1954) identified what he interpreted to be the outer cuticle on some dinosaur eggs from Laiyang. This conclusion was challenged by Chao and Chiang (1974) who identified the material as calcite (see also Hirsch & Quinn, 1990). Vianey-Liaud et al. (in Chapter 11) suggest that the calcite skin, also seen on the dinosaur eggshells from France, is the result of the decomposition of the egg's soft tissue. Ammonia, a by-product of such decay, would raise the local pH and result in the deposition of $CaCO_3$ The presence or absence of a cuticle on the dinosaur eggs is more than a minor point of debate because the presence or absence of the cuticle says a lot about the microenvironment of the nest. It was the apparent absence of such tissue on *Protoceratops* eggs that led van Straelen (1925) to argue that these eggs were laid in an arid environment.

Chow (1954) also claimed that the albumin and yolk of one uncrushed egg had been replaced by calcite (see Dong & Milner 1988, p. 102 for a color photograph). A detailed analysis, however, is needed to substantiate this claim.

The cuticle, shell membrane (an organic structure that covers the inside of the eggshell), scleroprotein matrix, and organic material inside the pore canals were identified in a Mongolian egg (Kolesnikov & Sochava, 1972). The identification of the cuticle is suspect (see discussion above).

Oxygen and carbon isotopes of dinosaur eggshell from France are discussed by Vianey-Liaud et al. in Chapter 11. The first study of these isotopes was conducted by Folinsbee et al. (1970), who raised the possiblity that isotopes might provide clues about the environment in which the egg-layers lived. Erben, Hoefs, and Wedepohl (1979) and Zhao, Yan, and Ye (1983) carried this further, suggesting that carbon isotopes might provide dietary clues. Dauphin (1990a), however, was not able to find a correlation between diet and eggshell composition.

Changes in the oxygen isotopes of eggshells were used by Zhao (1991) to indicate a possible climatic change toward the end of the Cretaceous. This climatic change, he argues, may be the cause for the extinction of the dinosaurs. Previously, Erben et al. (1979) had suggested thinning of dinosaur eggshell as a cause for the extinction. Paladino et al. (1989), on the other hand, suggest that the predominance of one sex among dinosaur hatchlings might have caused the dinosaur extinction.

A major problem with the use of isotopes from dinosaur eggshells is diagenesis, as pointed out by Dauphin and Jaeger (1990) and Dauphin (1992). For this and other reasons, Sabath (1991) urged caution about making environmental and dietary interpretations based solely on isotopic studies.

Physiology of dinosaur eggs

Physiological studies of dinosaur eggshell are not yet very common. The amniotic egg is thought to have freed tetrapods from having to reproduce in water by protecting the developing embryo from desiccation in a subaerial environment. Sochava (1969) took this one step farther by suggesting that the calcareous dinosaur egg developed in response to aridity during the Cretaceous. Such a hypothesis, Sochava argued, would explain the apparent rarity of Pre-Cretaceous dinosaur eggs; these eggs had a pliable shell that would rarely fossilize. Sochava's hypothesis is still advocated by Zhao in Chapter 12.

The use of negative evidence (the absence of eggshell) by Sochava and Zhao is not without danger, as indicated by Hirsch's description in Chapter 10 of eggs and eggshell from the Upper Jurassic Morrison Formation. Carruthers (1871) and van Straelen (1928) have described what they thought were leathery-shelled eggs, but the taphonomic loss of a calcareous shell that left an organic residue cannot be discounted at this time. We do not rule out the possibility of leathery dinosaur eggshell, but demonstration of its exsistence is difficult (see Summary for one possible method).

The conductance of water vapor and respiratory gas exchange through the pore system of dinosaur eggshell has been examined by Seymour (1980). Grigorescu et al. attempt to estimate these values for eggs described in Chapter 6. The correlation between pores, surface ornamentation, egg shape, position of the eggs in the nest, and type of nest (vegetation mounds or soil) is discussed by Moratalla and Powell in Chapter 3, and by Mikhailov et al. in Chapter 7.

The presence of resorption pits in the mammillae was used for the first time by Schwarz et al. (1961) to show that an embryo had developed. Such pits form in the mammillae of modern eggs as a result of the embryo drawing calcium out of the egg for bone formation.

Dinosaur eggshell pathologies

Most dinosaur egg studies, such as those in this volume, were conducted upon healthy eggshell. Sometimes, however, various structural irregularities, like extra spherulitic structures, can be incorporated into the shell layer during shell formation. The most important pathologic condition is the multilayered eggshell, caused by unintentional retention of the egg (Zhao, Chapter 12). First reported by Dughi and Sirugue (1958), these multilayered shells provide important clues about the reproductive organs of the dinosaurs. In birds, the formation and shelling of an egg takes place in different regions of the oviduct in assembly-line fashion; thus only one egg is laid at a time (Taylor 1970).

Sometimes, however, the shelled egg in birds may be retained unintentionally rather than being layed because of stress, illness, or environmental conditions. This may cause a reverse peristalsis which sends the shelled egg back up the oviduct to where it meets the next egg (not yet shelled) being formed. These two eggs move together back down the oviduct and a shell is formed around both eggs, resulting in an egg within an egg, or "ovum in ovo" (Romanoff & Romanoff, 1949). The retention of an egg in the oviduct may cause illness or death of the female bird. The term "ovum in ovo" has been used incorrectly for multilayered eggshells (e.g., Erben et al., 1979). As yet, ovum in ovo has not been observed in the fossil record.

In noncrocodilian reptiles, the eggs are formed and shelled within a single region of the oviduct, and the entire clutch is laid simultaneously (Aitken & Solomon, 1979). Crocodiles, on the other hand, are somewhat like birds in that formation and shelling of the eggs occur in different parts of the oviduct, but the entire clutch is laid at the same time (Palmer & Guillette, 1992). Unintentional retention of the eggs can occur for a variety of reasons, including a change in diet (Ewert, Firth, & Nelson, 1984). This retention can result in another egg membrane and layer of shell being deposited around the eggs. From this, we may conclude that dinosaurs did not have a avian-type reproductive system, but a reptilian-type. Ewert et al. (1984) also concluded from their study of turtles that multilayered eggs can occur if two or more clutches are laid per season. In addition, once retained eggs are laid, those at the bottom of a clutch (first laid) are more apt to be multilayered than those at the top. See Hirsch et al. (1989) for additional discussions.

Multiple layered eggshell is rare but widespread among different eggshell species. It does, however seem to be more common in *Megaloolithus* (Erben et al., 1979). This condition is fatal to the developing embryo because the pore canals of the layers are rarely aligned. This effectively blocks gas exchange.

Abnormally thin eggshell has also been reported by Erben et al. (1979), but contested by Penner (1985) who argues that different egg taxa are present. Some thinning is clearly diagenetic as discussed by Penner (1985) and Dauphin (1990b). Other pathological conditions have been reported by Zhao (1991). These include abnormal proportional thickness of the cone layer and columnar layer (i.e., part of the shell is too thick), misshapen calcite crystals in the columnar layer, and irregular spaces within the shell (but one of us, Hirsch, has evidence that this condition could sometimes be diagenetic). These pathologies are either rare or have not been recognized in other eggshells.

Effects of diagenesis

The study of eggshells must be undertaken carefully with the recognition that diagenetic overprinting can often occur as noted by Dauphin (1990a, 1990b, 1992) and Dauphin and Jaeger (1990), as well as Hirsch in Chapter 10 and Vianey-Liaud et al. in Chapter 11.

Certain features, such as a herringbone pattern or localized mineral deposits in the shell unit, have been described as part of the shell structure. It is now known that these and other features can distort or obliterate the true diagnostic features. Recrystallization of the eggshell calcium carbonate is a relatively common problem that can distort the exact size and shape of the prisms. Also, partial dissolution of the $CaCO_3$ can affect eggshell structures, especially the mammilla on the inner surface and the ornamentation on the outer surface. This dissolution can occur at any time and is controlled by local pH.

Nests and nesting behavior

The first possible dinosaur nest to be described was the clutch of eight or more eggs of *Oolithes bathonicae*. Buckman (1859) speculated that the eggs were laid reptile fashion in sand (i.e., buried eggs). Since the 1859 discovery, dinosaur nests have been found worldwide (see Carpenter & Alf, Chapter 1). But whether the eggs were buried in the ground as Buckman (1859) implied or on the surface as Dughi and Sirugue (1958) concluded has been the subject of much debate as discussed by Cousin et al. in Chapter 5.

As a result of the discovery of the many dinosaur nests, several different nesting patterns have been recognized as detailed by Moratalla and Powell in Chapter 3. Their study expands the earlier review by Coombs (1989). Moratalla and Powell note that eggs may be arranged in spirals, concentric circles, irregular clusters, arcs, parallel rows, or double rows. A variant of the parallel row not mentioned by them is the alternating parallel row. Such an arrangement is proposed by Zhao (1979b) for *Youngoolithus xiaguanensis*.

Colonial nesting grounds have been described by Young (1965) for sites in Laiyang, China, where the nests were about 2 m apart. In Nanxiong Basin, the nests are about 7 or 8 m apart, and cluster into "sets" of five or six nests. Young suggested that a single gravid female was responsibile for each "set." Horner (1982) described a regular spacing of *Maiasaura* and *Orodromeus* nests at the Willow Creek Anticline. The harsh conditions within a dinosaur nest are examined by Paul in Chapter 18.

Eggs and stratigraphy

The geographic distribution of eggshells in China was noted by Young (1965) who used them for correlation between various depositional basins. Chao and Chiang (1974) plotted the stratigraphic distribution of *Oolithes* near Laiyang and used the results to show the evolution of the various species within the basin. Zhao (1979a) discussed the use of egg species in biostratigraphy and regional correlations. The use of eggs for regional correlation has been accepted by Sabath (1991) and Mikhailov et al. in Chapter 7.

Discovery of baby dinosaur bones

The best-known baby dinosaurs are those of *Maiasaura* popularized by Horner (1984) and Horner and Gorman (1985, 1988). The underdevelopment of the joints suggested that the babies did not venture out of the nest, but that the adult(s) brought food to them. Parental care in dinosaurs had been predicted a year earlier by Case (1978). The subject has been examined by Coombs (1989). Lambert (1991) argues that altricial behavior only occurs in endotherms, therefore dinosaurs could not have been ectotherms. Paul expands on this argument in Chapter 18.

Other baby dinosaurs from Montana besides *Maiasaura* include an embryonic *Orodromeus* described briefly by Horner and Weishampel (1988). A new species of *Hypacrosaurus* is described from specimens of numerous babies and an adult by Horner and Currie in Chapter 21.

The baby *Maiasaura* were not the first baby dinosaurs to be described. In 1883, Marsh briefly described a "foetal" sauropod associated with the holotype of *Morosaurus* (= *Camarasaurus*) *grandis*. The specimen is illustrated for the first time by Carpenter and McIntosh in Chapter 17. It and the other sauropod babies described in Chapter 17 are too large to be embryos. A possible embryonic *Camarasaurus* is described by Britt and Naylor in Chapter 16. At present, this may be the only embryonic sauropod known despite the thousands of "sauropod" eggs known from France, Spain, Argentina, and India (contrary to Mohabey, 1987; see Chapter 17).

The distribution of dinosaur babies, along with some reconstructed skeletons, is presented in Chapter 1 by Carpenter and Alf. They report that embryonic dinosaurs are very rare, occurring mostly in Mongolia and the Two Medicine Formation of Montana. The first Mongolian embryonic dinosaur to be found may be in a "*Protoceratops*" egg collected by the American Museum of Natural History expedition at Bayn Dzak. The specimen had been on exhibit for many years, until one of us (Horner) studied the specimen but was unable to verify whether the white material was bone. Considering the poor fossilization of many Djadohkta bones, we cannot dismiss the possibility that it is embryonic bone.

A possible embryonic *Psittacosaurus* has been described by Coombs (1980, 1982). The association of pelvic and hind-limb material with ?anterior dorsals, ?posterior cervicals, and scapula (distal ends overlapped) would be expected in an embryo. This conclusion is based on an unhatched *Orodromeus makelai* embryo that shows the skull tucked down between the bent knees (see Horner & Weishampel, 1988, Fig. 1A). Overlap of the distal ends of the scapulae in the *Psittacosaurus* is probably due to collapse of the rib cage (see Chapter 1, Fig. 1.13, for skeletal reconstruction). That this specimen is an embryo would also explain the disparity in size between baby and adult as noted by Coombs (1982). The apparent absence of eggshell in the specimens could be due to loss by erosion (many of the bones were apparently lying loose when found), by sacrifice during collection or preparation in order to get the bones, by failure of the eggshell to fossilize well (see below), or by loss of the shell through dissolution by groundwater as indicated by "ghosts" of eggs or egg "steinkerns" (e.g., Djadohkta Formation, see Chapter 7).

Numerous other baby dinosaurs have been collected in Mongolia by Soviet, Polish, and Mongolian expeditions. An embryonic foot of an unknown dinosaur was found adhering to a small piece of dendroolithid eggshell (Sochava, 1972; Sabath, 1991). A baby *Protoceratops* skeleton has been collected and mounted for display next to an adult (see Rozhdestvensky, 1973, p.

36; see Chapter 1, Fig. 1.13B). Unfortunately, the specimen has never been described. A partial baby *Protoceratops* from Bayn Dzak, a baby *Bagaceratops* from Khermeen Tsav, and a baby *Breviceratops* (= ?*Protoceratops*) from Khulsan have been described by Maryanska and Osmólska (1975). Barsbold and Perle (1983) described a group of baby hadrosaurs and a baby *Protoceratops* from Toogreek. The hadrosaurs were found evenly spaced with a thin white powder around each of them (Currie, personal communication), suggesting that the babies were actually embryos and that the eggshells did not fossilize well. Phillipe Taquet (personal communication) suggests that the embryos are those of *Saurolophus*, a hadrosaur known from the younger Nemegt Formation; further study is needed to verify this identification. One of the few baby theropods known are the ingeniyids associated with an adult skeleton as described by Barsbold and Perle (1983). Other baby dinosaurs discussed in this volume include *Dryosaurus* by Carpenter in Chapter 19, an embryonic *Camptosaurus* by Chure, Turner, and Peterson in Chapter 20, and a nodosaurid ankylosaur by Jacobs et al. in Chapter 22. A nest full of baby dinosaurs would be a tempting source of food for many carnivores. Kirkland explores this possibility with terrestrial crocodiles in Chapter 9.

Growth of baby dinosaurs

The ontogenetic changes that accompany growth were examined briefly by Brown and Schlaikjer (1940), Rozhdestvensky (1965), Maryanska and Osmolska (1975), and Dodson (1976). These studies are based upon Mongolian specimens, especially of protoceratopsians. The ontogentic growth of a new species of *Hypacrosaurus* is discussed by Horner and Currie in Chapter 21.

Rate of growth is a new area of research, although it has been the subject of previous speculation (e.g., Case, 1978; Dunham et al., 1989). Ongoing histological research by one of us (Horner) on an entire growth series of the new *Hypacrosaurus* species should permit the first comprehensive growth curve for a dinosaur. Preliminary curves for another hadrosaur, *Maiasaura*, has been presented by Morell (1987) and used by Lockley in Chapter 23 for his population studies of footprints.

The implications of ontogeny for understanding the evolution of the Dinosauria is discussed in Chapter 14 by Weishampel and Horner. Paul, in Chapter 15, concludes that dinosaurs were r-strategists, with rapid rates of growth.

Taphonomy of eggshell and baby bones

The taphonomy of eggshells is still in its infancy, and little of this is devoted exclusively to dinosaur eggshells. One of us (Carpenter, 1982) presented a brief discussion on the effect local pH has on the preservation

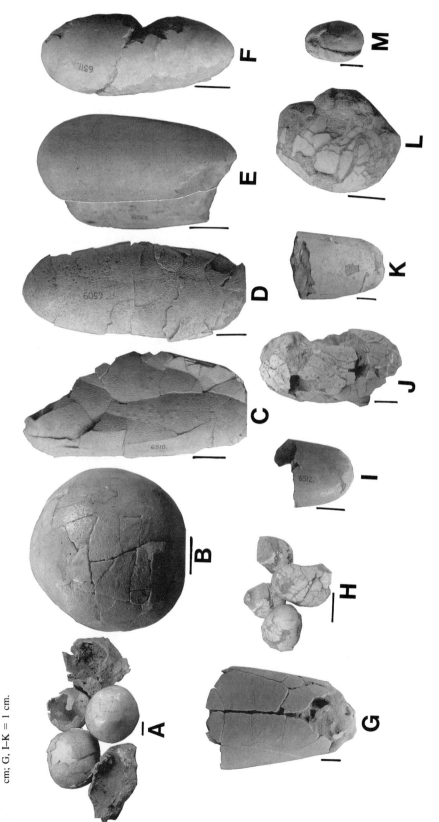

Introduction Figure 1. Some of the eggs recovered by the American Museum of Natural History expeditions to Shabarakh Usu (= Bayn Dzak), Mongolia, and Iren Dabasu (= Ermhot), China. **A.** Five or seven(?) eggs collected from Ermhot; three of the eggs have been partially or completely restored (AMNH field number 707). **B.** Enlargement of the restored egg in the front center of A (≈ 26.8 cm in circumference). A, B: *Paraspheroolithus irenensis* (Zhao, 1979a). **C.** *Elongatoolithus* sp. egg (AMNH 6510; cited in van Straelen, 1925). **D.** *Elongatoolithus* sp. egg (16.6 cm long, AMNH 6509). **E.** Two smooth eggs (most complete is 11.4 cm long, AMNH 6513). **F.** Smooth egg from clutch of five (11.3 cm long, AMNH 6511, cited in van Straelen, 1925). **G.** Hatched(?) smooth egg (AMNH 6505). **H.** Clutch of four "*Gobiopteryx minuta*" eggs (egg in foreground 4.1 cm tall, egg to left 4.3 cm long, egg in rear center 4.2 cm tall, AMNH 6642). **I.** Hatched(?) egg (AMNH 6512). **J.** *?Laevisoolithus sochavi* or large avian egg (AMNH 6507). **K.** *?Laevisoolithus sochavi* or large avian egg (egg in rear center 4.2 cm tall, partial egg to right >2.8 cm tall, AMNH 6657). **L.** Crushed *?Laevisoolithus sochavi* or large avian egg (14.4 cm long, 11.6 cm in circumference, AMNH 6652). **M.** Steinkern of a small egg (AMNH 6654) (see Sabath, 1991). Scale for A–F, H = 2 cm; G, I–K = 1 cm.

potential of dinosaur eggshell and baby bones. Tokaryk and Storer (1991) concluded from experimental evidence that eggshell can be transported a considerable distance with minimal abrasion.

In extant eggs, dissolution may actually occur in the nest due to microbial decay of nest vegetation and the production of acidic metabolites. Such dissolution may enhance gas exchange due to increased shell porosity, while at the same time making it easier for the hatchling to escape by weakening the shell. This type of dissolution produces very distinctive concentrically stepped erosion craters (Ferguson, 1981). As yet, such features have not been reported for dinosaur eggshells, although they may be predicted for those species having long incubation times.

With an unhatched egg, dissolution can begin shortly after the death of the egg or embryo. Decay of the organic material in the presence of oxygen produces carbon dioxide (CO_2) in large amounts making the calcium carbonate ($CaCO_3$) of the shell more soluble. If, however, access to air is restricted (e.g., burial of the nest by fluvial sediments), the process of decay is more complicated, and the effects on the eggshell is not always predictable (Krauskopf, 1979). Any CO_2 or hydrogen sulfide (H_2S) produced would lower the pH of the water in the vicinity of the egg, causing an increase in solubility. On the other hand, ammonia is often a byproduct of organic decay, and this would raise the local pH, resulting in the deposition of $CaCO_3$. In some instances, this ammonia may be the cause for the thin carbonate skin reported on eggshells (e.g., Vianey-Liaud et al, Chapter 11), mistakenly identified as the cuticle by Chow (1954).

Even if the egg hatches, shell fragments may still undergo dissolution before burial if the pH of the local environment is low (Carpenter, 1982; Hayward, Hirsch, & Robertson, 1991). After burial, dissolution can occur by hydrolysis involving groundwater (Krauskopf, 1979), or by carbonic acid (H_2CO_3) produced by CO_2 dissolved in rainwater. Carbonic acid is probably most responsible for dissolution of the fossilized eggshell that occurs after

weathering and erosion have returned the eggshell to near the ground surface.

The rate of dissolution is undoubtedly affected by how acidic conditions are, as well as by the thickness of the eggshell. In a similar vein, thin shells are more affected by diagenesis than thicker shells because of their greater surface to volume ratio.

The taphonomy of nesting grounds has only recently been examined. One of the first studies was made by Barsbold and Perle (1983) for sites at Toogreek, Mongolia. A more detailed study was presented by Winkler and Murry (1989) for a hypsilophodontid nesting ground in Texas. Hayward, Amlaner, and Young (1989) conducted a taphonomic study of an extant avian nesting colony buried by a volcanic ash fall, and their results should shed light on dinosaur nesting sites. In Chapter 8, Horner compares the taphonomy of an extant pelican nesting colony with those of two dinosaur nesting sites in Montana.

Egg collection by the Central Asiatic Expeditions

We feel it appropriate to close the Introduction by briefly discussing the historically important collection made by the Central Asiatic Expeditions of the American Museum of Natural History. Despite the importance of these eggs, they have been treated as curiosities, and only brief descriptions have appeared (e.g., van Straelen, 1925, 1928; Brown & Schlaikjer, 1940; Erben et al., 1979). We illustrate for the first time a representative sample of the eggs in Introduction-Figure 1. This sample includes eggs now referred to as *Paraspheroolithus irenensis* and to *Elongatoolithus* by Mikhailov et al. (Chapter 7) and Zhao (Chapter 12). Also shown are smooth protoceratopsian eggs, "*Gobiopteryx minuta*" eggs, possibly some *Laevisoolithus sochavi* eggs, and a single small steinkern of an egg Sabath (1991) has referred to Problematica.

The abundance of egg types from Bayn Dzak has resulted in confusion about which eggs are those of *Protoceratops*. Van Straelen unfortunately did not illustrate

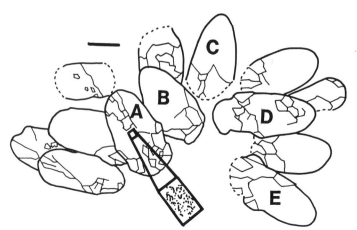

Introduction Figure 2. Nest of eggs (AMNH 6508 described by van Straelen (1925) as *?Protoceratops andrewsi*. Enlargement shows ramotubercular ornamentation in the equatorial portion. Most of the eggs are slightly crushed, thus exaggerating their width. Measurements for eggs: **A**. 14.9 cm long (partially telescoped), 7.7 cm wide; **B**. 14.7 cm long (partially telescoped), 6.7 cm wide; **C**. >14.3 cm long, 7.4 cm wide; **D**. 16.2 cm long, 6.7 cm wide (least crushed egg); **E**. 16 cm long, 7 cm wide. Dashed lines indicate portion still buried. Scale = 5 cm.

his second egg type nor did he adequately illustrate the eggs he considered as probably those of *Protoceratops*. To rectify the situation, we present an illustration (Introduction Fig. 2) of the nest from which van Straelen described and illustrated in thin section the egg he thought might be that of *Protoceratops andrewsi*. Photographs of the nest (AMNH 6508) in the field have appeared in Granger (1936, p. 24) and Brown and Schlaikjer (1940, Plate 12), after preparation (Colbert, 1961, Plate 93; 1983, p. 176), and as casts of the nest (Charig, 1979, p. 124; Norman, 1985, p. 133). Remarkably, Thulborn (1992) has correctly identified the nest of *Protoceratops* by the process of elimination.

The eggs of AMNH 6508 are paired, a condition Sabath (1991) thought was due to both oviducts participating in egg-laying. An enlargement of the equatorial portion of one egg shows ramotuberculate ornamentation (Mikhailov's 1991 terminology) composed of isolated hillocks and connected or joined hillocks. Such ornamentation is characteristic of Elongatoolithidae eggs (Chapter 7) and may require a new diagnosis of that egg family.

References

Aitken, R., & Solomon, S. 1976. Observations and the ultrastructure of the oviduct of the Costa Rican green turtle (*Chelonia mydas* L.). *Journal of Experimental Marine Biology and Ecology* 21: 75–90.

Andrews, R. 1932. *The New Conquest of Asia.* (New York: American Museum of Natural History).

Barsbold, R., & Perle, A. 1983. On the taphonomy of the joint burial of juvenile dinosaurs and some aspects of their ecology. *Transactions of the Soviet-Mongolian Paleontological Expedition* 24: 121–5.

Bonaparte, J. F., & Vince, M. 1979. El Hallazgo del primer nido de dinosaurios Triasicos (Saurischia, Prosauropoda), Triasico Superior de Patagonia, Argentina. *Ameghiniana* 16: 173–82.

Brown, B., & Schlaikjer, E. 1940. The structure and relationships of *Protoceratops*. *Annals of the New York Academy of Sciences* 40: 133–266.

Buckman, J. 1859. On some fossil reptilian eggs from the Great Oolite of Cirencester. *Quarter Journal of the Geological Society of London* 16: 107–10.

Carpenter, K. 1982. Baby dinosaurs from the Late Cretaceous Lance and Hell Creek formations and a description of a new species of theropod. *Contributions to Geology, University of Wyoming* 20: 123–34.

Carpenter, K., & Currie, P. 1990. *Dinosaur Systematics: Approaches and Perspectives.* (New York: Cambridge University Press).

Carruthers, W. 1871. On some supposed vegetable fossils. *Quarter Journal of the Geological Society of London* 27: 443–9.

Case, T. 1978. Speculations on the growth rate and reproduction of some dinosaurs. *Paleobiology* 4: 320–8.

Chao T.–K., & Chiang Y.–K. 1974. Microscopic studies on the dinosaurian egg–shells from Laiyang, Shantung Province. *Scientia Sinica* 17: 71–83.

Charig, A. 1979. *A New Look at the Dinosaurs.* (London: Heinemann Ltd.).

Chow, M. M. 1951. Notes on the Late Cretaceous dinosaurian remains and the fossil eggs from Laiyang Shantung. *Bulletin of the Geological Society of China* 31: 89–96.

1954. Fossil reptilian egg from Laiuang, Shantung, China. *Acta Palaeontologica Sinica* 2: 371–88.

1955. Additional notes on the microstructure of the supposed dinosaurian egg shells from Laiyang, Shantung. *Acta Scientia Sinica* 3: 523–6.

Colbert, E. 1961. *The Dinosaurs.* (New York: E. P. Dutton and Co.).

1983. *Dinosaurs: an Illustrated History.* (Maplewood: Red Dembner Enterprises Corp.).

Coombs, W. 1980. Juvenile ceratopsians from Mongolia – the smallest known dinosaur specimens. *Nature* 283: 380–1.

1982. Juvenile specimens of the ornithischian dinosaur *Psittacosaurus*. *Palaeontology* 25: 89–107.

1989. Modern analogs for dinosaur nesting and parental behavior. *In* J. Farlow, (ed.), *Paleobiology of the Dinosaurs. Geological Society of American Special Paper* 238: 21–53.

Dauphin, Y. 1990a. Microstructure et composition chimique des coquilles d'oeufs d'oiseaux et de reptiles. I. Oiseaux actuels. *Palaeontolographica* Abt. A 214: 1–12.

1990b. Incidence de l'état diagenetique des coquilles d'oeufs de dinosaures sur la reconnaissance des morphotypes – exemple du Bassin d'Aix en Provence. *Comptes Rendus de l'Académie des Sciences* 310: 849–54.

1992. Microstructure et composition chimique des coquilles d'oeufs d'oiseaux et de reptiles. II. Dinosauriens du Sud de la France. *Palaeontolographica* Abt. A 223: 1–17.

Dauphin, Y., & Jaeger, J.–J. 1990. Géochimie des oeufs de dinosaures du Bassin d'Aix en Provence (Cétacé): influence de la méthode d'analyse. *Neues Jahrbuch für Geologie und Palaeontologie Monatshefte* 8: 479–92.

Dodson, P. 1976. Quantitative aspects of relative growth and sexual dimorphism in *Protoceratops*. *Journal of Paleontology* 50: 929–40.

Dong Z., & Milner, A. 1988. *Dinosaurs from China.* (Beijing: China Ocean Press).

Dughi, R., & Sirugue, F. 1958. Observations sur les oeufs de dinosaures du bassin d'Aix–en–Provence: les oeufs à coquille bistratifées. *Comptes Rendus de l'Académie des Sciences* 246: 2271.

Dunham, A., Overall, K., Porter, W., & Forster, K. 1989. Implications of ecological energetics and biophysical and developmental constraints for life–history variation in dinosaurs. *In* J. Farlow, (ed.). *Paleobiology of the Dinosaurs. Geological Society of American Special Paper* 238: 1–19.

Erben, H. 1970. Ultrastrukturen und Mineralisation rezenter und fossiler Eischalen bei Vögeln und Reptilien. *Biomeralization Forschungsberichte* 1: 1–66.

1972. Ultrastrukturen und Dicke, der Wand pathologischer Eischalen. *Akademie der Wissenschaften und der Literatur* 6: 193–216.

Erben, H., Hoefs, J., & Wedepohl, K. 1979. Paleobiological and isotopic studies of eggshells from a declining dinosaur species. *Paleobiology* 5: 380–414.

Ewert, M., Firth, S., & Nelson, C. 1984. Normal and multiple eggshells in batagurine turtles and their implications for dinosaurs and other reptiles. *Canadian Journal of Zoology* 62: 1834–41.

Ferguson, M. 1981. Extrinsic microbial degradation of the alligator eggshell. *Science* 214: 1135–7.

Folinsbee, R., Fritz, P., Krouse, H., & Robblee, A. 1970. Carbon–13 and oxygen–18 in dinosaur, crocodile, and bird eggshells indicate environmental conditions. *Science* 168: 1353–6.

Gervais, P. 1877. Coquilles calcaires des oeufs. *Journal de Zoologies* 6: 88–96.

Gillette, D., & Lockley, M. 1989. *Dinosaur Tracks and Traces.* (New York: Cambridge University Press).

Granger, W. 1936. The story of the dinosaur eggs. *Natural History* 38: 21–5.

Hayward, J., Amlaner, C., & Young, K. 1989. Turning eggs to fossils: a natural experiment in taphonomy. *Journal of Vertebrate Paleontology* 9: 196–200.

Hayward, J., Hirsch, K., & Robertson, T. 1991. Rapid dissolution of avian eggshells buried by Mount St. Helens ash. *Palaios* 6: 174–8.

Hirsch, K., & Quinn, E. 1990. Eggs and eggshell fragments from the Upper Cretaceous Two Medicine Formation of Montana. *Journal of Vertebrate Paleontology* 10: 491–511.

Hirsch, K., Stadtman, K., Miller, W., & Madsen, J. 1989. Upper Jurassic dinosaur egg from Utah. *Science* 243: 1711–13.

Horner, J. 1982. Evidence of colonial nesting and 'site fidelity' among ornithischian dinosaurs. *Nature* 297: 675–6.

1984. The nesting behavior of dinosaurs. *Scientific American* 250: 130–7.

Horner, J., & Gorman, J. 1985. *Maia, a Dinosaur Grows Up.* (Bozeman: Museum of the Rockies).

1988. *Digging Dinosaurs.* (New York: Workman Press).

Horner, J., & Weishampel, D. 1988. A comparative embryological study of two ornithischian dinosaurs. *Nature* 332: 256–7.

Jensen, J. 1966. Dinosaur eggs from the Upper Cretaceous North Horn Formation of central Utah. *Brigham Young University Geological Studies* 13: 55–67.

Jepsen, G. 1964. Riddles of the terrible lizards. *American Scientist* 52: 227–46.

Kitching, J. 1979. Preliminary report on a clutch of six dinosaurian eggs from the Upper Triassic Elliot Formation, northern Orange Free State. *Palaeontologica Africana* 22: 41–5.

Kolesnikov, C., and Sochava, A. 1972. A paleobiochemical study of Cretaceous dinosaur eggshell from the Gobi. *Paleontological Journal* 7: 235–45.

Krauskopf, K. 1979. *Introduction to Geochemistry.* (New York: McGraw-Hill Book Co.).

Lambert, W. D. 1991. Altriciality and its implications for dinosaur thermoenergetic physiology. *Neues Jahrbuch für Geologie und Palaeontologie* 182: 73–84.

Lapparent, A. de 1947. Les dinosauriens du Crétacé Supérior de Midi de la France. *Mémoires de la Société Géologique de France* 56: 1-54.

Lapparent, A. de, & Aquirre, E. 1956. Algunes yacimientos de dinosaurios en el Cretécio superior de la Cuenca de Tremp. *Estudio Geologie* 12: 31–2.

Lapparent, A. de, & Zbyszewski, G. 1957. Les Dinosauriens du Portugal. *Mémoires du Service géologique de Portugal* 2: 1–63

Marsh, O. 1883. Principle characters of American Jurassic dinosaurs. Part VI. Restoration of *Brontosaurus. American Journal of Science* 26: 81–5.

Maryanska, T., & Osmolska, H. 1975. Protoceratopsidae (Dinosauria) of Asia. *Palaeontologica Polonica* 33: 133–81.

Mikhailov, K. 1991. Classification of fossil eggshells of amniotic vertebrates. *Acta Palaeontologica Polonica* 36: 193–238.

1992. The microstructure of avian and dinosaurian eggshell: phylogenetic implications. *In* K. Campbell (ed.) *Papers in Avian Physiology Honoring Pierce Brodkorb.* (Los Angeles: National History Museum of Los Angeles County), pp. 361–73.

Mohabey, D. 1987. Juvenile sauropod dinosaur from Upper Cretaceous Lameta Formation of Panchmahals District, India. *Journal Geological Society of India* 30: 210–16.

Mones, A. 1980. Nuevos elementos de la paleoherpetofauna del Uruguay (crocodilia y dinosauria). *Actas II Congresso Argentina Paleontologia et Biostratigrafia* 1: 265–77.

Morell, V. 1987. Announcing the birth of a heresy. *Discover* 8: 26–51.

Nesov, L. A. & Kaznyshkin, M. N. 1986. Discovery of a site in the USSR with remains of eggs of Early and Late Cretaceous dinosaurs. *Biological Sciences, Zoology* 9: 35–49.

Norman, D. 1985. *The Illustrated Encyclopedia of Dinosaurs.* (New York: Crescent Books).

Owen, R. 1841. Report on British fossil reptiles, Pt. II. *Report of the British Association for the Advancement of Science* 11:60–204.

Padian, K. 1986. *The Beginning of the Age of Dinosaurs.* (New York: Cambridge University Press).

Paladino, F., Dodson, P., Hammond, J., & Spotila, J. 1989. Temperature–dependent sex determination in dinosaurs? Implications for population dynamics and extinction. *In* J. Farlow, (ed.). *Paleobiology of the Dinosaurs. Geological Society of America Special Paper* 238: 63–70.

Palmer, B., & Guillette, L. 1992. Alligators provide evidence for the evolution of an Archosaurian mode of oviparity. *Biology of Reproduction* 46: 39–47.

Penner, M. 1985. The problem of dinosaur extinction. Contribution to the study of terminal Cretaceous eggshells from southeastern France. *Geobios* 18: 665–9.

Price, L. 1951. Un ovo de dinosaurio na formacao Bauru do Cretacio de Estado de Minas Gerais. *Departamento Nacional da Producao Mineral. Divisao de Geologia e Mineria. Notas Preliminares e Estudos* 53: 1–9.

Romanoff, A., & Romanoff, A. 1949. *The Avian Egg.* (New York; John Wiley and Sons).

Rozhdestvensky, A. 1965. Growth changes in Asian dinosaurs and some problems of their taxonomy. *Paleontologicheskii Zhurnal* 3: 95–109.

1973. *Animal Kingdom in Ancient Asia*. (Tokyo).

Sabath, K. 1991. Upper Cretaceous amniotic eggs from the Gobi Desert. *Acta Palaeontologica Polonica* 36(2): 151–92.

Sahni, A., & Gupta, V. 1982. Cretaceous egg shell fragments from Lameta Formation, Jabalapur, India. *Bulletin of the Indian Geological Association* 15: 85–8.

Schwarz, L., Fehse, F., Mueller, G., Anderson, F., & Sieck, F. 1961. Untersuchungen an Dinosaurier–Eischalen von Aix en Provence und der Mongolei (Sabarakh–Usu) *Zeitschrift fuer Wissenschaftliche Zoologie* 165: 344–79.

Seymour, R. 1980. Dinosaur eggs: the relationships between gas conductance throught the shell, water loss during incubation and clutch size. *Mémoires de la Société Géologique de France*. N. S. 139: 177–84.

Sigé, B. 1968. Dents de micromammifères et fragments decoquilles d'oeufs de dinosauriens dans la faune de Vertébrés du Crétacé supérieur de Laguna Umayo (Andes péruviennes). *Comptes Rendus de l'Académie des Sciences* (Sér. D) 267: 1495–8.

Sochava, A. 1969. Dinosaur eggs from the Upper Cretaceous of the Gobi Desert. *Paleontological Journal* 4: 517–27.

1971. Two types of eggshell in Senonian dinosaurs. *Paleontological Journal* 5: 353–61.

1972. The skeleton of an embryo in a dinosaur egg. *Paleontological Journal* 6: 527–31.

Srivastava, S., Mohabey, D., Sahni, A., & Pant, S. 1986. Upper Cretaceous dinosaur egg clutches from Kheda District (Gujarat, India). *Palaeontographica* Abt. A. 193: 219–33.

Straelen, V. van 1925. The microstructure of the dinosaurian egg–shells from the Cretaceous beds of Mongolia. *American Museum Novitates* 173: 1–4.

1928. Les oeufs de Reptile fossiles. *Palaeobiologica* 1: 295–312.

Straelen, V. van, & Denaeyer, E. 1923. Sur des oeufs fossiles du Crétacé supérieur de Rognac en Provence. *Bulletin Classe des Sociéte de L'Académie Royal Bde Belgique* 11: 14–26.

Swinton, W. E. 1950. Fossil eggs from Tanganyika. *The Illustrated London News* 217: 1082–83.

Taylor, T. 1970. How an eggshell is made. *In* Vertebrate Structures and Functions. (San Francisco: W.H. Freeman), pp. 371–7.

Thulborn, R. 1992. Nest of the dinosaur *Protoceratops*. *Lethaia* 25: 145–9.

Tokaryk, T., & Storer, J. 1991. Dinosaur eggshell fragments from Saskatchewan, and evaluation of potential distance of eggshell trasnport. *Journal of Vertebrate Paleontology* 11, Supplement to No. 3: 58A.

Voss–Foucart, M. F. 1968. Paleoproteines des coquilles fossiles d'oeufs de Dinosauriens de Créracé supéerior de Provence. *Comparative Biochemistry and Physiology* 24: 31–6.

Williams, D. L., Seymour, R. S., & Kérourio, P. 1984. Structure of fossil dinosaur egg shell from the Aix Bassin, France. *Palaeogeography, Palaeoclimatology, Palaeoecology* 45: 23–37.

Winkler, D. A., & Murry, P. A. 1989. Paleoecology and hypsilophodontid behavior at the Protor Lake dinosaur locality (Early Cretaceous), Texas. *In* J. O. Farlow, (ed.), *Paleobiology of the Dinosaurs. Geological Society of America, Special Paper*, 238: 55–61

Wyckoff, R. 1972. *The Biochemistry of Animal Fossils*. (Bristol: Scientechnica).

Young, C. C. 1954. Fossil reptilian eggs from Laiyang. *Acta Palaeontologica Sinica* 2: 371–88.

1965. Fossil eggs from Nanhsiung, Kwangtung and Kanchou, Kiangsi. *Vertebrata PalAsiatica* 9: 141–70.

Zhao Z. 1975. Microstructure of the dinosaurian eggshell of Nanhsiung, Kwangtung and Guangdong. *Vertebrata PalAsiatica* 13: 105–17.

1979a. Progress in the research of dinosaur eggs. *Mesozoic and Cenozoic Redbeds of South China* (Beijing: Science Press), pp. 330–40.

1979b. Discovery of the dinosaurian eggs and footprint from Neixiang County, Henan Province. *Vertebrata PalAsiatica* 17: 304–9.

1991. Extinction of the dinosaurs across the Cretaceous–Tertiary boundary in Nanxiong Basin, Guangdong Province. *Vertebrata PalAsiatica* 29: 1–20.

Zhao Z., Yan Z., & Ye L. 1983. Stable isotopic composition of oxygen and carbon in the dinosaur eggshells from Laiyang, Shandong Province. *Vertebrata PalAsiatica* 21: 204–9.

Distribution and history of collecting

1 Global distribution of dinosaur eggs, nests, and babies

KENNETH CARPENTER AND
KAREN ALF

Abstract

The geological and global distribution of dinosaur eggs, shell fragments, nests and baby bones, teeth and footprints is presented. The greatest number of localities are situated in Mongolia, China, France, India, and the United States. Most of these sites occur in Upper Cretaceous strata. Several skeletal reconstructions of baby and embryonic dinosaurs are also presented.

Introduction

Since their discovery in 1859, dinosaur eggs and eggshells have been found worldwide. The richest localities, however, occur in Aix-en-Provence in France, Laiyang and Nanxiong Counties in China, central India, southern Mongolia, and north central Montana in the United States. All of these sites occur in Upper Cretaceous strata, and most of these sediments were deposited in arid and semiarid environments. Elsewhere, one of us (Carpenter, 1982) suggested that this environmental distribution might reflect taphonomic bias.

Pre-Cretaceous eggshell is rare, leading Zhao (Chapter 12) to suggest that the calcareous egg was not widespread among the Dinosauria at this time. However, recent work in the Upper Jurassic Morrison Formation of the Western Interior suggests that this apparent near absence of pre-Cretaceous eggshell is due to the material not being recognized in the field.

Baby bones remain very rare globally and occur with appreciable frequency only in the Upper Cretaceous Two Medicine Formation of Montana (e.g., Horner & Makela 1979; Horner, 1982; Horner & Weishampel, 1988; Hirsch & Quinn, 1990) and in the Djadokhta Formation of Mongolia (least described). The oldest embryonic dinosaur bones co-occur with the oldest dinosaur eggshell in the Upper Triassic of Argentina. Similar associations of baby bone and eggshell occur in the Upper Jurassic Morrison Formation of Colorado, the Two Medicine Formation of Montana (U.S.A.) and Alberta (Canada), and the Djadohkta Formation of Mongolia. As yet, no bones have been reported associated with the hundreds of eggs from the Upper Cretaceous of France, China, and India. The fragmentary baby sauropod skeleton from India reported by Mohabey (1987) has been shown by Jain (1989) to be that of a booid snake, an identification accepted by Carpenter and McIntosh (Chapter 17).

Localities

The localition at which dinosaur eggs, eggshells, nests, baby remains, and baby footprints have been found are shown on maps (Figs. 1.1, 1.4, 1.7–1.12). The sites are presented by region from oldest to youngest.

Alaska and northern portion of North America (Fig. 1.1)

1. Sheep Creek, Wyoming. Morrison Formation, Kimmeridgian; *Apatosaurus excelsus* baby bones (Carpenter & McIntosh, Chapter 17).
2. Como Bluffs, Wyoming; Morrison Formation, Kimmeridgian; *Camarasaurus grandis* baby bones (Carpenter & McIntosh, Chapter 17).
3. Bridger, Montana; Cloverly Formation, Aptian–Albian; eggshells (Horner, personal communication).
4. Cashen Ranch, Montana; Cloverly Formation, Aptian–Albian; *Tenontosaurus tilleti* baby skeletons (Fig. 1.2A) (Forster 1990).
5. Wayan, Idaho; Wayan Formation, Albian; eggshells. (Dorr, 1985).
6. Thomas Fork Creek, Wyoming; Thomas Fork Formation, Albian; Ornithopod(?) eggshells (Dorr, 1985).
7. Dinosaur Provincial Park, Alberta; Judith River Formation, Campanian; eggshells and hadrosaur baby bones (Sternberg, 1955; Brinkman, 1986).

8. Crowsnest Pass, Alberta; Belly River Formation, Campanian; hadrosaur baby bones (Currie, personal communication).

9. Orion, Alberta; Judith River Formation, Campanian; hadrosaur baby bones and nest (Currie, personal communication).

10. Milk River, Alberta; Oldman Formation, Campanian; eggshells (Currie, personal communication).

11. Devil's Coulee, Alberta; Two Medicine Formation, Campanian; *Hypacrosaurus stebingeri* nest and baby skeletons (Fig. 1.3A); Horner and Currie, Chapter 21).

12. Salmond Ranch, Montana; Horsethief Formation, Campanian; eggshells (Bibler & Schmitt, 1986).

13. Fresno, Montana; Judith River Formation, Campanian; hadrosaur nest and baby bones (Carpenter, unpublished notes).

14. Judith River, Montana; Judith River Formation, Campanian; eggshells (Sahni, 1972).

15. Shawmut, Montana; Judith River Formation, Campanian; *Stegoceras* baby bones (Goodwin, personal communication).

16. Careless Creek, Montana; Judith River Formation, Campanian; Lambeosauridae baby bones (Fiorillo, 1987).

17. Landslide Butte, Montana; Two Medicine Formation, Campanian; hadrosaur nests and baby bones (Carpenter, unpublished notes).

18. Two Medicine River, Montana; Two Medicine Formation, Campanian; eggshells, hadrosaur nest, and baby bones (Carpenter, unpublished notes).

19. Blacktail Creek, Montana; Two Medicine Formation, Campanian; *Hypacrosaurus stebingeri* nest and baby bones, hadrosaur nests and baby bones (Horner & Currie, Chapter 21).

20. Dupuyer Creek, Montana; Two Medicine Formation, Campanian; Hadrosaur nest and baby bones, Lambeosauridae nest and baby bones, and Hypsilophodontidae baby bones (Carpenter, unpublished notes).

21. Choteau, Montana; Two Medicine Formation, Campanian; *Maiasaura peeblesorum* nests and baby skeletons (Fig. 1.3B), *Orodromeus makelai* nests and baby skeletons, and ornithoid and *?Troodon* sp. eggs. (Hirsch & Quinn, 1990).

22. Red Rock Locality, Montana; Two Medicine Formation, Campanian; *?Troodon* sp., *Maiasaura peeblesorum*, and *Orodromeus makelai* eggshells (Hirsch & Quinn, 1990).

23. Ocean Point, Alaska; Prince Creek Formation, Maastrichtian; *Edmontosaurus* sp. bones (Nelms, personal communication).

24. Cardston, Alberta; Willow Creek Formation, Maastrichtian; eggshells (Currie, personal communication).

25. Lethridge, Alberta; Saint Mary River Formation, Maastrichtian; eggshells (Currie, personal communication).

Figure 1.1. Dinosaur eggshell, nest, and baby bone locality map for Alaska (inset), Alberta, Saskatchewan, Montana, Idaho, Wyoming, and South Dakota.

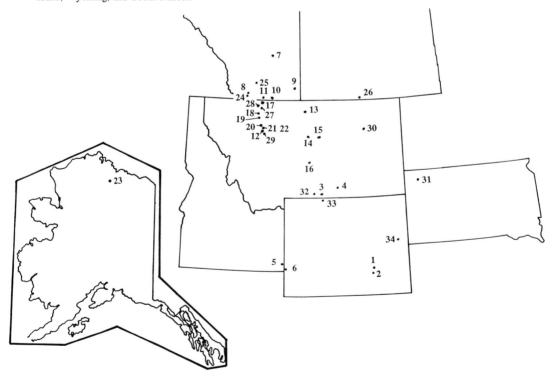

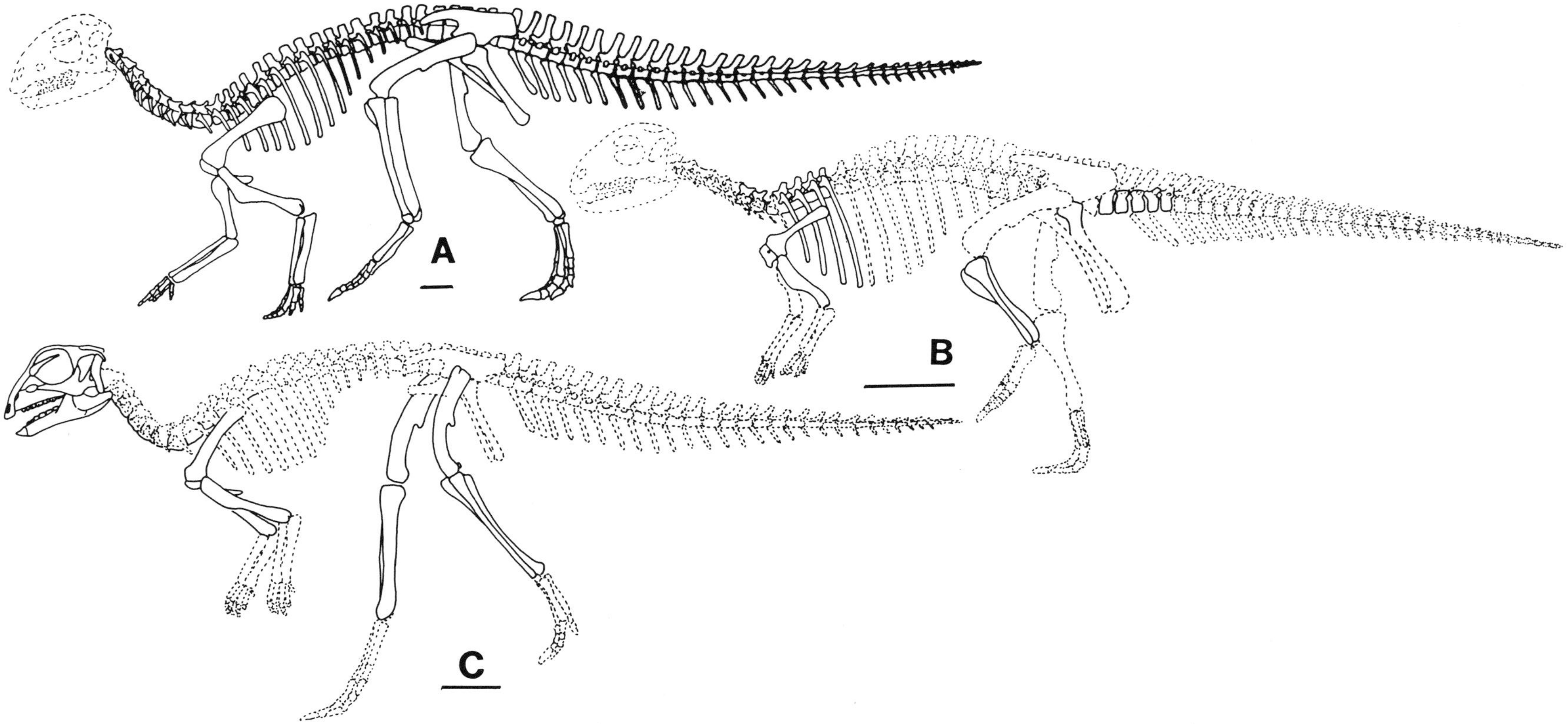

Figure 1.2. Reconstruction of baby ornithopod skeletons. A. *Tenontosaurus tilletti* (based on OU 10144). B. *Camptosaurus* sp. (based on DMN 15624). C. *Dryosaurus altus* (based on CMNH 11340). Scale for A, B = 4 cm; C = 2 cm.

26. Killdeer Badlands, Saskatchewan. Frenchman Formation, Maastrichtian; Ceratopsidae baby bone (Tokaryk, 1990).

27. Cut Bank, Montana; Saint Mary River Formation, Maastrichtian; nests (Weishampel, personal communication).

28. Browning, Montana; Saint Mary River Formation, Maastrichtian; hadrosaur baby bones and nests (Weishampel, personal communication).

29. Augusta, Montana; Saint Mary River Formation, Maastrichtian; nests (Weishampel, personal communication).

30. Fort Peck Reservoir, Montana; Hell Creek Formation, Maastrichtian; baby Dromaeosauridae teeth, baby *Aublysodon mirandus* teeth, baby *Tyrannosaurus* sp. tooth, baby Theropoda teeth, hadrosaur baby bones, baby ?Ankylosauridae teeth, and Ceratopsidae baby bone (Brown & Schlaikjer 1940; Carpenter, 1982 & unpublished notes).

31. Corson County, South Dakota; Hell Creek Formation, Maastrichtian; hadrosaur bones (Carpenter, unpublished notes).

32. Red Lodge, Montana; "Lance" Formation, Maastrichtian; eggshells (Jepsen, 1931).

33. Polecat Bench, Wyoming; "Lance" Formation, Maastrichtian; eggshells (Carpenter, 1982).

34. Lance Creek, Wyoming; Lance Formation, Maastrichtian; baby *Troodon* sp. jaw and teeth, baby Dromaeosauridae teeth, baby Theropoda teeth, baby *Aublysodon mirandus* teeth, baby *Thescelosaurus* sp. teeth, baby hadrosaur teeth, baby Ceratopsidae teeth, baby Ankylosauridae teeth. (Carpenter, 1982).

Southern portion of North America and northern Mexico (Fig. 1.4)

35. Ghost Ranch, New Mexico; Chinle Formation, Norian; *Coelophysis bauri* baby skeleton (Fig. 1.5A; Colbert 1989; Carpenter, unpublished notes).

36. Dinosaur National Monument, Utah; Brushy Basin Member, Morrison Formation, Kimmeridgian; *Dryosaurus altus* baby skeleton (Fig. 1.2C), *Camptosaurus* sp. baby skeleton (Fig. 1.2B), *Stegosaurus* sp. baby skeleton (Fig. 1.6A) (Carpenter, Chapter 19; Chure et al., Chapter 20; Galton, 1982).

37. Cleveland-Lloyd Quarry, Utah; Brushy Basin Member, Morrison Formation, Kimmeridgian; *Prismatoolithus coloradensis* egg (Hirsch et al., 1989; Hirsch, Chapter 10).

38. Fruita, Colorado; Morrison Formation, Kimmeridgian; *Prismatoolithus coloradensis* eggshells (Hirsch, Chapter 10).

39. Young Locality, Colorado; Salt Wash Member, Morrison Formation, Kimmeridgian; *Prismatoolithus coloradensis* egg, eggshells and nest (Hirsch, Chapter 10).

40. Dry Mesa, Colorado; Morrison Formation, Kimmeridgian; *Camarasaurus* sp. baby bone (Britt & Brinkman, Chapter 16).

41. Uravan Locality, Colorado; Brushy Basin Member, Morrison Formation, Kimmeridgian; *Prismatoolithus coloradensis* eggshells and *Dryosaurus altus* baby bones (Hirsch, Chapter 10; Scheetz, 1991).

42. Garden Park, Colorado; Morrison Formation, Kimmeridgian; *Prismatoolithus coloradensis* eggshells (Hirsch, Chapter 10).

Figure 1.3. Reconstruction of hadrosaur skeletons. **A.** *Hypacrosaurus stebingeri* (based on RTMP 89.79.52). **B.** *Maiasaura peeblesorum* embryo (based on MOR specimens). **C.** hadrosaur indet. (based on GI SPS 100/705). Scale = 2 cm.

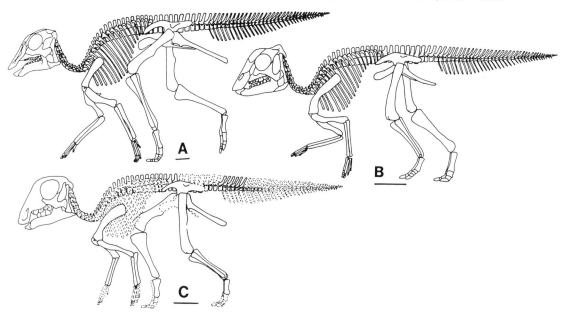

43. Kenton, Oklahoma; Morrison Formation, Kimmeridgian; *Apatosaurus* sp. baby bones (Fig. 1.5C) and *Camarasaurus* sp. baby bones (Carpenter & McIntosh, Chapter 17).
44. Coalville, Utah; Kelvin Formation, Berriassan–Albian; eggshells (Jensen, 1970).
45. Moore, Utah; Cedar Mountain Formation, Aptian–Albian; eggshells (Cifelli, personal communication).
46. Cedar Mountain, Utah; Cedar Mountain Formation, Aptian–Albian; eggshells (Cifelli, personal communication).
47. Castle Dale, Utah; Cedar Mountain Formation, Albian; Spheroolithidae eggshells (Jensen, 1970).
48. Proctor Lake, Texas; Twin Mountains Formation, Aptian; Hypsilophodontidae baby bones (Winkler & Murry, 1989).
49. Fort Worth, Texas; Paw Paw Formation, Albian; Nodosauridae baby bones (Fig. 1.6B; Jacobs et al., Chapter 22).
50. Wasatch Plateau, Utah; North Horn Formation, Maastrichtian; eggshells (Jensen, 1966).
51. Price, Utah; Blackhawk Formation, Campanian; Hadrosaur baby footprints (Carpenter, 1992).
52. Kaiparowitz Plateau, Utah; Wahweap and Kaiparowitz formations, Campanian; eggshells (Carpenter, unpublished notes).
53. San Juan Basin, New Mexico; Fruitland Formation, Campanian; eggshells (Wolberg & Bellis, 1989).
54. Harell Station, Alabama; Mooreville Chalk, Campanian; Dinosaur(?) egg (Dobie, 1978).
55. El Rosario, Mexico; El Gallo Formation, Campanian; eggshells (Hernandez, personal communication).

South America (Fig. 1.7)
56. Laguna La Colorada, Argentina; El Tranquilo Formation, Norian; *Mussaurus patagonicus* skeletons (Fig. 1.5B; Bonaparte & Vince, 1979).
57. Soriano, Uruguay; Mones-Ascencio Formation, Senonian; *Sphaeroolithus* (= *Sphaerouvum*) er-

beni and *Tacuaremboolithus* (= *Tacuarembouvum*) *oblongum* eggshells and nests(?) (Faccio, Chapter 4).
58. Algorta, Uruguay; Mones-Ascencio Formation, Senonian; eggshells (Faccio, Chapter 4).
59. Casa de Piedra, Argentina; Angostura Colorada Formation, Senonian; eggshells (Powell, 1987).
60. Pongo de Rentema, Peru; Bagua Formation, Campanian–Maastrichtian; eggshells (Mourier et al., 1988).
61. El Pintor, Peru; Bagua Formation, Campanian–Maastrichtian; eggshells (Mourier et al., 1988).
62. Fundo el Triunfo, Peru; Bagua Formation, Campanian–Maastrichtian; eggshells (Mourier et al., 1988).
63. Morerillo, Peru; Bagua Formation, Campanian–Maastrichtian; eggshells (Mourier et al., 1988).
64. Uberaba, Brazil; Bauru Formation, Campanian–Maastrichtian; eggshells (Price, 1951).
65. Laguna Umayo, Peru; Vilquechio Formation, Maastrichtian; eggshell (Kerourio & Sigé, 1984).
66. Moreno, Argentina; Allen Formation, Maastrichtian; Titanosauridae(?) eggshell (Powell, in press).
67. Ojo de Agua, Argentina; Allen Formation, Maastrichtian; Titanosauridae nest (Powell, in press).
68. General Roca, Argentina; Allen Formation, Maastrichtian; nest(?) (Powell, in press).

Europe (Fig. 1.8)
69. Cirencester, England; Great Oolite Formation, Bathonian; *Oolithus* (= *Oolithes*) *bathonicae* (dinosaur?) nest (van Straelen, 1928).
70. Stonesfield, England; Stonesfield Slate, Bathonian; *Oolithus* (= *Oolithes*) *sphaericus* (dinosaur?) eggs (van Straelen, 1928).
71. Petersborough, England; Oxford Clays, Oxfordian; Camptosauridae(?) egg (van Straelen, 1928).
72. Guimarota, Portugal; Unnamed formation, Kimmeridgian; eggshells (Hirsch, personal communication).

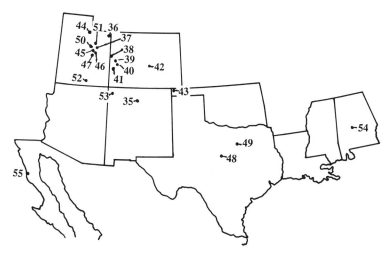

Figure 1.4. Dinosaur eggshell, nest, and baby bone locality map for Utah, Colorado, New Mexico, Oklahoma, Texas, Alabama, and Baja, Mexico.

Figure 1.5. Skeletal reconstruction of saurichians. **A.** *Coelophysis bauri* baby (based on MNA V3318). **B.** *Mussaurus patagonicus* embryo(?) (based on Fig. 1 in Bonaparte & Vince, 1979). **C.** *Apatosaurus* sp. baby (based on specimens in Carpenter and McIntosh, Chapter 17). Scale for A, B = 2 cm; C = 10.

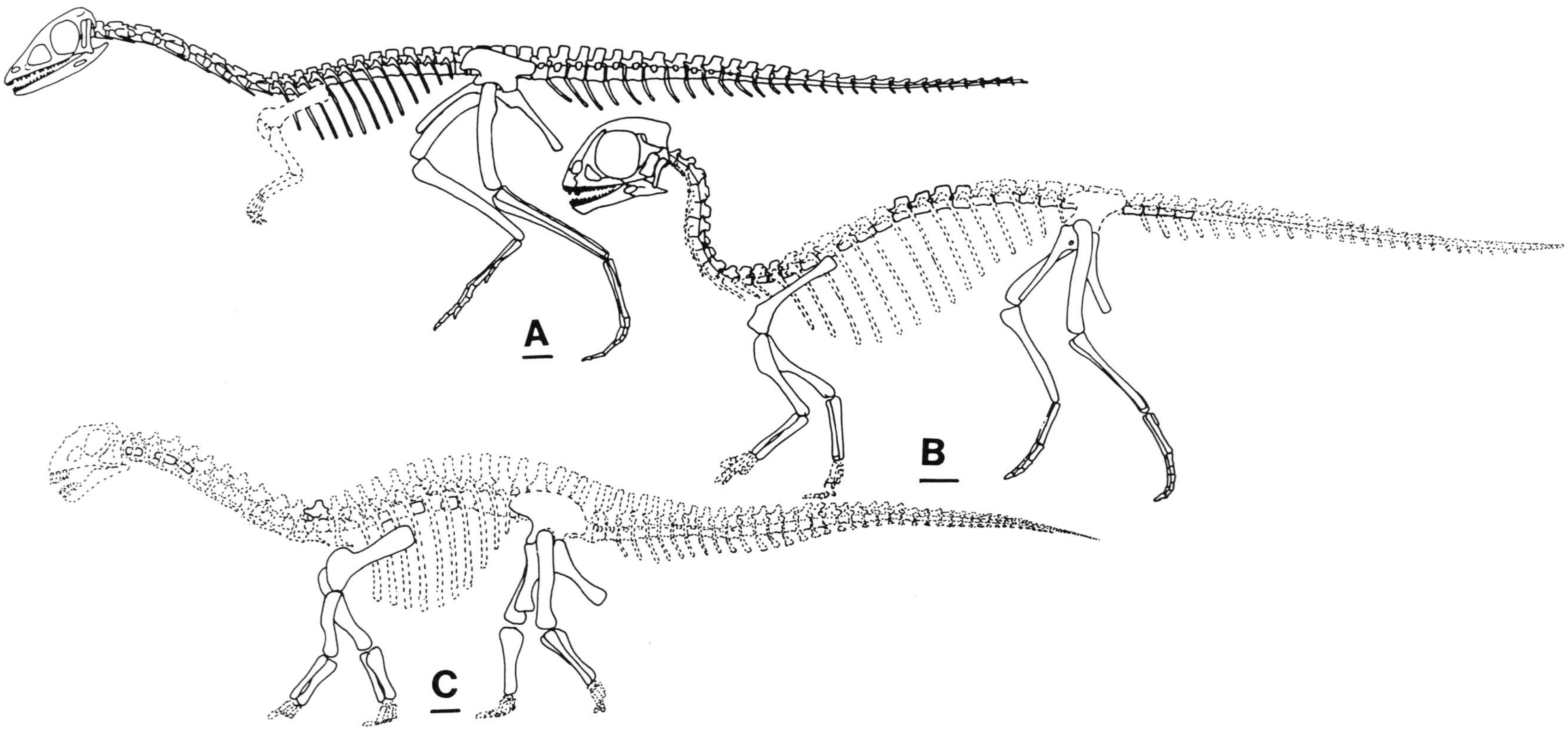

Figure 1.6. Skeletal reconstruction of thyreophors. **A.** *Stegosaurus* sp. (based on DNM 2438). **B.** Nodosaurid (based on specimen described by Jacobs et al. Chapter 22). C. *Pinacosaurus grangeri* (Carpenter, unpublished notes). Scale for A = 10 cm, B = 4 cm, C = 10 cm.

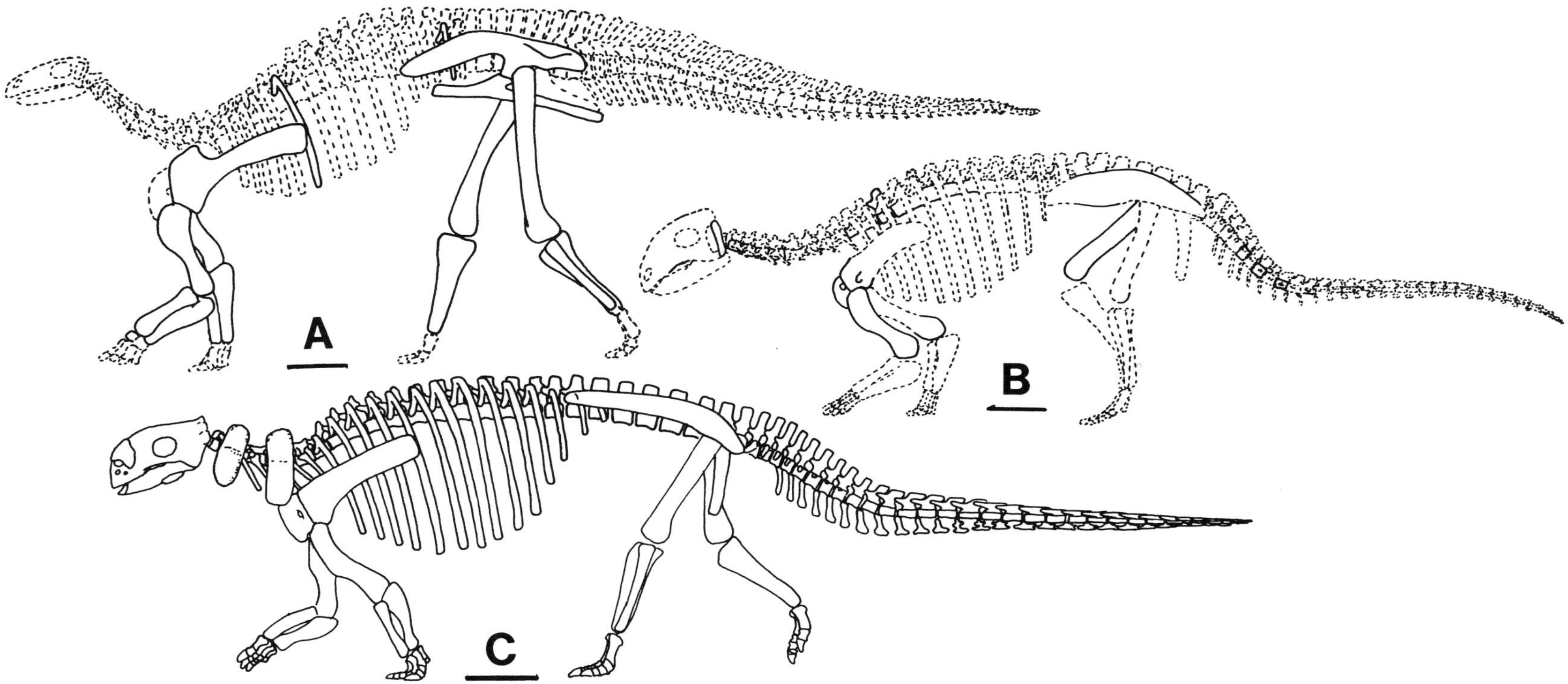

73. Peralta, Portugal; Unnamed formation, Kimmeridgian–Tithonian; Megaloolithidae and *Prismatoolithus*(?) sp. eggshells (Dantas et al., 1992).

74. Paimogo, Portugal; Unnamed formation, Kimmeridgian–Tithonian; Megaloolithidae eggshells (Dantas et al., 1992).

75. Valemitao, Portugal; Unnamed formation, Kimmeridgian–Tithonian; Megaloolithidae eggshells (Dantas et al., 1992).

76. Solnhofen, Germany; Solnhofen Limestone, Kimmeridgian; *Compsognathus*? eggs (Griffiths, 1992).

77. Galve, Spain; Las Zabachoras Beds, Barremian; eggshells (Kohring, 1990).

78. Isle of Wight, England; Wealden Formation, Lower Cretaceous; egg (probably not dinosaur, but included for thoroughness) (van Straelen, 1928).

79. Fox-Amphoux, France; unnamed formation, Maastrichtian; eggs (Lapparent, 1967).

80. Montmeyan, France; unnamed formation, Maastrichtian; eggs (Lapparent, 1967).

81. Sillans–Salernes, France; unnamed formation, Maastrichtian; eggs (Lapparent, 1967).

82. Rognette, France; unnamed formation, Maastrichtian; eggs. (Lapparent, 1967).

83. Rians, France; unnamed formation, Maastrichtian; eggs (Lapparent, 1967).

84. Pourcieux, France. Rognac Formation, Maastrichtian; nests (Williams, Seymour, & Kerouria, 1984).

85. Rousset, France; unnamed formation, Maastrichtian; eggs (Lapparent, 1967).

86. La Cairanne, France; unnamed formation, Maastrichtian; eggs (Penner, 1985).

87. Rogues-Hautes, France; unnamed formation, Maastrichtian; eggs (Lapparent, 1967).

88. Rognac, France; Rognac Formation, Maastrichtian; nest (Lapparent, 1967).

89. Eygalieres, France; unnamed formation, Maastrichtian; eggs (Lapparent, 1967).

90. Maussane, France. unnamed formation, Maastrichtian; eggs (Lapparent, 1967).

91. Argelliers, France; unnamed formation, Maastrichtian; eggs (Lapparent, 1967).

92. Montpellier, France; unnamed formation, Maastrichtian; eggs (Lapparent, 1967).

93. Villeveyrac, France; unnamed formation, Maastrichtian; eggs (Lapparent, 1967).

94. St. Chinian, France; unnamed formation, Maastrichtian; eggs (Lapparent, 1967).

95. Castigno, France; unnamed formation, Maastrichtian; eggs (Lapparent, 1967).

96. Fontfroide, France; Rognac Formation, Maastrichtian; eggshells (Penner, 1985).

97. St. André-de-Roquelongue, France; unnamed formation, Maastrichtian; eggs (Lapparent, 1967).

98. Vitrolles (= Velaux), France; Rognac Formation, Maastrichtian; nests (Cousin, Breton, & Gomez, 1987).

99. Albas, France; unnamed formation, Maastrichtian; eggs (Lapparent, 1967).

100. Orcau, Spain; Garumniense inferior, Maastrichtian; eggshells (Erben et al 1979).

101. Catalonia, Spain; Tremp Formation, Maastrichtian; eggshells (Buffetaut & Le Loeff, 1991).

102. Barranco de Bastus, Spain; Garumniense inferior, Maastrichtian; hadrosaur(?) eggshells (Erben, Hoefs, & Wedepohl, 1979; Kohring, 1989).

103. Barranco de la Munya, Spain; Garumniense inferior, Maastrichtian; eggshells (Erben et al., 1979).

104. Barranco de la Posa, Spain; Garumniense inferior, Maastrichtian; eggshells (Erben et al., 1979).

105. Quintanilla del Coco, Spain; Calizas de Lychnus Formation, Maastrichtian; eggshells (Moratalla, 1992).

106. Tustea, Romania; Densus–Ciula formation, Maastrichtian; Hadrosauridae nest (Grigorescu et al., Chapter 6).

107. Folkstone, England; unnamed formation, Upper Cretaceous; Egg (probably not dinosaur, but included for thoroughness) (van Straelen, 1928).

Eastern and southern portion of Africa (Fig. 1.9)

108. Rooidraai, South Africa; Elliot Formation, Early Jurassic; ?*Massospondylus* sp. nest and baby skeletons (Grine & Kitching, 1987).

109. Tendaguru, Tanzania; Tendaguru Formation, Kimmeridgian; Egg and eggshells. Swinton 1950.

Central Asia (Fig. 1.10)

110. Naryn River, Kirghizia; Khodzhaosmansk Formation, Aptian–Albian(?); nests (Nesov & Kaznachkin, 1986).

Figure 1.7. Dinosaur eggshell, nest, and baby bone locality map for Peru, Brazil, Uruguay, and Argentina.

111. Tayzhuzgen River, Kazakhstan; Manrakskaya Shiva, Senonian; *Macroolithus*? sp. and *Elongatoolithus* sp. eggs (Bazhanov, 1961; Sochava, 1971).
112. Shakhaftan, Kirghizia; Nichkesaisk Formation, Santonian–Campanian; eggshells (Nesov & Kaznachkin, 1986).
113. Arslanbob, Kirghizia; Ialovachsk Formation, Turonian–Santonian; eggshells (Nesov & Kaznachkin, 1986).
114. Charvak, Kirghizia; Ialovachsk Formation, Turonian–Santonian; eggshells (Nesov & Kaznachkin, 1986).
115. Tash Kumyr, Kirghizia; Ialovachsk and Nichkesaisk series, Turonian–Campanian; Spheroolithidae, Ovaloolithidae, Protoceratopsidae, and Elongatoolithidae eggs and eggshells (Nesov & Kaznyshkin, 1986).
116. Central Kazakhstan; unnamed formation, Upper Cretaceous(?); eggs (Nesov & Kaznyshkin, 1986).
117. Daugyztai, Uzbekistan; unnamed formation, Upper Cretaceous; Nest(?) (Nesov & Kaznyshkin, 1986).

118. Chimket, Kirghizia; unnamed formation, Upper Cretaceous(?); eggshells (Nesov & Kaznachkin, 1986).

India (Fig. 1.11)
119. Hathni River, India; Lameta Formation, Campanian–Maastrichtian; Sauropoda(?) eggshells and nests (Sahni et al., Chapter 13).
120. Anjar, India; Anjar Formation, Maastrichtian; Sauropoda and Ornithopoda nests (Ghevariya & Srikarni, 1990; Sahni et al., Chapter 13).
121. Dhori Dungri, India; Lameta Formation, Maastrichtian; eggshell (Sarkar, Bhattacharya, & Mohabey, 1991).
122. Khempur, India; Lameta Formation, Maastrichtian; nests (Srivastava et al., 1986).
123. Daulatpoira, India; Lameta Formation, Maastrichtian; nests (Sarkar et al., 1991).
124. Kevadiya, India; Lameta Formation, Maastrichtian; nests (Srivastava et al., 1986).
125. Rahioli, India; Lameta Formation, Maastrichtian; nests (Srivastava et al., 1986).

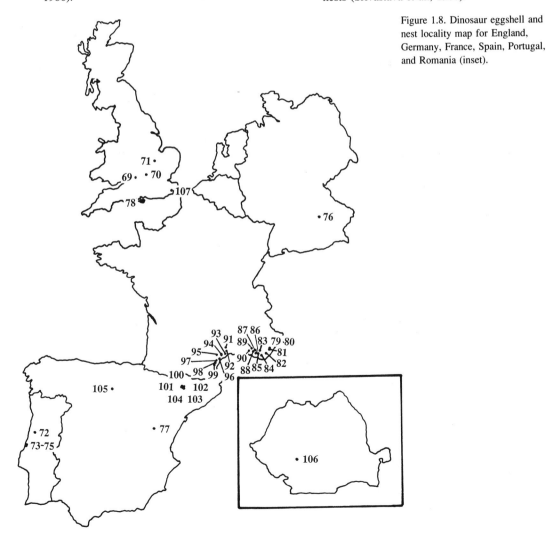

Figure 1.8. Dinosaur eggshell and nest locality map for England, Germany, France, Spain, Portugal, and Romania (inset).

126. Phensani, India;. Lameta Formation, Maastrichtian; nests (Mohabey, 1990).
127. Mirakheri, India; Lameta Formation, Maastrichtian; nests (Mohabey & Mathur, 1989).
128. Dholidhanti, India; Lameta Formation, Maastrichtian; eggs (Mohabey & Mathur, 1989).
129. Paori, India; Lameta Formation, Maastrichtian; nests (Mohabey & Mathur, 1989).
130. Waniawas, India; Lameta Formation, Maastrichtian; Sauropoda(?) nests (Srivastava et al., 1986).
131. Dhuvedia, India; Lameta Formation, Maastrichtian; nests (Srivastava et al., 1986).
132. Werasa, India; Lameta Formation, Maastrichtian; nests (Mohabey, 1990).
133. Balasinor, India; Lameta Formation, Maastrichtian; Sauropoda(?) nests (Srivastava et al., 1986).
134. Sonipur, India; Lameta Formation, Maastrichtian; nests (Sarkar et al., 1991).
135. Bara Simla and Chui Hills, India; Lameta Formation, Maastrichtian; Sauropoda(?) nests (Sahni et al., Chapter 13).
136. Lameta Ghat, India; Lameta Formation, Maastrichtian; nests (Jolly, Bajpai, & Srinlvasan, 1990).

137. Takli, India; Lameta Formation, Maastrichtian; Sauropoda eggshells (Sahni et al., Chapter 13).
138. Piraya, India; Lameta Formation, Maastrichtian; eggshells (Sahni et al., Chapter 13).
139. Tidakepar, India; Lameta Formation, Maastrichtian; nests (Mohabey & Udhoji, 1990).
140. Pisdura, India; Lameta Formation, Maastrichtian; eggshells (Sahni et al., Chapter 13).
141. Asifabad, India; Lameta Formation, Maastrichtian; Sauropoda(?) eggshells (Sahni et al., Chapter 13).
142. Bagh Caves, India; Lameta Formation, Maastrichtian; Sauropoda(?) eggshells and nests (Sahni et al., Chapter 13).

Mongolia and China (Fig. 1.12).

Spelling of Mongolian localities is based on Jerzykiewicz and Russell (1991)).

143. Kuren Tsav, Mongolia; unnamed formation, Neocomian; eggshells (Nesov & Kaznachkin, 1986).
144. Oshi Nur, Mongolia; Ondai Sayr, Lower Cretaceous; *Psittacosaurus mongoliensis* nest with baby bones (Fig. 1.13A; Coombs, 1982; Nesov & Kaznachkin, 1986).
145. Dushih Ula, Mongolia; Dushihin Formation, Lower Cretaceous; nests (Mikhailov et al., Chapter 7).
146. Builyasutuin Khuduk, Mongolia; Dushihin Formation, Aptian; Elongatoolithidae nest (Mikhalov et al., Chapter 7).
147. Ehr Chia Wu Tung (= Jianguanz), China; unnamed formation, Lower(?) Cretaceous; *Peishansaurus philemys* (= *Psittacosaurus*?) baby bones (Bohlin, 1953).
148. Ningxia Province, China; unnamed formation, Lower(?) Cretaceous; eggs (Young, 1979).

Figure 1.9. Dinosaur eggshell, nest, and baby bone locality maps for Tanzania and South Africa.

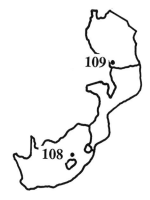

Figure 1.11. Dinosaur eggshell and nest locality map for India.

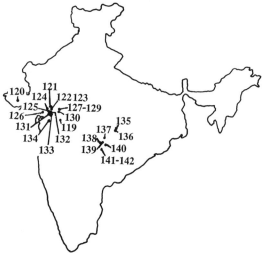

Figure 1.10. Dinosaur eggshell and nest locality map for Kirghizia, Uzbekistan, and Kazakhstan.

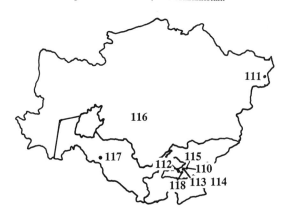

149. Xixia, China; unnamed formation. Lower Cretaceous; Faveoloolithidae, *Paraspheroolithus* cf. *P. irenensis, P.* sp. nov., and "*Placoolithus taohensis*" nests (Zhao, 1979a).
150. Shiregin Gashun, Mongolia; Bayn Shiren Formation, Cenomanian–Campanian; *Spheroolithus* sp. nest (Mikhailov et al., Chapter 7; referred to as Nemegt Formation).
151. Khara Khutul, Mongolia; unnamed formation, Cenomanian–Turonian; eggshells (Nesov & Kaznachkin, 1986).
152. Tel Ulan Ula, Mongolia; Bayn Shiren Formation, Cenomanian–Santonian; eggshells (Nesov & Kaznachkin, 1986).
153. Darigana, Mongolia; Bayn Shiren Formation, Cenomanian–Campanian; *Ovaloolithus* sp. eggs (Mikhailov et al., Chapter 7).
154. Ekhin Tukhum (= Ikh Eren?), Mongolia; Bayn Shiren Formation, Cenomanian–Santonian; eggshells (Nesov & Kaznachkin, 1986).
155. Moyogn Ulagiyn Khaets (= Mogoin Bulak?), Mongolia; Bayn Shiren Formation, Cenomanian–Campanian; *Ovaloolithus* sp. nests and hadrosaur baby bones (Mikhailov et al., Chapter 7).
156. Baynshin Tsav, Mongolia; Bayn Shiren Formation, Cenomanian–Campanian; eggshells (Nesov & Kaznachkin, 1986).
157. Baga Tarjach, Mongolia; Djadokhta Formation, Santonian–Campanian; Protoceratopsidae and *Spheroolithus* sp. nests (Mikhailov et al., Chapter 7).
158. Shiljust Ula, Mongolia; Barun Goyot Formation, Santonian–Campanian; *Dendroolithus* sp. nests (Mikhailov et al., Chapter 7).
159. Khermeen Tsav I and II, Mongolia; Barun Goyot Formation, Santonian–Campanian; *Ingenia yanshini* baby bone; *Bagaceratops rozhedestvenskyi* baby bones; Protoceratopsidae, *Faveoloolithus ningxiaensis*, and *Dendroolithus* sp. eggshells and nests (Maryanska & Osmolska 1975; Barsbold & Perle, 1983; Mikhailov et al., Chapter 7).
160. Undurshil Ula, Mongolia; Barun Goyot Formation, Santonian–Campanian. *Elongatoolithus* sp. eggs (Mikhailov et al., Chapter 7).
161. Bambu Khudu, Mongolia; Barun Goyot Formation, Campanian; eggshells (Nesov & Kaznachkin, 1986).
162. Udan Sayr, Mongolia; Djadokhta Formation, Campanian; Elongatoolithidae and *Ovaloolithus* sp. eggshells and nests (Mikhailov et al., Chapter 7).
163. Toogreek, Mongolia; Djadokhta Formation, Campanian; Protoceratopsidae nests and baby bones, Elongatoolithidae nests, and hadrosaur baby bones (Fig. 1.3C; Barsbold & Perle 1983; Mikhailov et al., Chapter 7).
164. Ologoy Ulan Tsav, Mongolia; Bayun Goyot Formation, Campanian; *Faveoloolithus* sp. nests (Mikhailov et al., Chapter 7).
165. Erenhot, China; Iren Dabasu Formation, Campanian(?); *Paraspheroolithus* cf. *P. irenensis* nest (Dong, Currie, & Russel, 1989).
166. Bagamod Khuduk, Mongolia; Bayun Goyot Formation, Campanian; nests (Mikhailov et al., Chapter 7).
167. Ikh Shunkht, Mongolia; Bayun Goyot Formation, Campanian; Protoceratopsidae, Elongatoolithidae, *Faveoloolithus* sp., and *Spheroolithus*

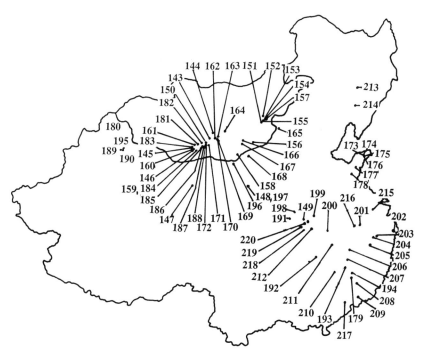

Figure 1.12. Dinosaur eggshell, nest, and baby bone locality map for Mongolia and China.

sp. eggshells and nests (Mikhailov et al., Chapter 7).

168. Bayan Manduhu, China; Djadokhta Formation, Campanian; *Protoceratops andrewsi* nests and baby bones and *Pinacosaurus grangeri* baby skeletons (Fig. 1.6C) (Dong et al., 1989).

169. Dzamyu Khond, Mongolia; Djadokhta Formation, Campanian; *Protoceratops andrewsi* baby skeletons (Barsbold & Perle, 1983).

170. Bayn Dzak, Mongolia; Djadokhta Formation, Campanian; Elongatoolithidae and *Spheroolithus* sp. nests; *Protoceratops andrewsi* nests and skeletons (Fig. 1.13B), and *Archaeornithoides deinosauriscus* baby bones (Maryanska & Osmolska, 1975; Elzanowski & Wellnhofer, 1992; Mikhailov et al., this volume).

171. Gilbent, Mongolia; Barun Goyot Formation, Campanian; eggs (Mikhailov et al., Chapter 7).

172. Khulsan, Mongolia; Barun Goyot, Campanian; *Dendroolithus* sp. eggshells and nests and *Breviceratops kozlowskii* baby bones (Maryanska & Osmolska, 1975; Mikhailov et al., Chapter 7).

173. Chiangchungting, China; Wangshi Formation, Campanian(?); *Spheroolithus chiangchiungtingensis* nests (Chao & Chiang, 1974).

174. Laiyang, China; Wangshi Formation, Campanian(?); *Spheroolithus chiangchiungtingensis*, ?*S. megadermus*, *Ovaloolithus laminadermus*, *O. chinkangkouensis*, and *Paraspheroolithus irenensis* eggs (Chao & Chiang, 1974).

175. Chaochun, China; Wangshi Formation, Upper Cretaceous; *Elongatoolithus* sp., *Spheroolithus* sp. and *Ovaloolithus* sp. nests (Chow, 1951).

176. Chinkangkou, China; Wangshi Formation, Upper Cretaceous; eggs (Young, 1965).

177. Jiaozhou, China; Wangshih Series, Cretaceous; Spheroolithidae, Ovaloolithidae and Elongatoolithidae eggs (Zhao, personal communication).

178. Zhucheng, China; Wangshi Formation, Upper Cretaceous; Spheroolithidae, Ovaloolithidae and Elongatoolithidae eggs (Zhao, personal communication).

179. Nanxiong, China; Yuanpu and Pingling formations, Campanian–Maastrichtian; *Macroolithus yaotunensis*, *M. rugustus*, *M.* sp. nov., *Elonga-*

Figure 1.13. Reconstruction of **A.** *Psittacosaurus mongoliensis* embryo (based on AMNH 6536). **B.** *Protoceratops andrewsi* (modified from Rozhdestvensky 1973, Figure 89). Scale for A = 2 cm; for B unknown.

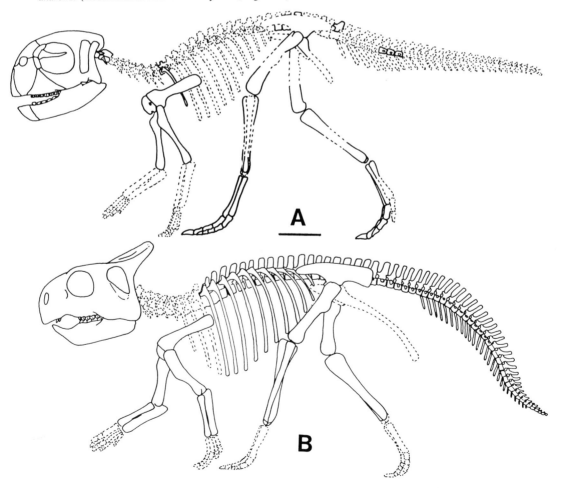

toolithus andrewsi, E. elongatus, E. sp. nov., *Nanshiungoolithus chuetienensis, Ovaloolithus* cf. *O. chinkangkouensis, O.* cf. *O. laminadermus, O* sp. nov., *Shixingoolithus erbeni,* and *Stromatoolithus pinglingensis* nests and eggs (Zhao et al., 1991, Chapter 12).

180. Ulungar, China; Donggou(?) Formation, Maastrichtian(?); eggs (Russell, unpublished notes).

181. Bugin Tsav, Mongolia; Nemegt Formation, Maastrichtian; *Laevisoolithus* sp. eggs (Mikhailov et al., Chapter 7).

182. Gurlin Tsav, Mongolia; Nemegt Formation, Maastrichtian; *Spheroolithus* sp. and Elongatoolithidae eggshells and nests (Mikhailov et al., Chapter 7).

183. Khaichin Ula I, Mongolia; Nemegt Formation, Maastrichtian; Elongatoolithidae, *Macroolithus*? sp. and *Subtilioolithus* sp. eggshells and nests (Mikhailov et al., Chapter 7).

184. Khermeen Tsav II, Mongolia; Nemegt Formation, Maastrichtian; Elongatoolithidae eggshells (Mikhailov et al., Chapter 7).

185. Tsagan Khushu, Mongolia; Nemegt Formation, Maastrichtian; Elongatoolithidae and *Spheroolithus* sp. nests (Mikhailov et al., Chapter 7).

186. Altan Ula III, IV, Mongolia; Nemegt Formation, Maastrichtian; Elongatoolithidae nests and Protoceratopsidae eggshells (Sabath 1991 Mikhailov et al., Chapter 7).

187. Naran Bulak, Mongolia; Nemegt Formation, Maastrichtian; eggshells (Nesov & Kaznachkin, 1986).

188. Nemegt, Mongolia; Nemegt Formation, Maastrichtian; *Laevisoolithus* sp. eggs (Mikhailov et al., Chapter 7).

189. Shenjinkou, China; Subash Formation, Maastrichtian; *Paraspheroolithus* sp. and *Elongatoolithus* sp. eggs (Hao & Guan, 1984).

190. Jiangjunmiao, China; Hongshaguan Formation, Maastrichtian; "*Protoceratops*" and *Paraspheroolithus* cf. *P. irenensis* nests (Young, 1965).

191. Shanyang Basin, China; Shanyang Formation, Maastrichtian; Spheroolithidae and Elongatoolithidae eggs (Zhao, personal communication).

192. Taoyuan and Dongting Basin, China; Fenshui'ao Formation, Maastrichtian; *Phaceloolithus hunanensis, Elongatoolithus magnus,* and *E.* sp. nests (Zeng & Zhang, 1979; Zhao, personal communication).

193. Chaling, China; Daijiaping Formation, Maastrichtian; *Elongatoolithus andrewsi* eggs, and Elongatoolithidae eggs (Young, 1965; Zhao, personal communication).

194. Taihe–Ganzhou area, China; Yuanpu Formation, Maastrichtian; *Macroolithus* sp. and *Nanhsiungoolithus chuetienensis* eggs (Young, 1965).

195. Qitai, China; Subashi Formation, Upper Cretaceous; *Elongatoolithus andrewsi* eggs (Young, 1965).

196. Tsondolein-Khuduk (= Ondor Mod?), China; Minke Formation, Upper Cretaceous; *Microceratops gobiensis* baby bones (Bohlin, 1953).

197. Alxa, China; unnamed formation, Upper Cretaceous; eggshells (Zhao & Ding, 1976).

198. Lingbao, China; Nanzhao Formation, Upper Cretaceous; Elongatoolithidae eggs (Zhao, personal communication).

199. Liguanqiao Basin, China; Hugang Formation, Upper Cretaceous; Elongatoolithidae eggs (Zhao, personal communication).

200. Anlu, China; Gong An Zhai Formation, Upper Cretaceous; *Dendroolithus wangdianensis* eggs (Zhao & Li, 1988).

201. Xuanzhou, China; Xuannan Formation, Upper Cretaceous; Elongatoolithidae eggs (Zhao, personal communication).

202. Tiantai, China; Laija B Formation, Upper Cretaceous; Faveoloolithidae and Spheroolithidae eggs (Mateer, 1989; Zhao, personal communication).

203. Gaotangshi, China; Quxian Formation, Upper Cretaceous; nests (Mateer, 1989).

204. Quzhou, China; Qujiang Group, Upper Cretaceous; Faveoloolithidae and Spheroolithidae eggs (Zhao, personal communication).

205. Anwen, China;. Xiaoyan Formation, Upper Cretaceous; eggs (Dong, 1980).

206. Gao'an, China; Qingfengqiao Formation, Upper Cretaceous; Elongatoolithidae eggs (Zhao, personal communication).

207. Xinyu, China; Qingfengqiao Formation, Upper Cretaceous; Elongatoolithidae eggs (Zhao, personal communication).

208. Heyuan, China; Nanxiong Group, Upper Cretaceous; Elongatoolithidae eggs (Zhao, personal communication).

209. Huizhou, China; Nanxiong Group, Upper Cretaceous; Spheroolithidae eggs (Zhao, personal communication).

210. Shiguguan, China; Hugang Formation, Upper Cretaceous; Elongatoolithidae eggs (Zhao, 1979b; personal communication).

211. Tsatzeyuanhsu, China; unnamed formation, Upper Cretaceous; eggshell (Young, 1965).

212. Xinyang, China; unnamed formation, Upper Cretaceous; Elongatoolithidae eggs (Zhao, personal communication).

213. Changchun, China; Quantou Formation, Cretaceous; Spheroolithidae eggs (Zhao, personal communication).

214. Quantou, China; Quantou Formation, Cretaceous; Spheroolithidae eggs (Zhao, personal communication).

215. Yixing, China; unnamed formation, Cretaceous; Elongatoolithidae eggs (Zhao, personal communication).

216. Anqing, China; unnamed formation, Cretaceous; Spheroolithidae eggs (Zhao, personal communication).

217. Guangzhou, China; Sanshui Formation, Cretaceous; Elongatoolithidae and Spheroolithidae eggs (Zhao, personal communication).

218. Xiaguan Basin, China; unnamed formation, Cretaceous; *Youngoolithus xiaguanensis* nests (Zhao, 1979b).

219. Yunxian, China; unnamed formation, Cretaceous; Dendroolithidae eggs (Zhao, personal communication).
220. Taohe Basin, China; Majiacun Formation, Cretaceous; Dendroolithidae and Dictyoolithidae eggs. (Zhao, personal communication).

Acknowledgment

We would like to express our gratitude to John R. Horner, Philip Currie, Richard Cifelli, and Zhao Zi-Kui for sharing their information on unpublished egg localities with us.

References

Barsbold, R., & Perle, A. 1983. On the taphonomy of the joint burial of juvenile dinosaurs and some aspects of their ecology. *Transactions of the Soviet-Mongolian Paleontological Expedition* 24: 121–5.

Bazhanov, V. 1961. First discovery of dinosaur eggshells in the USSR. *Trudy Zoologicheskogo Instituta, Akademiya Nauk Kazakstan SSR* 15: 177.

Bibler, C.J., & Schmitt, J. G. 1986. Barrier-island coastline deposition and paleogeographic implications of the Upper Cretaceous Horsethief Formation, northern Disturbed Belt, Montana. *The Mountain Geologist* 23: 113–27.

Bohlin, B. 1953. Fossil reptiles from Mongolia and Kansu. *The Sino-Swedish Expedition, Publication 37: VI. Vertebrate Palaeontology 6.* (Stockholm: Statens Etnografiska Museum).

Bonaparte, J. F., & Vince, M. 1979. El Hallazgo del primer nido de dinosaurios Triasicos, (Saurischia, Prosauropoda), Triasico Superior de Patagonia, Argentina. *Ameghiniana* 16: 173–82.

Brinkman, D. 1986. Microvertebrate sites: progress and prospects. *In Dinosaur Systematics Symposium, Field Trip Guidebook to Dinosaur Provincial Park* (Drumheller, Canada: Tyrrell Museum of Palaeontology), pp. 24–37.

Brown, B., & Schlaikjer, E. M. 1940. The origin of ceratopsian horn-cores. *American Museum Novitates* 1065: 1–7.

Buffetaut, E., & Le Loweff, J. 1991. Late Cretaceous dinosaur faunas of Europe: some correlation problems. *Cretaceous Research* 12: 159–76.

Carpenter, K. 1982. Baby dinosaurs from the Late Cretaceous Lance and Hell Creek formations and a description of a new species of theropod. *University of Wyoming Contributions to Geology* 20: 123–34.

1992. Behavior of hadrosaurs as interpreted from footprints in the "Mesaverde" Group (Campanian) of Colorado, Utah, and Wyoming. *University of Wyoming Contributions to Geology* 29: 81–96.

Chao T.-K., & Chiang Y.-K. 1974. Microscopic studies on the dinosaurian egg-shells from Laiyang, Shantung Province. *Scientia Sinica* 17: 71–83.

Chow, M. M. 1951. Notes on the Late Cretaceous dinosaurian remains and the fossil eggs from Laiyang Shantung. *Bulletin of the Geological Society of China* 31: 89–96.

Colbert, E. 1989. The Triassic dinosaur *Coelophysis. Museum of Northern Arizona Bulletin* 57: 1–160.

Coombs, W.P. 1982. Juvenile specimens of the ornithischian dinosaur *Psittacosaurus. Palaeontology* 25: 89–107.

Cousin, R., Breton, G., & Gomez, N. 1987. La campagne de "fouilles prliminaires" sur les lieux de ponte de dinosaures de Rousset–sur–Arc (Bouches–du–Rhône). *Bulletin trimestriel de la Société Géolique Normandie et des Amis Muséum du Havre* 74: 5–15.

Dantas, P. M., Moratalla, J. J., Hirsch, K. F., & Santos, V. F. 1992. Mesozoic reptile eggs from Portugal. New data. *Dinosaurs and Other Fossil Reptiles of Europe. Second Georges Cuvier Symposium.* Montbeliard, France.

Dobie, J. L. 1978. A fossil amniote egg from an Upper Cretaceous deposit (Mooreville Chalk of the Selma Group) in Alabama. *Copeia* 3: 460–4.

Dong Z. 1980. The dinosaurian faunas of China and their stratigraphic distribution. *Journal of Stratigraphy* 4: 256–63.

Dong Z., Currie, P. J., & Russell, D. A. 1989. The 1988 field program of the dinosaur project. *Vertebrata PalAsiatica* 27: 233–6.

Dorr, J. 1985. Newfound Early Cretaceous dinosaurs and other fossils in southeastern Idaho and westernmost Wyoming. *University of Michigan, Contributions from the Museum of Paleontology* 27: 73–85.

Elzanowski, A. and Wellnhofer, P. 1992. A new link between theropods and birds from the Cretaceous of Mongolia. *Nature* 359: 821–3.

Erben, H. K., Hoefs, J., & Wedepohl, K. H. 1979. Paleobiological and isotopic studies of eggshells from a declining dinosaur species. *Paleobiology* 5: 380–414.

Fiorillo, A.R. 1987. Significance of juvenile dinosaurs from Careless Creek Quarry (Judith River Formation), Wheatland County, Montana. *In P. J. Currie, & E. H. Koster (eds.)., Fourth Symposium on Mesozoic Terrestrial Ecosystems, Short Papers* (Drumheller, Canada: Tyrrell Museum of Palaeontology), 88–95.

Forster, C. A. 1990. Evidence for juvenile groups in the ornithopod dinosaur *Tenontosaurus tilletti* Ostrom. *Journal of Paleontology* 64: 164–5.

Galton, P. M. 1982. Juveniles of the stegosaurian dinosaur *Stegosaurus* from the Upper Jurassic of North America. *Journal of Vertebrate Paleontology* 2: 47–62.

Ghevariya, Z. G. and Srikarni, C. 1990. Anjar Formation, its fossils and their bearing on the extinction of dinosaurs. *In A. Sahni (ed.), Cretaceous Event Stratigraphy and the Correlation of the Indian Nonmarine Strata, Contributions from the Seminar cum Workshop I.G.C.P. 216 and 245.* (Chandigarh, India: Punjab University), pp. 106–9.

Griffiths, P. 1992. The question of *Compsognathus* eggs. *Dinosaurs and Other Fossil Reptiles of Europe. Second Georges Cuvier Symposium.* Montbeliard, France.

Grine, F. E., & Kitching, J. W. 1987. Scanning electron microscopy of early dinosaur egg shell structure: a comparison with other rigid sauropsid eggs. *Scanning Microscopy* 1: 615–30.

Hao Y., & Guan S. 1984. The lower–Upper Cretaceous and Cretaceous–Tertiary boundaries in China. *Bulletin of the Geological Society of Denmark* 33: 129–38.

Hirsch, K. F., & Quinn, B. 1990. Eggs and eggshell fragments from the Upper Cretaceous Two Medicine Formation of Montana. *Journal of Vertebrate Paleontology* 10: 491–511.

Hirsch, K. F., Stadtman, K. L., Miller, W. E. and Madsen, J. H. 1989. Upper Jurassic dinosaur egg from Utah. *Science* 243: 1711–13.

Horner, J. R. 1982. Evidence of colony nesting and "site fidelity" among ornithischian dinosaurs. *Nature* 297: 675–6.

Horner, J. R., & Makela, R. 1979. Nest of juveniles provides evidence of family structure among dinosaurs. *Nature* 82: 296–8.

Horner, J. R. and Weishampel, D. B. 1988. A comparative embryological study of two ornithischian dinosaurs. *Nature* 332: 256–7.

Jain, S. L. 1989. Recent dinosaur discoveries in India, including eggshells, nests and coprolites. *In* D. D. Gillette, & M. G. Lockley (eds.), *Dinosaur Tracks and Traces* (Cambridge: Cambridge University Press), pp. 99–108.

Jensen, J. A. 1966. Dinosaur eggs from the Upper Cretaceous North Horn Formation of central Utah. *Brigham Young University Geology Studies* 13: 55–67.

1970. Fossil eggs in the Lower Cretaceous of Utah. *Brigham Young University Geology Studies* 17: 51–65.

Jepsen, G. L. 1931. Dinosaur egg shell fragments from Montana. *Science* 73: 12–13.

Jerzykiewicz, T., & Russell, D. 1991. Late Mesozoic stratigraphy and vertebrates of the Gobi Basin. *Cretaceous Research* 12: 345–77.

Jolly, A., Bajpai, S., & Srinivasan, S. 1990. Indian sauropod nesting sites (Maastrichtian, Lameta Formation): a preliminary assessment of the taphonomic factors at Jabalpur, India. *In* A. Sahni (ed.), *Cretaceous Event Stratigraphy and the Correlation of the Indian Nonmarine Strata, Contributions from the Seminar cum Workshop I.G.C.P. 216 and 245* (Chandigarh, India: Punjab University), 78–81.

Kérourio, P., & Sigé, B. 1984. L'apport des coquilles d'oeufs de dinosaures de Laguna Umayo a l'âge de la Formation Vilquechico (Pérou) et à la compréhension de *Perutherium altiplanense. Newsletter in Stratigraphy* 13: 133–42.

Kohring, R. 1989. Fossile Eierschalen aus dem Garumnium (Maastrichtium) von Bastus (Provinz Lerida, NE–Spanien). *Berliner geowissenshaften Abhandlung* 106: 267–75.

Kohring, R. 1990. Fossile Reptil–Eischalen (Chelonia, Crocodilia, Dinosauria) aus dem unteren Barremium von Galve (Provinz Teruel, SE–Spanien). *Paläontologische Zeitschrift* 64: 329–44.

Lapparent, A.-F. de 1967. Les dinosaures de France. *Sciences* 51: 5–18.

Maryanska, T., & Osmolska, H. 1975. Protoceratopsidae (Dinosauria) of Asia. *Palaeontologica Polonica* 33: 133–81.

Mateer, N.J. 1989. Upper Cretaceous reptilian eggs from the Zhejiang Province, China. *In* D. Gillette & M. Lockley (eds.), *Dinosaur Tracks and Traces* (New York: Cambridge University Press), pp. 116–18.

Mohabey, D. M. 1987. Juvenile sauropod dinosaur from Upper Cretaceous Lameta Formation of Panchmahals District, Gujarat, India. *Journal of the Geological Society of India* 30: 210-6.

1990. Dinosaur eggs from Lameta Formation of western and central India: their occurrence and nesting behaviour. *In* A. Sahni (ed.), *Cretaceous Event Stratigraphy and the Correlation of the Indian Nonmarine Strata, Contributions from the Seminar cum Workshop I.G.C.P. 216 and 245* (Chandigarh, India: Punjab University), pp. 86–9.

Mohabey, D. M. & Mathur, U. B. 1989. Upper Cretaceous dinosaur eggs from new localities of Gujarat, India. *Journal of the Geological Society of India* 33: 32–7.

Mohabey, D. M. & Udhoji, S. G. 1990. Fossil occurrences and sedimentation of Lameta Formation of Nand Area, Maharashtra: palaeoenvironmental, palaeoecological and taphonomical implications. *In* A. Sahni (ed.), *Cretaceous Event Stratigraphy and the Correlation of the Indian Nonmarine Strata, Contributions from the Seminar cum Workshop I.G.C.P. 216 and 245.* (Chandigarh, India: Punjab University), pp. 75–7.

Moratalla, J. 1992. Dinosaurian and crocodilian eggshells. *In* C. Pol, A. D. Buscalioni, J. Carballeira, V. Frances, N. L. Martinex, B. Marandat, J. J. Moratalla, J. L. Sanz, B. Sigé, & J. Villatte. 1992. Reptiles and mammals from the Late Cretaceous new locality Quintanilla del Coco (Burgos Province, Spain). *Neues Jahrbuch für Geologie und Paläontologie, Abhandlungen* 184: 279–314.

Mourier, T., Bengtson, P., Bonhomme, M., Buge, E., Cappetta, H., Crochet, J.–Y., Feist, M., Hirsch, K. F., Moullade, M., Noblet, C., Pons, D., Rey, J., Sigé, B., Tambareau, Y. & Taquet, P. 1988. The Upper Cretaceous–Lower Tertiary marine to continental transition in the Bagua Basin, northern Peru. *Newsletters in Stratigraphy* 19: 143–77.

Nesov, L. A. & Kaznachkin, M. N. 1986. Discovery of a site in the USSR with remains of eggs of Early and Late Cretaceous dinosaurs. *Biological Sciences, Zoology* 9: 35–49.

Penner, M. M. 1985. The problem of dinosaur extinction. Contribution of the study of terminal Cretaceous eggshells from southeast France. *Geobios* 18: 665–9.

Powell, J. E. 1987. The Late Cretaceous fauna of Los Alamitos, Patagonia, Argentina. Part VI. The Titanosaurids. *Revista del Museo Argentino de Ciencias Naturales, Paleontología* 3: 147–53.

Powell, J. In press. Hallazgo de huevos asignables a Dinosaurios titanosáuridos (Saurischia, Sauropoda) de la provincia de Río Negro, Argentina.

Price, L. 1951. Un ovo de dinosaurio na formacao Bauru do Cretacio de Estado de Minas Gerais. *Departamento Nacional da Producao Mineral, Divisao de Geologia e Mineria, Notas Preliminares e Estudos* 53: 1–9.

Rozhdestvensky, A. K. 1973. *Animal Kingdom in Ancient Asia.* (Tokyo).

Sabath, K. 1991. Upper Cretaceous amniotic eggs from the Gobi Desert. *Acta Palaeontologica Polonica* 36: 151–92.

Sahni, A. 1972. The vertebrate fauna of the Judith River Formation, Montana. *American Museum of Natural History Bulletin* 147: 321–412.

Sarkar, A., Bhattacharya, S. K. & Mohabey, D. M. 1991. Stable-isotope analyses of dinosaur eggshells: paleoenvironmental implications. *Geology* 19: 1068–71.

Scheetz, R. 1991. Progress report of juvenile and embryonic *Dryosaurus* remains from the Upper Jurassic Morrison Formation of Colorado. *In* W. Averett (ed.) *Guidebook for Dinosaur Quarries and Tracksite Tour* (Grand Junction, CO: Grand Junction Geological Society).

Sochava, A. 1971. Two types of eggshell in Senonian dinosaurs. *Paleontological Journal* 3: 353–61.

Srivastava, S. Mohabey, D. M., Sahni, A., & Pant, S. C. 1986. Upper Cretaceous dinosaur egg clutches from Kheda District (Gujarat, India). *Paleontographica Abt. A* 193: 219–33.

Sternberg, C.M. 1955. A juvenile hadrosaur from the Oldman Formation of Alberta. *Annual Report, National Museum of Canada, Bulletin* 136: 120–2.

Straelen, V. van 1928. Les oeufs de reptiles fossiles. *Palaeontologica* 1: 295–317.

Swinton, W. E. 1950. Fossil eggs from Tanganyika. *The Illustrated London News* 217: 1082–3.

Tokaryk, T. 1990. A baby *Triceratops* or a very small adult dinosaur. *Saskatchewan Archaeological Society Newsletter* 11: 127–8.

Williams, D. L. G., Seymour, R. S., & Kerourio, P. 1984. Structure of fossil dinosaur eggshell from the Aix Basin, France. *Palaeogeography, Palaeoclimatology, Palaeoecology* 45: 23–37.

Winkler, D. A., & Murry, P. A. 1989. Paleoecology and hypsilophodontid behavior at the Protor Lake dinosaur locality (Early Cretaceous), Texas. *In* J. O. Farlow (ed.), *Paleobiology of the Dinosaurs. Geological Society of America, Special Paper*, 238: 55–61.

Wolberg, D. L. & Bellis, D. 1989. The discovery of dinosaur nests, eggs and tracks, Fruitland Formation, Late Cretaceous, San Juan Basin, New Mexico. *Geological Society of America Annual Meeting Abstracts*: A73.

Young, C. C. 1965. Fossil eggs from Nanhsiung, Kwangtung, and Kanchou, Kiangsi. *Vertebrata PalAsiatica* 9: 141–70.

1979. Note on an egg from Ninghsia. *Vertebrata PalAsiatica* 17: 35–6.

Zeng D., & Zhang, J. 1979. On the dinosaurian eggs from the western Dongting Basin, Hunan. *Verterbata PalAsiatica* 17: 131–6.

Zhao Z. 1979a. Progress in the research of dinosaur eggs. In *Mesozoic and Cenozoic Red Beds of South China* (Beijing: Science Press), pp. 330–40..

1979b. Discovery of the dinosaurian eggs and footprint from Neixiang County, Henan Province. *Vertebrata PalAsiatica* 17: 304–9

Zhao Z., & Ding S. 1976. Discovery of the dinosaurian eggshells from Alxa, Ningxia and its stratigraphic significance. *Vertebrata PalAsiatica* 14: 42–4

Zhao Z. & Li Z. 1988. A new structural type of the dinosaur eggs from Anlu County, Hubei Province. *Vertebrata PalAsiatica* 26: 107–15.

Zhao Z., Ye J., Li H., Zhao Z., & Yan Z. 1991. Extinction of the dinosaurs across the Cretaceous–Tertiary boundary in Nanxiong Basin, Guangdong Province. *Vertebrata PalAsiatica* 29: 1–20.

2 The discovery of dinosaur eggshells in nineteenth-century France

ERIC BUFFETAUT AND
JEAN LE LOEUFF

Abstract

Dinosaur eggshell fragments were discovered in the French Pyrenees as early as 1859 by the priest and geologist Jean-Jacques Pouech. Nearly complete eggs were later discovered in Provence by another geologist, Philippe Matheron (1869). These eggs were attributed to dinosaurs by Paul Gervais in 1877, but this identification was not accepted until the beginning of the twentieth century.

Introduction

One of the most striking and most widely publicized discoveries of the Central Asiatic Expeditions of the American Museum of Natural History was that of dinosaur eggs from the Upper Cretaceous of Mongolia. Roy Chapman Andrews (1932, p. 208), the expedition leader, told the story of the discovery at the "Flaming Cliffs" of Shabarakh Usu:

On July 13, 1923, George Olsen reported at dinner that he had found some fossil eggs. Inasmuch as the deposit was obviously Cretaceous and too early for large birds, we did not take his story very seriously. We felt quite certain that his so-called eggs would prove to be sandstone concretions or some other geological phenomena. Nevertheless, we were all curious enough to go with him to inspect his find. We saw a small sandstone ledge, beside which were lying three eggs partly broken. The brown striated shell was so egg-like that there could be no mistake. Granger finally said, "No dinosaur eggs have ever been found, but the reptiles probably did lay eggs. These must be dinosaur eggs. They can't be anything else."

Granger was wrong; dinosaur eggs had been found before in France, but this fact was not well publicized. The news of the discovery of dinosaur eggs in Mongolia quickly spread around the world, and newspapers and magazines claimed them as the first dinosaur eggs. In France, Andrews declared in the December, 1923, issue of the magazine *L'Illustration* that no one knew that dinosaurs laid eggs before the discovery of the eggs in Mongolia. A response soon came from French geologist and paleontologist Louis Joleaud of the University of Paris. Based on the discoveries of Matheron, he stated that French paleontologists had known of dinosaur eggs as early as 1869. These, not the Mongolian eggs, were the earliest dinosaur egg discoveries (Joleaud, 1924). Granger (1936, p. 21) admitted that

some fragments of what seemed to be reptilian egg shells were found at Rognac, southern France, in strata bearing dinosaur bones and there is a possibility that these are really bits of dinosaur eggs, but they may also belong to other contemporary reptiles.

Joleaud was correct about the priority of the French discoveries, but he was incorrect in that Matheron was not the first to discover dinosaur eggs in France, and furthermore the dinosaurian nature of the eggs had never been demonstrated.

Jean-Jacques Pouech: the forgotten pioneer

Jean-Jacques Pouech (1814–92) was a Roman Catholic priest and head of the Pamiers Seminary in southern France (Department of Ariège). He was also a pioneer in the geology, archeology, and paleontology of the Mas d'Azil region of the Pyrenees foothills of (Leclerc, 1983; Le Loeuff, 1991a). In the Upper Cretaceous strata of the area (now known to be Maastrichtian in age; see Le Loeuff, 1991b; Vianey-Liaud et al., Chapter 11), Pouech (1859, p. 403) described many "broken bones impregnated with iron hydrates and phosphates." Some other fossils, however, puzzled the priest-paleontogist. He wrote, the most remarkable are eggshell fragments of very great dimensions. At first, I thought that they could be integumentary plates of reptiles, but their constant thickness between two perfectly parallel surfaces, their fibrous structure, normal to the surfaces, and especially their regular curvature, definitely suggest that they are enormous eggshells, at least four times the volume of ostrich eggs. Their convex surface is granulated like an orange peel. The great development of the curvature, and consequently of the resulting volume, is the sole reason which has induced some doubt in my mind. All the other characters, especially the constant thickness (about 2 millimeters), suggests that of eggshells. Unfortunately, the fragments I possess are too small to provide at the present time, a rigorous demonstration, but I do not despair of discovering some others in the same locality (authors' translation).

In a footnote, Pouech estimated that the radius of the eggs was 18 cm, but he urged caution about this measurement. Nevertheless, from his description, Pouech was the first to discover and publish on dinosaur eggshells. Although he was convinced that he had eggshells, he did not connect them with dinosaurs. Considering the early date (1859), he may not have been familiar with the word "dinosaur."

In a second footnote added to Pouech's paper by d'Archiac, a paleontologist at the Muséum National d'Histoire Naturelle in Paris, he wrote, "the most competent scientists to whom the fragments were submitted did not confirm the author's hypothesis [about the nature of the fossils]" (authors' translation). As a result, Pouech changed his mind in a letter to d'Archiac dated May 20, 1859. This letter was quoted by d'Archiac in his own study of the fossils that Pouech had sent to the

Figure 2.1. Philippe Matheron (1807–1899) who discovered the first dinosaur eggs in the Upper Cretaceous of Provence. (Photograph courtesy of Roger Fournier, Musée d'Histoire Naturelle de Marseille.)

Muséum National d'Histoire Naturelle. Pouech wrote that after finding more fragments, he was no longer sure that they came from the eggs of "gigantic birds." He suggested, instead, that the enigmatic fragments could be the "solid parts of armadillos or chelonians" (d'Archiac, 1859, p. 803). D'Archiac added that "some people thought that they were crustacean tests," but that he felt that "their true origin was still uncertain" (authors' translation). Pouech's description of dinosaur eggshells was quickly forgotten until recent papers by Villatte, Taquet and Bilotte (1986), and by Bilotte et al. (1986).

Pouech's collection of fossil vertebrates was long considered lost, but in 1989 we luckily rediscovered it in the Collège Jean XXIII of Pamiers, where it had been carefully preserved (Buffetaut & Le Loeuff, 1989; Le Loeuff 1991a). Among the many reptile bones, we discovered a small box containing dinosaur eggshell fragments, including the one described by Pouech in his personal notes. He had used these eggshells to calculate the radius of the eggs. Pouech's eggshell fragments were collected near Brusquette Farm, close to Le Mas d'Azil. The entire area is now overgrown with vegetation, and few outcrops are visible. Yet, in 1989, we were able to collect a few dinosaur eggshell fragments from a clayey sediment in a ditch a few hundred meters from Brusquette Farm (Buffetaut & Le Loeuff, 1989).

Philippe Matheron and the "*Hypselosaurus priscus*" eggs

Ten years after Pouech's discovery of dinosaur eggshells, the geologist Philippe Matheron (Fig. 2.1) from Marseilles (Depéret, 1900) described the first dinosaurs from the Upper Cretaceous of Provence (southern France). He named an iguanodont from La Nerthe, *Rhabdodon priscus*, and a titanosaur from Rognac, *Hypselosaurus priscus*. Matheron correctly identified *Rhabdodon* as a dinosaur related to *Iguanodon*, but thought that *Hypselosaurus* was a gigantic crocodile. Matheron also mentioned the discovery of some very peculiar fossils:

With the bony fragments I have just mentioned were found two very enigmatic large fragments of a sphere or an ellipsoid, which have tried the patience of several paleontologists. All being considered, they seem to be eggshell fragments. Those eggs were even larger than those of a large bird, which has been named *Aepiornis* by Geoffroy Saint-Hilaire. Are the two above mentioned fragments the remains of two eggs of a giant bird, or are they remains of two hypselosaur eggs? The question still lacks an answer. I shall give a detailed description of these samples, together with drawings, at a later date (authors' translation).

Unfortunately, the detailed, illustrated description was never published. Nevertheless, Matheron's discovery did not go unnoticed, and almost 20 years later it formed the basis of the first study on the microstructure of dinosaur eggshells by Gervais (see below). Matheron also started the tradition of referring fossil eggs from southern France to *Hypselosaurus*. Some of Matheron's followers (e.g., Repelin, 1930) were not as cautious as he was, and referred the eggs from Provence to *Hypselosaurus priscus* solely because of their co-occurrence in the same formation.

Taxonomic identification: the work of Paul Gervais

Paul Gervais (1816–79) was a zoologist and paleontologist, and in 1877 he was head of the Laboratoire d'Anatomie Comparée at the Muséum National d'Histoire Naturelle in Paris. (For an account of Gervais's life and work, see Meunier, 1879.) Gervais was also one of the early leaders of vertebrate paleontology in France. In his study of Upper Cretaceous reptiles, Gervais was one of the first to recognize the occurrence of Owen's Dinosauria in France (Gervais, 1872, 1873, 1877a,b).

In the 1870s Gervais undertook a microscopic study of the eggshells collected by Matheron in the hopes of determining their origin. He compared thin sections of the eggshells with those of birds, turtles, tortoises, crocodiles, and geckos. He concluded that the mysterious eggs from Provence showed similarities with chelonian eggs, although their volume suggested that they were laid by *Hypselosaurus*. He wrote (Gervais 1877a, p. 163):

We are thus led to conclude:
1) that the large fossil eggs from the Rognac beds did not belong to a bird, but to a reptile of indeterminate position showing in the structure of its eggs uncontrovertible analogy with those of some Emydo-saurians.
2) that this reptile, if it was indeed Mr. Matheron's hypselosaur, as everything still leads us to suppose, had, at least in this respect, more resemblance with the chelonians than had been suggested on the basis of the still scanty parts of its skeleton which are hitherto known (authors' translation).

Gervais's final conclusion was cautious. Despite notable similarities in microstructure to the eggs of turtles, he refrained from claiming that Matheron's eggs had been laid by some gigantic species of turtle. He realized that nothing was known about the eggs of dinosaurs, a group which was known to be present in the "Garumnian" beds of southern France where the mysterious eggs had been found. He suggested that a better knowledge of the osteology of *Hypselosaurus* might well show it to have been a dinosaur, rather than a crocodile as Matheron suggested.

Gervais died in 1879, and his work on fossil eggs was all but forgotten. In 1885, Roule repeated Gervais's conclusions on the turtlelike aspect of the eggshells, but noted that their size suggested that they might be eggs of *Hypselosaurus* or *Rhabdodon*. It was not until 1923 that further research was done on the microstructure of the eggs from Provence. At that time, the Belgian scientists van Straelen and Denaeyer published a paper in which they stressed the microstructural resemblances between Matheron's eggs and those of birds. They concluded, however. that it was not possible to identify these fossil eggs.

Conclusions

Some of the richest dinosaur egg localities in the world are in the Upper Cretaceous of southern France. This fact was only recently appreciated because of the work of Dughi and Sirugue in the Aix-en-Provence Basin in the 1950s. Even though Pouech's discovery of dinosaur eggshells was made in the 1850s, it took almost three-quarters of a century until their occurrence in France was accepted (Joleaud 1924). In fact, the impetus for this was provided by the well-publicized discoveries of dinosaur eggs in Mongolia by the American Museum of Natural History. This drew the attention of both the public and the paleontologists to the existence of dinosaur eggs, something Pouech and Matheron did not do.

The reasons for this neglect may well be that they occurred too early. When Pouech made his first discoveries in the 1850s, very little was known about dinosaurs and their importance in Mesozoic ecosystems. The large eggs suggested giant birds rather than giant reptiles to the early French paleontologists. Moreover, the experts from the Paris museum were skeptical, and they failed to understand the importance of Pouech's finds. Not until Matheron found the co-occurrence of eggs and large reptile bones in Provence some years later, did the possible identity of the egg layers become known. However, Matheron thought that *Hypselosaurus*, the likely producer of the eggs, was a gigantic crocodile. Even as late as 1877, when Gervais published the first relatively detailed study of the dinosaur eggs, the affinity of *Hypselosaurus* was still uncertain. As a result, the large fossil eggs from southern France could not be assigned to the dinosaurs.

After Gervais's death in 1879, virtually no research was done on French dinosaur eggs until interest was rekindled by the Mongolian discoveries. Even then, their abundance and significance were not truly recognized until the 1950s. This delay in appreciation may be linked to a general lack of interest in dinosaurs on the part of French paleontologists during most of the first half of the twentieth century. A few papers were published on French dinosaurs during that period, but very little field work was done. (Lapparent's excavations in the Upper Cretaceous of Fox-Amphoux, started in 1939, were cut short by World War II.)

Finally, although the dinosaurian nature of most of the large fossil eggs from the Upper Cretaceous of southern France is now accepted, this is still largely based upon circumstantial evidence. Despite a large amount of work in the field and in the laboratory during the last 40 years, no direct association between eggs and dinosaur bones has yet been reported. As a result, the identity of the dinosaurs that laid the eggs is still uncertain. Usually the eggs are assigned to *Hypselosaurus* (which is itself a poorly defined genus), but this assignment is more the expression of a tradition that goes back to Matheron. Quite clearly, *Hypselosaurus* was not the only dinosaur living in southern France in the Late Cretaceous, and studies on eggshell microstructure suggest that several types of eggs are also present. However, until embryos or hatchlings are found in association with the eggs, much uncertainty will remain concerning the identity of the French dinosaur eggs.

References

Andrews, R. C. 1923. Les oeufs de dinosaure. *L'Illustration* 4217: 669.

1932. *The New Conquest of Asia.* (New York: American Museum of Natural History).

d'Archiac, E. J. A. 1859. Note sur les fossiles recueillis par M. Pouech dans le terrain tertiaire du département de l'Ariège. *Bulletin de la Société Geologique de France* 16: 783–814.

Bilotte, M., Duranthon, F., Clottes, P., & Raynaud, C. 1986. Gisements de dinosaures du nord-est des Pyrénées. *Actes du Colloque Les Dinosaures de la Chine à la France.* (Toulouse, France: Muséum de Toulouse), pp. 151–60.

Buffetaut, E., & Le Loeuff, J. 1989. La première découverte d'oeufs de dinosaures. *Pour la Science* 143: 22.

Depéret, C. 1900. Notice biographique sur Philippe Matheron (1807–1899). *Bulletin de la Société Géologique de France* 28: 511–26.

Gervais, P. 1872. Sur des ossements recueillis par M. Bleicher dans les dépôts lacustres de Villeveyrac. *Bulletin de la Société Géologique de France* 29: 306–7.

1873. Vertebre de l'*Hypselosaurus priscus*, trouvée à Pugère (Bouches du Rhône). *Journal de Zoologie* 2: 469–71.

1877a. Structure des coquilles calcaires des oeufs et caractères que l'on peut en tirer. *Journal de Zoologie* 6: 88–96.

1877b. De la structure des coquilles calcaires des oeufs et des caractères qu'on peut en tirer. *Comptes-Rendus de l'Académie des Sciences* 84: 159–65.

Granger, W. 1936. The story of the dinosaur eggs. *Natural History* 38: 21–5.

Joleaud, L. 1924. Oeufs de dinosauriens et d'oiseaux paleognathes fossiles. *La Feuille des Naturalistes* 3: 44–8.

Leclerc, G. 1983. Jean-Jacques Pouech archéologue d'après ses carnets. *Bulletin de la Société Arigeoise Sciences, Lettres et Arts* 1983: 143–58.

Le Loeuff, J. 1991a. L'Abbé Pouech et les dinosaures. *Géochronique* 38: 16.

1991b. Les vertébrés maastrichtiens du Mas d'Azil (Ariège, France): étude préliminaire de la collection Pouech. *Revue de Paléobiologie* 10: 61–7.

Matheron, P. 1869. Notice sur les reptiles fossiles des depots fluvio-lacustres crétacés du bassin a lignite de Fuveau. *Mémoires de l'Académie Impériale des Sciences, Belles Lettres et Arts de Marseille*: 345–79.

Meunier, S. 1879. Paul Gervais. *La Nature* 302: 225–6.

Pouech, J. J. 1859. Mémoire sur les terrains tertiaires de l'Ariège rapportés à une coupe transversale menée de Fossat à Aillères, passant par le Mas d'Azil, et projetée sur le méridien de ce lieu. *Bulletin de la Société Géologique de France* 16: 381–411.

Repelin, J. 1930. *Description Géologique succincte du Département des Bouches-du-Rhône.* (Marseille, Société Anonyme du Sémaphore de Marseille).

Roule, L. 1885. Recherches sur le terrain fluvio-lacustre inférieur de Provence. *Annales des Sciences Géologiques* 18: 1–138.

Straelen, V. van, & Denaeyer, M.E. 1923. Sur des oeufs fossiles du Crétacé superieur de Rognac en Provence. *Académie Royale de Begique, Bulletin de la Classe des Sciences* 9: 14–26.

Villatte, J., Taquet, P., & Bilotte, M., 1986. Nouveaux restes de dinosauriens dans le Crétacé terminal de l'anticlinal de Dreuilhe. État des connaissances dans le domaine sous-pyrénéen. *Actes du Colloque Les Dinosaures de la Chine à la France.* (Toulouse: Muséum de Toulouse), pp. 89–98.

II Nests

3 Dinosaur nesting patterns

J. J. MORATALLA AND J. E. POWELL

Abstract

We recognize two main egg-laying strategies among dinosaurs – clutches and linear patterns. Clutches may further be subdivided into three types, defined mainly by nest shape and egg distribution. The linear pattern of egg distribution may be typical of saurischian dinosaurs, and three types are recognized. Eggshell ornamentation may be bidirectional (with longitudinal ridges or striations) or multidirectional (with nodes). Bidirectional ornamentation is typical of eggs in which the long axis of the egg is subvertical in the nest. Multidirectional ornamentation is typical of eggs in clutches, or with the long axis of the egg horizontal. High gas conductance values of ornamented eggshells suggest that ornamentation might be an adaptation for increased gas exchange in buried eggs.

In titanosaurs, the relatively thick eggshell, compared to other dinosaur eggshell, could be an adaptation to nonparental care of the nest. Other dinosaurs may have had a more defined colonial nesting strategy, which included an annual return to the nesting site. This behavior may have been present in dinosaurs with both precocial and altricial hatchlings.

Introduction

Fossilized eggs suggest that dinosaurs were oviparous. The first report of dinosaur egg fragments appeared in France in 1859 (see Buffetaut and Le Loeuff, Chapter 2). Matheron (1869) attributed the fragments to the sauropod *Hypselosaurus* based on the thickness of the eggshell and presence of this sauropod in the same deposits. However, the first discovery of complete dinosaur eggs and egg clutches was made by the American Museum of Natural History expedition to Mongolia in 1923. Andrews (1927, 1932) and Brown and Schlaikjer (1940) reported not only eggs and nests, but also associated skeletal remains of *Protoceratops*. This associated material and an egg fragment with bone remains were used to identify the eggs as protoceratopsian. This identification is now in question because a definitive work on the microstructure of these eggshells has never been made, and the association of bones and eggs is equivocal (Hirsch, personal communication).

Problems in identification of the egg layer led Sochava (1969) to assign eggs to a morphotype. He identified the "*Protoceratops*" eggs as angusticanaliculate. However, some of the drawings and photographs (Sochava 1969, Fig. 2, Plate 1) show a ratite microstructure similar to that of ?*Troodon* eggshell from the Upper Cretaceous of Montana. It is possible, then, that some of the Mongolian eggs may belong to a theropod dinosaur. Pending a detailed restudy of the eggs from Mongolia, we refer to these elongated eggs as ?*Protoceratops*.

Dinosaur eggs and eggshells have been reported and described worldwide (see Carpenter and Alf, Chapter 1). Most of this material comes from Upper Cretaceous strata. This situation may result in a biased understanding of dinosaur nesting behavior because we are unable to document possible changes in dinosaur egg-laying strategies during the Mesozoic. It is even possible that the scarcity of eggs before the Upper Cretaceous itself indicates a different egg-laying behavior.

Because dinosaur eggs occur mostly in Upper Cretaceous strata, we must consider dinosaur nesting behavior almost exclusively in the Late Cretaceous and, of course, for only few species. Dodson (1990) estimates there are about 285 genera and 336 species of dinosaurs. In contrast, we only know the egg-laying strategies of ?*Protoceratops*, *Maiasaura*, *Orodromeus*, ?*Troodon*, and possibly the sauropod *Hypselosaurus*. These dinosaurs only comprise about 1.5% of the known genera. In addition, well-preserved nests are scarce and their excavation, let alone that of nesting areas, is not an easy task.

The study of dinosaur eggs must take into consideration the shell microstructure, egg macrostructure and the distribution pattern of the eggs within the nest. These three areas of study are strongly related because they control growth of the embryo. The egg provides all of the nutrients and conditions required for development of the embryo. The yolk and the albumen provide it food, and the eggshell protects the embryo from the external environment (microorganisms, temperature, humidity, etc.), while allowing gas exchange with the environment [water vapor, carbon dioxide (CO_2), and oxygen (O_2)]. The nesting behavior of the adult keeps these environmental conditions in equilibrium.

Eggshell microstructure

The amniotic eggshell is composed of calcium carbonate in the form of aragonite (turtles) or calcite (other amniotic vertebrates). The carbonate material is organized into closely packed spherulitic or prismatic shell units. The shape, size, and arrangement of these shell units are variable, depending on the vertebrate group (Fig. 3.1). The classification of eggs is based on shell morphology and pore canals. As yet, there is no universal system of egg nomenclature, and several classifications have been proposed (Sochava, 1969; Erben, Hoefs, & Wedepohl, 1979; Penner 1983, 1985; Williams, Seymour & Kérourio, 1984). We follow the classification proposed by Hirsch and Quinn (1990) based on Mikhailov (1987a, 1987b). In this classification there are several basic organizational groups: geckonoid, testudoid, crocodiloid, dinosauroid and ornithoid (Fig. 3.1).

The ornithoid type of shell includes theropod and bird eggshells (Fig. 3.2). The more diverse dinosauroid group may be subdivided into spherulitic and prismatic types (Fig. 3.2). Spherulitic dinosaur eggshells show several organizational patterns that reflect the shell units and pore canals. These structures form the basis for further subdivision of eggs into the following three morphotypes: tubospherulitic, prolatospherulitic, and multispherulitic (Fig. 3.2). The diversity of tubospherulitic eggs suggests that there are also several different types that might be distinguished in this group.

Dinosaur egg-laying strategies

The egg-laying strategies of dinosaurs were probably diverse, although at present we can identify only

two main types: clutched (nest) and linear. Moreover, the clutched type can be subdivided on the basis of how the eggs were distributed into concentric circles, spirals, or inverted cones. The linear type also occurs in different patterns of parallel rows or arcs.

Saurischians
Prosauropods

The oldest dinosaur nest is from the Upper Triassic of Patagonia, Argentina (Bonaparte & Vince 1979). Eggshell fragments were found in close association with hatchling skeletons of *Mussaurus*. Unfortunately, there was no evidence of the nest type or the arrangement of the eggs within the nest. The discovery included five hatchlings of the same size in association with eggshell fragments, which suggested that all the remains were in a nest and that this prosauropod might have exhibited altricial behavior.

Another prosauropod nest was reported by Kitching (1979) from the Lower Jurassic Elliot Formation in South Africa. The clutch consisted of six partial eggs about 6.5 cm in length and 5.5 cm wide. The eggs and associated bones suggested that they represent a nest. The occurrence of the prosauropod *Massospondylus* bones in the area and of juvenile bones in the nest, led Kitching to attribute the eggs to a prosauropod dinosaur. The eggshells were relatively thin (0.2–0.3 mm), and their outer surface was smooth (Grine & Kitching, 1987). The photographs showing the microstructure of the shell are not clear, but the shell may be spherulitic with shell units closely interlocked and probably fused. Unfortunately, the pore canals were not described, nor were polarized micrographs published, so no classification of the eggshell is now possible.

Sauropods

The eggs that are referred to sauropods are spherical or subspherical in shape with a nodular surface and a spherulitic microstructure. This type of egg has been

Figure 3.1. Morphology of shell units in different amniotic taxa. **A.** Geckonoid. **B.** Testudoid. **C.** Crocodiloid. **D.** Dinosauroid. **E.** Ornithoid.

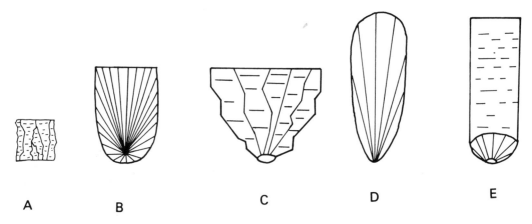

A   B   C   D   E

discovered worldwide, including Spain, France, South America and India (see Vianey-Liaud et al., Chapter 11; Cousin et al., Chapter 5, Sahni et al., Chapter 13). Despite their abundance, no identifiable embryos or hatchlings have been discovered in close association with eggs. Nevertheless, they have been attributed to the titanosaurs based upon (1) the large size of the eggs (5,500 ml), (2) the presence of titanosaur bones in the same strata, (3) the apparent absence of this type of eggshell where skeletal remains of titanosaurs are not present. Furthermore, Mohabey (1987) reported juvenile bone remains in close association with this eggshell type (however, see Carpenter and McIntosh, Chapter 17). The thick shell of the sauropod egg would protect the embryo from small predators because the shell would be difficult to break from the outside.

Titanosaur dinosaurs have several different egg-laying strategies. The first is a circular pattern of six to eight eggs in random distribution. This nest shape is conical in cross section (Fig. 3.3; Kérourio, 1981) and may have been excavated with the forefeet, possibly with the enlarged ungual phalanx of digit I. Such nests have been discovered in the Tremp Basin of Spain (Fig. 3.4; de Lapparent, 1958).

A variant of the conical titanosaur nest was discovered in Balasinor (Rahioli Village, Gujarat) and Bara Simla (Jabalpur) in India. These nests were rounded shallow pits (Mohabey, 1984) with three-to-six randomly distributed eggs. Some nesting areas of spherulitic titanosaur eggshells suggest a colonial nesting strategy, with the clutches well separated or close together (Jain, 1989). We do not yet know whether each clutch was laid by one female or whether one female was responsible for several clutches. The eggs are spher-

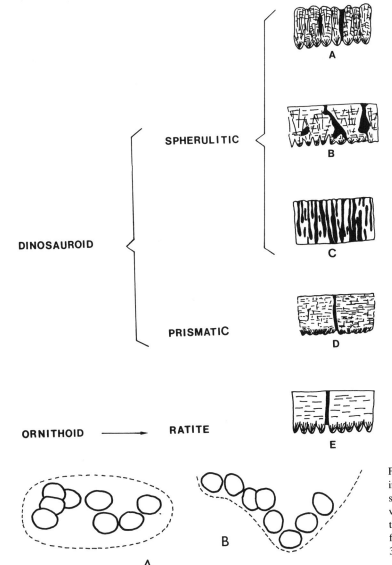

Figure 3.2. Morphotypes of dinosaur eggshells.
A. Tubospherulitic.
B. Prolatospherulitic.
C. Multispherulitic.
D: Prismatic.
E: Ratite. (Based on Sochava, 1969, and Hirsch & Quinn, 1990.)

Figure 3.3. Clutch-type titanosaur nest in which the eggs were deposited into a small excavation in the ground. **A.** Top view. **B.** Profile view. Distribution of the eggs is probably random. (Redrawn from Kérourio, 1981.) See also Figures 3.4 and 3.5.

Figure 3.4. Titanosaur egg clutch from the Upper Cretaceous of the Tremp Basin (Lérida, Spain). Scale = 80 cm.

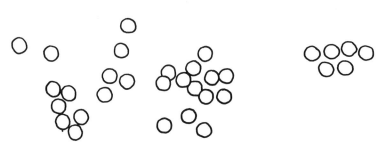

Figure 3.5. Aerial view showing the distribution of titanosaur egg clutches at Salitral Ojo de Agua, Río Negro Province, Argentina. (Redrawn from Powell, in press.)

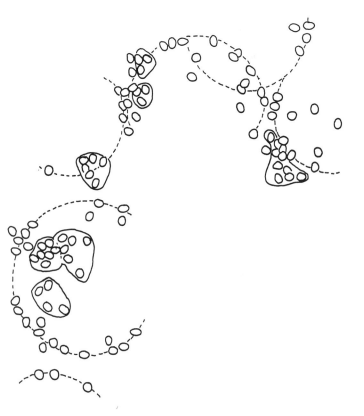

Figure 3.6. Arc-type titanosaur egg pattern in which the squatting females turned around while laying eggs. Diameter of the arcs are proportional to body size. Upper Cretaceous of Rennes-le-Château (France). Redrawn from Beetschen (1985).

ical or subspherical in shape, and the eggshell is of the spherulitic or tubospherulitic type. Several variants of this eggshell microstructure have been described (Vianey-Liaud, Jain, & Sahni et al., 1987; Chapter 11).

Several groups of eggs thought to be nests were recovered and mapped at Salitral Moreno, south of General Roca, Río Negro Province, Argentina (Powell, in press). The egg groups exhibit an irregular pattern of egg-laying. Some groups consist of two levels of circular, randomly distributed eggs, while others are aligned in an irregular pattern. The nests, containing up to 12 eggs, are close together (Fig. 3.5). This closeness and the large size of the adults might exclude the possibility of direct parental care of eggs and hatchlings. Although based on negative evidence, the absence of skeletal remains of babies and juveniles in the nest bearing layer and the presence of many hatched eggs suggest a precocial behavior for this species.

At two sites, the superposition of two levels of eggs were found, suggesting an annual return to the nesting site area. The eggs are spherical and have a diameter of about 18 cm. The shells are very thick (5 mm) and have a multispherulitic structure consisting of very long, thin, crystalline units. These units are usually divided toward the external surface in a dichotomonic arrangement (Powell, 1985, In press). These eggs are similar to those found in Uruguay (Mones, 1980) and Peru (Vildoso-Morales, 1991).

The second type of titanosaur egg-laying strategy has been observed at a site in Rennes-le-Château, France. The eggs are not in clutches, but in arcs, which if connected, would form circles containing fifteen to twenty eggs (Fig. 3.6; Beetschen, 1985; Breton, Fournier, & Watté, 1986; Coombs, 1989). The radius of the circles is 1.3–1.7 m and may correspond to the turning radius of a squatting, egg-laying female. Variations in the circle diameter could be due to the different sizes of the females. These arcs of eggs are not isolated but occur in overlapping groups (Breton et al., 1986). Clutches of six to eight eggs sometimes occur between these circles (Fig. 3.6) and have been attributed to the ornithopod *Rhabdodon* (Breton et al., 1986; Cousin et al., Chapter 5). However, because titanosaurs also laid eggs in clutches, a detailed analysis is needed to correlate eggshell structure, egg shape, and nesting pattern to separate the nesting strategies of *Rhabdodon* and the titanosaurs.

The third egg-laying strategy among titanosaurid dinosaurs was the linear pattern. Dughi and Sirugue (1966) described eggs arranged in up to five linear rows. However, Kérourio (1981) was unable to substantiate this in the more than fifty localities he studied. Moreover, this linear egg-laying pattern has never been reported in other regions in which titanosaur eggs have been found, such as South America and India. This suggests that either the linear pattern was very rare among titanosaurs or that the eggs were not laid by titanosaurs.

Linear rows of eggs were described by Grigorescu et al. (1990) at a site in the Hateg Basin of Transilvania (Romania). The eggs are subspherical in shape with a diameter of about 15 cm. The eggshell is tubospherulitic and is 2.3 to -2.4 mm thick. The eggs were arranged in four linear rows, each containing two or four eggs. The distance between two rows was 0.5 m (Grigorescu et al., 1990). Weishampel, Grigorescu, and Norman (1991; Grigorescu et al., Chapter 6) identified the eggs as hadrosaurid based on embryonic remains found nearby. If the eggs are those of hadrosaurs, and the eggshells are tubospherulitic, then hadrosaurs had two different eggshell types, the other being prolatospherolitic (Hirsch & Quinn, 1990). At present, we prefer to wait for further analyses to shed light on this apparent discrepancy.

Theropods

Theropod eggs have been questionably identified from the Upper Cretaceous of Montana (Horner 1987; Hirsch & Quinn, 1990). These eggs, attributed to ?*Troodon*, are elongated and have a length of about 12 cm. The eggshell has a ratite microstructure and is relatively thin, 1.2–1.3 mm (Hirsch & Quinn, 1990). The eggs were found in the same outcrops as hypsilophodontid nests (Horner, 1984) and occurred in two parallel linear rows with the eggs deposited side by side (Fig. 3.7). This pattern suggests that the eggs were rearranged by the female after egg-laying.

Kurzanov and Mikhailov (1989) reported on a theropod nest from the Lower Cretaceous of Builjasutuin, Mongolia. The eggshell microstructure is of the ratite type, similar to that of ?*Troodon* eggshell. No complete eggs or large shell fragments were found, and the distribution of the eggs within the nest is unknown.

Ornithischians
?*Protoceratops*
Eggshells, eggs, nests, and associated bone remains identified as *Protoceratops* were first reported from the Upper Cretaceous of Mongolia in 1923. The eggs are elongated, with an enlarged blunt end. The surface, except at the poles, has ridges arranged in subparallel and longitudinal patterns (Brown & Schlaikjer, 1940).

The ?*Protoceratops* nest was a circular depres-

Figure 3.7. Two parallel linear rows referred to ?*Troodon*. Upper Cretaceous of Montana. Redrawn from Horner (1987).

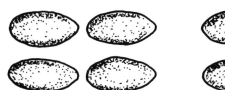

sion with a diameter of about 1–1.5 m. The shape of the nest suggests that it was excavated by the female, although it is possible that a natural depression in the ground was used. The eggs were almost vertically oriented with the blunt end upwards. The number of eggs in a nest is not known, but the best preserved nests suggest about thirty to thirty-five eggs (Brown & Schlaikjer, 1940).

The eggs were arranged in a circular pattern, with up to 3 concentric circles (Fig. 3.8). These circles were not at the same level; the central circle was at the lowest level (Brown & Schlaikjer, 1940). It would seem that it would be difficult for *Protoceratops*, or any quadrupedal dinosaur, to lay or arrange the eggs in neat concentric circles. This would be easier for a bipedal animal, such as a theropod dinosaur, where the "hands" could manipulate the eggs.

The relatively complex egg arrangement in concentric circles is not present in extant animals, excepting possibly in megapodid birds (Coombs, 1989). These birds lay their eggs vertically or subvertically, but not in concentric circles. The concentric dinosaur egg-distribution pattern was relatively complex and sophisticated. It could have been an adaptation for using the minimal amount of space for a nest without the eggs being in contact with one another. It could also have been possible that the temperature of the three different circles and levels were different, making temperature-dependent sex control possible.

Dinosaur eggs and nests similar to ?*Protoceratops* have been reported in several Upper Cretaceous localities in China (Young, 1954, 1965; Zhao, Chapter 12). These have been called *Elongoolithus* (= *Oolithes*) *elongatus*. Unfortunately, there are no associated bone remains that permit accurate identification of these eggs. The nest shape and distribution of the eggs are similar to those of ?*Protoceratops* (Fig. 3.9). The similar age and proximity of both the Mongolian and Chinese sites suggest that these eggs could have been laid by the same type of dinosaur or closely related species.

Maiasaura

The hadrosaur *Maiasaura peeblesorum*, from the Upper Cretaceous Two Medicine Formation in Montana, laid its eggs in a bowl-shaped nest about 2 m in diameter and 0.75 m deep (Fig. 3.10; Horner & Makela, 1979). Vegetation was used to cover the eggs to incubate them (Coombs, 1989). The shape of the eggs is not very well preserved, but may be elongated, with one end larger. The eggs are 10–12 cm long and 7–9 cm wide (Hirsch & Quinn, 1990). Their external surface has ridges oriented more or less longitudinally. The pattern and number of the eggs within the nest are unknown. Horner and Makela (1979) reported a nest with eleven hatchlings, suggesting at least this many eggs had been present (Fig. 3.10). Tooth wear of the hatchlings suggests that the parents had fed them in the nest, implying there was some parental care provided by these dinosaurs.

An assemblage of eight hadrosaur nests, some of them with hatchling remains, was reported by Horner (1982). The nests were at least 7 m apart, coinciding approximately with the length of an adult *Maiasaura*. This spacing suggests that the site was a colonial nesting ground, where the separation between the nests permitted the adults to move through the colony. This type of nesting strategy could have reduced nest predation because of shared vigilance by the adults.

Orodromeus

The eggs of the hypsilophodontid *Orodromeus makelai* were slightly more elongated than those of *Maiasaura*. One end of the egg was larger than the other. The length was about 15 cm. The external surface was smooth, with faint longitudinal striations (Hirsch & Quinn, 1990).

The nests were circular and about 1 m in diam-

Figure 3.8. Nest of (?)*Protoceratops* from the Upper Cretaceous of Mongolia. The eggs seem to be arranged in up to three concentric circles. (Redrawn from a photograph in Colbert, 1983.)

eter. The eggs were arranged almost vertically, with the pointed end buried in the sediment. The axis of the blunt end was often oriented through a middle point situated over the nesting structure. Some of the nests showed a spiral arrangement of the eggs (Fig. 3.11). Horner (1984) suggests that the standard number of eggs in a nest was twelve, and that nests with twenty-four eggs could be the result of two successive egg-laying episodes from the same or different females. The spiral arrangement of the eggs was peculiar and could be due to manipulation of the eggs with the hands or beak. This suggests that a relatively complex behavior was exhibited by this dinosaur.

As with hadrosaur nests, *Orodromeus* nests are not isolated but in groups that suggest colonial nesting. The distance between the nests is about 3 m, coinciding more or less, with the length of an adult *Orodromeus*. There is some evidence to suggest that use of the nesting site continued over several years (Horner, 1982). The structure of the nests is not very well known. Horner and Gorman (1988) suggest that *Orodromeus* did not excavate mounds or hollows. Instead, they laid their eggs in a spiral clutch (Fig. 3.11). Nevertheless, the spiral pattern of the eggs and the distance between the clutches suggest that these dinosaurs carefully tended their eggs (Horner & Gorman, 1988).

Upper Jurassic dinosaur nest

Dinosaur nests from earlier than the Upper Cretaceous are very scarce in the fossil record (see Zhao, Chapter 12). However, Hirsch, Young, and Armstrong (1987) reported a nestlike structure from the Upper Jurassic Morrison Formation in Colorado. The nest structure was not well preserved but appeared to be bowl-shaped. Numerous shell fragments and only a single very compressed egg were recovered. As a result, nothing is known of the egg-distribution pattern. The eggshell fragments are very abundant and have a prismatic microstructure with a smooth outer surface (Hirsch et al., 1987; Hirsch, Chapter 10).

Discussion

It is important to determine whether there is any correlation among eggshell structure (surface ornamentation and microstructure), the egg (shape and size), the nest (shape and egg distribution pattern), and nesting

behavior. The interaction of these elements was probably very important for the proper development of the embryo.

Eggshell structure is very important for embryo respiration, with gas exchange occurring through the pore canals. The main gases involved in this exchange are CO_2, O_2, and water vapor. The number and size of the pores determine the exchange capability and/or conductance of the eggshell. Conductance estimates for dinosaur eggshell gives values about eight to sixteen times that of bird eggs (Seymour 1980). This suggests different environmental conditions, such as high humidity and CO_2, and probably low O_2 (Seymour, 1980). Therefore, a dinosaur egg incubated in the open air could have serious problems with water loss, except in a very humid environment. As a consequence, the environment for dinosaur eggs was most likely underground or within a nesting mound. The various known dinosaur nesting structures support this hypothesis (Mikhailov et al., Chapter 7).

Another important eggshell structure that is predominant in Cretaceous dinosaur eggs is the surface ornamentation provided by nodes, ridges, or striations. This contrasts with the mostly smooth eggshells from the Triassic (Bonaparte & Vince, 1979), Lower Jurassic

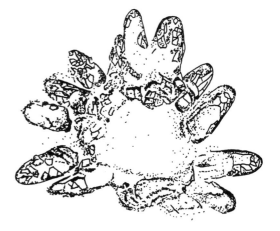

Figure 3.9. Nest of *Elongoolithus elongatus* from the Upper Cretaceous of Chinkangkou (Laiyang Province, China). The morphology of these eggs and nest seems to be similar to that of (?)*Protoceratops* from Mongolia. Egg length is about 11–15 cm. (Redrawn from Young 1954.)

Figure 3.10. *Maiasaura* nest with skeletal remains from the Upper Cretaceous Two Medicine Formation in Montana. Diameter of the nest, where the bones were recovered, is about 2 m. (Redrawn from Horner & Makela, 1979.)

(Kitching, 1979; Grine & Kitching, 1987) and Upper Jurassic (Hirsch et al., 1987). The sole exceptions are rare eggshells with nodes from the Morrison (see Hirsch, Chapter 10). By the Early Cretaceous, eggshells with a nodular external surface appear in greater abundance in Mongolian sites (Kurzanov & Mikhailov, 1989).

The ornamentation of nodes and ridges is formed by the external ends of the crystal units forming the shell. This ornamentation could have had several variations and purposes. For one, it could have strengthened portions of the shell without unduly increasing the length of the pore canals to the point where respiration becomes difficult. This is because the pores occurred not on the tops of the nodes or ridges, but in the valley where the shell is thinnest. Furthermore, strength was achieved without increasing overall shell thickness to the point where the embryo could not rupture the shell. The shell remained thinnest between the ornamentation where the pores also occur.

Another purpose of eggshell ornamentation could have been for underground nests. The external ornamentation could have increased egg placement in the interstitial space between the eggshell and the surrounding substrate and thus improve gas exchange. Eggshell sculpture is not always a necessary feature for nesting below ground because crocodiles and turtles, which bury their eggs, have smooth eggshells. Nevertheless, Mellon (1982) suggests that the longitudinal ornamentation of a vertically placed egg was an adaptation for the removal of CO_2. The ornamentation would have allowed the gas to go up through the interstitial spaces between the ridges. In a horizontally aligned egg, this ornamentation would have caused the accumulation of CO_2 at the poles (Mellon, 1982). This problem could

have been avoided by nodular ornamentation because the CO_2 could have escaped between the nodes.

The various ornamentation styles could also have enhanced O_2 and water vapor exchange by serving as the main pathways for gas circulation between the eggshell and the surrounding substrate. We therefore call the longitudinal arrangement bidirectional ornamentation and the nodular type multidirectional ornamentation.

The existence of nestlike structures (clutches) in the Upper Triassic suggests that nest-building and possible parental care were developed by that time. Therefore, the linear pattern of egg-laying in theropod and titanosaur dinosaurs could be considered as a derived character among saurischian dinosaurs. On the other hand, subsurface nests could be the main reason for the greater abundance of fossil dinosaur eggs in the Cretaceous, especially in the Upper Cretaceous, than in pre-Cretaceous strata. This behavior, of course, would improve the possibilities of eggshell fossilization.

Conclusions

1. Dinosaur egg-laying and nesting behavior were variable among dinosaurs, with two main nesting patterns: circular and linear.
2. Ornithischian dinosaurs made nests with the eggs deposited in a concentric circle or a spiral. The eggs were vertically oriented with the blunt end upwards.
3. The complex egg arrangement in ornithischians could have been an adaptation to use the minimum space necessary for a nest without the eggs touching one another. This suggests that these animals had the capability for recognizing geometrical patterns.
4. Saurischian dinosaurs had more variable nesting behavior. Prosauropods probably made nests. Sauropods, on the other hand, laid eggs in circular clutches, parallel rows, or arcs. Theropods also show a linear pattern or a nestlike structure. The eggs of theropods and sauropods may have been deposited on the ground horizontally and randomly.
5. The most common egg-laying pattern of titanosaur dinosaurs was the circular clutch. The scarcity of linear patterns suggests that parallel rows and arcs were very rare in titanosaur egg-laying behaviors.
6. The ornamentation of the eggshell surface was a typical character of Cretaceous dinosaur eggs. On the contrary, most described Triassic and Jurassic eggshells lack any kind of ornamentation.
7. The eggshell ornamentation could have been an adaptation for subsurface nesting by providing interstitial spaces between the eggshell and the substrate for gas exchange.
8. The two main types of eggshell ornamentation

Figure 3.11. Spiral clutch of *Orodromeus* eggs. Upper Cretaceous Two Medicine Formation in Montana. Eggs are slightly tilted toward the center, and the blunt ends point toward a point situated over the nest. (Redrawn from Horner, 1987.)

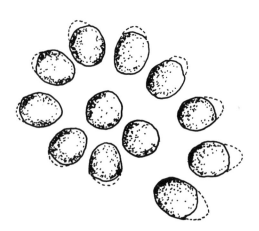

correlate with nesting patterns. (1) Bidirectional ornamentation (longitudinal with ridges or striations) is typical of eggs that were arranged vertically or subvertically in circular nests. (2) Multidirectional (nodular) ornamentation is typical of eggs that were deposited randomly in circular clutches or in a linear pattern such as linear rows or arcs.

9. The relatively greater thickness of titanosaur eggshells compared to other dinosaur eggshells could have been an adaptation for lack of parental care or supervision of the nest, and for incubation by the sun; hatchlings were probably precocial.
10. The behavior of constructing nests by dinosaurs probably developed since the Triassic.
11. Some dinosaurs, such as sauropods, hadrosaurs, and hypsilophodontids, had some type of colonial nesting strategy, including a cyclical return to the nesting site. This colonial behavior is present in dinosaurs with both precocial (hypsilophodontids and probably titanosaurs) and altricial (hadrosaurs) hatchlings.
12. The linear egg-laying pattern seems to have been a derived character for saurischian dinosaurs.

Acknowledgments

We thank Iberdrola for financial support and Karl F. Hirsch for critically reading the manuscript and making valuable suggestions.

References

Andrews, R. C. 1927. Auf der Faehrte der Urmenschen. *Abenteuer und Entdeckungen dreier Expeditionen in die Mongolische Wueste.* (Leipzig: F. A. Brockhaus).
1932. *The new conquest of Central Asia, a narrative of the explorations of the Central Asiatic Expeditions in Mongolia and China, 1921–1930. Natural History of Central Asia,* vol. 1. (New York: American Museum of Natural History).

Beetschen, J. C. 1985. Sur les niveaux à coquilles d'oeufs de Dinosauriens de la région de Rennes-le-Château (Aude). *Actes du Colloque Les Dinosaures de la Chine à la France.* (Toulouse, France: Muséum de Toulouse), pp. 113–126.

Bonaparte, J. F., & Vince, M. 1979. El hallazgo del primer nido de Dinosaurios triásicos, (Saurischia, Prosauropoda), Triásico superior de Patagonia, Argentina. *Ameghiniana* 1(2): 173–82.

Breton, G., Fournier, R., & Watté, J. P. 1986. Le lieu de ponte de dinosaures de Rennes-le-Château (Aude); premiers résultats de la campagne de fouilles 1984. *Annuales du Muséum du Havre* 32: 1–13.

Brown, B., & Schlaikjer, E. M. 1940. The structure and relationships of *Protoceratops. Annals of the New York Academy of Science* 40(3): 133–266.

Colbert, E. H. 1983. *Dinosaurs, an Illustrated History.* (Maplewood, N.J.: Hammond).

Coombs, W. P. 1989. Modern analogs for dinosaur nesting and parental behavior. *Geological Society of America, Special Paper* 238: 21–53.

Dodson, P. 1990. Estimating dinosaur diversity. *Journal of Vertebrate Paleontology* 10(supplement to No. 3): 21A.

Dughi, R., & Sirugue, F. 1966. Sur la fossilization des oeufs de dinosaures. *Comptes rendus des séance de l'Académie des Sciences* 262: 2330–2.

Erben, H. K., Hoefs, J., & Wedepohl, K. H. 1979. Paleobiological and isotopic studies of eggshells from a declining dinosaur species. *Paleobiology* 5(4): 380–414.

Grigorescu, D., Seclamen, M., Norman, D. B., & Weishampel, D. B. 1990. Dinosaur eggs from Romania. *Nature* 346: 417.

Grine, F. E., & Kitching, J. W. 1987. Scanning electron microscopy of early dinosaur egg shell structure: a comparison with other rigid sauropsid eggs. *Scanning Microscopy* 1(2): 615–30.

Hirsch, K. F., & Quinn, B. 1990. Eggs and eggshell fragments from the Upper Cretaceous Two Medicine Formation of Montana. *Journal of Vertebrate Paleontology* 10(4): 491–511.

Hirsch, K. F., Young, R. G., & Armstrong, H. J. 1987. Eggshell fragments from the Jurassic Morrison Formation of Colorado. In *Dinosaur Triangle Paleontological Field Trip Guidebook* (Grand Junction, CO: Museum of Western Colorado).

Horner, J. R. 1982. Evidence of colonial nesting and "site fidelity" among ornithischian dinosaurs. *Nature* 297: 675–6.
1984. The nesting behavior of dinosaurs. *Science American* 250(4): 130–7.
1987. Ecologic and behavioral implications derived from a dinosaur nesting site. *In* S. Czerkas & E. Olson (eds.), *Dinosaurs Past and Present.* (Los Angeles. CA: Natural History Museum of Los Angeles County).

Horner, J. R., & Gorman, J. 1988. *Digging Dinosaurs.* (New York: Workman Publishing)

Horner, J. R., & Makela, R. 1979. Nest of juveniles provides evidence of family structure among dinosaurs. *Nature* 282: 296–8.

Jain, S. L. 1989. Recent dinosaur discoveries in India, including eggshells, nests and coprolites. *In* D. Gillette & M. Lockley (eds.), *Dinosaur Tracks and Traces.* (New York: Cambridge University Press).

Kérourio, P. 1981. Nouvelles observations sur le mode de nidification et de ponte chez les dinosauriens du Crétacé terminal du Midi de la France. *Comptes rendus sommaire des séance de la Société Géologique de France* 1: 25–8.

Kitching, J. W. 1979. Preliminary report on a clutch of six Dinosaurian eggs from the Upper Triassic Elliot Formation, Northern Orange Free State. *Paleontographica Africana* 22: 41–5.

Kurzanov, S. M., & Mikhailov, K. E. 1989. Dinosaur eggshells from the Lower Cretaceous of Mongolia. *In* D. Gillette & M. Lockley (eds.), *Dinosaur Tracks and Traces.* (New York: Cambridge University Press).

Lapparent, A. F. de 1958. Découverte d'un gisement d'oeufs de Dinosauriens dans le Crétacé supérieur du bassin de Tremp (Province de Lérida, Espagne). *Comptes rendus des séance de l'Académie des Sciences* 247: 247.

Matheron, M. P. 1869. Notice sur les reptiles fossiles des dépôts fluvio-lacustres Crétacés du bassin à lignite de Fuveau. *Mémoires de l'Academie Imperiale des Sciences, Belles Lettres et Arts de Marseille*: 345–79.

Mellon, R. M. 1982. *Behavioral Implications of Dinosaur Nesting Patterns*. Unpublished Senior Thesis, Princeton University, New Jersey

Mikhailov, K. E. 1987a. Some aspects of the structure of the shell of the egg. *Paleontological Journal* 3: 54–61.

　　1987b. The principal structure of the avian eggshell: data of SEM studies. *Acta Zoologia Cracoviensa* 30: 53–70.

Mohabey, D. M. 1984. The study of dinosaurian eggs from Infratrappean limestone in Kheda District, Gujarat. *Journal Geological Society of India* 25(6): 329–37.

　　1987. Juvenile sauropod dinosaur from Upper Cretaceous Lameta Formation of Panchmahals District, Gujarat, India. *Journal Geological Society of India* 30: 210–16.

Mones, A. 1980. Nuevos elementos de la herpetofauna del Uruguay (Crocodylia y Dinosauria). *Actas del Congreso Argentino de Paleontologia y Bioestratigrafia* 1: 265–77.

Penner, M. M. 1983. *Contribution à l'étude de la microstructure des coquilles d'oeufs de dinosaures du Crétacé supérieur dans le Bassin d'Aix-en-Provence (France): Application Biostratigraphique*. Unpublished Ph.D. Thesis, L'Université Pierre et Marie Curie, Paris.

　　1985. The problem of dinosaur extinction. Contribution of the study of terminal Cretaceous eggshells from Southeast France. *Geobios* 18(5): 665–70.

Powell, J. E. 1985. Hallazgo de nidadas de huevos de Dinosaurios (Sauropoda-Titanosauridae) del Cretácico superior del Salitral Ojo de Agua, Provincia de Río Negro. *II. Jornadas argentinas de Paleontologa de Vertebrados (resumen)*: 15.

　　In press. Hallazgo de huevos asignables a Dinosaurios titanosuridos (Saurischia, Sauropoda) de la Provincia de Río Negro, Argentina.

Seymour, R. S. 1980. Dinosaur eggs: the relationships between gas conductance through the shell, water loss during incubation and clutch size. *Mémoires de la Societe Géologique de France* N.S. 139: 177–84.

Sochava, A. V. 1969. Dinosaur eggs from the Upper Cretaceous of the Gobi Desert. *Paleontological Journal* 4: 517–27.

Vianey-Liaud, M., Jain, S. L., & Sahni, A. 1987. Dinosaur eggshells (Saurischia) from the Late Cretaceous intertrappean and Lameta Formations (Deccan, India). *Journal of Vertebrate Paleontology* 7(4): 408–24.

Vildoso-Morales, C. A. 1991. Tetrápodos del miembro inferior de la Formacin Bagua (Cretácico tardío-Paleoceno) del norte peruano. *VII. Jornadas argentinas de Paleontologa de Vertebrados. Resúmenes*: 251–52.

Weishampel, D., Grigorescu, D., & Norman, D. 1991. The dinosaurs of Transylvania. *National Geographic Research and Exploration* 7: 196–215.

Williams, D., Seymour, R., & Kérourio, P. 1984. Structure of fossil Dinosaur eggshell from the Aix Basin, France. *Palaeogeography, Palaeoclimatology, Palaeoecology* 45: 23–37.

Young, C. C. 1954. Fossil Reptilian eggs from Laiyang, Shantung, China. *Scientia Sinica* 3: 505–22.

　　1965. Fossil eggs from Nanhsiung, Kwangtung and Kanchou, Kiangsi. *Vertebrata PalAsiatica* 9: 141–70.

4 Dinosaurian eggs from the Upper Cretaceous of Uruguay

GUILLERMO FACCIO

Abstract

A nesting site with abundant dinosaurian eggs and egg-shell fragments and another locality containing eggshell fragments are described from the Upper Cretaceous of Uruguay. At the first locality – the Soriano Site – the eggs have a spherical shape (17–20 cm diameter) and show variability in eggshell thickness (2.5–5 mm) and in the sculpturing of the outer surface. The eggshell material was found at different stratigraphic levels. The tops of the eggs are crushed in with the internal surfaces of the shell fragments facing upwards. The eggs lie very close to each other; some are in contact and some superimpose one another, suggesting that the eggs were laid in a shallow pit. The second locality – the Algorta Site – is composed only of eggshell fragments that have been transported by water. The microstructure of the specimens of the two localities is of the multicanaliculate morphotype.

Introduction

A large variety of egg material has been found worldwide from the Upper Cretaceous (e.g., France, Spain, Russia, Mongolia, China, India, and the United States; see Carpenter & Alf, Chapter 1). A few dinosaur eggshell localities have been reported from South America. Bonaparte and Vince (1979) have reported dinosaur eggs from the Upper Triassic of Patagonia, Argentina, as the oldest from South America. They described two eggs and a nest with several incomplete skeletons of a juvenile Prosauropoda. The eggshell structure has not yet been described in detail. Also from Argentina, Frenguelli (1951) described an isolated egg from the Upper Cretaceous and attributed it possibly to a dinosaur. Powell (in press) described two egg localities from the lower Maastrichtian Allen Formation at Río Negro in Patagonia. He has also reported eggshell fragments in the Campanian Los Alamitos Formation (Powell 1987).

In Brazil, a dinosaur egg has been reported from the Bauru Formation (Price, 1951). Eggshells have been discovered in the Vilquechico Formation in Peru and were referred to the Upper Cretaceous by Sigé (1968).

These eggshells were described by Kérourio and Sigé (1984) who noted two shell types – a tubocanaliculate structure that is dinosaurian and an ornithoid structure that is possibly dinosaurian. These identifications conflict with Van Valen's (1988) assignment of the Vilquechico Formation to the Lower Paleocene rather than Cretaceous. Three types of eggshells were also described by Hirsch in Mourier et al. (1988) from the Bagua Formation (Upper Cretaceous) of Peru. These included a large mamillae type, a vermicular type and a thick gecko-like type.

Mones (1980) has described two kinds of eggs from the Upper Cretaceous of Uruguay based on a parataxonomical classification. The first one, *Sphaerovum erbeni*, is assigned to the titanosaurids based on its spherical shape and large size of the eggs. The second parataxon, *Tacuarembouvum oblongun*, is attributed to the Ornithischia, because of its ellipsoid shape. Upper Cretaceous dinosaurian eggs from the Soriano Site have been described by Faccio, Ford, and Gancio (1990). I now provide additional information about these two egg localities in Uruguay.

Geological setting

The two egg localities, Soriano Site and Algorta Site, are about 150 km apart (Fig. 4.1). Although both sites occur in the Mercedes Formation, they differ taphonomically. The Soriano Site consists of nests with more or less complete eggs and eggshell fragments. The Algorta Site is composed of eggshell fragments of a single structural type. The matrix at both localities is quite similar: fine- to medium-grained sandstone, regular sorting, quartzitic, with clayey siliceous cement, massive, and rose colored (Ford, 1988). Diagenesis has destroyed the original sedimentation structures.

Four genera of sauropods of the family Titanosauridae (*Argyrosaurus*, *Antarctosaurus*, *Laplatasaurus*, and *Titanosaurus*) were described by von Huene

(1929) from strata that could correspond to the Mercedes Formation. This material, however, was not found in situ. The sandstone that contained the dinosaurian eggs has been included in the Mercedes Formation by Ford and Gancio (1988). The age of the Mercedes Formation is Senonian (Upper Cretaceous), based on its relation with other lithostratigraphic units (Sprechman, Bossi, & Da Silva, 1981; Bossi & Navarro, 1990). The Guichon Formation underlies the Mercedes Formation and contains two species of Crocodilian: *Uruguaysuchus aznarezi* and *Uruguaysuchus terrai* (Mesosuchia, Notosuchia, Uruguaysuchidae) described by Rusconi (1933), and are believed to be Cenomanian. Von Huene (1934) described the teeth of an ornithischian, which Bonaparte (1978) assigned to the Iguanodontidae. He suggested that one of them was of chronological significance indicating an Upper Senonian age.

Egg localities
Soriano Site

This site is composed of abundant eggs and eggshell fragments in an area of about 20 m × 40 m. The twenty seven eggs that have been collected from this site are spherical and have a diameter of 17–20 cm. The thickness of the shell varies from 2.5 to 5 mm, but is mostly 5 mm. The stratigraphic level that contained the eggs and eggshell fragments is about 60 cm thick. The sediments that contained the eggs and eggshell frag-

ments were partially covered by soil, which was a major obstacle in the extraction of the material, as was the silicification of the sediment. As a result, not every egg could be collected.

The eggs were found very close to each other; some were in contact with other eggs, and some were superimposed on one another (see Fig. 4.2A,D). The top of the eggs are broken, while the lower portions are well preserved (Figs. 4.2A–D; 4.3A–D). The fragments from the tops lay within the lower portions of the eggs. Most of the fragments were situated with their internal surfaces facing upward (Fig. 4.2B,C; 4.3A–D). Eggshell fragments were also scattered around the eggs. These showed no orientation and pointed in all directions (Fig. 4.3E). Bioturbation occurs in the sediments at this level. A small bone fragment was found in association with the eggs (Fig. 4.3F); its edges are well rounded.

Another egg-bearing level of Soriano occurred 150 cm higher that the specimen described above. The eggshell fragments were of the same structural morphotype. Small bone fragments a few centimeters in length were also found. Microscopic examination of these bones reveal haversian and interstitial systems (Fig. 4.5E), indicating that bone remodeling had occurred. Hence, the bone fragments are from mature individuals. The mode of occurrence of these fossil eggs at the Soriano Site and their state of preservation suggest an autochthonous accumulation.

Figure 4.1. Maps of South America and Uruguay showing dinosaurian egg and eggshell localities – Soriano Site and the Algorta Site.

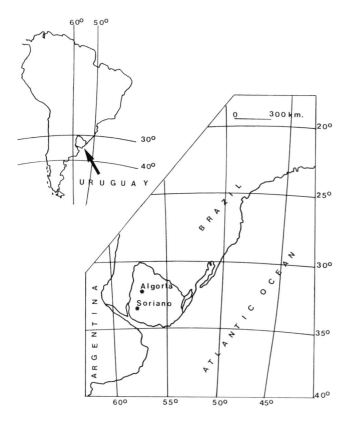

Figure 4.2. Soriano Site. **A.** Nest with four crushed eggs. Eggs are in contact with one another, and two are even superimposed on one another. Lower portions are well preserved, while upper portions are crushed (MPAB 4075). **B, C.** Eggshell fragments from the top of the eggs lie within the lower portions filled with sediment. **D.** Nest with eggs very close to each other and superimposing one another (MPAB 4081). Scale: A, bar = 5 cm; B, bar = 2.1 cm; C, bar = 5 cm; D, 2.5 cm.

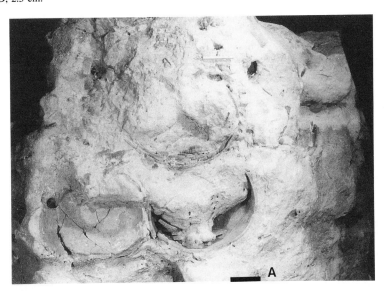

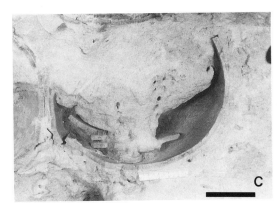

Algorta Site

This site is composed of only a concentration of eggshell fragments of a single structural type. The fragments were found in a stratigraphic unit 10–20 cm thick and 200 m in length. The thickness of the eggshell fragments is about 5 mm. The distribution of the eggshell fragments suggest that they were briefly transported by water.

Eggshell morphology

Outer surface

The outer surface of the Soriano Site specimens is sculptured with nodes clustered very close to each other (Fig. 4.4A–D). The nodes are numerous and are almost 0.8 mm high. In some eggs the nodes are elongated and are up to 1.3 mm high. In some specimens these elongated nodes are oriented in the same direction

Figure 4.3. Dinosaurian eggs from Soriano Site. **A–D.** Well-preserved, sediment-filled lower portions and crushed upper portions of eggs. **E.** Eggshell fragments surrounding eggs (arrows), MPAB 4079. **F.** Crushed eggs in association with a small bone fragment (arrow); note well-rounded edges. A, B: Museo Nacional de Historia Natural, 1; C: MPAB 4076; D: MPAB 4078. Scale: A, bar = 2.5 cm; B, bar = 5 cm; C, bar = 2.5 cm; D, bar = 5 cm.

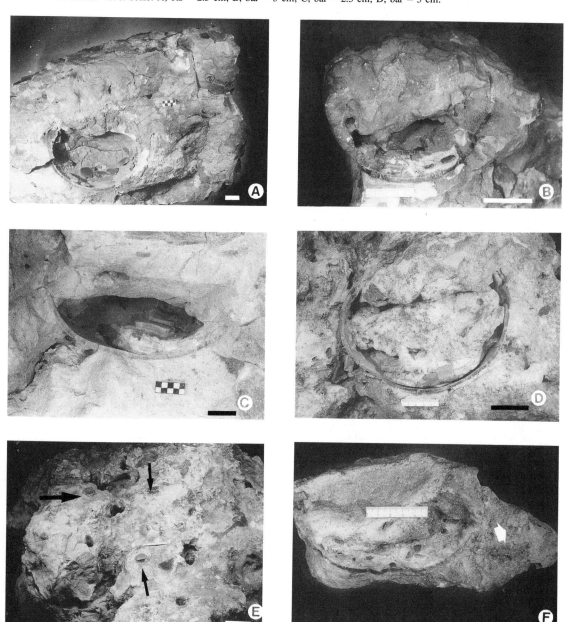

Figure 4.4. **A–D.** Scanning electron micrograph (SEM) of outer surface of eggshell. Note variability in sculpturing. **E., F.** Inner surface of eggshell. Note craters in the mammillae in F. Scale: A–E = 1,000 μm; F, bar = 100 μm.

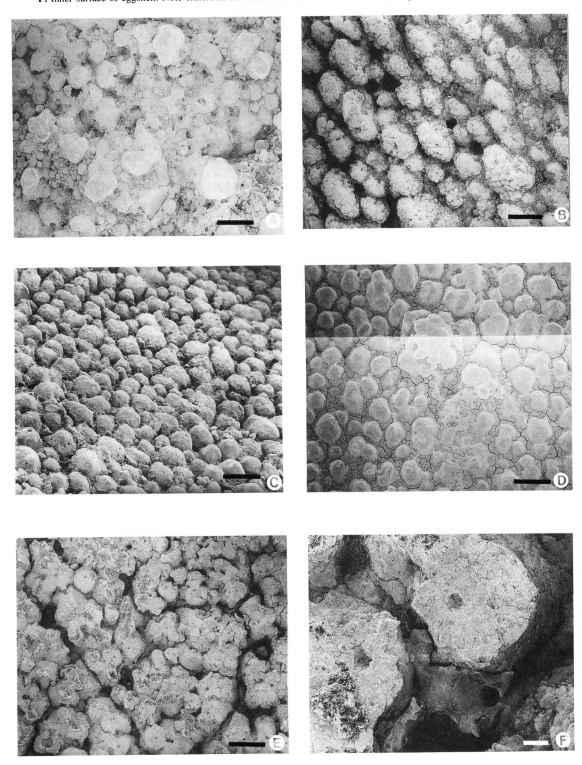

(Fig. 4.4B). In other specimens the nodes are round (Fig. 4.4C), or are overgrown and fused (Fig. 4.4D). Secondary mineral deposits have not been observed on the outer surface of the shell.

Inner shell surface

Specimens from Soriano Site have been replaced by silica (chalcedony). This has not preserved most of the internal structures, and the base of the growth units is generally not well preserved. However, some craters (Fig. 4.4E,F) are visible that could have been caused by the resorption of calcium by the embryo. It is also possible that silicification and diagenesis could have caused these structures. The growth units on the inner surface of the shell are 0.39 to 0.64 mm in diameter.

Radial view and pore system

The shell consists of two layers: a very thin mammillary layer and a thick continuous shell layer. Thin sections between cross polars show that the shell is substituted by chalcedony, with a microfibrous structure (Fig. 4.5D). The shell columns are usually branched and have growth striations (Fig. 4.5B,C).

Pore canals are numerous and their distribution is shown in Figure 4.5A. They are usually filled by chalcedony, which differs from that replacing the unit layers in that the spherulitic microfibers are more developed (Fig. 4.5D). The pores are 0.3–0.4 mm wide.

The structure of the shell layer and the shape and size of the egg are similar to the multicanalicular specimens from Mongolia described by Sochava (1969) and Erben, Hoefs, and Wedepohl (1979).

Discussion and conclusions

The eggs from Uruguay are similar in diameter and spherical shape to those described by Mones (1980) as *Sphaerovum erbeni*. Unfortunately, diagenesis has completely obliterated the eggshell structure. The dinosaurian eggs from the Río Negro localities in Argentina are also similar to the dinosaurian eggs from the Soriano and Algorta sites. These Argentine eggs are from the Allen Formation (Powell, in press) and Los Alamitos Formation (Powell, 1987).

Radial views of thin sections from the Soriano and Algorta eggs identifies these eggshells as belonging to the multicanaliculate morphotype. This type of eggshell was originally described for Mongolian eggs by Sochava (1969) and Erben, Hoefs, & Wedepolh, (1979) (Hirsch, personal communication).

Recently, Mikhailov (1991) proposed a classification for the fossil eggshells of amniotic vertebrates. He accepted the parataxonomic classification of Zhao (1975, 1978, 1979a,b) for the lower taxonomic categories of family, genus, and species. Mikhailov based his diagnoses of the families on the basic type, type of pore system, type of ornamentation, shape of the egg, and range of shell thickness. According to Mikhailov's clas-

sification, multicanaliculate eggshells are now included in the family Faveoloolithidae (Zhao & Ding, 1976), and he suggested a filispherulitic morphotype for this type of eggshell. However, the Uruguayan specimens differ from those of Mongolia in that the shell columns are more pronounced, are branched and have arched growth striations. The outer surface is nodose, and the thickness of the eggshell is greater.

Multicanaliculate eggs have been assigned to the Sauropoda because of their size and shape. Indeed, their large volume and spherical shape suggest that they were deposited by large dinosaurs with a large pelvic cavity, such as the titanosaurid sauropods. However, positive identification must be based on identifiable embryonic remains in eggs or hatchlings within the nests (Hirsch & Packard, 1987).

The top of the eggs are crushed inward, while the lower portions are intact, which suggests that hatching had occurred. Hatched eggs have been noted by Horner (1982, 1984, 1987) and Horner and Gorman (1988) among Ornithischian dinosaurs. However, it is possible that the top portion of the Uruguayan eggs could have caved in due to sediment pressure after being buried. If so, then the occurrence of the eggshells within the eggs with their internal shell surfaces facing upward could be by chance.

Judging by the very close arrangement of the eggs at the Soriano Site where they are in contact and sometimes superimpose one another, the eggs probably were laid in a shallow pit. Burial of eggs by dinosaurs in a shallow pits has been proposed by Mohabey (1984) at Kheda (locality 2), Erben et al. (1979) for the eggs of Corbieres (Albas section), and Kérourio (1981) for the eggs of Aix-en-Provence, Bouches-du-Rhône. The eggs at these localities are spherical, similar to the eggs found at the Soriano Site.

Seymour (1979) proposed that the high porosity of dinosaur eggs indicated a low oxygen, high humidity nest environment, such as would occur underground. The gas conductance capacity for the eggs from France was estimated to be eight to twenty-four times that of avian eggs incubated by body contact (Williams, Seymour, & Kérourio, 1984; Vianey-Liaud et al., Chapter 11), while that for the multicanalicular eggs (?sauropod) from the Gobi Desert was estimated to be 100 times larger (Seymour & Ackerman, 1980). These results suggest that the multicanaliculate pore system of the Uruguayan eggshell indicate a very humid nest environment for the Soriano Site.

The presence of eggshell at two stratigraphic levels at the Soriano Site suggests recurrent nesting at the site (although in the upper level only eggshell fragments and small bone fragments were found). The mode of occurrence of these fossil eggs and their state of preservation suggest an autochthonous accumulation. The presence of bioturbation indicates that the eggs were laid in a paleosol.

Figure 4.5. Scanning electron micrographs (SEM) of eggs and eggshells. **A.** Radial view of Soriano Site eggs, showing distribution of pore canals. **B.** Radial view of Algorta eggshell fragments showing growth striations (arrows). **C.** Radial view showing the columns. **D.** Radial view between cross polars showing that shell replaced by chalcedony and that spherulitic forms (arrow) are more developed in the pores than in the unit layers of the shell. **E.** Thin section of bone fragments showing the haversian and interstitial systems (arrows). Scale: A, bar = 1,000 gmm; B, bar = 100 μm; C, bar = 1,000 μm; D, bar = 0.151 mm; E, bar = 100 μm.

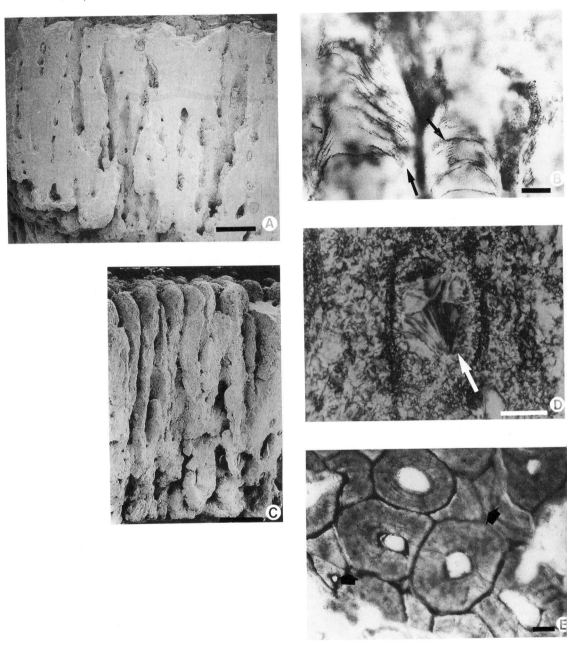

Acknowledgments

I would like to express my thanks to K. F. Hirsch, J. Bonaparte, J. Montana, J. Bossi and to I. Ford and F. Gancio, who discovered the eggshell material. The SEM photographs were taken at the Laboratory of Scanning Electronic Microscopy, U. B. A.

References

Bonaparte, J. F. 1978. El Mesozóico de America del Sur y sus tetrapodos. *Opera Lilloana* 26: 1–596.

Bonaparte, J. F., & Vince, M. 1979. El hallazgo del primer nido de dinosaurios Triásico (Saurischia Prosauropoda), Triásico Superior de Patagonia. *Ameghiniana* 16: 173–82.

Bossi, J., & Navarro, R. 1990. *Geologia del Uruguay*, vols. 1 & 2. (Montevideo: Universidad de la Republica, Departamento de Publicaciones).

Erben, H.K., Hoefs, J., & Wedepohl, K. H. 1979. Paleobiological and isotopic studies of eggshells from a declining dinosaur species. *Paleobiology* 5: 380–414.

Faccio, G., Ford, I., Gancio, F. 1990. Primer registro fosil in situ de huevos de dinosaurios del Cretácico Superior del Uruguay (Fm. Mercedes). *Universidad de la Republica, Facultad de Agronomia, Boletin de Investigacion* 26.

Frenguelli, J. 1951. Un huevo fosil del Rocanense. *Asociacion Geologica Argentina, Revista* 6(2): 108–12.

Ford, I. 1988. Areniscas con huevos de dinosaurios (Biozona Informal): Posible definicion de una nueva formacion en la columna estratigrafica uruguaya. *Panel de Geologia del Litoral. Reunion de Geologia del Uruguay* 1988: 54–6.

Ford, I., & Gancio, F. 1988. Asociacion Caolinita Montmorillonita en un paleosuelo del terciario Inferior del Uruguay. *Universidad de la Republica Facultad de Agromonia. Boletin de Investigacion* 12.

Hirsch, K., & Packard, M. J. 1987. Review of fossil eggs and their shell structure. *Scanning Microscropy* 1: 383–400.

Horner, J. R. 1982. Evidence of colonial nesting and 'site fidelity' among ornithischian dinosaurs. *Nature* 297: 675–6.

1984. The nesting behavior of dinosaurs. *Scientific American* 250: 130–7.

1987. Ecologic and behavioral implications derived from a dinosaur nesting site. *In* S. J. Czerkas & E. C. Olsen (eds.), *Dinosaurs Past and Presents*, Vol. ll. (Los Angeles: Natural History Museum of Los Angeles County), pp. 51–63.

Horner, J. R. and Gorman, J.. 1988. *Digging Dinosaurs*. (New York: Workman Publishing).

Huene, F. von. 1929. Terrestriche Oberkreide in Uruguay. *Centralblatt für Mineralogie, Geologie und Palaeontologie* 8: 107–12.

1934. Neue Saurier-Zahne aus der Kreide von Uruguay. *Zentralblatt für Mineralogie und Palaeontologie* 1934: 182–9.

Kérourio, P. 1981 Nouvelles observations sur le mode de nidification et de ponte chez les dinosauriens de Crétacé terminal du Midi de la France. *Comptes rendus sommaire des séance de la Société Géologique de France* 1: 25–8.

Kérourio, P., & Sigé, B. 1984. L'apport des coquilles d'ouefs de dinosaures de Laguna Umayo à l'âge de la Formation Vilquechico (Pérou) et à la comprehension de *Perutherium altiplanense*. *Newsletters in Stratigraphy* 13: 133–42.

Mikhailov, K. E. 1991. Classification of fossil eggshells of amniotic vertebrates. *Acta Paleontologica Polonica* 36(2): 193–238.

Mohabey, D. M. 1984. The study of dinosaurian eggs from Infratrappean Limestone of Kheda Distric, Gujarat. *Journal of the Geological Society of India* 25: 329–37.

Mones, A. 1980. Nuevos elementos de la paleoherpetofauna del Uruguay (Crocodilia y Dinosauria). *Actas del Congreso Argentino de Paleontología y Bioestratigrafía* 1: 265–77.

Mourier, T., Bengtson, P., Bonhomme, M., Buge, E., Cappetta, H., Crochet, J. Y., Feist, M., Hirsch, K. F., Jaillard, E., Laubacher, G., Lefranc, J. P., Moullade, M., Noblet, C., Pons, D., Rey,J., Sig, B., Tambareau, Y., & Taquet, P. 1988. The Upper Cretaceous-Lower Tertiary marine to continental transition in the Bagua basin, northern Peru. Paleontology, biostratigraphy, radiometry, correlations. *Newsletters in Stratigraphy* 19(3): 143–77.

Powell, J. (1987). The Late Cretaceous fauna of Los Alamitos, Patagonia Argentina. Part Vl. The Titanosaurids. *Revista del Museo Argentino de Ciencias Naturales "Bernardino Rivadavia," Paleontología.* lll, (3).

In press. Hallazgo de huevos asignables a dinosaurios Titanosáuridos (Saurischia, Sauropoda) de la Provincia de Río Negro, Argentina. *Acta Zoologica Lilloana.*

Price, L.I. 1951. Un ovo de dinosaurio na formacao Bauru do Cretácico do Estado de Minas Gerais. *Departamento Nacional da Producao Mineral, Divisao de Geologia e Mineria, Notas Preliminares e Estudos* 53: 1–9.

Rusconi, C. 1933. Sobre reptiles Cretaceos del Uruguay (*Uruguaysuchus aznarezi* n.g. n.sp.) y sus relaciones con los notosuquidos de la Patagonia. *Instituto de Geologia y perforaciones del Uruguay, Boletin* 30.

Seymour, R. S. 1979. Dinosaur eggs: gas conductance through the shell, water loss during incubation and clutch size. *Paleobiology* 5: 1–11

Seymour, R. S., & Ackerman, R. A., 1980. Adaptations to underground nesting in birds and reptiles. *American Zoologist* 20: 437–47.

Sigé, B. 1968. Dents de micromammifères et fragments de coquilles d'oeufs de dinosauriens dans la faune de Vertébrés du Crétacé supérieur de Laguna Umayo (Andes péruviennes). *Comptes rendus des seances de l'Academie des Sciences* (D) 267: 1495–8.

Sochava, A. V. 1969. Dinosaur eggs from the Upper Cretaceous of the Gobi Desert. *Paleontological Journal* 4: 517–27.

Sprechman, P., Bossi, J., & Da Silva, J. 1981. Cuencas del Jurasico y Cretácico del Uruguay. *Cuencas Sedimetarias del Jurasico y Cretacico de America del Sur* 1: 239–70.

Valen, L. Van 1988. Paleocene dinosaur or Cretaceous ungulate in South America. *Evolutionary Monograph* 10: 1–79.

Williams, D. G. L., Seymour, R. S., & Krourio, P. 1984. Structure of fossil dinosaur eggshell from the Aix Basin, France. *Palaeogeography, Palaeoclimatology, Palaeoecology* 45: 23–37.

Zhao Z. 1975. The microstructure of dinosaurian egg-shells of Nanhsiung, Kwangtung. *Vertebrata PalAsiatica* 13: 105–17.

———. 1978. A preliminary investigation on the thinning of the dinosaurian eggshells of Late Cretaceous and some related problems. *Vertebrata PalAsiatica* 16: 314–21.

———. 1979a. Progress in the research of dinosaur eggs in China. In *Mesozoic and Cenozoic "Red Beds" in South China* (Beijing, China: Science Publishing Co.), pp. 329–40.

———. 1979b. Discovery of the dinosaurian eggs and footprints from Neixiang County, Henan Province. *Vertebrata PalAsiatica* 17: 304–9.

Zhao Z., & Ding, S. H. 1976. Discovery of the dinosaurian eggshells from Alxa, Ningxia and its stratigraphical meaning. *Vertebrata PalAsiatica* 14: 42–5.

5 Dinosaur egglaying and nesting in France

RÉMI COUSIN, GÉRARD BRETON,
ROGER FOURNIER, AND
JEAN-PIERRE WATTÉ

Abstract

The first dinosaur eggs were found in 1859, but were not recognized as belonging to dinosaurs until the early twentieth century. Since the 1960s, microstructural analysis of the egg shells and a geometrical analysis of the clutches have been used in France. All clutches and eggs come from continental Maastrichtian deposits (= Bégudien and Rognacien) in the south of France. The main sites for eggs are Aix-en-Provence Basin, Rennes-le-Château, Campagne-sur-Aude and Saint-André-de-Roquelongue. One problem yet to be determined with clutches is whether or not they were originally buried. The studies of Dughi, Sirugue, Erben, and Kérourio are compared to those of Cousin and Breton, especially about the exceptional nesting site at Rennes-le-Château. Any attempt to associate clutches and dinosaur taxa remains highly hypothetical. An appendix gives a list of the French collections housing French dinosaur eggs.

Introduction

Dinosaur egg sites from the Upper Cretaceous of southern France have never provided eggs and embryonic bones together. Therefore, the ten or more different egg or shell types cannot be identified to specific species of dinosaurs known from the same strata. The studies of eggshell in France stress two approaches – microstructural analysis and the organic and isotopic chemistry of the shell developed by Vianey-Liaud et al. (Chapter 11). Our study emphasizes field studies, the analysis of the size and shape of the eggs, and the geometry of clutches.

History of research on French dinosaur eggs and nests

As early as 1859, Abbé Jean-Jacques Pouech published a report of his paleontological research in the vicinity of Mas-d'Azil (Ariège). There, he found, with a few badly preserved bones, some "very large egg shells" (see Buffetaut & Le Loeuff, Chapter 2).

Then, in 1869 Philippe Matheron, "the Father of Provençal geology" (Fournier, 1983), found near Rognac (Bouches-du-Rhône), two enigmatic egg fragments associated with bones attributed to the big herbivorous sauropod *Hypselosaurus priscus* (Matheron, 1869). Matheron (1869) speculated whether the two "large segments of sphere or ellipsoid represent the remains of two eggs of a gigantic bird, or are the remains of two eggs of *Hypselosaurus*."

Gervais (1877) made a microscopic study of the eggshells and compared their microstructure with those of different birds, tortoises, crocodiles, and geckos. He stated that Matheron's shells were most similar to those of the chelonians. He concluded, "I cannot assert that we are dealing here with the eggs of some gigantic species belonging to the order of the tortoises, rather than some reptile of a different group, whose affinities with chelonians remain uncertain. This reservation is necessary because we completely do not know the characteristics of the dinosaurs' egg." We must make it clear that, in 1877, *Hypselosaurus* was not recognized as a dinosaur.

The existence of dinosaur eggs was finally accepted in 1922 after their discovery by an expedition of the American Museum of Natural History in Central Asia. The next year, Maurice Derognat, a Provençal amateur geologist, collected eggshell fragments in the Rognacian beds near Velaux and Rognac (Bouches-du-Rhône) (Derognat, 1935). A microstructural study of these shells was conducted by Victor Van Straelen, who attributed the eggshells to *Hypselosaurus priscus* (Van Straelen & Denaeyer, 1923).

In 1930, 2 km north east of Rousset-sur-Arc, Pierre Chamoux, a vine grower, discovered a whole egg while plowing the sandy marls of Calcaire de Rognac. This was the third whole dinosaur egg known up to that time. Since 1939, the number of egg discoveries in

France has increased. Abbé A.-F. de Lapparent prospected around Fox-Amphoux (Var), Roque-Haute and Eygalières (Bouches-du-Rhône), and Castigno and Saint-Chinian (Hérault). De Lapparent (1947) concluded that these areas only had the eggs of one species of dinosaur, *Hypselosaurus priscus*.

Several complete eggs were collected by Madame Frigara, an amateur geologist, from the Marnes Rutilantes near Rousset-sur-Arc (Bouches-du-Rhône). The Marnes Rutilantes was previously considered as Eocene, and her discovery led to a reassessment of the stratigraphy of the Upper Cretaceous continental deposits (de Lapparent, 1957). Dughi and Sirugue studied the Cretaceous beds of the Aix-en-Provence Basin, and discovered many complete eggs organized in "nests" at Les Grands-Creux (Dughi & Sirugue, 1957a). Unlike de Lapparent, Dughi and Sirugue concluded that at least five egg types were present.

Since the 1960s, the number of egg discoveries in France has increased. Many new sites have been recorded, particularly in the northeast Pyrénées. Paleontologists are now trying to describe the mode of egg laying and nesting of the dinosaurs at these sites (e.g., Vianey-Liaud et al., Chapter 11).

Stratigraphy and paleogeography

Upper Cretaceous continental deposits (lacustrine and fluvial) are widespread in the Pyreneo-Provençal and northern Iberic areas. Several authors have tried to correlate the stages of the continental formations with the marine stages as follows:

Plaziat (1970a, b): Maastrichtian = Bégudien + Rognacien

Babinot et al. (1983): Dano-Montian = Vitrollian; Maastrichtian = Bégudien + Rognacien; Campanian = Valdonnian = Fuvélien

Westphal (1989): Maastrichtian = Rognacien; Campanian = Fuvélien + Rognacian; Uppermost Santonian = Valdonian

The biostratigraphical correlation (Table 5.1) used here is based upon the correlations of Babinot et al. (1983).

Provence

The Upper Senonian (= Coniacian, Santonian, Campanian, and Maastrichtian of North American usage) is well developed in the Aix-en-Provence Basin. The sediments are lacustrine and fluvial. Other less important outcrops, are found in the Alpilles, the Beausset syncline, and north of the Var department. The deposits of the Aix-en-Provence Basin, which can locally be more than 1,000 m thick, have been used as a reference series. The continental stages defined from this series during the last century include:

Rognacian with *Bauxia*, *Lychnus*, and reptiles
Begudian with physas and charophytes
Fuvélian with unios and corbicules
Valdonian with *Campylostylus galloprovincialis*

The base of the deposit correspond to the beginning of the Valdonian according to Matheron's (1878) definition. The top of the deposits is defined by the last dinosaurs, clavatoracea algae, and mollusca with Cretaceous affinities. Eggs and rare bones of dinosaurs are mainly found in the "Argilites inférieures" and "Argilites supérieures" formations below and above the "Barre calcaire de Rognac" at the localities in Rognac,

Table 5.1. *Stratigraphic correlation between the marine and continental series; Upper Cretaceous, southeastern France*

Age (My)	Period	Marine stage	Continental equivalent stage
65		Maastrichtian	Rognacien Bégudien
	Upper Cretaceous	Campanian	Fuvélien Valdonian
100		Santonian Coniacian Turonian Cenomanian	

Notes: Valdonnien: Matheron, 1878, from Valdonne (Bouches-du-Rhône); brackish facies. Fuvélien: Matheron, 1878, from Fuveau (Bouches-du-Rhône); lacustrine, lignitous facies. Bégudien: Villot, 1883, from La Bégude near Fuveau (Bouches-du-Rhône); continental facies. Rognacien: Villot, 1883, from Rognac (Bouches-du-Rhône); continental facies.

Velaux, Rousset, Roques-Hautes, and Saint-Cer. At Fos-sur-Mer, Begudian red marls and argilites have produced dinosaur eggshell fragments (Fig. 5.1).

Bas-Languedoc

Authors of older papers only recognized the Begudo-Rognacian, with its mollusk and reptile faunas at Montpellier, Villeveyrac, Saint-Chinian, and Cobières. Freytet (1970) recognized the Fuvélien, which has dinosaurs, first in the Cobières, then at Villeveyrac and Saint-Chinian. Near Montpellier in the outcrops at Clapiers and Jacou, bones and whole eggs were collected from a coarse Rognacian sandstone (Mattauer & Thaler, 1961). Similar fossils were also collected at Argelliers, 20 km west-northwest of Montpellier.

Alaric, Mouthoumet, Plantaurel, and Petites Pyrénées

The sub-Pyrenean trough and shelf were infilled by deltaic sediments, followed by continental deposits at Grés d'Alet and Grés de Labarre. This infilling is diachronous (Campanian to Maastrichtian) from the east and north toward the west and south (Figs. 5.2, 5.3, 5.4).

Northeast Spain

Lagoonal and continental sediments were deposited at the end of the Maastrichtian (Rognacian) in northeastern Spain (Masriera & Ullastre, 1981, 1983). The "Grés à Reptiles" have produced Cretaceous dinosaur remains. However, the age of the overlaying calcareous bed is still undetermined (Fig. 5.5).

North Cantabric and North Iberic ranges

The "Grés à Reptiles" formation is widespread from Basse-Provence to northwestern Spain through the north side of the Pyrenees. It is diachronous, being Fuvelian in the Bas-Languedoc, Begudo-Rognacian in Provence, and Rognacian in Catalogne and northwest Spain.

The marine province regressed northward and

Figure 5.1. Composite section of the fluviolacustrine Upper Cretaceous of Provence.

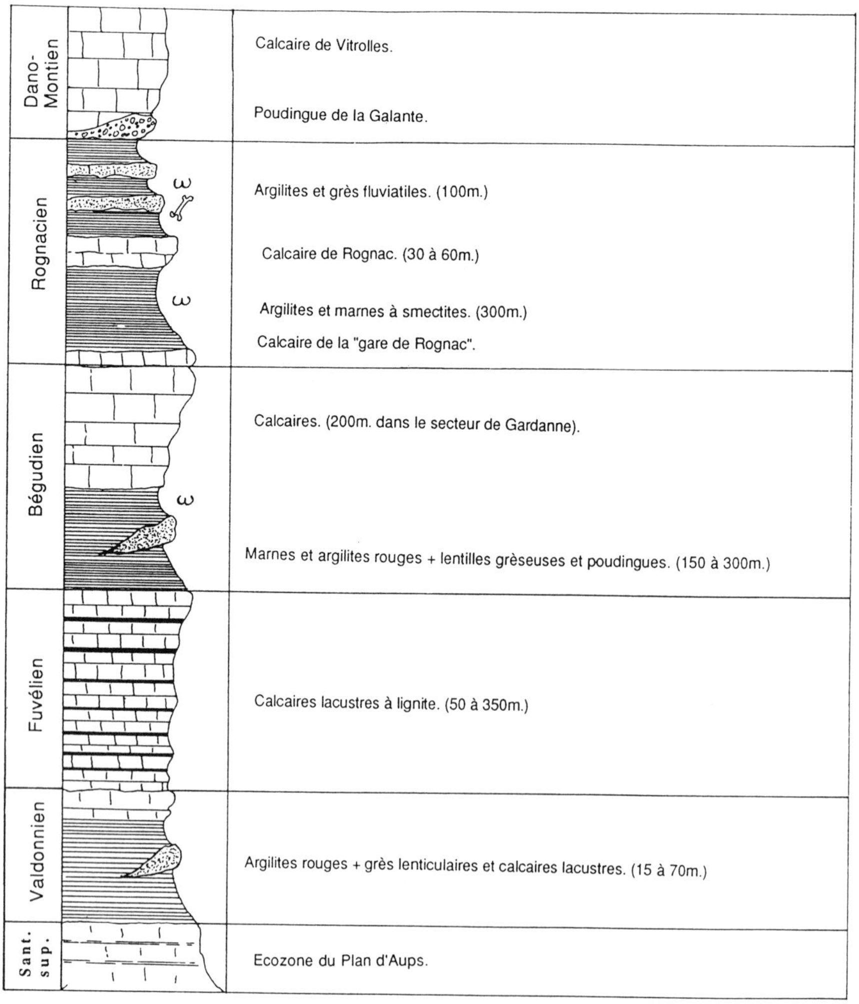

was replaced by diachronous continental deposits of Campanian (Iberic ranges) to Upper Maastrichtian in age (Cantabric ranges). De Lapparent (1957) discovered a site with dinosaurs at Cubilla (Soria Province). More recently, Floquet (1979) identifies it as probably of Maastrichtian age.

Geographical and stratigraphical distribution of dinosaur egg sites

The main sites of dinosaur egg discoveries in southeastern France are shown in Figure 5.6. Deposits with dinosaur eggs vary from mudstone and siltstones to sandstone and coarse conglomeratic channels (Fig. 5.7). The dinosaurs laid their eggs on muddy, sandy, or marshy ground along the shores of rivers or lakes.

Main French dinosaur egg sites

Aix-en-Provence Basin

The most famous French sites are those at the foot of the Montagne-Sainte-Victoire, near Aix-en-Provence (Bouches-du-Rhône) – Roquehaute (= Les Grands-Creux) at Beaurecueil. The outcrops are very extensive with easy access, and are very fossiliferous

(de Lapparent, 1947; Dughi & Sirugue, 1957a,b; Kérourio, 1978). The sites have been collected by many "tourists" and are now protected by a decree passed in 1964.

Between a little stream, which runs parallel to the contact between the Paleocene Breche du Tholonet and the Calcaire de Rognac, is a 100-m section in the Upper Maastrichtian Argiles et Grés à Reptiles (= Lower Rognacian) (Fig. 5.8). These sediments are pellites interspersed with decimeter thick, lenticular units of lighter colored, fine to conglomeratic sandstones. Several beds show hydromorphic pedogenesis that led to a redistribution of the iron compounds as indicated by mottling of the sediments. These beds, interpreted as paleosols, contain carbonated nodules and burrows with horizontal bioglyphs. Dinosaur eggs are abundant in these paleosols, where they are sometimes found with internal molds of small terrestrial gastropods. Lithic gravel, with clasts of millimeter to centimeter in size, are common at all the levels of the section. In the sandstones beds, reworked fragments of dinosaur egg shells are abundant.

During the planning of an on-site museum in 1987, we began an exploratory excavation of a paleosol

Figure 5.2. Composite section of the Upper Cretaceous of the Alaric area. (From Bilotte, 1985.)

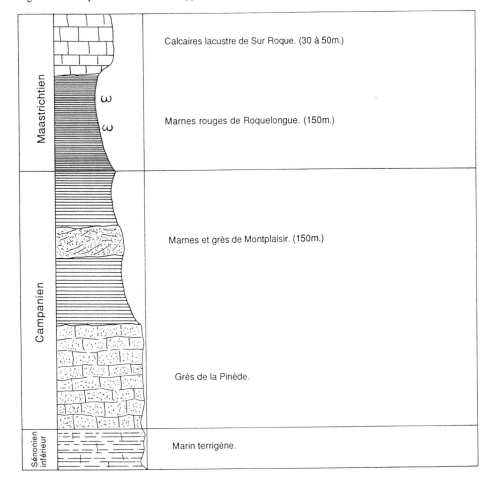

Calcaires lacustre de Sur Roque. (30 à 50m.)

Maastrichtien

Marnes rouges de Roquelongue. (150m.)

Marnes et grès de Montplaisir. (150m.)

Campanien

Grès de la Pinède.

Sénonien inférieur

Marin terrigène.

containing many eggs. The aim was to prepare and display a nesting site as it had been left by newly hatched dinosaurs. Our preliminary excavation, at Rousset-sur-Arc (Bouches-du-Rhône) (Cousin et al., 1987), used the same technique of horizontal excavation as is used in archaeology. Using very fine tools, we were only able to excavate the soft, upper 10 cm of sediment. These sediments corresponded to either a thin, soft alluvial layer and/or to a weathered zone. Below this level the sandstone was too hard to excavate.

Rennes-le-Château (Aude)

Caillaud (1968) described this site which lies northeast of the Pyrenees (Fig. 5.9). It is the westernmost known in France to contain whole eggs in situ. The exposures are smaller than those in Provence. Unlike the site at Roquehaute, the nests at Rennes-Le-Château are exceptionally well preserved. Here, it has been possible to excavate horizontally with fine tools. More than 115 hatched eggs were unearthed from the variegated soft marl (Begudo-rognacian Marnes Rouges Inférieures Formation) during an excavation made in 1984 by the natural history museums of Le Havre and Marseille (Breton, Fournier, and Watté, 1986a,b;

Cousin et al., 1989). This site is described in detail later in this chapter.

Campagne-sur-Aude (Aude)

Located a few kilometers west of Rennes-le-Château, Campagne-sur-Aude was first described by Clottes and Raynaud (1983). A recent excavation yielded a large number of bones, making it one of the richest sites in the Upper Cretaceous of France (Buffetaut & Le Loeuff, 1989b). Numerous eggs are also present in the Begudo-rognacian Grés et Marnes Rouges Inférieures. Clottes and Raynaud (1983) identified two nesting areas that are different from the bone-bearing areas. The eggs are visible in cross section in sandstone blocks that have fallen down the slopes. Clottes and Raynaud (1983) counted thirty-four unhatched eggs. However, we made a mold of three hatched eggs from this site, and found fragments of egg shell from the ''hatching window'' at the bottom of the eggs. The ''hatching window'' had been in the upper part of the eggs (Fig. 5.10) where they had been covered by matrix.

Saint-André-de-Roquelongue (Aude)

This site (Figs. 5.11, 5.12, & 5.13) is located in

Figure 5.3. Composite section of the Upper Cretaceous of the Mouthoumet area. (From Bilotte, 1985.)

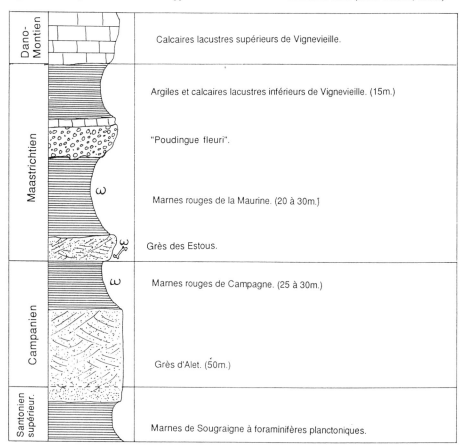

the western Cobières, and has been described by Freytet (1965). The eggs are in a paleosol, which is distinguished from the surrounding Rognacian Marnes Rouges de Roquelongue, by the occurrence of vertical calcareous concretions 1 to 2 cm in diameter which outline roots.

This site has been involved in a debate concerning whether or not the eggs had been buried. Freytet (1965) observed eggs 60 cm beneath the surface of the paleosol and concluded that the dinosaurs laid their eggs in a nest dug 60 cm deep. He also suggested that incubation and pedogenesis were synchronous. We have also observed eggs 50 to 60 cm beneath the surface of the paleosol. However, in 1987, erosion exposed eggs at the top of the paleosol (Fig. 5.11). These eggs are apparently hatched. The upper part of the shell, interpreted as the "hatching window," lies scattered at the bottom of the egg. Some eggs are coated with calcium carbonate, which resembles the vertical calcareous concretions (Fig. 5.13). In some cases, the bottom of the shell and the remains of the "hatching window" lying at the bottom of the shell are coated with carbonate. This indicates that the calcium carbonate was deposited after hatching.

We conclude that the surface upon which the eggs were deposited did not have time to undergo pedogenesis because soon after hatching the eggs were rapidly buried by 60 cm of sediment. Pedogenesis could only occur during a long pause in sedimentation so soil development could occur, with roots and associated concretions overgrowing the buried eggs.

Other dinosaur egg sites

Other smaller and lesser known sites have also provided dinosaur bones and eggs. The Fox-Amphoux site, in the Var department, was first collected from by de Lapparent in 1939, and more recently by Philippe Taquet. Numerous dinosaur taxa are represented, including *Hypselosaurus*, *Rhabdodon*, "*Megalosaurus*," an ankylosaur, and a hadrosaur. Le Loeuff et al. (1989) reported the discovery of the posterior part of a titanosaurid skull in the Var department, but did not indicate the exact location for fear of damage to the site. Dinosaur eggs and bones have also been found in the areas

Figure 5.4. Composite section of the Upper Cretaceous of the "Petites Pyrénées" area. (From Babinot, et al., 1983.)

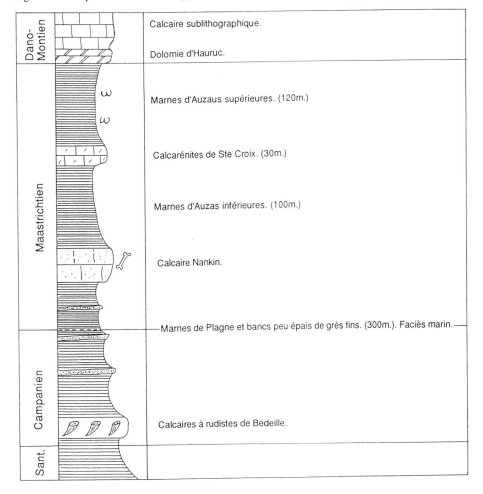

surrounding Montpellier (Hérault), Clapiers, Jacou, and Argelliers (Mattauer & Thaler, 1961).

Numerous eggs and some bones in sandstones were first discovered by Depéret (1900), 22 km from Béziers (Hérault) at Saint-Chinian. A group of two eggs is at the Natural History Museum in Le Havre. Numerous eggs in clutches have been collected at Albas (Aude) and described first by Plaziat (1961). These nests typically contain four, apparently hatched, eggs grouped in a 30 cm radius. Many of them have become misshapened due to compaction. One complete egg is ellipsoidal, and measures 17 × 15 cm. A deep nest has been described from the same site by Erben, Hoefs, & Wedepohl (1979).

Previous studies on dinosaur clutches and egg physiology

Some of the earliest studies on clutches of dinosaur eggs were by Dughi and Sirugue (1958a,b, 1966, 1976). They concluded from their research at Aix-en-Provence Basin and synclines of the Var department that

Dinosaurs did not lay their eggs in sludge, but laid them subaerially, on the firm ground of river banks and marshes, most often on the alluvium left after river flooding that was produced by the tectonic reactivation of the basin headwaters where they lived. (authors' translation)

Their hypothesis that the nests were exposed to the open air is based upon (1) field observations and (2) the fact that the eggs are amniotic, which indicates they belonged to air-breathing organisms. However, the egg-laying habits of certain present-day reptiles indicate that amniotic eggs can be buried in clutches.

Dughi and Sirugue suggested that a clutch was deposited by a crouching female dinosaur who laid one to five eggs directly upon the surface of the ground. The dinosaur then advanced one step, again laid one to five eggs, and so on. A complete clutch consisted of one to five longitudinal rows of eggs. These clutches laid in rows had previously been interpreted as isolated nests of four to five eggs. Dughi and Sirugue pointed out that a whole clutch comprising several rows of eggs could only be observed if erosion cut parallel through all the eggs, rather than by oblique or perpendicular cuts. The geometry of the clutches might have been disturbed by deposition of sediment or by land movement.

Dughi and Sirugue (1958a) originally suggested that the eggs were rapidly buried by sediment, which infilled the eggs through a hole produced by damage or by hatching of the young. Later, they suggested that infilling occurred following the rupture of the eggs caused by putrefaction (Dughi & Sirugue, 1976). This

Figure 5.5. Composite section of the Upper Cretaceous of Catalogne (northern Spain) (From Masriera & Ullastre, 1983.)

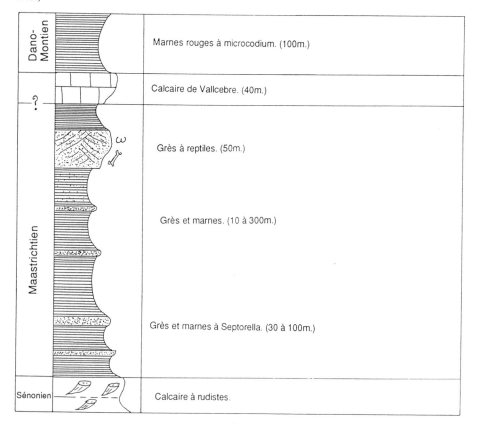

hypothesis implies that the eggs never hatched. Our excavations at Rennes-le-Chteau, however, suggest that the hole in the egg is the "hatching window," which would indicate that the eggs had hatched (Breton et al., 1986a). Our research concludes that

In the Aix-en-Provence Basin, the number of eggs increases upwards in the lower Vitrollian [sensu Matheron = Upper Maastrichtian]. The presence of eggs presupposes the presence of beasts, and the presence of beasts presupposes the presence of fertile eggs.

There is, therefore, a limitation to the hypothesis by Dughi and Sirugue (1966) of dinosaurs laying infertile eggs.

Dughi and Sirugue's hypothesis of open air nests, supported by Thaler (1965) and Ginsburg (1980), has been questioned by Erben et al. (1979) and Kérourio (1981, 1987), who noted that an open air clutch would have been scattered by fluvial action on the floodplain. The possibility that the eggs might have been disturbed (turned upside down) had been considered by Dughi and

Sirugue (1966). Kérourio commented that Dughi and Sirugue's assumption that the topography was flat where the eggs were laid was not true, and that two clutches located at the same level today were not necessarily contemporaneous.

The first buried nests to be discovered in France were found near Albas (Aude). Here, Erben et al. (1979) observed eggs at the bottom of a deep, narrow pit. Two years later, Kérourio (1981) described several buried clutches in the central part of the Arc Syncline (Aix-en-Provence Basin) (Fig. 5.14). One of these, located 1 km north of Rousset-sur-Arc (Bouches-du-Rhone), was excavated and yielded eight large eggs. The eggs appear to delineate a cone 70 cm deep, with an opening 120 cm long and 70 cm wide (Fig. 5.14A,B). No other clutch was excavated because the matrix was too hard. The eggs, however, were drawn in plan view. Kérourio noted similarities between these buried clutches to the one previously described by Erben et al. (1979), and noted that they differed from the "structured" clutches from the Upper Cretaceous of Mongolia and China (Young,

Figure 5.6. Main dinosaur egg sites in southeastern France. **1.** Fox Amphoux (Var). **2.** Rousset-sur-Arc (Bouches-du-Rhône). **3.** Les Grand-Creux, commune de Beaurecueil, Roquehaute (Bouches-du-Rhône). **4.** Rognac (Bouches-du-Rhône). **5.** Saint-Chinian (Hérault). **6.** Saint-André-de-Roquelongue (Aude). **7.** Albas (Aude). **8.** Rennes-le-Château (Aude). **9.** Campagne-sur-Aude (Aude).

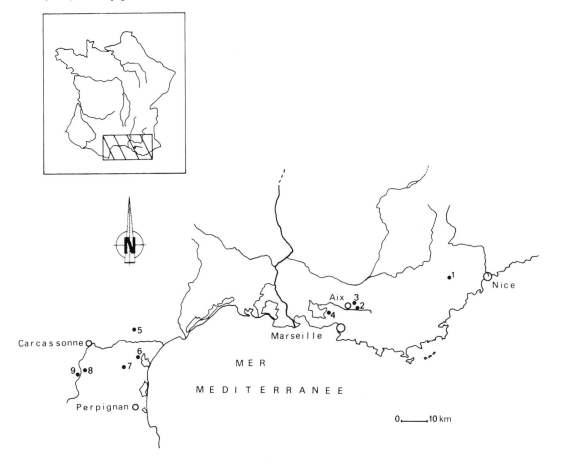

1965; Zhao & Chiang, 1974) and Montana (U.S.A.) (Horner, 1984).

Kérourio (1981) also noted a large number of isolated eggs, from which he concluded that an unknown species of dinosaur laid only single eggs. He evidently never observed clutches in rows like those described by Dughi and Sirugue (1958b, 1976). Nevertheless, Kérourio's plan view of the ''nest'' at Rousset shows three eggs in an arcuate row (labeled 1, 2, 3 in Fig. 5.14A). Because the matrix filling the nest can not be differentiated from the surrounding sediment, we do not rule out the possibility that these eggs do not belong to the clutch; additionally, the limits of the nest cannot be determined. From this it is difficult to assert that (1) the eight eggs, distributed through 70 cm of matrix, cannot belong to successive clutches, or (2) there are no eggs outside the excavation that cannot belong to one of these successive clutches. Perhaps expansion of Kérourio's excavation would reveal an altogether different organization of the clutches, with juxtaposition of rows and clusters similar to the ones we described at Rennes-le-Chteau. With this caution, the clutch shown in Figure 5.14C must be interpreted carefully.

A clutch collected at Puyloubier (Fig. 5.14D), and curated in the Muséum du Havre is homogeneous and is here interpreted as a single cluster of eggs. The two rows of three eggs (labeled 1, 2, 3 and 4, 5, 6 in Fig. 5.14D) is misleading. In fact, all six eggs are very eroded, and the diameter of the preserved portions is four times less than the greatest diameter of the original eggs. When laid, the eggs were in greater proximity than they appear today.

The physiology of the dinosaur eggshell has been used by Kérourio to suggest that dinosaur eggs were incubated beneath a layer of earth and/or plants. This interpretation is based upon the gas conductance of the shell as determined by Seymour (1979). In present-day reptiles, conductance (i.e., permeability of the shell to oxygen, carbon dioxide, and metabolic water) is high because the eggs are incubated in a reducing environment, which can either be under plant debris for crocodilians, or in the earth for chelonians. Gas conductance was determined by Kérourio in shells collected from four different types of whole eggs from the Aix Basin (Williams, Seymour, & Kérourio, 1984). This study showed that the values were eight to twenty-four times greater than those of avian eggs of the same size. From this, Kérourio concluded that incubation of the dinosaur eggs was similar to that of present-day reptiles, whose eggs are buried by the female.

Nesting site at Rennes-le-Château (Aude)

In April 1984, the natural history museums of Marseilles and Le Havre combined their expertise, staff, and physical resources for a 50 m² excavation in the Upper Maastrichtian continental deposits (Rognacian) at Rennes-le-Château (Aude) (Figs. 5.15–5.18). Because the matrix was a soft red marl, we excavated horizon-

Figure 5.7. Channel fill in a Rognacian sandstone at Châteauneuf-le-Rouge (Bouches-du-Rhône).

tally with very fine tools at 10 cm intervals that were recorded as surfaces A, B, C, D, beneath a grid (Watt, Fournier and Breton 1986; Watt, 1989). Because of the matrix, egg shells were extensively fragmented, evidently during compaction of the sediment. Most of the fragments were only a few square centimeters in size and rarely reconsolidated. Isolated fragments were very abundant in layer A (several hundred per square meter), and very rare in layers B, C and D. This suggests that dispersal of the fragments in layer A was due to recent pedogenesis and associated bioturbation. Dispersal of the shell during the Rognacian, whether by dinosaurs or

by fluvial action that deposited sediment on the eggs, was quite limited.

In contrast to Dughi and Sirugue's observations on the Provencal material, we noted that the opening of the eggs was always directed upward, and we interpreted this as a ''hatching window.'' When the egg hatched, the irregular shaped ''window'' was pushed upwards and fell back into the egg shell, where the fragments lay concave side upwards. In a few cases, we have observed the ''window'' fragment lying subvertically, and, in one instance, it lay broken partly inside and partly outside the egg (Fig. 5.10). Of the more than 115 eggs that were excavated, only two

Figure 5.8. Lithostratigraphy of the Rognacian sequence at Les Grands-Creux (= Roquehaute), Beaurecueil (Bouches-du-Rhône). (Sections by Breton and Fournier, July 12, 1990.) **1.** Red and yellow variegated argilites. **2.** Fine-grained, well-sorted, white sandstone, bioturbated at the top. **3.** Red argilites with gravel centimeters in diameter. **4.** Fine-grained, white sandstone, with small-grained conglomeratic lenses; bed thickness variable. **5.** Wine-red argilites with large-grained gravel, becoming a variegated paleosol upward. **6.** Discontinuous bed of white sandstone with small-grained conglomeratic lenses. **7.** Variegated bed with calcareous concretions. **8.** Lens of bioturbated sandstone. **9.** Reddish argilites with gravel, becoming a paleosol upward with small gastropods and whole eggs (34–36 m). **10.** Lenticular bed of white, fine-grained, well-sorted sandstone. **11.** Reddish argilites, becoming a paleosol upward, with numerous shell fragments and whole eggs. **12.** Wine-red argilites, very fine grained, becoming a paleosol upwards with thick egg fragments. **13.** Variegated paleosol, with several egg fragments. **14.** Paleosol, no eggshell. **15.** Red argilites with some concretions, becoming a paleosol upward. **16.** Red argilites with some concretions, becoming a paleosol upward. **17.** Fine red argilites. **18.** Bed of fine-grained, crossbedded sandstone with conglomerate at the base. **19.** Argilites, variegated and reddish at the base, and yellow-green at the top, capped by the paleosol of bed 20. **20.** Massive, white, calcareous paleosol. **21.** Paleosol with whole eggs. **22.** Pink calcareous paleosol, with whole eggs at the base. **23, 24.** Grey marls, becoming a paleosol upward. **25.** Covered. **26.** Thick, calcareous bed = Calcaire de Rognac.

Figure 5.9. Lithological sequence from the Begudo-rognacian to the Thanetian at Rennes-le-Château (Aude). (Section by Fournier.)

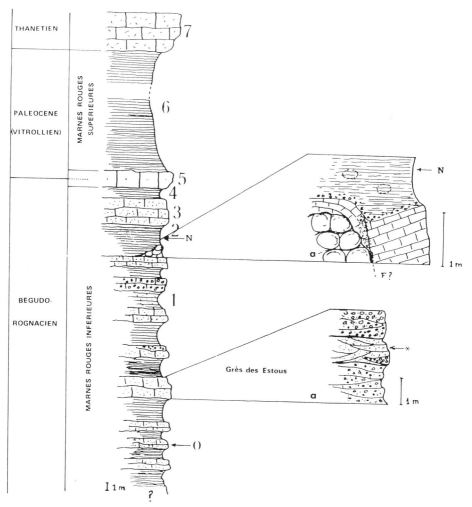

Figure 5.10. Three hatched eggs in a Rognacian sandstone at Campagne-sur-Aude (Aude).

had not hatched. The shell microstructure of these eggs did not show any pathological features. Nor were embryonic bones recovered from them, possibly because the eggs were not fertile.

Horizontal distribution of eggs

Because of the difficulty of identifying paleosols or paleosurfaces (other than the contact point with the bottom of the egg) and because of the probable irregular topography of the paleosurfaces, we group on the same plane all the eggs, regardless of their 10 cm stratigraphic level designation (Fig. 5.16). Two different patterns of laying are clearly evident from the resulting map.

1. Arcuate rows of two, three or four eggs (Figs. 5.16 & 5.17). By connecting these arcs, the eggs may belong to larger, circular groups of up to fourteen eggs. The radii of these groups range from 1.30 to 1.70 m. Apparently, the axes of the eggs were randomly oriented about the circle. All the eggs of a circular group are believed to have been laid at the same time. This type of egg grouping is similar to the rows described by Dughi and Sirugue for the Aix Basin, although the arrangement is more obvious here.

2. Clusters of three or up to possibly eight eggs with no particular arrangement. These clusters look like the "nests" described by Kérourio (1981; Fig. 5.14A,B), but their vertical distribution does not exceed 35 cm.

Other eggs are isolated and could have been laid singly (a third type of egg-laying or a partial arc type). The eggs could also belong to a stratigraphically higher group that has since been removed by erosion (or previous collecting!), or else be lying outside the limits of the excavation.

Depth of eggs laid

The detailed horizontal excavation enabled us to obtain a view of the horizontal and vertical distribution of the eggs. Without such a spatial perspective, many errors would have arisen. For example, collection FG 11 (Fig. 5.18) shows a group of nine eggs laid 30 cm deep. A vertical section suggests that the eggs were laid in a nest dug into the soil, like those described by Kérourio (1981). However, excavation of this group with fine tools showed that three egg levels were superposed, indicating at least three successive episodes of egg-laying. From this and similar analyses of the clutches at Rennes-le-Château, our conclusion excludes the possibility that the eggs were buried. This is in marked contrast to the conclusion of egg burial reached by Kérourio based on gas conductance of dinosaur eggshells. At present, we are unable to explain this contradiction between the evidence for buried eggs and for eggs in an open nest. It would be profitable to extend this type of excavation technique to the other French sites, but only at Rennes-le-Château does the topography and soft matrix make this possible.

Egg-layers and egg-laying behavior

The taxonomic identity of any egg is extremely hypothetical without its close association with skeletal remains. Nevertheless, we tentatively suggest the identity for two of the egg-layers on the basis of their skeletal abundance in the Upper Cretaceous of southern France. These are the titanosaurid sauropod *Hypselosaurus priscus* Matheron and the iguanodontid ornithopod *Rhabdodon priscus* Matheron.

We attribute the circles of eggs (type 1 above) to

Figure 5.11. Profile of a paleosol with dinosaur eggs at Saint-André-de-Roquelongue (Aude). **1, 2, 3, 4, 5.** Complete eggs. **6, 7.** Egg fragments. **A.** Variegated marls with polyhedral structure and cylindrical calcareous concretions. Many isolated eggshell fragments occur between the base of the paleosol and the 100-cm level. **B.** Sandstone lenses with layers of blunted shell fragments. (Section by Cousin, November 27, 1987.)

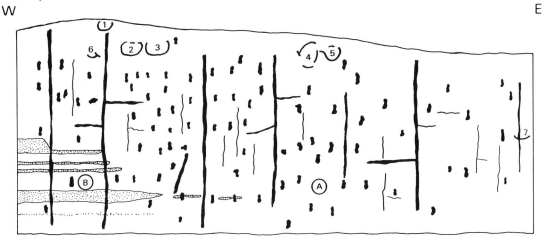

W   E

20 cm

Hypselosaurus. The eggs could have been deposited by a single female that turned while laying, trampling the center of the circle with her hind legs. The radius of this circle is thus equal to the horizontal distance between the hind feet and the cloaca. The eggs deposited in such circles could only be laid from an erect or semi-erect bipedal stance because a squatting position is excluded by functional morphology. This large sauropod (12 m long according to Rolland, 1980) had front legs much shorter (154 cm) than its hind legs (190 cm; de Lapparent 1947).

The difference in size between front and rear limbs indicates the possibility of bipedal posture; however, it is doubtful that the females assumed this position while egg-laying because the cloaca would be 1.5 m above the ground, too far for an egg to fall safely. By squatting, the eggs would exit closer to the ground. Testing this hypothesis may be done by comparing the hind leg measurements and estimated position of the cloaca with the radius of the egg circles. Such comparisons actually suggest that the female assumed a half crouching position with the hind legs slightly bent. Variations in the radii of the egg circles (e.g., 1.30, 1.50, and 1.70 m) indicate different sized females in the herd, probably as a result of continual growth after sexual maturity.

The other eggs arranged in clusters (type 2 above) might be, as implied by our hypothesis, attributed to *Rhabdodon priscus* or a sauropod distinct from *Hypselosaurus priscus.*

Taxonomy of egg-layers

Bones collected from the Upper Cretaceous of southern France are not very common but belong to at least six taxa, four of which are known at the specific level (Buffetaut, Méchin, & Salessy, 1988; Buffetaut 1989; Villate, Taquet, & Bilotte, 1986). These are:

Suborder Sauropoda	*Hypselosaurus priscus* Matheron
	Titanosaurus indicus Lydekker
Suborder Ornithopoda	*Rhabdodon priscus* Matheron
	Orthomerus sp.
Suborder Theropoda	"*Megalosaurus*" sp.
Suborder Ankylosauria	*Struthiosaurus* sp.

In addition, a skull of a hadrosaur has also been collected from the Begudian Sandstones of the Aix basin.

According to Dughi and Sirugue (1976), dinosaur

Figure 5.12. Detail of Figure 5.11 showing eggs. **A.** Group of eggs at the top of the paleosol. Eggs 1 and 3 are open upward and are thus presumed hatched. Egg 2 looks complete, but shell fragments within the egg suggests that the opening is behind the section plane. **B.** Two eggs with a "double bottom," which might be a "hatching window" that has fallen into the egg after hatching. **C.** Bottom of an egg containing the "hatching window" (the upper part of the egg may have been destroyed by erosion).

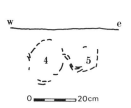

Figure 5.13. Concretion-coated egg from the paleosol shown in Figure 5.11, collected 60 cm beneath the top of the paleosol. (Breton & Fournier, August 21, 1986.)

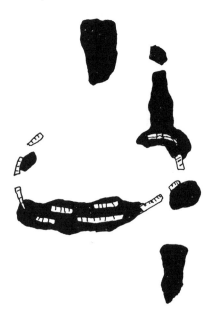

egg ''species'' include at least ten types based on morphologic, morphometric, and microstructural criteria. These authors further tried to assign these egg types to various dinosaurs by reconstructing the diversity and relative abundance of the dinosaurs by stage.

During the Begudian, the dinosaur population consisted mainly of the herbivorous and aquatic sauropods, along with a few ornithopods and very few theropods. In the Lower Rognacian, the terrestrial and carnivorous theropods were briefly the dominant group before they died out together with sauropods at the beginning of the Upper Rognacian. Finally, the herbivorous and terrestrial ornithopods remained until they died out at the end of the Cretaceous.

From this information, Dughi and Sirugue (1976) concluded that sauropods had

medium-sized, ellipsoidal eggs, with thin and delicate shells, light brown in color, with simple [pore] canals, regular[ly spaced] external tubercles, crystal units [with a] low, top-shape, and big, brownish crystals.'' [Theropods have] ''medium-sized, ellipsoidal eggs, with brown, grey or sometimes blackish shells, [that are] thick and sturdy, the tubercles of which are diversely distributed, with long, top-shaped crystal units [that are] often few or poorly contiguous, with big crystals and a matrix concentrically arranged, with branched aeration canals. [The ornithopod eggs] are small, medium-sized, or are amongst the largest. They are either ellipsoidal (small and medium eggs) or subspherical, and their shells, which bear tubercles, are of various thickness. Sometimes white, but often stained by the surrounding sediment, they can always be distinguished by the section of their shell, which is glassy and white or whitish.

Except for these brief descriptions, Dughi and Sirugue did not adequately describe or illustrate the egg

species, and have been criticized for this by Aujard and Kérourio (1986) and Beetschen (1986). An attempt to replicate Dughi and Sirugue's study was made by Penner (1983), but he could only identify three types of monolayered eggshells which he assigned to sauropods. Dauphin (1990) argued that without adequate evidence no real association between clutches and reptile taxa is possible today for the eggs found at French sites. We agree with this, especially because the number of egg shell types may be greater than the number of dinosaurian taxa known today.

Conclusion

Excavating with archaeological methods and field studies provides valuable information on nesting, laying, and hatching of dinosaur eggs in prehistoric France. At Rennes-le-Château, the clutches were apparently not buried, despite the high gas conductance of the eggshell. Reexamination of other previously described eggs as buried clutches has prompted to us to believe that all were clutches laid upon the surface of the ground.

Acknowledgments

The authors are very much indebted to the voluntary workers who made the excavation at Rennes-le-Château in 1984. They also want to thank M. Vianey-Liaud, and the two reviewers, Karl F. Hirsch and Walter P. Coombs, Jr. for their help and for the improvements they suggested.

References

Aujard, C., & Kérourio, P. 1986. Les oeufs de dinosaures. *Archeologia* 215: 66–75.

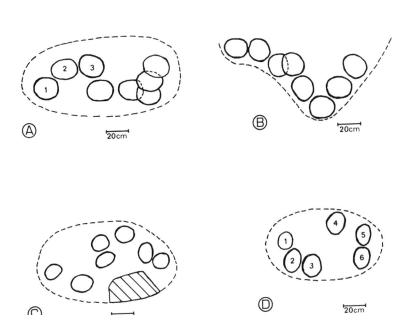

Figure 5.14. Dinosaur ''nests'' from the uppermost Cretaceous of Aix basin, southeastern France. **A.** Plan view of a clutch found in the Upper Rognacian at Rousset (Bouches-du-Rhône). Dashed line indicates the possible limits of the ''nest.'' **B.** Vertical section of the nest drawn in A. **C.** Plan view of a clutch (not excavated) in the Lower Rognacian at Les Grands-Creux (= Roquehaute), Beaurecueil (Bouches-du-Rhône). Dashed line indicates the possible limits of the ''nest.'' Area destroyed shown by hatched lines. **D.** Plan view of a clutch from the Upper Rognacian, at St-Ser, Puyloubier (Bouches-du-Rhône). Dashed line indicates the possible limits of the ''nest.'' Clutch depicted has been removed and is housed in the Natural History Museum, Le Havre (Seine-Maritime) (MHNH 2272). (Redrawn from Kérourio, 1981.)

Babinot, J.-F., Freytet, P., Amiot, M., Bilotte, M., Broin, F. de, Colombo, F., Durand, J.-P., Floquet, M., Gayet, M., Lange-Badré, B., Masriera, A., Massieu, M., Medus, J., Tambareau, Y., Ullastre, J., & Villate, J. 1983. Le Sénonien supérieur continental de la France méridionale et de l'Espagne septentrionale: état des connaissances biostratigraphiques. Colloque sur le Sénonien. *Géologie méditerranéenne* 10 (3–4): 245–68.

Beetschen, J.-C. 1986. Sur les niveaux à coquilles d'oeufs de dinosaures de la région de Rennes-le-Château (Aude). *Actes du Colloque Les Dinosaures de la Chine à la France.* (Toulouse: Muséum de Toulouse), pp. 114–26.

Bilotte, M. 1985. Le Crétacé supérieur des plates-formes est-pyrénéennes. Doctoral thesis. *Strata* 5 (Ser. 2): 1–438.

Breton, G., Fournier, R., & Watté, J.-P. 1986a. Le lieu de ponte de dinosaures de Rennes-le-Château (Aude); premiers résultats de la campagne de fouilles 1984.

Annales du Muséum du Havre 32: 1–13.

1986b. Le lieu de ponte de Rennes-le-Château (Aude): premiers résultats de la campagne de fouille 1984. *Actes du Colloque Les Dinosaures de la Chine à la France.* (Toulouse: Muséum de Toulouse), pp. 127–140.

Buffetaut, E. 1989. Les dinosaures du Crétacé terminal du Sud de la France: données tirées des restes osseux. *Cahier de la Réserve Géologique de Haute-Provence* 1: 17–22.

Buffetaut, E., & Le Loeuff, J. 1989a. 100 ans; La première découverte d'oeufs de dinosaures. *Pour la Science* 143: 22.

1989b. Les dinosaures du Crétacé supérieur des Corbières, France. *Géochronique* 32: 4.

Buffetaut, E., Méchin, P., & Salessy, A. 1988. Un dinosaure théropode d'affinités gondwaniennes dans le Crétacé supérieur de Provence. *Comptes rendus des séances de l'Académie des Sciences*, Ser. II 306: 153–8.

Figure 5.15. Hatched egg from grid G-9 bc (see Fig. 5.16) of the 1984 excavation at Rennes-le-Château (Aude). Part of the ''hatching window'' is in a subvertical position; the other part lies in bottom of the egg.

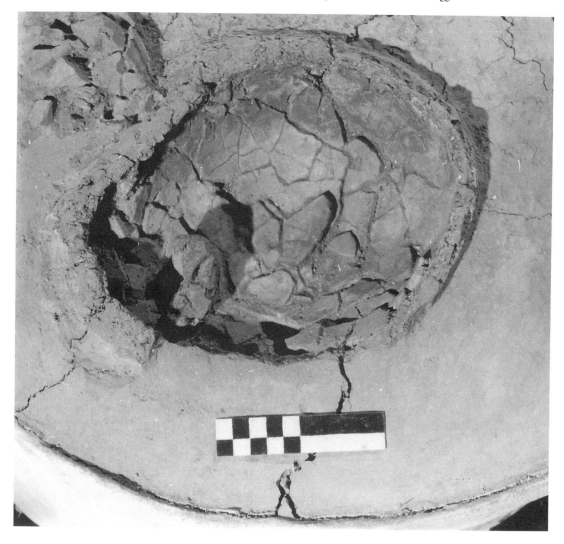

Figure 5.16. Dinosaur nesting ground in the Upper Maastrichtian at Rennes-le-Château (Aude). The site was excavated in April, 1984, and shows the distribution of eggs. Stippled eggs belong to an arc group. Unstippled eggs belong to a clutch (surrounded with thick line). Also shown are isolated eggs or eggs not assigned to any group.

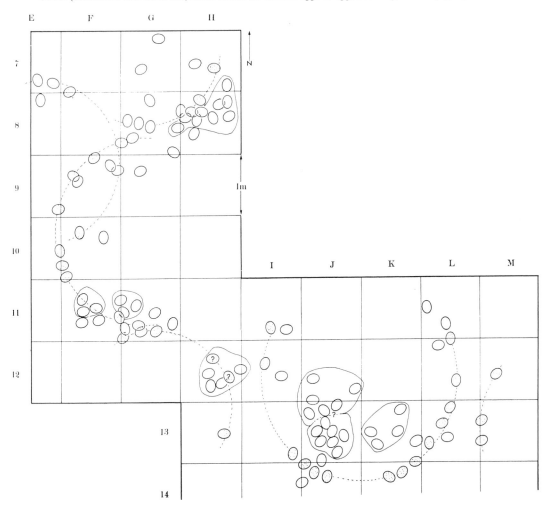

Figure 5.17. Arcuate row of hatched eggs shown in Figure 5.16 in grids I-13 and J-14.

Caillaud, P. 1968. Extension, vers Rennes-le-Château (Aude), des gisements d'oeufs de Dinosaures dans le Crétacé supérieur des Corbières. *Comptes rendus sommaires des séances de la Société Géologique de France* 1: 111–12.

Clottes, P., & Raynaud, C. 1983. Le gisement à dinosauriens de Campagne-sur-Aude-Esperaza. Observations préliminaires. Premiers resultats. *Bulletin de la Société d'Etudes Scientifiques de l'Aude* 83: 5–14.

Cousin, R., Breton, G., Fournier, R., & Gomez, N. 1987. La campagne de "fouilles préliminaires" sur les lieux de ponte de dinosaures de Rousset-sur-Arc (Bouches-du-Rhône). *Bulletin trimestriel de la Société Géologique de Normandie et des Amis du Muséum du Havre* 74: 5–15.

Cousin, R., Breton, G., Fournier, R., & Watté, J.-P. 1989. Dinosaur egg-laying and nesting: the case of an Upper Maastrichtian site at Rennes-le-Château (Aude, France). *Historical Biology* 2: 157–67.

Dauphin, Y. 1990. Comparative microstructural studies of egg shells. 1. Dinosaurs of Southern France. *Revue de Paléobiologie* 9: 121–37.

Déperet, M. 1900. Compte rendu de la séance du 5 mars 1900 présidée par L.-A. de Lapparent, président. *Bulletin de la Société Géologique de France*, Ser. 3, 28 (1): 107–8.

Derognat, M. 1935. Les oeufs fossiles de dinosaures. *Association française pour l'Avancement des Sciences*, Nantes: 295–6.

Dughi, R., & Sirugue, F. 1957a. Les oeufs de Dinosauriens du Bassin d'Aix-en-Provence. *Comptes rendus des séances de l'Académie des Sciences* 245: 707–10.

1957b. La limite supérieure des gisements d'oeufs de dinosauriens dans le bassin d'Aix-en-Provence. *Comptes rendus des séances de l'Académie des Sciences* 245: 907–9.

Figure 5.18. Three superimposed egg levels at the egg-nesting ground at Rennes-le-Château (Aude). Sequence of events: (1) deposition of sediment 1; laying and hatching of "alpha" (α) eggs. (2) Sediment 2 completely buries the "alpha" eggs. (3) Erosion of upper part of sediment 2, leaves only the bottoms of the "alpha" (A) eggs and removes nearly all dispersed shell fragments that might have been present. (4) Laying and hatching of "beta" (β) eggs on surface of sediment 2. (5) Sediment 3 completely buries the "beta" eggs. (6) Laying and hatching of "gamma" (γ) eggs. (7) Sediment 4 completely buries "gamma" eggs. (8) Erosion and recent bioturbation remove all but the bottoms of the "gamma" eggs, but numerous shell fragments remain dispersed in the soil.

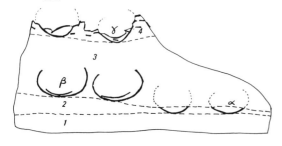

1958a. Observations sur les oeufs de dinosaures du bassin d'Aix-en-Provence: les oeufs à coquilles bistratifiées. *Comptes rendus des séances de l'Académie des Sciences* 246: 2271–4.

1958b. Les oeufs de dinosaures du Bassin d'Aix-en-Provence: les oeufs du Bégudien. *Comptes rendus des séances de l'Académie des Sciences* 246: 2386–8.

1966. Sur la fossilisation des oeufs de Dinosaures. *Comptes rendus des séances de l'Académie des Sciences*, Ser. D, 262: 2330–2.

1976. L'extinction des dinosaures à la lumière des gisements d'oeufs du Crétacé terminal du Sud de la France, principalement dans le bassin d'Aix-en-Provence. *Paléobiologie continentale* 7: 1–39.

Erben, H., Hoefs, J., & Wedepohl, K. 1979. Paleobiological and isotopic studies of egg shells from a declining dinosaur species. *Paleobiology* 5: 380–414.

Floquet, M. 1979. La série carbonatée Coniacian supérieur a Santonian dans la région de Soria (Chaînes ibériques septentrionales): analyse et interprétations. *Cuadernos Geologica Iberica* 5: 365–83.

Fournier, R. 1983. *Catalogue de l'exposition Matheron Ph. 1807–1899: 60 ans de géologie provençale.* (Marseille: Musée d'Histoire Naturelle).

Freytet, P. 1965. Découverte d'oeufs de Dinosaures à Saint-André-de-Roquelongue (Aude). *Bulletin de la Société d'Etudes Scientifiques de l'Aude* 65: 121–4.

1970. *Les dépôts continentaux et marins du Crétacé supérieur et des couches de passage a l'Eocène en Languedoc.* Unpublished Ph.D. Thesis (Orsay).

Gervais, P. 1877. De la structure des coquilles calcaires des oeufs et des caractéristiques que l'on peut en tirer. *Comptes rendus des séances de l'Académie des Sciences* 84: 159–65.

Ginsburg, L. 1980. Les gisements à oeufs de Dinosaures du Crétacé terminal du Midi de la France et la physiologie des Ornithopodes. *Mémoires de la Société Géologique de France* N.S., 139: 109–10.

Horner, J. 1984. Les oeufs et les nids de Dinosaures. *Pour la Science* 80: 60–8.

Kérourio, P. 1978. Les dinosaures de provence. *Science AM* 6: 6–13.

1981. Nouvelles observations sur le mode de nidification et de ponte chez les dinosauriens du Crétacé terminal du Midi de la France. *Compte rendu sommaire des séances de la Société Géologique de France* 1: 25–8.

1987. Les nids de dinosaures en Provence. *La Recherche* 18, 185: 256–7.

Lapparent, A.-F. de. 1947. Les Dinosauriens du Crétacé supérieur du Midi de la France. *Mémoires de la Société Géologique de France* 26(6): 1–54.

1957. Les oeufs de Dinosauriens fossiles de Rousset (Bouches-du-Rhône). *Comptes rendus des séances de l'Académie des Sciences* 245: 546–8.

Le Loeuff, J., Buffetaut, E., Mchin, P., & Méchin-Salessy, A. 1989. Un arrière-crâne de dinosaure titanosauridé (Saurischia Sauropoda) dans le Crétacé supérieur du Var (Provence, France). *Comptes rendus des séances de l'Académie des Sciences*, Ser. II, 309: 851–7.

Masriera, A., & Ullastre, J. 1981. Contribution des minéraux lourds à la lithostratigraphie du Crétacé terminal des

Pyrénées catalanes et quelques conséquences paléo-géographiques et tectoniques. *Comptes rendus des séances de l'Académie des Sciences* 293: 179–82.

—— 1983. Essai de synthèse stratigraphique des couches continentales de la fin du Crétacé des Pyrénées catalanes (N.E. de l'Espagne); Colloque sur le Sénonien. *Géologie méditerranéenne* 10: 283–90.

Matheron, P. 1869. Notice sur les reptiles fossiles des dépôts fluvio-lacustres crétacés du bassin à lignite de Fuveau. *Bulletin de la Société Géologique de France*, Ser. 2, 26: 781.

—— 1878. *Recherches paléontologiques dans le Midi de la France. 15ème partie. Terrain tertiaire:* (Marseilles: Published by author).

Mattauer, M., & Thaler, L. 1961. Découverte d'oeufs et d'os de Dinosaures dans le Crétacé terminal des environs de Montpellier (Hérault). *Comptes rendus sommaires des séances de la Société Géologique de France* 7: 7–9.

Penner, M.-M. 1983. Contribution à l'étude de la microstructure des coquilles d'oeufs de dinosaures du Crétacé supérieur dans le Bassin d'Aix-en-Provence (France): Application biostratigraphique. Doctoral thesis. Mémoires des Sciences de la Terre, Université Pierre-et-Marie-Curie 83–52: 1–234.

Plaziat, J.-C. 1961. Présence d'oeufs de dinosaures dans le Crétacé supérieur des Corbières et existence d'un niveau marin dans le Thanétien aux environs d'Albas (Aude). *Compte rendu sommaire des séances de la Société Géologique de France* 7: 196–7.

—— 1970a. Conséquences stratigraphiques de l'interstratification du Rognacian dans le Maastrichtien supérieur d'Alava (Espagne). *Comptes rendus des séances de l'Académie des Sciences* 270: 2768–71.

—— 1970b. La limite crétacé-tertiaire en Alava méridionale (Pays basque espagnol): le Rognacian n'y est pas l'équivalent continental du Danien. *Comptes rendus des séances de l'Académie des Sciences* 3: 77–8.

Pouech, J.-J. 1859. Mémoire sur les terrains tertiaires de l'Ariège rapportés à une coupe transversale menée du Fossat à Aillères, passant par le Mas d'Azil et projetée sur le méridien de ce lieu. *Bulletin de la Société Géologique de France*, Ser. 2, 16: 381–411.

Rolland, C. 1980. Révision des ossements de dinosauriens de la collection Matheron au Muséum du Palais-Longchamp (Marseille). *Rapport de DEA: Géologie des formations sédimentaires.* (Marseilles: Université de Provence St Charles).

Seymour, R.-S. 1979. Dinosaur eggs: gas conductance through the shell, water loss during incubation and clutch size. *Paleobiology* 5: 1–11.

Straelen, V. van, & Denaeyer, M.E. 1923. Sur les oeufs fossiles du Crétacé supérieur de Rognac en Provence. *Bulletin de l'Académie Royale Belge, Section Science* 9: 14–26.

Thaler, L. 1965. Les oeufs de dinosaures du Midi de la France livrent le secret de leur extinction. *Science Progrès – La Nature* 45: 41–8.

Villate, J., Taquet, P., & Bilotte, M. 1986. Nouveaux restes de dinosauriens dans le Crétacé terminal de l'anticlinal de Dreuilhe. Etat des connaissances dans le domaine sous-pyrénéen. *Actes du Colloque Les*

Dinosaures de la Chine à la France. (Toulouse: Museum de Toulouse), pp. 89–98.

Villot, L. 1883. Etude sur le bassin de Fuveau et sur un grand travail à y exécuter. *Annales des Mines* 3: 5–66.

Watté, J.-P. 1989. Archéologues normands et recherche paléontologique sur un lieu de ponte de dinosaures de l'Aude. *Bulletin de la Commission Départementale des Antiquités de la Seine-Maritime* 35: 141–3.

Watté, J.-P., Fournier, R., & Breton, G. 1986. Coopération archéologues-paléontologues: Plusieurs musées s'associent pour fouiller un lieu de ponte de dinosaures. *Bulletin de l'Association Philomatique d'Alsace Lorraine* No. h.s.: 81–6.

Westphal, M. 1989. Magnétostratigraphie du Crétacé supérieur du bassin d'Aix-en-Provence. Etude préliminaire. *Cahier Réserve Géologique de Haute-Provence* 1: 28–33.

Williams, D. L., Seymour, R. S., & Kérourio, P. 1984. Structure of fossil dinosaur egg shell from the Aix Basin, France. *Palaeogeography, Palaeoclimatology, Palaeoecology* 45: 23–37.

Young, C. C. 1965. Fossil eggs from Nanhsiung, Kwangtoung and Kanchou, Kiangsi. *Vertebrata PalAsiatica* 9: 141–70.

Zhao T. K., & Chiang Y. K. 1974. Microscopic studies on the dinosaurian egg shells from Laiyang, Shantung Provinces. *Scientia Sinica* 17: 63–74.

Appendix 5.1. Survey of dinosaur egg collections in France

Even though most research on dinosaur eggs has been carried out by Provençal paleontologists (e.g., Matheron, Dughi, Sirugue, and Kérourio), the natural history museums in southern France house few specimens. The largest collection is housed at the Natural History Museum of Le Havre (Seine-Maritime) as a result of a change in field techniques used since 1983 (see above).

Aix-en-Provence Museum (Bouches-du-Rhône)

Eggs collected at Roquehaute (Beaurecueil, Bouches-du-Rhône)

1. Group of two eggs and part of another (9 × 14 cm, 9 × 17 cm, 15 cm), prepared bottom-side up on a plaster shell. From a red sandstone. Beige color with a finely ornamented shell.

2. Crushed egg, exposed in section (14 cm). From a red sandstone. Whitish color, with a fine shell.

3. Two eggs (17 × 12 cm, 16 × 12 cm), collected by Dughi and Sirugue (1957) north of Roquehaute farm. Whitish color with a finely tuberculated shell (thickness-1.5 mm).

4. Part of an egg collected in 1957, probably by Dughi and Sirugue. Whitish color with a coarsely tuberculated shell (thickness-2.4–2.6 mm).

Eggs collected at Rousset-sur-Arc (Bouches-du-Rhône)

1. A spherical egg (19 × 19 cm), with a grey, smooth shell.

2. Egg (19 × 17 cm) with a blunt shell.

Eggs collected at Châteauneuf-le-Rouge (Bouches-du-Rhône)

An egg labeled "*Hypselosaurus*" (22 × 19 cm) in sandstone. White shell with local large tubercles.

Unlabeled eggs

1. Egg in gravelly breccia with a brown, worn shell. Obvious tubercles on a shell fragment.
2. Egg, prepared bottom side up (17 × 17 cm) from a sandy paleosol; smooth shell (thickness ca. 2 mm).
3. Part of an egg (15 × 10 cm) from a rusty colored, micaceous clay; grey, thin shell (ca. 1.4–1.6 mm) with many fine tubercles.

Marseilles Museum (Bouches-du-Rhône)

Eggs collected at Rousset-sur-Arc (Bouches-du-Rhône)

1. Unhatched egg (12 × 10 cm) filled with calcite crystals, collected in 1975; tuberculated thick shell.
2. Group of three eggs (18 × 17 cm; 21 cm; 22 cm), prepared bottom upward, collected by Kérourio (1981); smooth and rather thick shell.
3. Group of two eggs (18 × 14 cm; 16 × 15 cm), same data as in No. 2.

Eggs collected at Châteauneuf-le-Rouge (Bouches-du-Rhône)

Group of two eggs and bottom of a third one, located 8 cm higher than the other two; from a sandstone; collected in 1980 and prepared by Romelli; weathered shell (thickness: 1.8–1.9 mm).

Eggs collected at Rennes-le-Château (Aude), 1984

Two groups, the first of which is still unprepared. The other displays two hatched eggs.

Saint-Benoit Geology Center at Digne Alpes-de-Haute-Provence)

Four eggs uncovered by road work in 1987 were collected at La Bégude (Rousset-sur-Arc, Bouches-du-Rhône). The stratigraphical location was Bed No. 2 in Rognacian sandstone (Cousin et al., 1987). The collection was made by G. Breton of the Natural History Museum of Le Havre.

1. A complete egg, slightly crushed (21 × 17 cm).
2. An incomplete, misshapen egg (18.5 × 13 cm).
3. A very misshapen egg (17.5 × 13.5 cm).
4. Internal mold of half an egg (20 cm).

Esperaza Museum (Aude)

This museum devoted to dinosaurs will soon display a part of the regional discoveries, mainly bones collected at Campagne-sur-Aude (Aude) (Buffetaut & Le Loeuff, 1989b). Two groups of eggs from the 1984 excavation at Rennes-le-Château (prepared bottom-side up) will also be on display.

Rennes-le-Château Museum (Aude)

One egg collected and prepared during the field work (1984) has been deposited in this small village museum.

Le Havre Museum (Seine-Maritime)

Eggs collected at Rennes-le-Château (Aude)

1. Seventeen groups, with a total of about forty eggs, from the excavation made in 1984 by the museums of Marseilles and Le Havre. All are prepared as originally found.
2. A 10 m² surface, displaying 20 eggs, cast during the same excavation by R. Millo and J.-P. Hauchecorne.
3. Three groups with a total of five eggs collected at the same site in 1983 during preliminary field work. All are prepared bottom-side up.
4. Four other groups, which are still in preparation, collected from another site at Rennes-le-Château. This site was excavated in 1988 by the staff of the Museum of Le Havre. No complete and well preserved eggs were found; all were crushed, and their fragments scattered.

Eggs collected at Saint-Chinian (Hérault)

Two apparently complete eggs are shown in a very hard red sandstone. These eggs were not collected in situ.

Eggs collected at Puyloubier (Bouches-du-Rhône)

A clutch of six eggs, in a whitish Rognacian sandstone, was proposed as a "nest" by Kérourio (1981, Fig. D). The average size of the eggs is 18 × 15 cm.

Laboratory of Vertebrate Paleontology, University of Montpellier II (Hérault)

Eggs collected at Argelliers – Montarnaud (Hérault)

One spherical and almost complete egg and four blocks with portions of eggs of the same type are housed in this laboratory.

Eggs collected at Saint-Chinian (Hérault)

Eight complete or fragmentary ellipsoidal eggs from this site are housed in this laboratory.

6 Late Maastrichtian dinosaur eggs from the Haţeg Basin (Romania)

DAN GRIGORESCU,
DAVID WEISHAMPEL,
DAVID NORMAN, M. SECLAMEN,
M. RUSU, A. BALTRES,
AND V. TEODORESCU

Abstract

The Haţeg Basin in western Romania has principally been known for the Upper Cretaceous dinosaur material it has produced. A newly discovered nesting site with complete eggs and near-term embryonic or hatchling specimens is described here. Subspherical eggs were laid in paired rows, each containing two, three, or four eggs. Based on the embryonic/hatchling skeletal remains, parentage of these eggs is thought to be the hadrosaurid *Telmatosaurus transsylvanicus*. Eggshell fabric is tubospherulitic. Pore geometry and density, eggshell thickness, and local sedimentology associated with the nests suggest that levels of humidity within the nest were high (85–90 percent) and that incubation was on the order of 50 to 60 days.

Introduction

The Upper Cretaceous fluviolacustrine deposits of the Haţeg Basin (Transylvanian region, Romania; Fig. 6.1) have been known for almost a century, particularly for their rich dinosaur, crocodilian, and turtle faunas. These fossils most commonly come from "fossiliferous pockets," concentrations in which the prevailing preservation is of disarticulated skull and postcranial elements. The dinosaurs include at least seven major taxa, among them sauropods, small and large theropods, ornithopods, and ankylosaurs. As such, the assemblage is very similar to those of Upper Cretaceous deposits elsewhere in the Tethyan region, including northern Spain, southern France, and Austria (Buffetaut & Le Loeuff, 1991; Le Loeuff, 1991). The principal collections of the Haţeg fauna were made before World War I by Franz Baron Nopcsa, who discussed the systematic, paleobiologic, and paleogeographic aspects of this fauna in more than a dozen papers published between 1897 and 1929 (see Weishampel & Reif, 1984; Weishampel, Grigorescu, & Norman, 1991). The largest collection of material described by Nopcsa is now housed at the Natural History Museum in London. A smaller collection, provided by Kadić O. and studied by Nopcsa, is in the Hungarian Geological Institute in Budapest.

More recently, new collections from these same Haţeg deposits have been made by one of us (Grigorescu) and students from the University of Bucharest, helped by a local villager, Doinel Vulc and Ion Groza from the Deva County Museum. The expeditions, begun 14 years ago, provide evidence of new small and large theropods, as well as multituberculate mammals (Grigorescu, 1983, 1984; Grigorescu & Hahn, 1987). A systematic revision of the entire dinosaur fauna from the Haţeg Basin is presently being undertaken by Weishampel, Norman, and Grigorescu (viz., Weishampel et al., 1991).

Until recently, only skeletal material, principally of dinosaurs, was known from the Haţeg localities. This is in contrast to the Provence region of southern France, from which, in addition to the skeletal remains of some of the same Transylvanian taxa, dinosaur eggs and eggshell fragments have been reported for more than 125 years (Dughi & Sirugue, 1957; Breton, Fournier, & Watté, 1986). This situation changed recently with the discovery of a nesting site by N. Ghinescu near the village of Tustea, Romania (Grigorescu et al., 1990b). This locality, in the northern part of the Haţeg Basin, produced not only hatched and unhatched eggs, but also embryonic and hatchling skeletal material.

Litho- and chronostratigraphy of the dinosaur-bearing deposits of the Haţeg Basin

The Haţeg region constitutes a tectonic basin that began forming at the end of the Cretaceous by the Laramian compression (the Laramian Orogeny that was instrumental in producing the Carpathian Mountains). The basin itself is filled predominantly with as much as 4,000 m of nonmarine molasse deposits (Fig. 6.1).

The first phase of sedimentation began in the late Maastrichtian and continued into the early Paleogene. Two partially correlative formations are recognized from this sequence: the Densuş–Ciula and Sînpetru formations (Grigorescu et al., 1990a). Although they have several paleontological aspects (palynofloras, fresh water gastropod, and dinosaur assemblages) in common, these lithostratigraphic units can be clearly distinguished by their sedimentologic features.

Outcropping in the northern part of the Haţeg Basin, the Densuş–Ciula Formation consists mainly of coarse clastic rocks (conglomerates and sandstones) interbedded with siltstones and marls. The color of these beds varies from gray and greenish-gray to red and purplish-blue. The lower member of the Densuş–Ciula Formation is characterized by a mixture of terrigenous and volcanoclastic sediments, which themselves are interbedded with andesitic and rhyolitic tuffs. Dinosaur remains are absent from this member. Instead, a rich subtropical flora is represented by leaf impressions that are found on the surface of the fine-grained tuffaceous rocks. This flora includes not only ferns (*Asplenium foersteri*, *Phyllites* sp.), but also palms (*Palmophyllum longirachis*) and dicotyledon angiosperms of Senonian age.

The middle member of the Densuş–Ciula Formation consists mainly of coarse clastic rocks (including cross-bedded conglomerates and sandstones) with interbedded lacustrine limestones and thin coal layers. A palynological assemblage within this member is dominated by *Normapolles* and *Postnormapolles*. In addition, the assemblage includes *Pseudopapillopollis praesubhercynicus*, an index species for the Maastrichtian stage of the Upper Cretaceous. The dinosaur nesting site is located in the upper part of this middle member (Fig. 6.2).

The upper member of the Densuş–Ciula Formation is lithologically similar to the middle member, but differs in its absence of both dinosaur bones and interbedded volcanoclastic units. The upper member gives no indication of its age from palynological or other paleontological materials, but its stratigraphic position (i.e., above upper Maastrichtian and below middle Miocene sediments) suggests a Paleogene age. Thus, when considering the stratigraphic position of the lower member of the Densuş–Ciula Formation above the lower Maastrichtian flysch deposits, the whole formation is most likely middle Maastrichtian to lower Paleogene in age.

In contrast to the Densuş–Ciula Formation, the Sînpetru Formation outcrops in the central part of the Haţeg Basin. It consists of heterotopic facies, the deposition of which was probably diachronous in its upper part with the more northerly Densuş–Ciula Formation. The two appear to interfinger in the central and western parts of the basin.

The 2500 m thick Sînpetru Formation is almost completely devoid of coarse volcanoclastic sediments, made up instead mainly of clastic rocks of wide-ranging grain sizes. Subunits within the formation form cyclic fluviolacustrine sequences. Within the middle part of the Sînpetru Formation, there is a palynoflora nearly identical to that described above from the middle member of the Densuş–Ciula Formation. Dinosaur and other amniote remains (e.g., turtles, crocodilians) are frequently recovered from fossiliferous pockets scattered throughout the thickness of the Sînpetru Formation, including its highest level which

Figure 6.1. Geographic position of the Haţeg Basin, Romania, with geologic (A) and geographic (B) maps of the Haţeg Basin. 1 Proterozoic and Paleozoic metamorphic schists. 2. Cretaceous marine deposits. 3. Densuş–Ciula Formation [(?)middle–upper Maastrichtian–lower Paleogene continental deposits]. 4. Sînpetru Formation [(?)middle-upper Maastrichtian continental deposits]. 5. Upper Paleogene–lower Miocene continental deposits. 6. Middle Miocene marine deposits. 7. Upper Miocene brackish deposits. 8. Quaternary deposits. 9. Dinosaur egg site. (?): possible Cretaceous–Tertiary boundary.

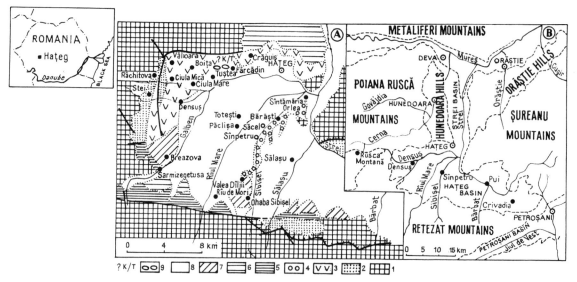

Stratigraphic setting of the nesting horizon

The 1988 discovery of dinosaur eggs from the Haṭeg Basin was the fortuitous result of a rock fall along a 10 m high vertical escarpment outside the village of Tustea. This fall revealed the remains of fourteen eggs (Fig. 6.3), including an almost complete specimen. The remainder consisted of large crushed remains and small eggshell fragments. All were collected soon after the rock fall. In addition, the remains of another four eggs (including another complete egg) eroded from the site over the following year and were collected in 1989. The specimens, split into polyhedral fragments, were found in a 0.5 m thick, red, massive, silty mudstone. This unit is intercalated between a lower 1.2-m thick, reddish mudstone that contains small calcretes, plant rootlets, and vertical burrows, and an upper 4 m thick cross-bedded conglomerate with a friable matrix (Fig. 6.2).

As indicated by their distribution in the rocks, the subspherical eggs were apparently lying in linear rows, each egg located close to another. The eggs were arranged vertically in two closely superposed levels, each containing two groups of two, three, or four eggs. The groups were spaced about a half meter from one another (Fig. 6.4). The "left" and "right" superposed groups may represent two contemporaneous clutches deposited by the same female, or they may belong to two clutches laid in two different years. We feel that the closeness of the egg groups provides support for the first alternative. On the "right" side, large isolated fragments of another egg were found beneath the first row.

Most of the eggs from Haṭeg are represented only by their lower halves. These are often more or less compressed and fractured. Upper surfaces, when preserved, are collapsed; some of the upper shell fragments are preserved within the internal matrix. The two eggs that preserve an unbroken upper surface are unhatched (Fig. 6.3A). These eggs found in 1989 were arranged in a single row, apparently behind the eggs from the second level of the "left" side of the outcrop (Fig. 6.4B). The vertical orientation of the bedding plane and the unstable conglomerate that overlies the fossiliferous level made excavation of the locality dangerous. As a consequence, it was impossible to determine the geometric structure of the clutches.

Dinosaur bones within the mudstone that contain these eggs consist of few adult hadrosaurid limbbone fragments. In addition, a few well-preserved, but small limb elements were collected from the nesting site. These come from a near-term embryo or young

hatchling, probably of the hadrosaurid *Telmatosaurus transsylvanicus* (viz., Weishampel, Norman, & Grigorescu, in press). Another small element, a complete tibia of a small theropod, was also found in the vicinity.

Finally, preliminary computed-tomographic (CT) analysis indicates possible embryonic material concentrated at one end of one of the unhatched eggs (Fig. 6.5). At present, we are unable to comment on the details and significance of these inclusions, but are hopeful that continued CT analyses, among other techniques, will further elucidate the nature of these possible embryonic remains.

The eggs and eggshells

Egg shape and dimensions

All of the hatched eggs are more or less compressed and fractured by burial, so exact egg shape and size are difficulty to ascertain. However, the two unhatched eggs appear to be only slightly deformed, thus preserving their original subspherical and slightly ellipsoidal morphology. The three axes of the best preserved specimen (Fig. 6.3A) are 7.5, 5.6, and 6.0 cm long, respectively. The volume of this egg is 981 cm^3.

Figure 6.2. Lithologic sequence with dinosaur eggs and embryonic bones in the upper part of the middle member of the Densuş–Ciula Formation. **1.** Coarse, poorly cemented, grayish sandstone. **2.** Pinkish silty marls with calcretes and plant rootlets. **3.** Red, massive, silty mudstone with dinosaur egg levels (indicated by the arrow). **4.** Grayish, poorly cemented, cross-bedded conglomerate. **5.** Variegated (reddish and grayish) silty marls with calcretes. **6.** Soil.

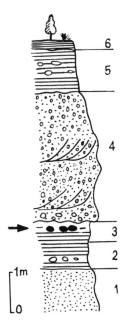

Fabric and composition of the eggshell

Eggshell thickness ranges from 2.1 to 2.7 mm (most common thickness = 2.3–2.4 mm). The outer surface is covered by an irregular pattern of fine, hemispheric tubercles raised 0.15 mm to 0.20 mm from their base (Fig. 6.6A). Each tubercle is isolated from others or else grouped into small clusters that sometimes coalesce at their bases. Irregularly distributed pore openings among the tubercles are commonly plugged by sparry calcite. Internally, mammillary bases lack the cratering indicative of resorption of calcium carbonate by the de-

veloping embryo (Fig. 6.6B, C, D; Hirsch & Quinn, 1990). Between the mammillae, the shell is pitted by irregular interstices. These funnel-shaped interstices are connected with the proximal part of the pores and penetrate deeply between the wedge-shaped shell units, as shown in Figures 6.8 and 6.9.

Microstructurally, the eggshell is tubospherulitic, agreeing well with the dinosauroid–spherulitic morphotype described by Hirsch and Quinn (1990) and Mikhailov (1987, 1991). In plane light, radial sections of the eggshell reveal thin wedge-shaped shell units composed

Figure 6.3. Dinosaur eggs from the Haţeg Basin. **A.** Unhatched egg. **B.** Two closely positioned hatched eggs. **C, D.** Two hatched eggs from the upper ''left'' level (see Fig. 6.4).

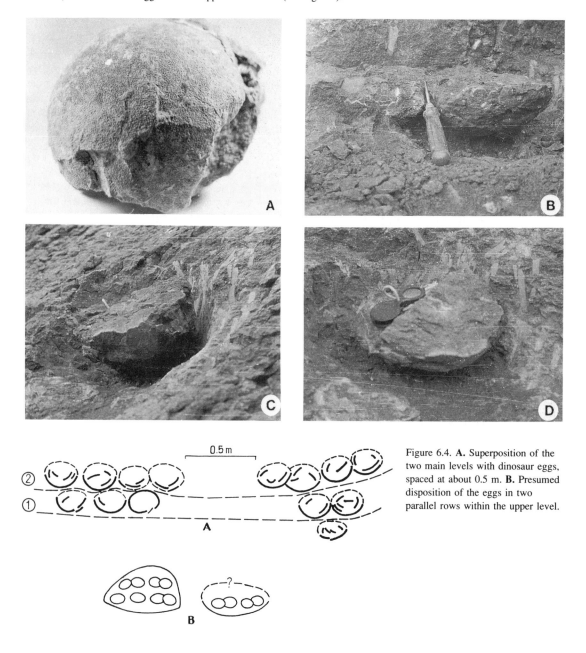

0.5 m

Figure 6.4. **A.** Superposition of the two main levels with dinosaur eggs, spaced at about 0.5 m. **B.** Presumed disposition of the eggs in two parallel rows within the upper level.

of aggregates of tabular calcite crystallites (Erben, 1970; Mikhailov, 1991). These wedges radiate upward from a line of small, closely spaced mammillae (Figs. 6.7 & 6.8). The c-axis of the calcite is oriented parallel to the long axes of the wedges, with parallel extinctions for each wedge. Between the mammillae are thin, discontinuous seams of dark, opaque organic matter (Fig. 6.10). Finely, upwardly arched growth lines pass through adjacent bundles as light-and-dark, brown or orange bands (Fig. 6.7A). These appear to consist of diffuse organic matter that was deposited at fluctuating rates during eggshell growth. Each band comprises four or five fine, parallel striae (Fig. 6.13). Paired striation sets resembling a herringbone pattern are sometimes present (Fig. 6.14).

In tangential section, eggshells exhibit a polygonal mosaic that corresponds either to the boundaries between the closely-packed columnar bundles or to discontinuous growth lines that result in a concentric pattern (Fig. 6.11). Acetate peels show a finer network within this mosaic, which represent the ragged intercrystalline boundaries of calcite. Under crossed nicols, this xenotopic calcite mosaic shows a composite to weakly undulatory extinction (Fig. 6.12). In radial section, the bundles do not extinguish uniformly, showing instead a sweeping extinction pattern (Fig. 6.7B). A schematic cross-section of the columnar bundles at different levels within the shell is provided in Figure 6.15.

Electron microscopy and x-ray diffraction study of the eggshell

Details of shell texture and carbonate crystalline structure are available from electron microscopy and x-ray diffraction analyses. The x-ray diffraction spectra obtained from powdered eggshell indicates pure, rhombohedral calcite. Figure 6.16A shows the hexagonal reflection system and measured relative intensities of powdered eggshell. The spectra obtained from tangentially and radially sectioned eggshell are shown in Figures 6.16B and C, respectively. In Figure 6.16B, the intensity of reflections are high for (006) and low for (110). In radial section (Fig. 6.16C), the intensity of reflections are very high for (110) and (202) and insignificant for (006). Therefore, the trigonal axis of the calcite crystals is oriented along their long axes.

Each thin, calcite wedge is in fact a crystallite, which together with its neighbors, is grouped into radial bundles that diverge from perpendicular to the surface of the eggshell by 30° to 40°. The long dimension of each crystallite is 1 to 2μm (determined from scanning electron micrographs). The number of crystallite defects is high (approximately $5 \times 10^{10}/cm^2$), most likely the consequence of dispersed, fine-grained organic matter and by crystal dislocations. Scanning electron micrographs also reveal an amorphous organic residue (0.1 to 1 μm) between the calcite crystallites. The total organic matter in the eggshell is 0.113 percent and the kerogen content is 0.78 percent.

Pore morphology and dimensions

Precise eggshell pore geometry is difficult to determine, in large part because of the intricate form and shape of many of these pores. However, such precision is obviously important for quantitative analyses of both physical and biological characteristics of the eggs. Because such parameters as pore length and diameter determine important biological functions of the egg, we present estimated ranges and extremes for many of these values.

In the eggshells from Transylvania, there are two types of voids: closed (i.e., not directly communicating between the external and internal surfaces of the shell; represented by the broken line in Fig. 6.13), and open (connecting inner and outer shell surfaces; hatched regions in Fig. 6.13). We define a pore as a type "b" void. These pores provide connection between the embryo and the outside environment, and account for both mass transport and heat exchange. Type "a" voids also contribute to heat exchange by conduction, but they do not contribute directly to gas and vapor (mass) transport

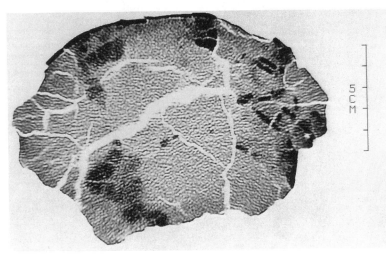

Figure 6.5. Computed-tomography (CT) scan approximately midway through one of the unhatched eggs from the Hațeg Basin. CT analyses were done on a Siemens Scann Tomograph. We tentatively identify the collection of long, narrow inclusions on the right side of the egg as embryonic skeletal elements.

Figure 6.6. **A.** Scanning electron micrographs of the external surface of the eggshell showing tuberculate surface. **B.** Internal surface of the eggshell of an unhatched egg. Note the absence of the cratered mammillae. **C.** Transverse thin section of the eggshell (unpolarized light). **D.** The same specimen in polarized light.

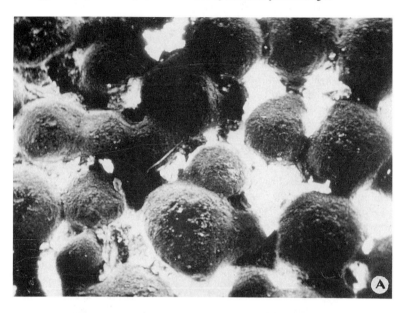

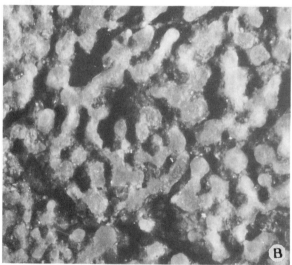

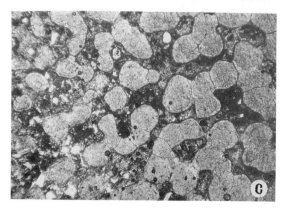

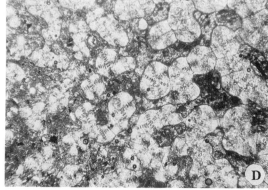

Figure 6.7. Radial section of an eggshell showing columnar bundles of calcite radiating from opaque centers (mammillae, c). Fine, arched growth lines (g) pass between adjacent bundles. **A.** Plane-polarized light. **B.** Crossed nicols. In B, the bundles are in various extinction positions, giving a sweeping extinction appearance.

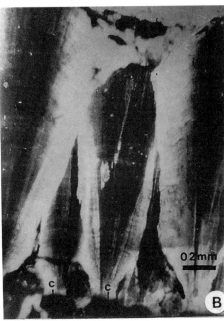

Figure 6.8. **A.** Scanning electron micrograph of a polished and etched radial section of part of an eggshell showing the internal mold of a pore (p). Note the boundary between adjacent bundles (cr) and the opaque spherulitic centers (mammillae, c) at the origin of each calcite bundle. **B.** The same specimen in thin section, plane-polarized light.

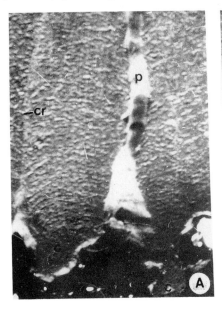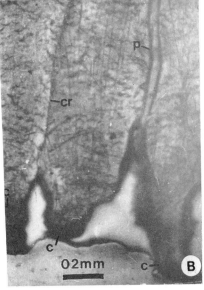

through the shell and therefore are not considered pores.

A series of transverse and tangential sections was made through the shell of the Transylvanian eggs to estimate the form and dimensions of pores. Tangential sections were taken parallel to the outer surface of the egg at approximately 0.2 mm intervals. Sections were successively polished and etched and acetate peels were taken from these sections. These peels were then examined under the microscope and were also photographically enlarged as negative peel prints on high-contrast paper.

The pores of the Transylvanian eggs are not always simple cylindrical tubes, but can often be considerably contorted and sometimes bifurcated (viz., Fig. 6.17, which provides evidence of variation in diameter and form along the length of the pore). Because of the twisted pathway of some pores, pore length may be considerably greater than the thickness of the shell itself.

As a consequence, pore length can vary from 2.3 to 4.6 mm for a 2.3-mm thick eggshell. Mean pore length appears to be 3.45 mm.

Statistical distribution of the pore diameter

Table 6.1 provides the principal statistics for each of the twelve sections (in order of increasing depths). Because the pores in the first two sections ($z = 1, 2$) and for the last ($z = 12$) may have been enlarged by diagenesis, we consequently did not use these replicas for the measurements that follow. Over the remaining sections, 2,095 pores were counted and measured. The number of pores/unit area (pore density) is highly correlated with mean pore diameter, so there are few large pores, but many small ones (Fig. 6.18; Table 6.2). With logarithmic transformation, pore number is a linear

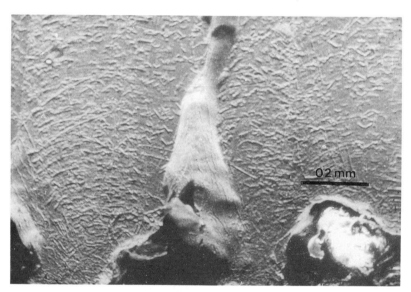

Figure 6.9. Scanning electron micrograph showing in detail the funnel-shaped internal mold of an interstice passing to the proximal part of the pore shown in Figure 6.6A. Faint striae are visible along the outer surface of a removed bundle.

0.2 mm

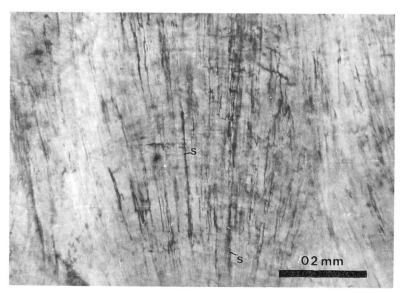

Figure 6.10. Discontinuous seams of opaque organic matter (s) delineating the calcite aggregates. Radial section in plane-polarized light.

0.2 mm

function of the mean pore diameter under linear regression methods:

$$n = 241.8^{(-7.88d)};$$

where n = number of pores and d = mean pore diameter, expressed in millimeters.

For sections obtained from the middle of the eggshell (and hence probably more representative of original pore geometry), mean pore diameter is 0.14 mm and mean pore area (ap; from its circular approximation) is 0.0258 mm^2. For the entire egg, total estimated pore area is:

$$Ap = ap \times n \times At = 26.64 \text{ cm}^2$$

where At is eggshell area (457 cm^2).

Water-vapor exchange through the shell

Exchange of water between egg and environment is obviously vital during embryonic incubation. Rates of water-vapor exchange that are either too high or too low could be fatal to the enclosed embryo. To calculated this water vapor exchange rate for the Transylvanian eggs, we begin with the relationships among conductance (G) of the eggshell, temperature, water vapor diffusion rate, and pore area and length (Ar et al., 1974):

Figure 6.11. **A, B.** Tangential sections of calcite bundles and pores (p), showing a polygonal pattern (cr) marking the boundaries between the columnar bundles and ghosts of growth rings. Note in B the position of pores between adjacent bundles. Acetate peels after polished and etched surfaces.

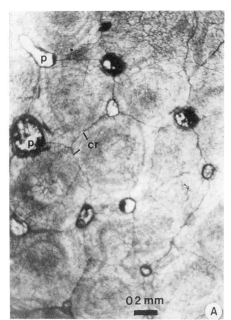
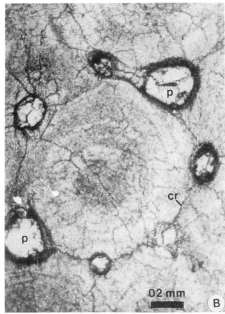

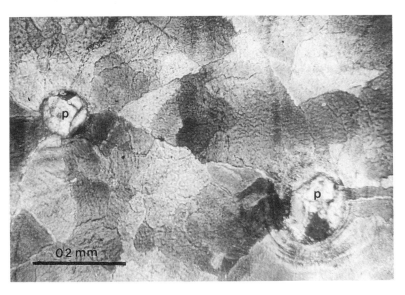

Figure 6.12. Xenotopic mosaic of calcite crystals showing composite to weakly undulatory extinction. Fine, discontinuous growth rings are centered around the circular transverse section of a pore (p). Crossed nicols.

Figure 6.13. Schematic drawing of the eggshell in the radial view. Cross-hatches represent the pores and some interstices.

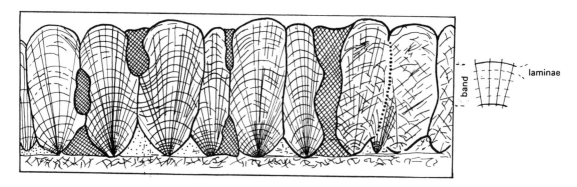

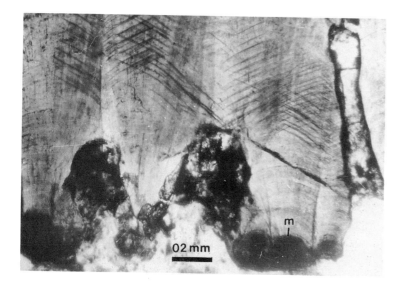

Figure 6.14. Radial section of the internal surface of an eggshell showing paired striation sets producing a herringbone pattern, deeply penetrating interstices and dark mammillae (m). Thin section in plane-polarized light.

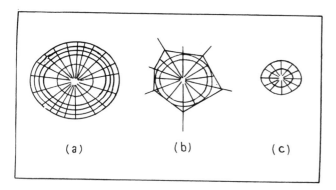

(a)   (b)   (c)

Figure 6.15. **A–C.** Schematic cross sections showing the circular outlines of columnar bundles in their proximal (C) and distal (A) parts, in contrast to the polygonal outline of their middle part (B).

$$G_{H_2O} = C/RT \times D_{H_2O} \times Ap/L$$

where $C = 0.56 \times 10^9$ s $\times$ mg/day/mol, a conversion constant; $R = 6.24 \times 10^4$ cm^3 mmHg/mole/K, the gas constant; $T = 303$ K ($= 30°$ C); $D = 0.292$ cm^2/s, the diffusion coefficient of water vapor in air at $3°$ C; Ap = total pore area of the egg (cm^2); and L = pore length (cm). An incubation temperature of 30°C is used

Figure 6.16. **A–C.** X-ray diffraction pattern of calcite powder of the eggshell (A), X-ray diffraction spectra of calcite obtained from polished tangential surface of the eggshell (B) and from polished radial section (C).

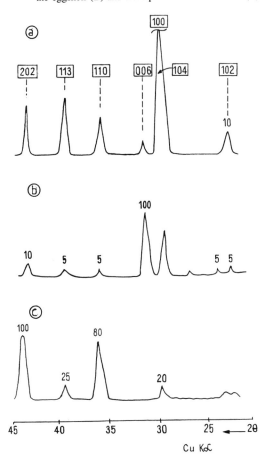

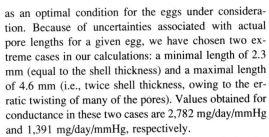

as an optimal condition for the eggs under consideration. Because of uncertainties associated with actual pore lengths for a given egg, we have chosen two extreme cases in our calculations: a minimal length of 2.3 mm (equal to the shell thickness) and a maximal length of 4.6 mm (i.e., twice shell thickness, owing to the erratic twisting of many of the pores). Values obtained for conductance in these two cases are 2,782 mg/day/mmHg and 1,391 mg/day/mmHg, respectively.

Mass loss of the egg and its contents during incubation consists predominantly of water loss. Total water loss (M) depends on water vapor conductance (G), difference in water vapor pressure across the shell (dp), and length of time of incubation (I):

$$M = G \times dp \times I$$

Conductance is the same as in previous calculations, but incubation time and vapor-pressure differences are uncertain. We suggest that a reasonable guess for incubation time of these eggs is between 10 and 100 days (viz., Ferguson, 1985, for crocodilians; Wealty and Baptista, 1988, for birds), and internal vapor pressure should be the saturation pressure at incubation temperature. This pressure (measured in mmHg) is:

$$P_{sat} = 27.06 - 1.542t + 0.0564t^2$$

where t is temperature expressed in °C.

According to Seymour (1979), a 20 percent water loss could likely be tolerated by a dinosaur embryo within an egg. For modern avian eggs, for example, this loss is no more than 15 to 20 percent. With a total mass (W) and a maximum mass loss of 0.2 W, the ambient water vapor pressure can be calculated from:

$$P_{ext} = P_{sat} - 0.2W/(G \times I)$$

from which relative humidity (r) of the surrounding medium is:

$$r = 100\, P_{ext}/P_{sat}$$

From this final calculation, it is likely that the relative humidity of the incubation site must have been high, probably between 85 and 97 percent over an incubation period greater than 30 days, even for the case of minimum conductance ($G = 1.391$).

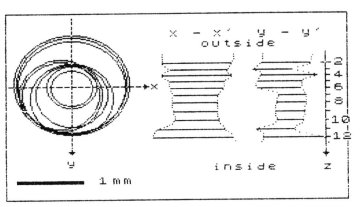

Figure 6.17. Variation of diameter and form of a pore through the entire 2.3 mm thickness of the eggshell. Left side: horizontal projection. Right side: vertical section. 2-12: number of the level of successive serial section from the outer to inner eggshell surfaces.

Significance of multistratified laminae of eggshells and of Transylvanian nesting environment

Multistratified growth lines or laminae are common structural features of the dinosaur eggshells from the Hațeg Basin. These laminae are also found in other fossil eggshells, especially among those of the angusticanaliculate morphotype (Erben, Hoefs, & Wedepohl, 1979). However, they appear to be otherwise absent in many other shell types. It is possible that these differences are biologically real. However, we suspect that growth lines were originally present in many or perhaps all hard-shelled eggs. Their absence in many fossils may well be the product of diagenetic recrystallization of eggshell calcite (Hirsch, personal communication). As primary structural elements of the eggshell, growth laminae appear to have formed before the calcite aggregates, because the opaque arrays of laminae traverse the aggregates. The latter likely crystallized from an amorphous mixture of bicarbonate gel and organic matter. We suspect that variation of the proportion of this gel and organics across the thickness of the shell may be a function of maternal circulatory carbon dioxide (CO_2) concentrations. This concentration, as well as the rate of carbonate precipitation in the shell, is controlled by such physiological functions as maternal respiration and pulse. If these relationships prove to be true, then the approximately 200 growth lines arranged in about fifty bands may well indicate the number of maternal heart beats or breaths during the development of the eggshell.

The eggs themselves appear to have been buried superficially in fine sandy sediments, which were then rapidly overlain by a thick blanket of coarse sediments. Successful incubation of the eggs in this environment required that water loss from the eggs, itself a function of shell conductance and sediment humidity, should not

Table 6.1. *Pore measurement on serial eggshell sections*

z	Investigated Area (cm²)	Number of pores	Pore density (cm⁻²)	Mean pore diameter (cm)	Total pore area (cm²)	Area occupied by (%) pores
3	1.07	106	98.9	0.386	0.0395	3.690
4	1.07	159	148.5	0.213	0.0180	1.685
5	1.43	244	170.7	0.173	0.0182	1.277
6	1.43	233	163	0.162	0.0153	1.069
7	1.43	221	154.6	0.137	0.0104	0.725
8	1.43	252	176.3	0.153	0.0147	1.031
9	1.43	272	190.3	0.133	0.0120	0.841
10	1.43	296	207.1	0.133	0.0131	0.915
11	1.43	312	218.3	0.167	0.0217	1.521

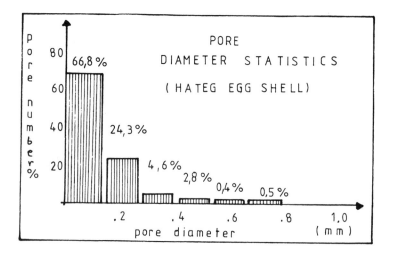

Figure 6.18. Relationship between pore frequency and diameter in Transylvanian dinosaur eggshell.

Table 6.2. Pore frequency and diameters of serial
eggshell sections

Mean pore diameter (cm)	Number of pores	Total pore number (%)	Mean pore area (mm²)	Calculated number of pores
0.07	1,400	66.83	0.00385	1,392.83
0.20	510	24.34	0.0314	500.04
0.33	97	4.63	0.1386	179.52
0.46	60	2.86	0.166	64.45
0.61	7	0.33	0.292	19.76
0.75	11	0.52	0.442	6.56

have exceeded 20 to 25 percent of egg weight. For eggs
with an internal temperature of about 30°C and the conductance values derived in this chapter, water loss
would have been much higher than 25 percent when the
surrounding sediments were completely dry. In contrast,
the best conditions for embryonic development require
50 to 60 days of incubation and 85 to 95 percent sediment humidity. A nest formed in such humid, warm,
fine-grained sand has at least two principal advantages.
First, such a nest maintains high and constant levels of
humidity. Second, thermal stability, with evaporation
and condensation mediating changes in temperature, is
slowed throughout the incubation period. Finally, it is
important to note that the paleoecophysiology of the
Transylvanian eggs and nests documented here is supported by the clay–mineralogic analyses conducted by
Radan (in Grigorescu et al., 1990a). These studies indicate a predominance of smectite, a group of clay minerals formed by the diagenesis of the volcanic ashes in
humid conditions. The much rarer calcretes within the
local section containing the nests suggest a limited degree of aridity within the region.

Acknowledgments

We thank the editors of this volume for the opportunity
to report on the eggs and nests newly recovered from the Hațeg
Basin of Romania. We also thank P. J. Currie, J. R. Horner,
and L. M. Witmer for information and discussions on embryonic and hatchling dinosaurs. This research was supported by
funds from the Romanian Academy of Science.

References

Ar, A., Paganelli, C. V., Reeves, R. B., Greene, D. G., &
    Rahn, H. 1974. The avian egg: water vapor conduc-
    tance, shell, thickness, and functional pore area. *Con-
    dor* 76: 153–8.
Breton, G., Fournier, R., & Watté, J.-P. 1986. Le lieu de
    ponte de dinosaures de Rennes-le-Château (Aude).
    Premiers resultats de la campagne de fouilles 1984.
    Annales Museum du Havre 32: 1–13.
Buffetaut, E., & Le Loeuff, J. 1991. Late Cretaceous dino-
saur faunas of Europe: some correlation problems.
    Cretaceous Research 12: 159–76.
Dughi, R., & Sirugue, F. 1957. Les oeufs de Dinosauriens du
    basin d'Aix-en-Provence. *Compte rendu de l'Acadé-
    mie de Science, Paris* 245: 707–10.
Erben, H. K. 1970. Ultrastrukturen und Mineralisation rezen-
    ter und fossiler Eischalen bei Voegeln und Reptilien.
    Biomeneralisation 1: 1–66.
Erben, H. K., Hoefs, J., & Wedepohl, K. H. 1979. Paleobio-
    logical and isotopic studies of eggshells from a de-
    clining dinosaur species. *Paleobiology* 5: 380–414.
Ferguson, M. W. J. 1985. Reproductive biology and embry-
    ology of the crocodilians. *In* C. Gans, F. Billet, & P.
    F. A. Maderson (eds.). *Biology of the Reptilia, 14A*
    (New York: John Wiley & Sons), pp. 329–491.
Grigorescu, D. 1983. A stratigraphic, taphonomic and paleo-
    ecologic approach to a "forgotten land": the dino-
    saur-bearing deposits from the Hațeg Basin
    (Transylvania, Romania). *Acta Palaeontologica Po-
    lonica* 28: 103–21.
    1984. New tetrapod groups in the Maastrichtian of the
    Hațeg Basin: coelurosaurians and multituberculates. *In*
    W.-E. Reif & F. Westphal (eds.). *Third Symposium on
    Mesozoic Terrestrial Ecosystems, Short Papers*. (Tü-
    bingen: Attempto Verlag), pp. 99–104.
Grigorescu, D., & Hahn, G. 1987. The first multituberculate
    teeth from the Upper Cretaceous of Europe (Roma-
    nia). *Geologica et Palaeontologica* 21: 237–43.
Grigorescu, D., Avram, E., Pop, G., Lupu, M., Anastasiu, N.,
    & Radan, S. 1990a. *Guide to excursions. Interna-
    tional Symposium I.G.C.P. Projects 245 and 262*.
    (Bucharest: Institute of Geology and Geophysics).
Grigorescu, D., Seclaman, M., Norman, D. B., & Weisham-
    pel, D. B. 1990b. Dinosaur eggs from Romania. *Na-
    ture* 346: 417.
Hirsch, K. F., & Quinn, B. 1990. Eggs and eggshell frag-
    ments from the Upper Cretaceous Two Medicine For-
    mation of Montana. *Journal of Vertebrate
    Paleontology* 10: 491–511.
Le Loeuff, J. 1991. The Campano-Maastrichtian vertebrate
    faunas of southern Europe and their relationships with
    other faunas in the world; palaeobiogeographical im-
    plications. *Cretaceous Research* 12: 93–114.
Mikhailov, K. E. 1987. Some aspects of structure of the shell
    of the egg. *Paleontological Journal* 21: 54–61.
    1991. Classification of fossil eggshells of amniote verte-
    brates. *Acta Palaeontologica Polonica* 36: 193–238.
Seymour, R. S. 1979. Dinosaur eggs: gas conductance
    through the shell, water loss during incubation and
    clutch size. *Paleobiology* 5: 1–11.
Weishampel, D. B., Grigorescu, D., & Norman, D. B. 1991.
    The dinosaurs of Transylvania. *National Geographic
    Research and Exploration* 7: 196–215.
Weishampel, D. B., Norman, D. B., & Grigorescu, D. In
    press. *Telmatosaurus transsylvanicus* from the Late
    Cretaceous of Romania: the most basal hadrosaurid
    dinosaur.
Weishampel, D. B., & Reif, W.-E. 1984. The work of Franz
    Baron Nopcsa (1877–1933): dinosaurs, evolution, and
    theoretical tectonics. *Jahrbuch der geologischen Bun-
    desanstalt, Wien* 127: 187–203.
Wealty, J. C., & Baptista, L. 1988. *The Life of Birds* (New
    York: Saunders College Publishing).

7 Eggs and nests from the Cretaceous of Mongolia

KONSTANTIN MIKHAILOV,
KAROL SABATH, AND
SERGEY KURZANOV

Abstract

Fossil nests, eggs, and eggshells are reviewed from more than fifteen Upper Cretaceous localities in the Gobi Desert, Mongolia. These sites are in the Bayn Shireh (Cenomanian–Santonian), Barun Goyot and Djadokhta (Santonian–Campanian), and Nemegt (Maastrichtian) formations. Mass accumulations of eggs occur at Bayn-Dzak, Toogreekiyn-Shireh, Khermiyn-Tsav, Khulsan, Ologoy-Ulan-Tsav, and Gooreeliyn-Tsav. For each type of egg, a short history of the discovery is given, followed by a description of egg morphology, shell structure, and nest taphonomy. The eggshells are organized according to a parataxonomic classification (egg families Faveoloolithidae, Dendroolithidae, Spheroolithidae, Ovaloolithidae, Elongatoolithidae, Laevisoolithidae, Subtiliolithidae, eggs of protoceratopsid dinosaurs and gobipterygiform birds). Also discussed are the taxonomy and paleobiology (including paleophysiology and nesting behavior) of the Cretaceous beach-nesting communities, as well as the stratigraphic implications of the fossil egg assemblages.

Introduction

The Central Asiatic Expedition of the American Museum of Natural History discovered dinosaurian eggs during the 1920s in the Upper Cretaceous beds of "Shabarakh-Usu" (= Bayn-Dzak) in the Gobi Desert, Mongolia (Andrews, 1932, pp. 162–3, 207–14). The microstructure of the shells of these eggs was studied by Van Straelen (1925), Schwartz et al. (1961), and Erben (1970).

The Mongolian Paleontological Expeditions of the USSR Academy of Sciences (1946–1949), the Joint Soviet–Mongolian Paleontological and Geological Expeditions (since 1969), and the Polish–Mongolian Paleontological Expeditions (1963–1971) have expanded the number of known Cretaceous egg-bearing localities in the Gobi area of Mongolia (Fig. 7.1). For historical details about these and other paleontological discoveries, see Efremov (1949), Kielan-Jaworowska and Dovchin (1969), Sochava (1969), Kielan-Jaworowska and Barsbold (1972), and Shuvalov (1982).

The remarkable diversity of fossil eggs in the Gobi Desert makes it the "classic site" for paleooology. The material collected after World War II is housed in Moscow at the Paleontological Institute of the Russian Academy of Sciences, in Ulan Bator at the Laboratory of Paleontology of the Geological Institute of the Mongolian People's Republic Academy of Sciences and the State Museum of Regional Studies, and in Warsaw at the Institute of Paleobiology of the Polish Academy of Sciences. The collection in Warsaw has recently been described (Sabath, 1991a). The Ulan Bator and Moscow collections, first studied by Sochava (1969, 1971, 1972), have received less attention, except for the most intriguing specimens (Kurzanov & Mikhailov, 1989). Recently, the Moscow collection has been used to construct a general classification of fossil oological remains based on eggshell ultrastructure and microstructure (Mikhailov, 1991a).

Materials and Methods

Most of the specimens to be described are housed in Moscow and Warsaw, with supplemental observations made by one of us (Mikhailov) on the egg collection in Ulan Bator. These supplemental observations were used for general grouping of fossil eggs. Primarily, we focus on the morphological characters of the eggs and clutches, their systematic interpretation, the reconstruction of nesting conditions, and the sedimentological features of dinosaur nesting sites. The results of studies on the structure and development of eggshells (Mikhailov, 1987, 1991a,b) permit us to group fossil eggs into types differing in their eggshell histostructure. The techniques used to study the eggshells have been described elsewhere (Mikhailov, 1987, 1991a), as has the new terminology (Kurzanov & Mikhailov, 1989; Mikhailov, 1991a; Sabath 1991a). The paleophysiological estimates are modified from Sabath (1991a). The water vapor conductance coefficient used for the G_{H_2O} given below is 1.56 instead of the erroneous 0.56 used previously (Sabath, 1991a, after Seymour, 1979). Accordingly, previous estimates are approximately one-third the new values.

Abbreviations

JSMGE – Joint Soviet–Mongolian Geological Expedition(s)

JSMPE – Joint Soviet–Mongolian Paleontological Expedition(s)

MPE – Mongolian Paleontological Expedition(s) of the USSR Academy of Sciences

PMPE – Polish–Mongolian Paleontological Expedition(s)

A – total functional pore area (cm²)

A_s – eggshell surface area (cm²)

E - elongation coefficient of egg (length/equatorial diameter)

G_{H_2O} – water vapor conductance of the eggshell [mg (day Torr)⁻¹]

L – shell thickness (mm)

N – number of pores

V – egg volume (cm³)

Mongolian egg-bearing localities

The major egg-bearing localities in southern Mongolia are confined to the area along the boundary between Omnogov aimak (South Gobi Province) and the Ovor Khangai and Bayan Khongor aimaks. Only a few egg-bearing localities are known from the Dornogov (East Gobi) aimak. Isolated eggshell fragments and isolated eggs occur in most of the Upper Cretaceous Gobi localities. The southern group of localities is better known. They lie in the Nemegt Basin and southeast of the Arts-Bogdo Mountains (Fig. 7.1). These localities are: Khermiyn-Tsav ("Khermeen Tsav"), Gooreeliyn-Tsav, Boogiyn-Tsav, Khaichin-Ula I, Nemegt, Khulsan, Toogreekiyn-Shireh (=Toogreekiyn-Us = "Toogreeg"), Bayn-Dzak ("Shabarakh Usu"), and Udan-Sayr. To the northeast is the Ologoy-Ulan-Tsav locality. The eastern localities are scattered over a larger area and include Mogoyn-Ulagiyn-Khaets (Tel-Ulan-Ula Mountains), Baga-Tariach, Ikh-Shunkht (Bayshiyn-Tsav Basin), and Shiljust-Ula (Borzongiyn-Gobi).

Geological data relating to the sites are available elsewhere (Gradziński, Kaźmierczak, & Lefeld, 1969; Gradziński, 1970; Lefeld, 1971; Gradziński & Jerzykiewicz, 1972; Sochava, 1975; Shuvalov, 1975, 1982). Therefore, we shall only present a brief summary of the general stratigraphical and paleogeographical information on these sites. Most localities occur in the Bayn Shireh (Cenomanian–Santonian), Barun Goyot (Santonian–Campanian) or Nemegt (Maastrichtian) formations or their biostratigraphic equivalents. Only Mogoyn-Ulagiyn-Khaets and Dariganga in the eastern Gobi occur in the Bayn Shireh Formation.

The upper beds of the Bayn Shireh Formation are heterogeneous, consisting of red clay and sandstones intercalated with gray sands. These strata probably represent lacustrine and alluvial deposits where the hydrodynamic flow regime varied. Some of the lacustrine deposits appear to have been predominantly warm brackish water (up to 20°C), indicating a relatively humid climate (Stankevitch, 1982).

Figure 7.1. Cretaceous egg-bearing localities in the Mongolian Gobi Desert. Locality symbols are designated by formation or their biostratigraphical equivalents. Southwestern group: **1.** Dushih–Ula, **2.** Khermiyn–Tsav I and II, **3.** Tsagan–Khushu, **4.** Khaichin–Ula I, **5.** Boogiyn–Tsav, **6.** Gooriliyn–Tsav, **7.** Sheregiyn–Gashoon, **8.** Altan–Ula III, **9.** Nemegt, **10.** Khulsan, **11.** Gilbent, **12.** Builjasutuin–Khuduk, **13.** Udan–Sayr, **14.** Toogreekiyn–Shireh, **15.** Bayn–Dzak, **16.** Ologoy–Ulan–Tsav. Eastern group: **17.** Ikh–Shunkht, **18.** Baga–Mod–Khuduk, **19.** Mogoyn–Ulagiyn–Khaets, **20.** Baga–Tariach (and Dariganga eastward). **21.** Shiljust–Ula in southernmost Mongolia (Borzongiyn Gobi).

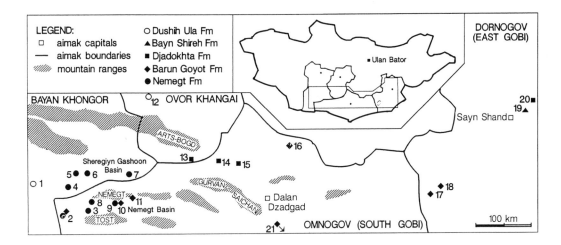

Most of the egg localities occur in the Barun Goyot Formation, which overlies the Djadokhta Formation (Gradziński, Kielan-Jaworowska, & Maryańska, 1977; Osmólska, 1980; Barsbold, 1983). Lithologically, the two formations are similar and consist of red well-sorted sandstones, intercalated with clay and gravel (except the gray beds at Toogreekiyn-Shireh). The facies diversity of the Barun Goyot Formation consists of dune and interdune deposits, playas, and ephemeral stream deposits (Gradziński & Jerzykiewicz, 1974; Ryszard Gradziński, personal communication to Sabath in 1991). In the Djadokhta Formation, eolian sandstone (dunes and intermittent pond deposits) dominate, especially in the stratotype outcrop at Bayn-Dzak and at Toogreekiyn-Shireh. Caliche horizons mark sedimentation gaps under arid conditions. Interpreting the Djadokhta is to some extent obscured by secondary features resulting from much later infiltration of groundwater, which has cemented the sand with calcitic matrix. Although lithologically similar, the Djadokhta and Barun Goyot formations have a different faunal composition (there are some common or similar lizard and dromaeosaurid genera, however). If this difference is explained by facies distinction alone, then the faunal division arose only because of different local paleogeographical situations.

The Djadokhta and Barun Goyot formations have been assumed by Russian geologists to have had a similar environmental setting in a large intracontinental basin with a complex drainage system. This consisted of numerous islands, bars and shallows, and a complicated shoreline. Fluctuations of water level would result in large seasonal changes in the area submerged by strongly carbonated and brackish water (Shuvalov, 1982). The bottom sands exposed during droughts were redeposited by winds, forming eolian facies. The water temperature, as high as 26°C (Kolesnikov, 1982), is consistent with the postulated hot and arid climate for this period. Proluvial conglomerates and gravels occur in some localities, associated with ''islands'' of older, mainly Paleozoic, rocks bordered by Cenozoic fault systems. [For an alternative interpretation, see Gradziński & Jerzykiewicz, 1974 – editors].

The outcrops of the Nemegt Formation are undoubtedly coeval, formed under predominantly fluvial (Gradziński, 1970; Sochava, 1975) or limnic (Shuvalov, 1982) conditions. Both authors agree that a temperature decrease followed the Barun Goyot/Djadokhta temperature maximum, thereby increasing fluvial sedimentation. Furthermore, the abundance of cross-bedded channel sediments, and the taphonomy of dinosaur skeletons (prolonged decay and subaerial drying on the ground, followed by quick burial) is inconsistent with the presence of a huge, single lake (Stankevitch, 1982; Shuvalov, 1982). The rapid burial can be explained by alternating dry and rainy seasons. Despite a decrease in salinity, the lakes were still brackish because of evaporation.

At the Ologoy-Ulan-Tsav locality, abundant sauropod eggs have been found, which belong to the typical Barun Goyot egg assemblage (see last section below). Therefore, the Lower Cretaceous age of Ologoy-Ulan-Tsav (Sochava, 1975; Shuvalov, 1982) is far from proven. Indisputable Lower Cretaceous eggshells are thus far only known from Builjasutuin-Khuduk (Kurzanov & Mikhailov, 1989) and Dushih-Ula (Shuvalov, 1982).

Mass accumulations of fossil eggs occur mostly in the red beds of the Barun Goyot Formation (including the Djadokhta Formation) and rarely in the gray beds (Toogreekiyn-Shireh). The Nemegt beds contain only separate eggs or clutches and much eggshell debris. Along with the eggs, dinosaur skeletal remains were found in some localities (Bayn–Dzak, Toogreekiyn-Shireh, Khermiyn-Tsav). Other red bed localities (Ikh-Shunkht, Mogoyn-Ulagiyn-Khaets, Khaichin-Ula I) have not yielded any bones.

Description of eggs and eggshells

All of the Cretaceous dinosaurian and avian eggs and eggshells can be classified into a dozen parataxonomic families (Mikhailov, 1991a), of which nine families are known from the Mongolian Gobi Desert. Because many fossils have not been formally described and named, we designate groups and subgroups at the generic/specific level by the first letter of the family name followed by a successive number. Distinction of families is based on eggshell histostructure, and, for lower-rank groups, on egg morphology (shape, size, ornamentation, shell thickness) (Figs. 7.2–7.4).

Faveoloolithidae Zhao and Ding (1976) (?Sauropoda)

Eggshell histostructure

The eggshell is the basic dinosauroid type, of the filispherulithic morphotype, with a multicanaliculate pore system (Figs. 7.2A, 7.5C,D).

F1: *Faveoloolithus ningxiaensis* (Zhao & Ding, 1976) (Figs. 7.2A, 7.3A, 7.5A–D).

Material

Whole and fragmentary clutches (PIN 2970/1, and PIN uncatalogued; numerous in Ulan Bator collection), whole eggs and eggshell debris (PIN 4225-1 and PIN uncatalogued; ZPALA MgOv-III/18b).

Morphology

Eggs subspherical in shape; 150–165 mm long ($V = 900$–$1,200$ cm³); eggshell thick ($L = 2.1$–2.6 mm over the whole egg, main range: 2.2–2.4 mm); outer surface smooth (or rough if eroded). Due to the high porosity of the eggshell, the diffusive conductance exceeds that of all other known rigid-shelled eggs by more than an order of magnitude [lumen of pores occupies almost half of a tangential section area: $A_s = 470$ cm²,

$A = 200$ cm^2, $N = 650,000$, $G_{H_2O} = 21,000$ mg (day Torr)$^{-1}$].

Nest structure

Cluster of two to three layers of eggs with up to fifteen eggs in each; nest diameter: 1–1.5 m (Fig. 7.4E).

Occurrence

Ologoy-Ulan-Tsav (?Upper Cretaceous), Khermiyn-Tsav (biostratigraphic equivalent of the Barun Goyot Formation), and Ikh-Shunkht (Barun Goyot Formation).

History, taphonomy, and paleobiology

Clutches and eggs found at Ologoy-Ulan-Tsav by JSMGE (Shuvalov, 1975; Figs. 19, 31) were originally described as eggs with a multicanaliculate shell (Sochava, 1969, Plate 11:1). In 1987, a clutch and some broken eggs were collected at Ikh-Shunkht by JSMPE (PIN collection). Fragments of eggshell have been identified in the materials from Khermiyn-Tsav (ZPAL). All localities (except for Ologoy-Ulan-Tsav?) belong to the Barun Goyot Formation or its biostratigraphic equivalents. A clutch of *F. ningxiaensis* eggs was also found in the Upper Cretaceous of China (Zhao & Ding, 1976).

The age of the Ologoy-Ulan-Tsav locality is uncertain (see above). The outcrops are of red and red-brown clays interbedded with variously grained sands, showing coarse oblique bedding, or with conglomerates and gravels. Carbonate concretions are common. The alluvial (and occasional proluvial) deposits formed near channel bars of a river, indicating periodical drops in water level (Sochava, 1975). The eggs and clutches only occur in the conglomerates and gravels, indicating that the animals nested at the end of the rainy season when the gravel bars were exposed.

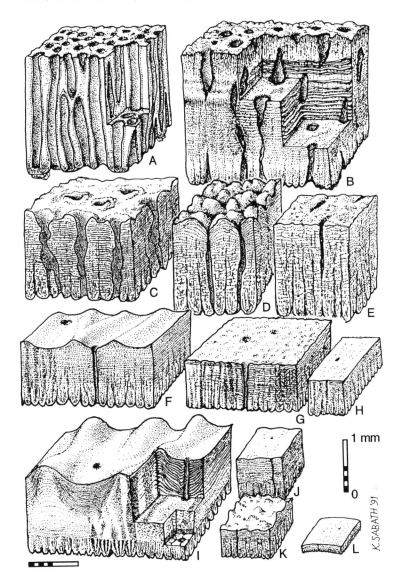

Figure 7.2. Diagrammatic comparison of the histostructure and surface ornamentation of dinosaurian eggshells from Mongolia. Oblique views of the eggshell fragments, with radial fracture exposed, are drawn to scale. Extinction patterns (in polarized light) are superimposed onto the front surface of F, G, H, and I to show the similarities and differences between protoceratopsid and elongatoolithid eggshells. **A.** Multicanaliculate faveoloolithid eggshell. **B–D.** Prolatocanaliculate eggshells: dendroolithid (B), spheroolithid (C), ovaloolithid (O1 group, D). **E.** Rimocanaliculate ovaloolithid (O2) eggshell. **F–H.** Angusticanaliculate protoceratopsid eggshells: P2 (F), P1 (G), and P3 (H). **I–L.** Angusticanaliculate ornithoid eggshells (equatorial zone of egg): elongatoolithid (E1: I), *Laevisoolithus* (J), *Subtiliolithus* (K), and smaller "*Gobipteryx*" eggs (G1 group; L). In D and E the horizontal scale is compressed by approximately 1.5 to show the pore and ornamentation patterns. (A, B, F, G, H, I, and L after Sabath, 1991a.)

1 mm

0

K. SABATH '91

Eggshells are always recrystallized and impregnated with iron salts; the pore canals are filled with secondary calcite (e.g., Fig. 7.5C). The lack of resorption craters at the base of the spherulites shows that some eggs died before incubation, possibly due to flooding. The Ologoy-Ulan-Tsav site reveals mass egg laying by many females, year after year at the same sites along sand and gravel beaches. The large diffusive permeability of the eggshells suggests an extremely humid incubation environment (Seymour, 1979; Sabath, 1991a). This, together with the primitive nest structure (chelonianlike clusters of eggs), suggests underground incubation of the *Faveoloolithus* eggs.

No traces of embryos or hatchlings have been found, so a definitive taxonomic assignment of the eggs is not possible. However, the large size of eggs (comparable to *Megaloolithus* from the Maastrichtian of the Aix-en-Provence Basin, France; see Vianey-Liaud et al., Chapter 11) and the mode of nesting and incubation conditions suggests that faveoloolithid eggs belong to sauropod dinosaurs. In fact, the few bones found at Ologoy-Ulan-Tsav belong to an unidentified sauropod (Efremov, 1949).

In China, besides subspherical *Faveoloolithus* eggs, a clutch of larger ellipsoid eggs with the same shell histostructure (Faveoloolithidae: *Youngoolithus xiaguanensis*) has been found in association with a tridactyl footprint (Zhao, 1979). On the basis of the print, the eggs might be those of an ornithopod (Sabath 1991a), but it is also possible that the footprint could have been left by an animal unrelated to the eggs.

Dendroolithidae Zhao (1988 (?Sauropoda or ?Ornithopoda)

Eggshell histostructure

Eggshells are of the basic dinosauroid type and of the dendrospherulithic morphotype. They have a pro-latocanaliculate pore system with the pores interconnected and forming a network of microcanals among the prisms (Figs. 7.2B, 7.5E, 7.6D). Within the single Mongolian "genus," *Dendroolithus*, variants (D1–D3) can be distinguished, differing in size and shell thickness (Figs. 7.3B,D, 7.6A–D & Table 7.1).

Material

An egg and eggshell fragments (PIN uncatalogued) from a clutch of eggs housed in Ulan Bator. Half an egg and eggshell fragments (ZPAL MgOv-I/16, MgOv-III/17,18,20); eggshell debris (PIN 3142-454, 466, 475).

Morphology

Eggs are subspherical or slightly oval, smaller than *Faveoloolithus* with thick eggshell (ca. 2 mm). Surface is rougher than in faveoloolithid eggs; sometimes a microshagreen pattern is formed by separate fine prisms of spherulites. The pore orifices are large; eggshell is less porous than those of *Faveoloolithus* and *Megaloolithus* eggs, or of ?titanosaurid eggs from Los Alamitos, Argentina (Sabath, 1991a); $As = 400$ cm², $A = 4.5$ cm², $N = 32,000$, $G_{H_2O} = 1500$ mg (day Torr)⁻¹.

Nest structure

Irregular cluster of eggs, resembling a single layer of a *Faveoloolithus* clutch (Fig. 7.4D).

Occurrence

Shiljust-Ula, Khermiyn-Tsav (biostratigraphic equivalent of the Barun Goyot Formation), and Khulsan (Barun Goyot Formation).

History, taphonomy, and paleobiology

After *Dendroolithus* eggs were described from the Upper Cretaceous of China (Zhao, 1988), fragments

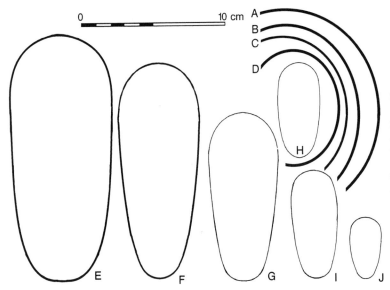

Figure 7.3. Comparison of the sizes and shapes of selected Mongolian fossil eggs. Shell thickness is shown roughly proportional to scale. **A.** *Faveoloolithus* (F1). **B, D.** *Dendroolithus* (D2 & D1, respectively). **C.** *Spheroolithus* (S1). **E.** ?*Macroolithus* (E1). **F.** Larger protoceratopsid eggs (P1 & P2 groups). **G.** Smaller protoceratopsid egg (P3). **H.** *Laevisoolithus*. **I–J.** "*Gobipteryx*" eggs (G2 & G1, respectively). (Modified from Sabath, 1991a.)

0 _____ 10 cm

of eggshell were identified in the PIN and ZPAL collections from Khermiyn-Tsav and Khulsan. In 1989, a whole clutch and an egg plus numerous fragments from another clutch were found in the red beds (Barun Goyot Formation) in southernmost Mongolia (Shiljust-Ula locality, Borzongiyn Gobi). Our field observations (Mikhailov and Kurzanov) revealed that the burial conditions of dendroolithid eggs at Shiljust-Ula were similar to those of *Faveoloolithus* eggs at Ologoy-Ulan-Tsav and Ikh-Shunkht. The iron stained eggs occur in red, coarse-grained sands and gravels associated with a zone containing large, limnic, brackish water mollusks. This lithology indicates ponds on an alluvial plain. Two clutches were 6 m apart. The babies apparently hatched through a round opening in the egg (Fig. 7.6D). The taphonomy of the eggs and the well developed pore system of the eggs indicate underground incubation in humid conditions (sand beaches along a shore). The hatchlings (like those of *Faveoloolithus*) may have been amphibious.

The taxonomic affinity of the eggs is unknown, but based on the eggshell structure and the mode of nesting, could be of sauropod or ?ornithopod dinosaur.

Spheroolithidae Zhao (1979) (Ornithopoda: possibly some hadrosaurs).

Eggshell histostructure

The shell is the dinosauroid basic type of the prolatospherulithic morphotype, with a prolatocanaliculate pore system (Figs. 7.2C & 7.8E,H). Two groups of spheroolithids are known from Mongolia and are referred to as S1 and S2.

Group S1: *Spheroolithus* sp. (= "*Oolithes spheroides*" Young, 1954, redescribed by Zhao, 1978; Figs. 7.3C, 7.7, & 7.8A,E).

Material

Whole clutches (PIN 4216-400; some also in Ulan Bator), eggs and shell fragments (PIN 2970/3, 4216-401).

Morphology

Eggs are large (80–100 × 70–80 mm), spherical and slightly ellipsoid ($E = 1.1$–1.2), with the shell 1.1 to 2.2 mm thick over the whole egg (main range, 1.4–2.0 mm). The smooth outer surface exhibits a fine sagenotuberculate pattern. The porosity is lower than in *Faveoloolithus*, *Megaloolithus*, and perhaps *Dendroolithus* eggs suggesting drier incubating conditions.

Nest structure

Clutch PIN 2970/3 is single-layered and does not show the circular arrangement of the eggs in contrast

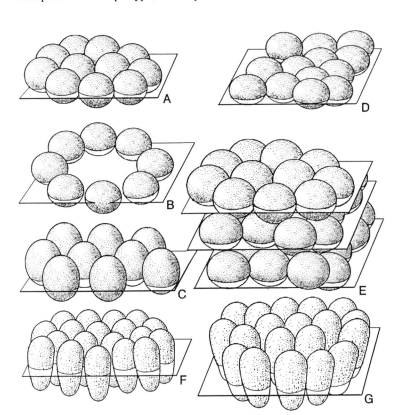

Figure 7.4. Diagrammatic comparison of clutch patterns of Asiatic dinosaur nests (horizontal planes are for reference only). A, B. *Spheroolithus* (S1) from Gooriliyn-Tsav (A) and Laiyang, China (B). C. Clutch of ?S2 eggs from Shiljust-Ula. D. *Dendroolithus* from Shiljust-Ula; E. *Faveoloolithus* (Ologoy-Ulan-Tsav). F. Protoceratopsid clutch. G. Elongatoolithid clutch.

Table 7.1. Subdivision of Mongolian
Dendroolithus eggs

Sub-group	Shell thickness (mm)		Egg size (mm)	Egg volume (cm³)
D1	1.5–3	(2.0–2.7)	70 × 60	150–200
D2	1.5–3	(2.0–2.7)	115 × 108	650–800
D3	1.5–3.8	(2.7–3.5)	>120 × >110?	>800?

with some of the clutches found in China (compare Figs.
7.7A,B & 7.4A,B).

Occurrence
Gooreeliyn-Tsav, Sheregiyn-Gashoon (Nemegt
Formation).

History and taphonomy
Numerous clutches of *Spheroolithus* eggs were
found at Laiyang, Shantung Province, China, during the

Figure 7.5. Faveoloolithid (A–D) and dendroolithid (E) fossils. A. *Faveoloolithus ningxiaensis*, upper layer of a clutch
(PIN 4225-1, Ologoy-Ulan-Tsav, egg diameters about 15 cm). **B–E.** Eggshell structure: multicanaliculate pore system
(B, C; ZPAL MgOv-III/18b) and (C, D) filispherulithic morphotype of vertical shell units (su) in faveoloolithids,
prolatocanaliculate pore system and dendrospherulithic morphotype (E) in dendroolithids. Transverse (B) and radial (C)
thin section with reflected light; radial views of freestanding eggshells (D, E), scale bar = 1 mm, shell thickness ca. 2
mm; pc – pore canals). (B, C after Sabath, 1991a.)

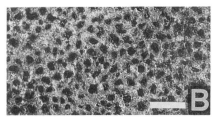
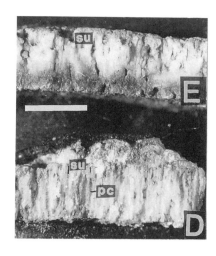

1960s (Young, 1954, Chow, 1954). The JSMGE and JSMPE later discovered spheroolithid eggs in Mongolia. These eggs, now housed in Moscow and Ulan Bator, were assigned by Sochava (1969, Plate 11,4) to the prolatocanaliculate group of eggshells.

Unlike *Faveoloolithus* and *Dendroolithus*, *Spheroolithus* eggs have been found in the white and dark gray sands, not in the red beds. The eggs from Gooreeliyn-Tsav are gray, and those from Sheregiyn-Gashoon are black (*Spheroolithus* eggs from Laiyang, China, are dark purple). The eggs are unhatched, and probably lack remains of ossified embryos (no tomography or X-ray studies have been made, however). *Spheroolithus* fossils are rare in Mongolia, but, in China, they are the most common group of dinosaur eggs (85 percent of the well-preserved specimens; Chao & Chiang, 1974, p. 82).

Group S2: The specimens have not described previously (Figs. 7.8B,H, & 7.9A).

Material

An egg (PIN 4228-2) plus abundant eggshell debris (PIN 522-400, 614-606, 607; ZPAL MgOv-III/17, 18b) and one ?complete clutch in Ulan-Bator (an egg with shell fragments from this clutch: PIN, uncatalogued).

Morphology

The eggs are subspherical or elliptical in shape, and relatively large compared to most eggs: 90×70 mm ($A^s = 200$ cm^2, $V = 300$ cm^3). The eggshell thickness varies from 1.0 to 1.6 mm over the egg (as in other

Figure 7.6. Dendroolithidae. **A, B.** *Dendroolithus* sp. (D1 group), different view of an egg (PIN, uncatalogued, Shiljust–Ula); **C, D.** ?*Dendroolithus* sp. (D2 group; ZPAL MgOv-I/16, Khermiyn-Tsav), outer view (C), scale bar = 1 cm, and radial thin section through the eggshell with transmitted light (D), scale bar = 1 mm (for abbreviations see caption for Fig. 7.5D,E). (D after Sabath, 1991a.)

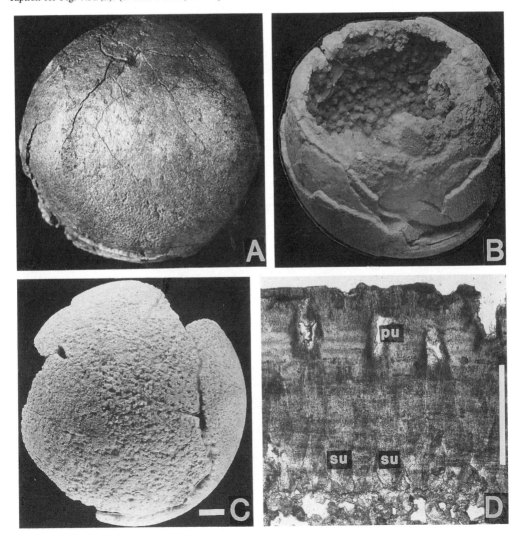

spheroidal eggs, there was a thinner eggshell area, perhaps serving as a "hatching window" from which the hatchling emerged), main range: 1.2–1.5 mm. The surface sculpture is similar to the sagenotuberculate ornamentation of the *Maiasaura* type (i.e., with a comparatively large, roughly polygonal mesh of protuberances; some of them surround orifices of prolatocanaliculate pores). The diffusive permeability is similar to that of *Spheroolithus* (S1 group), or perhaps greater (G_{H_2O} = 2,000 mg (day Torr)$^{-1}$). The egg morphology and shell structure strongly resemble those of the *Maiasaura* eggs (Mikhailov 1991a).

Occurrence

Bayn-Dzak (Djadokhta Formation), Baga-Tariach, Ikh-Shunkht, ?Shiljust-Ula (?Barun Goyot Formation), Tsagan-Khushu (Nemegt Formation).

History and taphonomy

Abundant eggshell debris was collected at Bayn-Dzak in the 1940s, and recently at Tsagan-Khushu. At the Shiljust-Ula locality (Borzongiyn-Gobi), a whole clutch of ?spheroolithid eggs similar to the S2 group was found in 1989 by the JSMPE (housed in Ulan Bator). The elliptical eggs were arranged with their long

Figure 7.7. Spheroolithidae (S1 group). **A.** Clutch of *Spheroolithus* sp. (PIN 4216-400, Gooriliyn–Tsav). **B.** Clutch of *Spheroolithus chiangchiungtingensis* (V-731; Laiyang, China; housed in PIN), eggs 7–8 cm long.

axis subvertical, were separated by sandy matrix, and formed one layer. The nest was elongated, or there were two very closely placed clutches (Fig. 7.4C).

The Shiljust-Ula clutch was found on the weathered surface of the red badlands (?Barun-Goyot Formation). Eggshell fragments were also found at Tsagan-Khushu and Khermiyn-Tsav. At Tsagan-Kushu, they eroded from the white beds of the Nemegt Formation into sayrs (canyons) where they were found. At Khermiyn-Tsav, the eggshells were collected on the gravel tops of hills. Poorly preserved fragments of the eggshells, evidently redeposited, were collected from the red sands at Bayn-Dzak (white beds at Bayn-Dzak are apparently not fluvial).

The S1 group (*Spheroolithus*) and S2 group fossils are found at different localities of the Nemegt Basin. Both groups are quite common in Central Asia (although in Mongolia, they are outnumbered by protoceratopsid and elongatoolithid eggs). Eggshell histostructure identifies these eggshells as belonging to hadrosaurs (Mikhailov, 1991a), which prevailed in the Late Cretaceous of Central Asia (Kurzanov, 1989; Zhen et al., 1985). The *Spheroolithus* eggs from China (Laiyang) have also been attributed to hadrosaurs; the eggs

of *S. irenensis* were found in close association with *Bactrosaurus* remains (Chao & Chiang, 1974, p. 81). The pattern of surface ornamentation (especially in the S2 group), pore system, and structure of eggs and clutches are similar to *Maiasaura peeblesorum* eggs (Horner & Makela, 1979; Hirsch & Quinn, 1990); their prolatocanaliculate pore systems are also functionally similar. We can only speculate on the possible *Maiasaura*-like nesting habits (vegetation-covered nesting mounds, altricial juveniles) of the Asiatic hadrosaurs.

Ovaloolithidae Mikhailov (1991a) (?Ornithopoda)
Eggshell histostructure
The shell is of the dinosauroid basic type, angustispherulithic morphotype, and with a pore system of mixed rimocanaliculate and angusticanaliculate types (Figs. 7.2D,E & 7.8F,G).

The Mongolian ovaloolithid eggs form two groups, O1 and O2, and do not compare well with the Chinese genus *Ovaloolithus* (comparisons with eggshell from clutch V-736, "*Oolithes chinkangkouensis*" from Laiyang). This genus was established by Zhao (1978,

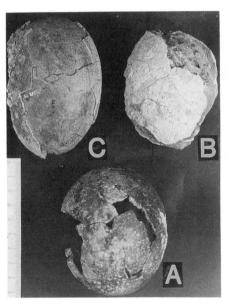

Figure 7.8. Spheroolithidae (A,B,E,H) and Ovaloolithidae (C,D,F,G). **A.** *Spheroolithus* sp. (S1 group; PIN 2970-3, Sheregiyn-Gashoon). **B.** Egg of S2 group (PIN 4228-2, Baga-Tariach); **C.** Egg of O2 group (PIN 4231-1, Altan–Ula–III); scale bar = 90 mm. **D.** Limb bones of a dinosaur embryo on the inner surface of eggshell, O1 group (PIN 2970-7, Mogoyn–Ulagiyn–Khaets). Eggshell structure, radial view (E-G): prolatospherulithic morphotype (E–H), angustispherulithic morphotype (F,G). **E, F.** Freestanding eggshell, scanning electron microscope (SEM). **G, H.** Thin section using transmitted light; shell unit (su); shell thickness (E–H) ca. 1.8–2.0 mm.

1979) for some eggs originally referred to *Oolithes spheroides* by Young (1954).

Group O1 (Figs. 7.9C & 7.10B).

Material

A large fragment from an egg (PIN 2970/7) and numerous eggshell fragments, partly from badly eroded nests (PIN 3225-150, 151, 152).

Morphology

The eggs were ?subspherical, or slightly elliptical in shape ($E = 1.2–1.3$?), 70–80 mm wide (only equatorial diameters of the eggs are known). Although the surface has sagenotuberculate ornamentation with small, irregularly arranged swellings, protruding nodes and hillocks, and a hieroglyphic pore pattern among them (Fig. 7.9C). The shell thickness varies over the egg from 1.4 to 3.0 mm (the lower values refer to the "hatching window"); the main range is 2.2–3.0 mm.

Clutch structure

Like the Spheroolithidae S1 group, eggs formed a single-layered, subcircular cluster.

Occurrence

Mogoyn-Ulagiyn-Khaets, Dariganga (Bayn-Shireh Formation).

History, taphonomy, and paleobiology

All material of the O1 group was collected by JSMGE and JSMPE in the eastern Gobi. Material from the dark purplish-red deposits of Mogoyn-Ulagiyn-Khaets was described by Sochava (1969), who assigned it to the prolatocanaliculate type of eggshell. In one broken egg (PIN 2970/9) metatarsal bones of a dinosaur embryo prove that dinosaurs were oviparous (Sochava, 1972). Mass burials of dinosaur nests were discovered at Mogoyn-Ulagiyn-Khaets in 1989. The large area (ca. 200 × 50 m) of weathered badlands (dark pink sand) at the base of sandstone cliffs revealed multiple damaged nests (shell fragments bordering the eggs where once a nest existed). The arrangement of nests and lithology suggests colonial nesting of the dinosaurs near shore. The histostructure and superficial features of the shells, as well as the arrangement of the eggs in clutches, are similar to the Spheroolithidae and tentative suggest assignment to ornithopods, perhaps hadrosaurs.

Remarks

The four small bones associated with PIN 2970/9 (Fig. 7.8D) were originally attributed by Sochava (1972) to a protoceratopsian dinosaur. One of us (Kurzanov) has reexamined the specimen and found that the bone originally identified as metatarsal III is actually not a metapodial bone. The diaphyses preserved do not allow for more exact identification of the specimen than

as dinosaurian, but not protoceratopsian. Protoceratopsians laid distinctly different types of eggs, with an entirely different shell histostructure (Mikhailov, 1991a; see also below).

Group O2 (Figs. 7.8C,F,G; 7.9B, & 7.10A).

Material

Whole egg (PIN 4231-1) and eggshell fragments (PIN 3097-502).

Morphology, history, and taphonomy

The single whole egg from Altan-Ula III is elliptical (105 × 74 mm). The surface is smooth with a pore pattern similar to that of hypsilophodontid eggs (see Hirsch & Quinn, 1990; i.e., a "dinornithoid" pattern of short grooves with two to three pores oriented longitudinally). Fragments of eggshell with the same pattern of ornamentation were collected in 1985 from yellow sands at Udan-Sayr (north of Nemegt Basin). Judging from the curvature of the fragments, the shape of the egg was elliptical. Shell thickness in the Altan-Ula egg is 1.5–1.8 mm and 1.1–1.8 mm (main range, 1.3–1.6 mm) in the Udan–Sayr eggshells.

Occurrence

Udan-Sayr (Djadokhta Formation), Altan-Ula III (Nemegt Formation).

Eggs of protoceratopsian dinosaurs (Protoceratopsidae)

Eggshell histostructure

Eggshell is of the dinosauroid basic type and of prismatic morphotype, with an angusticanaliculate pore system (Figs. 7.2F-H, 7.3F,G, & 7.12E,F).

We distinguish three groups of protoceratopsid eggs – P1, P2, and P3 – differing in size, eggshell thickness, and ornamentation (Sabath, 1991a).

P1 Group (larger smooth-shelled protoceratopsid eggs) (Figs. 7.2G, 7.3F, 7.11A).

Material

Complete and partial clutches (PIN 614-58, PIN uncatalogued; ZPAL MgOv-II/2,2a,3a,3b, ?some clutches in Ulan Bator), numerous eggs with eggshell debris (PIN 614-601, PIN 603; ZPAL MgOv-II/20, 21).

Morphology

Eggs of medium-to-large size (40–50 × 110–120 mm), slightly asymmetrical (ovoid) and elongated ($E = 2.3–2.5$) in shape. The surface is smooth, but sometimes faint parallel striation is visible in the equatorial part of the egg. The pore canals are narrow and often filled with secondary calcite. The main range of shell thickness is 0.6–0.7 mm in the equatorial part of egg, 0.3 mm on the pointed end, and up to 1.2 mm on the blunt end. The eggshell diffusive properties are A_s = 230 cm^2, A = 0.35 cm^2, N = 7,000, G_{H_2O} = 70 mg (day Torr)$^{-1}$.

Nest structure

About twenty to thirty (perhaps up to 40) eggs are arranged vertically (sometimes in two circles) around the center (clutch diameter, 0.5 m) (Fig. 7.4F). The eggs are often preserved in pairs, possibly as a result of the simultaneous action of both oviducts and the gluing together of the eggs with a mucous secretion.

Occurrence

Bayn-Dzak, Toogreekiyn-Shireh (Djadokhta Formation), Khermiyn-Tsav (biostratigraphic equivalent of the Barun Goyot Formation), Ikh-Shunkht (?Barun Goyot Formation).

P2 Group (protoceratopsid eggs with prominent ornamentation) (Figs. 7.2F, 7.3F, 7.12A-D, & 7.13C).

Material

Whole and partial clutches (PIN 3142-496 – 19 eggs; ZPAL MgOv-II/23 – 19 eggs, MgOv-II/4, 5, 8, 10, 22 – ca 20 eggs, and MgOv-II/1a-f – 7 eggs), (some clutches in Ulan Bator); numerous whole eggs and their parts (PIN 3142-415, 429, 447–453, 455, 489, 495; ZPAL MgOv-I/5, 7–10, 12–15, 17, 20, 23, 24, 26, 27, MgOv-III/1-8, 19).

Morphology

Eggs similar to the P1 group, but larger: 50–57 × 130–150 mm (egg lengths differ by 20 mm within a single clutch, e.g., PIN 3142-496), V = 220 cm^3. Ornamentation is present in the equatorial area of the egg linearituberculate (tiny long ridges, oriented parallel to the long axis of an egg); the poles lack the ornamentation. The main range of shell thickness is 0.6–0.7 mm, 0.3 mm on the sharp end, and up to 1.4 mm on the blunt end. Egg diffusion rate: A_s = 230 cm^2, A = 0.5 cm^2, N = 9,000, G_{H_2O} = 90 mg (day Torr)−1.

Nest structure

Clutch arrangement is similar to the P1 group, but the eggs often occur in oblique or subhorizontal position.

Remarks

The "protoceratopsid" eggs from Bayn-Dzak are of similar shape, but have a branching ornamentation pattern near the poles and nodes on the poles. The eggs

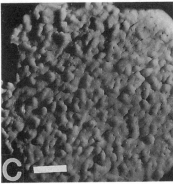

Figure 7.9. Outer surface of spheroolithid (A) and ovaloolithid (B,C) eggshells. **A.** *Maiasaura*-like sagenotuberculate ornamentation, S2 group (PIN 614–600, Bayn–Dzak); scale bar = 10 mm. Longitudinally oriented grooves are typical of the rimocanaliculate pore system (B,C) and tiny nodes and protuberances of the sagenotuberculate ornamentation (C). **B.** O2 group; **C.** O1 group (detail of Fig. 7.10); scale bar = 1 mm.

belong to the Elongatoolithidae (the eggshell histostructure is, however, obscured by recrystallization).

Occurrence

Bayn-Dzak (Djadokhta Formation), Khermiyn-Tsav (red beds; biostratigraphic equivalent of the Barun Goyot Formation), and Khulsan (Barun Goyot Formation).

P3 Group (smaller, smooth shelled protoceratopsid eggs) (Figs. 7.2H, 7.3G, & 7.11B)

Material

Part of a clutch (PIN 4228-1, 4 eggs), incomplete eggs and their polar parts (PIN 3143-121, 122,123,

ZPAL MgOv-I/1,10,25; MgOv-III/9); clutches from Ulan Bator.

Morphology

The shape and surface of the eggs are similar to the P1 group, but a little smaller: 42–?50 × 100–?110 mm; $V = 100$ cm^3; the diffusion is $A_s = 130$ cm^2, $N = 2600$, G_{H_2O}lt80 mg (day Torr)$^{-1}$. The shell thickness is 0.3–0.7 mm. Eggs are always preserved in an upright position.

Occurrence

Toogreekiyn-Shireh, Baga-Tariach (Djadokhta Formation), Khermiyn-Tsav (biostratigraphic equivalent

Figure 7.10. Outer surface of ovaloolithid eggshells. **A.** O2 group (PIN 3097-502, Udan–Sayr), note transition from almost smooth surface to the complete pattern of a fine sagenotuberculate ornamentation ("aberrant" fragments in lower right corner), similar to O1 group (B). **B.** O1 group (PIN 3225-150, Mogoyn-Ulagiyn-Khaets) (scale in millimeters; see enlargements in Fig. 7.10B,C).

of the Barun Goyot Formation), and Khulsan (Barun Goyot Formation).

History, taphonomy, and paleobiology of P1–P3 groups

These are the most famous fossil eggs from Mongolia, having been found by the Central Asiatic Expeditions of AMNH in "Shabarakh-Usu" (= Bayn Dzak; Andrews, 1932, pp. 162–3, 207–14). The eggs were attributed to the ceratopsian dinosaur *Protoceratops andrewsi* because its bones and skeletons were common in the same fossil-bearing strata (e.g., Thulborn, 1991). Although claims referring to identification of embryos inside the eggs (Andrews, 1932, p. 210) proved to be unfounded (Elżanowski, 1981), the referral of these elongated eggs having a prismatic shell morphotype to protoceratopsians is plausible. Three other dinosaur nesting sites have been found (Toogreekiyn-Shireh,

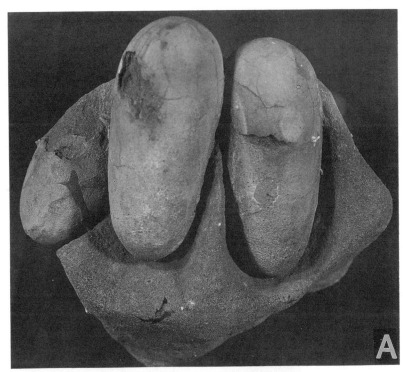

Figure 7.11. Protoceratopsid eggs. **A.** P1 group (PIN 614-58/1-3, Bayn-Dzak), typical eggs of *Protoceratops andrewsi.* **B.** P3 group (PIN 4228-1, Baga-Tariach). Eggs are shown in their original upright position.

Khermiyn-Tsav, and Khulsan), where these eggs co-occur with articulated skeletons of juvenile and adult protoceratopsians. Bone fragments of an undetermined protoceratopsian and eggs of this type have been found at Ikh-Shunkht and Baga-Tariach. All of these localities are in the Barun Goyot and Djadokhta formations. Sochava's (1969) doubts about the protoceratopsid affinities of these eggs arose from confusion between the superficially similar protoceratopsid (P2 group) and elongatoolithid eggs from Bayn-Dzak (see also "Remarks" section under Ovaloolithidae).

The differences among the P1, P2, and P3 groups may not be taxonomically significant and may only reflect inter- or intrapopulation variability. At Toogreekiyn-Shireh, only *Protoceratops andrewsi* bones and smooth protoceratopsid eggs (P3 and P1 groups) have been found. At Khermiyn-Tsav and Khulsan, there are abundant bones of *Breviceratops kozlovskii* and sculptured eggs of the P2 group (rarely any P1 group). At Bayn-Dzak, the situation is more complicated because only *P. andrewsi* skeletal remains are known, while both smooth (P1 and ?P3 group) and sculptured (P2 group)

protoceratopsid eggs have been found (plus very similarly sculptured elongatoolithid eggs attributable to theropods).

Protoceratopsid eggs occur mainly in red sands (Bayn-Dzak, Ikh-Shunkht) and sandstones (Khermiyn-Tsav and Khulsan), and less often in yellow-white sands (Toogreekiyn-Shireh, Baga-Tariach). Lithological and taphonomical evidence point to periodic colonial nesting of protoceratopsians at the same places along the streams and ponds. The P1 and P3 eggs are always found in vertical position (blunt end up). This suggests to one of us, (Sabath, 1991a,b) to infer the sand-nesting mode for the protoceratopsians. The thinner shell of P3 eggs indicates more humid incubation conditions with low oxygen partial pressure (see Calder 1978).

Elongatoolithidae Zhao (1975) (?Theropoda)
Eggshell histostructure

The eggshell is of the ornithoid basic type, "ratite" morphotype (ratio between the single and mammillary layer from 3:1 to 5:1) and has an angusticanaliculate pore system (Figs. 7.2I, 7.13B, & 7.14A,B). The Mongolian

Figure 7.12. Protoceratopsid eggs of P2 group. Clutches from Khermiyn–Tsav (A, PIN 3142-496, possibly laid by *Breviceratops kozlovskii*), and from Bayn-Dzak (B, ZPAL MgOv-II/2, prepared from below, scale bar = 10 cm). Note asymmetry of both ends of the egg (C, ZPAL MgOv-II/23) and the fine linearituberculate ornamentation (C, D; scale = 10 mm). Radial thin sections reveal differences in prismatic morphotype structure between thick (E) and thin (F) shell fragments from a single egg (scale bar = 0.1 mm). (B, C after Sabath, 1991a.)

elongatoolithid eggs belong to two main groups, E1 and E2.

E1 group includes *Elongatoolithus* and *Macroolithus* Zhao (1975) (Figs. 7.3E, 7.13A, 7.14A,B, & 7.15A–C).

Material
Incomplete clutch of five eggs (PIN 3143-125), crushed clutch and egg (PIN: 4216-400, 3142-400), small parts of eggs and shell debris (PIN 614-59, 604, 605, 610, 611; 2970/2-1; 3097-500; 4229-1; 4230-1, 4, 5; 522-401; ZPAL MgOv-I/2; MgOv-II/10,11,12,14,16; MgOv-III/18a-c); some clutches also in Ulan Bator.

Morphology
Medium to large sized eggs, superficially similar to the ornamented protoceratopsid eggs (P2), but slightly longer and wider (length = 68–70 × 150–170 mm, V = 280–00 cm^3, E = 2.1–2.5, elliptical in shape without apparent asymmetry); usually preserved flattened. The surface is always sculptured, with the ornamentation pattern consisting of distinct nodes and short ridges that look like "fused" nodes. These ridges are arranged longitudinally to the long axis of the egg, are parallel to one another or bifurcate in the equatorial part, and meander or form vortex patterns closer to poles.

Both poles are always ornamented with distinct nodes (discretituberculate ornamentation), and in this they differ markedly from the fine continuous ridges of P2 eggshells. In transverse sections of eggshell, the growth (accretion) lines undulate along with the outer sculptured surface. The pores open into the valleys between the ridges. The diffusive properties of the eggshell are: G_{H_2O} = 250 mg (day Torr)$^{-1}$, A = 320 cm^2, A = 1.3 cm^2, N = 17,500. Eggshell thickness varies regularly over the whole egg, with the lower pole (pointed end with the thinnest shell) and the upper pole (thickest-shelled) differing by up to 0.7–0.8 mm. E1 group eggs can be subdivided into five groups that differ in eggshell thickness and egg size (if known) as shown in Table 7.2.

Nest structure
At least twenty, perhaps more than thirty eggs per clutch, usually inclined away from the center of the nest and crushed, but sometimes standing upright (Fig. 7.4J).

Occurrence
Bayn-Dzak, Udan-Sayr, Toogreekiyn-Shireh (Djadokhta Formation); Ikh-Shunkht, Undurshil-Ula (Barun Goyot Formation); Khaichin-Ula I (white beds of the Nemegt Formation), Gooreeliyn-Tsav, Tsagan-Khushu, and Khermiyn-Tsav (white beds of the Nemegt

Figure 7.13. Elongatoolithidae, E1 group. **A.** Clutch of ?*Elongatoolithus* sp. (PIN 3143-126, Toogreekiyn–Shireh, eggs about 15 cm long). Elongatoolithid eggshells belong to the ornithoid basic type and ratite morphotype. **B.** Thin radial section under polarized light, scale bar = 1 mm. Note the undulating growth lines following the outer ornamentation, although in this specimen the upper (outer) surface has been eroded. **C.** Protoceratopsid egg, P2 group (for comparison). Abbreviations: see Fig. 7.14. (B after Sabath 1991a.)

Figure 7.14. Structure of elongatoolithid (**A, B**) and protoceratopsid (**C**) eggshells in radial view (freestanding eggshell, SEM; see also Figs. 7.12E,F & 7.13B). A,B. Note ultrastructural difference between the lower mammillary layer (ml) and the upper spongy layer (sl) typical of the ornithoid basic type. There are no vertical prisms in the upper ("single") layer in the ratite morphotype; surface ornamentation reflects the undulation of the accretion lines in the shell. The only layer of the protoceratopsid eggshell (C, dinosauroid basic type, prismatic morphotype) corresponds to the mammillary layer of A and B, and is clearly divided into prismatic shell units (su). Scale bar = 0.1 mm.

Formation including ?uppermost Maastrichtian beds – pink beds of Tsagan-Khushu).

E2 Group: unnamed genus (Kurzanov & Mikhailov, 1989; Fig. 7.15D).

Material
Eggshell debris from a ?nest (PIN 4227-1).

Morphology
Eggs are strongly elongated (judging from shell-fragment curvature) with linearituberculate ornamentation in the equatorial area, mainly as longitudinally oriented tiny heteromorphic nodes, rarely short ridges. The size of the eggs is unknown. The eggshell is 0.8–1.2 mm thick, including the nodes.

Occurrence
Builjasutuin-Khuduk (Lower Cretaceous, Dushih-Ula Formation).

History, taxonomy, and paleobiology of E1 and E2 groups
The genera *Macroolithus* and *Elongatoolithus* are among the most common fossil oological remains in Central Asia. They were established by Zhao (1975) for clutches of *Oolithes elongatus* found during the early 1950s in Shantung Province (Laiyang), China (Young, 1954), and later in Guandung and Jiangxi Provinces (Young, 1965).

Elongatoolithid eggs and shells (?*Elongatoolithus* sp.) were first found in Bayn-Dzak by the Central Asiatic Expeditions of the AMNH in the 1920s (and later found by MPE, JSMPE, and PMPE). The eggs were, however, confused with protoceratopsid eggs because of their strong superficial similarity to the latter. This confusion led Bakker (1971) and Case (1978) to assume that *Protoceratops* laid eggs 20 cm long with a mass of 0.5 kg, twice that for real *Protoceratops* eggs. This error resulted in faulty calculations for the growth rate and physiology of *Protoceratops*.

Partial elongatoolithid (?*Macroolithus*) eggs and shell debris collected at Khaichin–Ula I were described as angusticanaliculate eggshell (Sochava, 1969, Plate. 1, pp. 6–8; actually the illustrations of angusticanaliculate eggshell in Sochava should be referred to elongatoolithid and not to protoceratopsid). Abundant elongatoolithid eggshells (very similar to the Khaichin-Ula ?*Macroolithus* eggshells) were collected during the late 1950s in the Zaissan Basin, Kazakhstan (Bazhanov, 1961).

At present, the Elongatoolithidae are known from at least ten Upper Cretaceous and one Lower Cretaceous locality in Mongolia, especially the Barun Goyot Formation (including the Djadokhta Formation) and Nemegt Formation. Most specimens come from gray, white, and yellow sandstones, and only few from red sandstones.

The eggshell structure of the Elongatoolithidae (ornithoid basic type) is unique among known groups of dinosaurian eggs because of its birdlike type of eggshell. It is implausible to assign this type of eggshell to ornithopod or sauropod dinosaurs because all positively identified eggshells of those groups (hadrosaurs, e.g., *Maiasaura*; ceratopsians, e.g., *Protoceratops*; hypsilophodonts, e.g., *Orodromeus*; sauropods, e.g., *Hypselosaurus*) belong to the dinosauroid basic type (Mikhailov, 1991a,b).

Because elongatoolithid specimens are very common in the Upper Cretaceous deposits of Central Asia, they should belong to one of the widespread groups of Upper Cretaceous dinosaurs other than protoceratopsids and hadrosaurs. Such a group can only be the theropods (Kurzanov & Mikhailov, 1989). This suggestion recently received strong support based on an egg from the Upper Cretaceous of Montana and referred to ?*Troodon* (Theropoda: Troodontidae) by Hirsch & Quinn (1990). The egg possesses a histostructure very similar to the Mongolian elongatoolithids (Mikhailov 1991a). The presence of a well-defined mammillary layer and a single (= spongy, prismatic, continuous) layer, as in bird eggs (especially ratites) confirms the close kinship between theropods and birds. The nest structure (probably a vegetation mound; Sabath, 1991a) and evidence of altriciality and feeding by parents (Kurzanov & Mikhailov, 1989) are also consistent with the theropod affinities of the eggs.

Laevisoolithidae Mikhailov (1991) (?Aves or ?Theropoda)

Eggshell histostructure

The shell is the ornithoid basic type and "ratite" morphotype (the ratio between the single layer and mammillary layer is 2:1.5), with an angusticanaliculate pore system (Figs. 7.2J, & 7.17A).

A single type of laevisoolithid eggs is known: *Laevisoolithus sochavai* Mikhailov (1991a) (Figs. 7.3H, 7.16C, & 7.17A).

Material

Two partly collapsed eggs (PIN 2970/5, ZPAL MgOv-I/3) and the polar end of an egg (ZPAL MgOv-II/9-e).

Table 7.2. Subdivision of Mongolian elongatoolithid eggs

Subgroup	Shell thickness (mm)		Egg size (mm)
	Equatorial area	Pole	
E1-1	0.4–0.7	1.0–1.2	150–170 × 60–70
E1-2	0.7–0.8	?	?
E1-3	0.9–1.1	?	?
E1-4	1.2–1.7	1.7–1.9	150–170 × 60–70
E1-5	1.2–1.7	1.7–2.2	?>170 × > 70

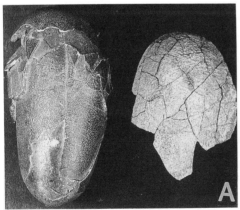

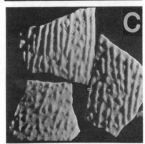

Figure 7.15. Elongatoolithid fossils. **A.** Cast of an ?*Elongatoolithus* egg (left) and the blunt end of ?*Macroolithus* egg (right; PIN 4230-1, Khaichin-Ula I). Eggshell surface shows a discretituberculate (**B**) ornamentation on poles, or linearituberculate (**C, D**) pattern near the equatorial zone of the egg; **B, C.** S1 group (PIN 2970-8, Zaissan). **D.** S2 group (PIN 4227-1, Builjasutuin-Khuduk). Fragment of protoceratopsid eggshell **E.** (P2 group) shows another variant of the linearituberculate ornamentation without tubercles. Scale bar = 10 mm.

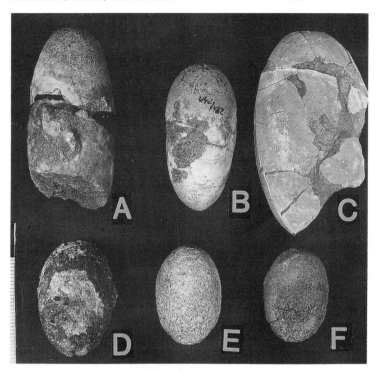

Figure 7.16. Fossil eggs (A–C) and egglike concretions (D–F). **A.** Larger "*Gobipteryx*" egg (PIN uncatalogued, Khulsan). **B.** Smaller "*Gobipteryx*" egg (PIN 3142-487, Khermiyn-Tsav). **C.** *Laevisoolithus sochavai* egg (PIN 2970-5, Boogiyn–Tsav). **D–F.** Egglike concretions from Toogreekiyn–Shireh and Bayn–Dzak (scale bar in millimeters).

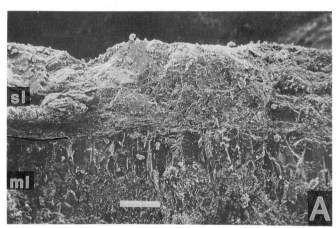

Figure 7.17. Shell histostructure of a laevisoolithid egg (**A**) and of an gobipterygiform bird egg (**B**); radial view, freestanding eggshell, SEM (bar 0.1 mm; for abbreviations see Fig. 7.14A,B).

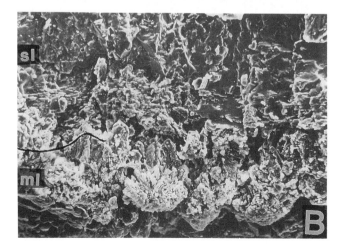

Morphology

Medium-to-small ellipsoid eggs (about 70–71 × 35–40 mm; $E = 1.9$; $V > 45$ cm^3), having a very smooth outer surface. The eggshell measures 0.7–0.9 mm thick, but the actual thickness may not exceed 0.7 mm because the outer layer is secondarily crystallized.

Occurrence

?Bayn-Dzak (Djadokhta Formation), and Boogiyn-Tsav, Nemegt (Nemegt Formation).

History, taxonomic identity, and paleobiology

The eggs were first found at Boogiyn-Tsav by JSMGE and (PIN 2970/5) were preliminarily described by Sochava (1969, Plate 11,5). Recently, laevisoolithid specimens were identified in the ZPAL collection (Sabath, 1991a). The two almost complete eggs came from the Nemegt Formation outcrops at two localities, Boogiyn-Tsav and Nemegt, less than 100 km apart. Identification of the eroded polar end of an egg from Bayn-Dzak (Djadokhta Formation) is tentative. The taxonomic identity of the laevisoolithid eggs is unknown,

Figure 7.18. Eggshell of *Subtiliolithus microtuberculatum* (PIN 4230-3, Khaichin-Ula I). **A.** Surface with microtuberculate ornamentation. **B.** Eggshell structure in radial view (freestanding eggshell, SEM, bar = 0.1 mm); note disproportionally thick mammillary layer.

but the eggshell histostructure suggests that they were laid either by birds, or by some small theropods.

Subtiliolithidae (Mikhailov, 1991a) (?Aves or small ?Theropoda)

Eggshell histostructure

The shell is of the ornithoid basic type and "ratite" morphotype (but the ratio between single layer and mammillary layer is opposite that in the typical ornithoid eggshell, from 1:2 up to 1:3). The pore system is angusticanaliculate (Figs. 7.2K & 7.18B).

Only one type of subtiliolithid eggs is known, *Subtiliolithus microtuberculatum* Mikhailov (1991a) (Fig. 7.18).

Material

Broken eggshells belonging to several clutches (PIN 4230 – 3,7,8) embedded in a sandy matrix.

Morphology

The size and shape of the eggs are still undetermined. The shell surface is smooth or microtuberculate, and the shell is thin (0.3–0.6 mm, mainly 0.3–0.4 mm).

Taphonomy and paleobiology

Clusters of eggshell fragments were found by JSMPE in 1981–1983 in the red beds (almost barren of skeletal fossils) at the Khaichin-Ula I locality in the Nemegt Formation. Local lithology is typical for red beds (see the section on localities). As with the protoceratopsid eggs, the shell is whitish in color. In one case, the distribution of eggshells in about 30-mm diameter clusters clearly define eggs of a clutch. Furthermore, the distribution of the eggshell clutches indicates colonial nesting. The identity of the egg layer is unclear. The histostructure of the eggshell suggest avian or possibly theropod origin.

Eggs of volant palaeognathous birds (Gobipterygiformes)

Eggshell histostructure

The eggshell is of the ornithoid basic type and prismatic ("neognatheous") morphotype (ratio between spongy and mammillary layer is "normal" – 2:1), with an angusticanaliculate pore system (Figs. 7.2L, 7.17B).

Two groups of these eggs, G1 and G2, are known from Mongolia.

G1 group: smaller "*Gobipteryx*" eggs (Figs. 7.3J, 7.16B, & 7.17B).

Material

Numerous intact and broken eggs (PIN 3142 – 401–410, 416–445, 456–459, 460, 462–468, 471–473, 482–494; ZPAL MgOv-III/10-14).

Morphology

Small eggs (30–46 × 20–24 mm, $V = 7$–12 cm³) of elongated and slightly asymmetrical (ovoid) shape ($E = 1.8$–2.0). The surface is smooth, and the eggshell thin – 0.1–0.2 mm.

Occurrence

Khermiyn-Tsav (red beds; biostratigraphic equivalent of the Barun Goyot Formation) and Khulsan, Gilbent (Barun Goyot Formation).

G2 group: larger "*Gobipteryx*" eggs (Figs. 7.3I & 7.16A).

Material

A few eggs (PIN 551-1002; ZPAL MgOv-I/19,21,25, MgOv-II/6,7,25).

Morphology

Very similar in appearance to eggs of the G1 group, but larger (53–70 × 26–30 mm, $V = 35$–40 cm³) and also with a thicker eggshell (0.2–0.4 mm).

Occurrence

Bayn-Dzak (Djadokhta Formation), and Khulsan (Barun Goyot Formation).

History, taphonomy, and paleobiology (G1 and G2)

Numerous eggs of both types were found by members of PMPE in the mid-1960s in the red beds of Khermiyn–Tsav and Khulsan. Additional eggs were later repeatedly collected by PMPE and JSMPE at these localities. The eggs were preliminaryly described by Ełżanowski (1981), and the eggshell histostructure was studied by Mikhailov (1991a). In some G1 eggs fairly complete skeletons of well developed embryos of volant palaeognathous birds were found (Ełżanowski, 1981; Kurochkin, 1988). A few hind limb bone fragments (of adults) were found associated with the G2 eggs (Sabath, 1991a). E. N. Kurochkin (personal communication to Mikhailov in 1990) disagrees with Ełżanowski's (1981) identification of the embryos in the G1 eggs from the

Gilbent locality (Ełżanowski, 1977) to *Gobipteryx minuta* and believes that the embryos belong to another genus of volant paleognathous bird.

The abundance of G1 eggs from different stratigraphic levels at Khermiyn-Tsav demonstrates the site fidelity of the birds. The nesting ground spread along lake and river margins. Good preservation of egg shape (many unbroken ones) is interpreted as due to underground incubation (Sabath, 1991a). Although such a mode of nesting exists among birds (e.g., Megapodidae: *Megapodius freycinet*), an alternative scenario may also explain the taphonomical setting at Khermiyn-Tsav. The eggs were primarily arranged in clutches (Sabath, 1991a; e.g., ZPAL MgOv-II/7) at Bayn-Dzak. However, at Khermiyn-Tsav and Khulsan, no clutches occur. Instead isolated eggs are irregularly dispersed over the slopes in vertical or subvertical position. This suggests regular flooding of the nesting colonies with the unhatched eggs floating in a vertical position. As the water level dropped, the eggs would slowly sink and be embedded in the soft bottom matrix.

The shells of most eggs are strongly recrystallized. The pore canals are filled with secondary calcite, and the pores are fused and unidentifiable. These diagenetic changes resulted from the action of water.

Eggs of turtles (Fig. 7.19)

Eggshell histostructure

The shell is of testudoid basic type and testudoid morphotype, with a pore system consisting of widely separated, simple tubes.

Material

Two eggs dissected in half (PIN 4225-2,3). In one egg a well-developed embryo has been detected (Fig. 7.19).

Morphology and taphonomy

The eggs are spherical, small (35–40 mm) and smooth-shelled. The eggshell is 0.3–0.4 mm thick. The eggs were found by JSMGE in the red deposits at the

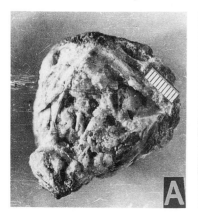

Figure 7.19. Two views (A,B) of a fossil chelonian egg with an embryo (PIN 4225-2, Ologoy–Ulan–Tsav; scale bar = 10 mm).

Ologoy-Ulan-Tsav locality (?Upper Cretaceous) in beds with eggs of *Faveoloolithus*.

Problematica (Figs. 7.16D, & 7.17B,C)

At some localities (Khermiyn-Tsav, Khulsan, Bayn-Dzak, Toogreekiyn-Shireh), numerous small, ovoid and ellipsoid sandstone forms have been collected by JSMPE, JSMGE, and PMPE (housed in PIN and ZPAL). Although they were often considered as "?casts of turtle eggs" (Sochava, 1969, Figs. 1: 2,3) or "?soft-shelled eggs of turtles, crocodiles or lizards" (Sabath 1991a), most of them should be referred to as inorganic concretions (nodules). During fieldwork by JSMPE in 1985–1989, piles of hundreds of such concretions (30–50 mm long), some possessing a typical ovoid shape, were observed in different places at Bayn-Dzak. The concretions also fall from the sand walls at Toogree-kiyn-Shireh.

Some elliptical forms from Khermiyn-Tsav and Khulsan might be lacertid eggs. Among eight small, sandy, elongated forms (PIN 3142-411), 11–13 × 6–7 mm in size, one is hollow and has a thin mineralized shell (hardly structured, but with clear irregular vertical

blocks), as well as a characteristic oval "window" in the central area, through which the juvenile could have hatched (Fig. 7.20). These ovoid forms are comparable to the eggs of geckoes. Larger sandstone forms (23–27 × 14–15 mm) from Khermiyn-Tsav with traces of ?mineralized shell are housed in Warsaw (ZPAL MgOv-II/9a-c, 17a,b).

Nesting ecology of dinosaurs

The taphonomical settings and the geographical and stratigraphical distribution of different groups of fossil eggs and shells correlate with the nesting behavior of the egg-laying dinosaurs. Protoceratopsid, faveolool-ithid (?sauropods) and dendroolithid eggs, as well as those of gobipterygiform birds, commonly occur as whole eggs or clutches of unhatched eggs. They are usu-ally found in fine-grained sandstones and occasionally in gray and yellow sands (Toogreekiyn-Shireh). Fav-eoloolithid eggs have also been preserved in coarse-grained red sandstones and sometimes in gravels (Ologoy-Ulan-Tsav and upper levels at Ikh-Shunkht) where chelonian eggs are also present.

Dendroolithid egg clutches occur just above the

Figure 7.20. Egglike problematica from Khermiyn-Tsav (a smaller "*Gobipteryx*" egg is shown for comparison in the upper left corner). The far left one in the middle row has an amorphic mineralized shell. The lowest row comprises ?lizard eggs (PIN 3142-411); the specimen first from the left has a hollow, mineralized shell with a "hatching window."

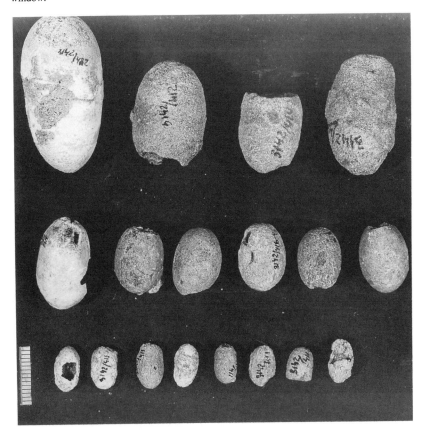

zone with lacustrine bivalves at Shiljust–Ula. The most porous eggshells, Faveoloolithidae and Dendroolithidae, are strongly impregnated with iron salts, thus acquiring a red color. A layer of secondary calcite crystals often appears on the inner surface as well. The shells of protoceratopsids and "*Gobipteryx*" retain a more natural beige color, but show massive recrystallization and bear traces of water dissolution. In addition, the pore canals are filled with secondary calcite and thus are hardly discernible from the shell. Piles or scattered eggshell are rare, although small fragments showing dissolution are frequently dispersed over a large area. This suggests that pieces of shell from hatched eggs were blown or washed into a water channel where most of them dissolved. Such a taphonomical bias suggests nesting on beaches or other half-submerged spots.

The nest taphonomy at Ologoy-Ulan-Tsav, Khermiyn-Tsav, Khulsan, and Bayn-Dzak points to gregarious nesting grounds at favorable sites for long periods. Protoceratopsids and gobipterygiform birds gathered at the breeding sites to form nesting colonies. Sauropods (Faveoloolithidae) appeared on beaches only for egglaying, although some of them might have fed nearby (isolated bones are found in Ologoy-Ulan-Tsav). Juvenile sauropods might have sought safety in water. The composition of the egg assemblages in the main beach-nesting sites is given in Table 7.3.

Unlike the eggs listed above, whole elongatoolithid eggs are very rarely found. Eggshell debris is found scattered around the nest with little evidence of water transport. Resorption craters at the bases of mammillae indicate hatching had occurred. The abundance of successfully hatched eggs may reflect a different nesting habit of these dinosaurs compared to the other dinosaurs. They might have preferred nesting areas less prone to flooding during the incubation season, or they may have had better hatching success due to more parental care. Perhaps the parents were also less susceptible to predation and displayed more birdlike parental behavior (Kurzanov & Mikhailov, 1989; Sabath 1991a).

The only evidence for colonial nesting by the layers of elongatoolithid eggs is at Tsagan-Khushu. Elsewhere the nests were apparently solitary. Furthermore, in contrast to the eggs of typical shore-nesters (sauropods, protoceratopsids, and *Gobipteryx*-like birds), most elongatoolithid eggshells are well preserved. The lumen of the pore canals is distinct; none show recrystallization or dissolution of the shell fragment edges. These observations hold true not only for elongatoolithid eggs from Tsagan-Khushu, Khaichin-Ula I, Gooreeliyn-Tsav, and Udan-Sayr, but also for similar eggs from Zaissan Basin, Kazakhstan. Again, this indicates that the eggs were not flooded even after hatching. Fragments of broken eggs were simply buried within the sand around the nest, and some were scattered by the wind. The nest must have been far from the water margin, possibly on the upper bank terraces.

Some eggs of ?*Elongatoolithus* from the red sands of Bayn-Dzak and a single clutch from the gray sands of Toogreekiyn-Shireh suggest flooding in situ. At Bayn Dzak, their shell structure is altered, which makes them confusingly similar to protoceratopsian eggs. These clutches of elongatoolithid eggs were found among abundant protoceratopsid nests. Similarly, at the Lower Cretaceous locality Builjasutuin-Khuduk, a possible theropod nest was found with numerous remnants of psittacosaurs (Kurzanov & Mikhailov, 1989). In a similar fashion, the eggs attributed to ?*Troodon* were found among numerous eggs of hypsilophodontid dinosaurs in Montana (Horner, 1987). Perhaps some carnivores nested on the shore near the colonies of small herbivores, which would provide "fast-food" for the theropods and their hatchlings. This strategy, however, increased risk of flooding of the nest.

Eggs of spheroolithids (S2 group which includes hadrosaurs and possibly some other ornithopods) often occur as shell debris. The eggshells, however, were apparently transported far from the nest sites (e.g., Tsagan-Khushu, Khermiyn-Tsav, and Bayn-Dzak). The surface and edges of fragments have a "melted" look because of their dissolution in water. At the Chinese localities (Laiyang, Shantung), numerous whole spheroolithid and ovaloolithid eggs have been found (Young, 1954). Mass burials of hatched ovaloolithid eggs have also been found at Mogoyn-Ulagiyn-Khaets in Mongolia.

The lower preservation potential of whole spheroolithid and ovaloolithid eggs compared to the protoceratopsid and faveoloolithid eggs may be due to the better protection of the nests by parents. Parental care among hadrosaurs has been demonstrated by Horner (1982, 1987). Spheroolithid and ovaloolithid eggs have never been found with protoceratopsid and faveoloolithid eggs, implying a spatial separation of their nesting sites.

Three basic modes of nesting can be distinguished from the dinosaur egg sites in Mongolia: a typical underground (hole) nest (chelonianlike type: faveoloolithid and dendroolithid eggs, Fig. 7.4A–E), and two types of mound nests, where the nest is built out of sand or vegetation (Figs. 7.4F,G & 7.21). Each mode results in a different nest taphonomy (Sabath, 1991a,b). The hole-nests contain more spherical eggs, without much asymmetry, because they are randomly clustered in the hole.

Sand-mound nests contain elongated eggs with their long axis oriented vertically (Fig. 7.21G). This is exemplified by the smooth-shelled eggs of protoceratopsids and gobiopterygid birds. Moreover, the lower, pointed end of the egg is usually intact and not flattened or crushed. Such preservation suggests that the eggs were laid vertically in a sand nest and, after burial, the hollow interior was later filled with sediment through the upper pole of the egg (Fig. 7.21H, based on ZPAL MgOv-II/3). If the fossil nest is covered with plaster of

Paris in the field, and then prepared from below, the result is like that shown in Figures 7.12B and 7.21I.

If the nest was made of vegetation, then after burial and decay of the nest material the eggs would fall over and the pressure of the overlying sediment would flatten each egg perpendicularly to its long axis (Fig. 7.21D–F). Such flattening is typical of most ornamented elongated eggs, even though their shells are thicker than those of the smooth-shelled eggs of the same size. The ornamented eggs thus lie almost flat, with a typical arrangement being like that of MgOv-II/23 (Fig. 7.21A).

The rotting vegetation would produce a lot of carbon dioxide and lower the partial pressure of oxygen. The embryos would face respiratory problems more severe than those in the sand nests. Perhaps the ornamentation aided in the exchange of respiratory gases. The

pores always open into the valleys between the ornamentation where they would be protected from plugging by nesting material. The parallel and bifurcating valleys would serve to convect the respiration gases upward (Fig. 7.21B,C).

The taphonomical differences between nests of two types of elongated eggs from Mongolia has also been noticed by Thulborn (1992), who proposed that in one type of nest (designated type A), several concentric circles of subvertically laid eggs alternate with a covering of soil around a mound, thus forming a concentric pile. His type B nest consisted of eggs deposited along the rim of a craterlike depression. It is not believed that the female manipulated the eggs into position.

In contrast to Sabath (1991a, b), Thulborn assumes the same material was used in both nests. Fur-

Table 7.3. *Egg and skeletal assemblages typical for Upper Cretaceous formations of Mongolian Gobi*

Formation	Eggs (systematic interpretation)	Skeletal remains
Bayn Shireh	O1 (ornithopods) (d)	Hadrosaurs (d)
		Sauropods
		Segnosaurids
		Tyrannosaurids
		Ornithomimids
		Crocodiles
Djadokhta	P1–P3 (protoceratopsids) (d)	Protoceratopsids (d)
	E1–4 (theropods)	Avimimids (sd)
	G2 ("*Gobipteryx*")	Sauropods
	S2 (hadrosaurs)	Oviraptorids
		Ankylosaurs
		Tyrannosaurids (rare)
		Dromaeosaurids
		Hadrosaurs
		Crocodiles
Barun Goyot	P1–P3 (protoceratopsids) (d)	Protoceratopsids (d)
	G1–G2 ("*Gobipteryx*") (sd)	Ankylosaurs (sd)
	F1 (sauropods)	Oviraptorids
	D1–D3 (?sauropods)	Tyrannosaurids
	O2 (ornithopods)	
	E1–2, 4 (theropods)	
	S2 (hadrosaurs)	
Nemegt	E1–5 (theropods)	Tyrannosaurids (d)
	S1 (hadrosaurs)	Hadrosaurs (d)
	S2 (hadrosaurs)	Ankylosaurs (?sd)
	O2 (ornithopods)	Sauropods
	E1–4 (theropods)	Deinocheirids
	Laevisoolithus (?birds)	Dromaeosaurids
	Subtiliolithus (?birds)	Therizinosaurids
		Pachycephalosaurs

Notes: d, dominant; sd, subdominant.

Figure 7.21. Comparison of ornamented and smooth-shelled elongated eggs, including nest structure and taphonomy of the nests. See text for explanation. (Modified from Sabath, 1991a,b).

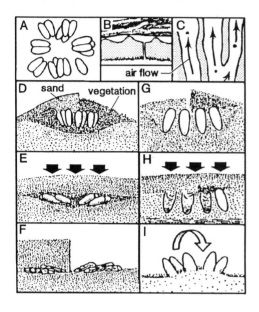

thermore, he does not take into account differences between the two egg types in ornamentation or porosity, the fact that the eggs are usually found in pairs, nor differences in the frequency of crushing. These omissions are due to his basing his interpretation on photographs, rather than a study of original material. Nevertheless, while his nest building scenarios remain plausible, his taxonomic conclusions are questionable. Thulborn based his taxonomic identification solely on the arrangement of the eggs in the nest. He seems unaware of the micro- and ultrastructural studies of elongated eggs conducted over the past 50 years that have demonstrated the existence of several different taxa.

Stratigraphy of Cretaceous egg assemblages

The Bayn Shireh Formation has yielded distinctive ovaloolithid eggs of the O1 group from a single locality (Mogoyn-Ulagiyn-Khaets). The Djadokhta Formation and Barun Goyot Formation eggshell assemblages are very similar (Table 7.4), as were probably the nesting habits of the egg-laying animals. Protoceratopsid eggs (P1–P3 groups) clearly dominate in most localities (six out of eight). Eggs of gobipterygiform birds

Table 7.4. Distribution of Cretaceous eggs in Mongolian localities

Locality	Formation, beds	Eggs (main – bold type)
Builjasutuin-Khuduk	Dushih Ula, red	E2
Mogoyn-Ulagiyn-Khaets	Bayn Shireh, purple	**O1**
Dariganga	Bayn Shireh, purple	O1
Bayn-Dzak	Djadokhta, red	**P1, P2**, E1–4, S2, G2
Toogreekiyn-Shireh	Djadokhta, yellowish	**P1, P3**, E1–4
Baga-Tariach	Djadokhta, red	P3, S2
Udan-Sayr	Djadokhta, yellowish	O2, E1–2
Khermiyn-Tsav	Barun Goyot, red	**G1, P2**, P1, P3, G2, D1–D3, ?F1, ?S2
	Nemegt, gray	S2, E1
Ikh-Shunkht	Barun Goyot, red	P1, F1, E1, S2
Khulsan	Barun Goyot, red	**G1, P2**, P3, G2, D2
Shiljust-Ula	Barun Goyot, red	D1, ?S2
Ologoy-Ulan-Tsav	?U. Cretaceous, red	**F1**, chelonian eggs
Gooriliyn-Tsav	Nemegt, gray	**S1, E1**
Khaichin-Ula I	Nemegt, gray	**E1–5**
Sheregiyn-Gashoon	Nemegt, gray	S1
Tsagan-Khushu	Nemegt, gray	E1, ?S2
Altan-Ula IV	Nemegt, rose	**E1**
Nemegt	Nemegt, rose	O1, *Laevisoolithus*
Boogiyn-Tsav	Nemegt, gray	*Laevisoolithus*

(G1 and G2 groups) were found at three localities, and were abundant only at two of them (Table 7.3). Elongatoolithid eggs are represented by the E1–E4 groups (?*Elongatoolithus*). Faveoloolithid and dendroolithid specimens are also common in both formations. Based on the taxonomic similarities of the eggs between the Djadokhta and Barun Goyot formations, we refer to the eggshell assemblages as a "protoceratopsid nesting assemblage." This supports the idea of stratigraphically unifying the Djadokhta and Barun Goyot formations.

At Khermiyn-Tsav and Ikh-Shunkht, *Faveoloolithus ningxiaensis* (F1 group) eggs co-occur with protoceratopsid nests (Table 7.5) and indicate an Upper Cretaceous age for the Barun Goyot Formation. The presence of *F. ningxiaensis* eggs also suggests an Upper Cretaceous age for the outcrops at Ologoy-Ulan-Tsav (?Barun Goyot Formation) previously thought to be Lower Cretaceous in age.

The egg assemblage of the Nemegt Formation (Table 7.4) comprises elongatoolithid eggshells of the E1–5 group (?*Macroolithus*) and spheroolithid eggs of the S1 group (*Spheroolithus*), which dominate in all localities. In China, *Spheroolithus* eggs are common in the Wangshih series in Laiyang (Chao & Chiang, 1974). *Laevisoolithus* and *Subtiliolithus* eggs are rare in the Nemegt Formation, and protoceratopsid, faveoloolithid, and gobipterygiform eggs are absent.

The Nemegt assemblage suggests nesting far from water. Most egg remains occur in white-yellow-green and gray sands. The exception is the eggshells of *Subtiliolithus* known only from the red sands at Khaichin-Ula I. No evidence for shore-nesting is known, in marked contrast to the Barun Goyot formation. This is probably not due to phylogenetical changes and extinction of the Barun Goyot fauna, but to different local paleogeographical and climatical conditions.

Conclusions

The spectacular discoveries of Gobi dinosaur nests in the 1920s were the first widely recognized proof for hard shelled eggs belonging to dinosaurs (actually, the first dinosaur eggshells were collected in France half

Table 7.5. *Late Cretaceous dinosaur nesting communities of Mongolia*

Nesting sites (age)	Supposed egg-laying animals (egg groups)	Predominant skeletal remains
Mogoyn-Ulagiyn–Khaets (Cenom.–Santonian)	Ornithopods (?hadrosaurs) (O1)	Hadrosaurs
Bayn–Dzak (Santonian–Campanian)	Protoceratopsids (P1-P3)	*Protoceratops*
	Gobipterygiform birds (G2)	Theropods
	Hadrosaurs (S2)	Ankylosaurs
	Theropods (E1-4)	Turtles, lizards
Toogreekiyn-Shireh (Santonian–Campanian)	Protoceratopsids (P1, P3)	*Protoceratops*
	Theropods (E1–4)	*Velociraptor*
		Hadrosaurs (juv.)
		Lizards
Khermiyn-Tsav (Santonian–Campanian)	Gobipterygiform birds (G1, G2)	*Breviceratops*
	Protoceratopsids (P1, P2)	Theropods
	Sauropods (F1)	Sauropods
	?Sauropods (D2, D3)	Pachycephalosaur
	Theropods (E1–4)	Crocodiles
	Lizards	Turtles, lizards
Khulsan (Santonian–Campanian)	Protoceratopsid (P1)	*Breviceratops*
	Sauropods (F1)	Pachycephalosaur
	Hadrosaurs (S2)	Ankylosaurs
	Theropods (E1)	Theropods
		Lizards
Ologoy-Ulan-Tsav (?Upper Cretaceous)	Sauropods (F1)	Sauropods
	Turtles	

a century earlier by Abbé Pouech, see Buffetaut & Le Loeuff, 1989; Chapter 2). Our work shows that the Upper Cretaceous deposits of the Mongolian Gobi remain important for dinosaur egg research. Although more egg types have been described from this area than from any other comparable territory in the world, we believe that still more fascinating material awaits future workers. Our speculations about the taphonomy, paleophysiology, and nesting behavior of the dinosaurs, as well as the taxonomical inferences about the egg-layers, may provide some ideas to be tested when applied to other fossil eggs.

References

Andrews, R. C. 1932. *The New Conquest of Asia.* (New York: American Museum of Natural History).

Bakker, R. T. 1971. Dinosaur physiology and the origin of mammals. *Evolution* 25: 636–58.

Barsbold, R. 1983. Khishchnye dinozavry mela Mongolii. *Trudy Sovmestnykh Sovetsko-Mongolskikh Paleontologicheskikh Ekspeditsiy* 19: 1–117.

Bazhanov, V. S. 1961. Pervoe mestonakhozhdenie skorlupi yaits dinozavrov v SSSR. *Trudy Instituta zoologii Kazakhskoi SSR* 15: 177–81.

Buffetaut, E., & Le Loeuff, J. 1989. La première découverte d'oeufs de dinosaures. *Pour la Science* 143: 22.

Calder, W. A. III 1978. The kiwi: a case of compensating divergences from allometric predictions. In J. Piiper (ed.), *Respiratory functions in birds, adult and embryonic.* (Berlin: Springer).

Case, T. J. 1978. Speculations on the growth rate and reproduction of some dinosaurs. *Paleobiology* 4: 320–8.

Chao T. K., & Chiang Y. K. 1974. Microscopic studies on the dinosaurian egg-shells from Laiyang, Shantung Province. *Scientia Sinica* 17: 73–83.

Chow, M. M. 1954. Additional notes on the microstructure of the supposed dinosaurian egg shells from Laiyang, Shantung. *Scientia Sinica* 3: 523–6.

Efremov, I. A. 1949. Predvaritelnye rezultaty rabot pervoi Mongolskoi paleontologicheskoi expeditsii AN SSSR 1946 g. *Trudy Mongolskoi komissii AN SSSR* 38: 1–49.

Elżanowski, A. 1977. Skulls of *Gobipteryx* (Aves) from the Upper Cretaceous of Mongolia. *In* Z. Kielan-Jaworowska (ed.), *Results of the Polish-Mongolian Paleontological Expeditions* 7, *Palaeontologia Polonica* 37 Warsaw & Crakow, Poland: Państwowe Wydawnictwo Naukowe).

1981. Embryonic bird skeletons from the Late Cretaceous of Mongolia. *In* Z. Kielan-Jaworowska (ed.), *Results of the Polish-Mongolian Paleontological Expeditions* 9, *Palaeontologia Polonica* 42 (Warsaw & Crakow: Państwowe Wydawnictwo Naukowe).

Erben, H. K. 1970. Ultrastrukturen und Mineralisation rezenter und fossiler Eischalen bei Vögeln und Reptilien. *Biomineralisation* 7: 28–36.

Gradziński, R. 1970. Sedimentation of dinosaur-bearing Upper Cretaceous deposits of the Nemegt Basin, Gobi Desert. *In* Z. Kielan-Jaworowska (ed.), *Results of the Polish-Mongolian Paleontological Expeditions* 2, *Palaeontologia Polonica* 21 (Warsaw & Crakow: Państwowe Wydawnictwo Naukowe).

Gradziński, R., & Jerzykiewicz, T. 1972. Additional geographical and geological data from the Polish-Mongolian Paleontological Expeditions. *In* Z. Kielan-Jaworowska (ed.), *Results of the Polish-Mongolian Paleontological Expeditions* 4, *Palaeontologia Polonica* 27 (Warsaw & Crakow: Państwowe Wydawnictwo Naukowe).

1974. Dinosaur- and mammal-bearing aeolian and associated deposits of the Upper Cretaceous in the Gobi Desert (Mongolia). *Sedimentary Geology* 12: 249–78.

Gradziński, R., Kaźmierczak, J., & Lefeld, J. 1969. Geographical and geological data from the Polish-Mongolian Paleontological Expeditions. *In* Z. Kielan-Jaworowska (ed.), *Results of the Polish-Mongolian Paleontological Expeditions* 1, *Palaeontologia Polonica* 19 (Warsaw & Crakow: Państwowe Wydawnictwo Naukowe).

Gradziński, R., Kielan-Jaworowska, Z., & Maryaska, T. 1977. Upper Cretaceous Djadokhta, Barun Goyot and Nemegt formations of Mongolia, including remarks on previous subdivisions. *Acta Geologica Polonica* 27: 281–318.

Hirsch, K. F., & Quinn, B. 1990. Eggs and eggshell fragments from the Upper Cretaceous Two Medicine Formation of Montana. *Journal of Vertebrate Paleontology* 10: 491–511.

Horner, J. R. 1982. Evidence of colonial nesting and "site fidelity" among ornithischian dinosaurs. *Nature* 297: 675–6.

1987. Ecologic and behavioral information derived from a dinosaur nesting site. In S. J. Czerkas, & E. C. Olson (eds.) *Dinosaur Past and Present*, vol. 2 (Seattle: Washington University Press).

Horner, J., & Makela, R. 1979. Nest of juveniles provide evidence of family structure among dinosaurs. *Nature* 282: 296–8.

Kielan-Jaworowska, Z., & Barsbold, R. 1972. Narrative of the Polish-Mongolian Paleontological expedition 1967–1971, *In* Z. Kielan-Jaworowska (ed.), *Results of the Polish-Mongolian Paleontological Expeditions* 4, *Palaeontologia Polonica* 27 (Warsaw & Crakow: Państwowe Wydawnictwo Naukowe).

Kielan-Jaworowska, Z., & Dovchin, N. 1969. Narrative of the Polish-Mongolian Paleontological expedition 1963–1965, *In* Z. Kielan-Jaworowska (ed.), *Results of the Polish-Mongolian Paleontological Expeditions* 1, *Palaeontologia Polonica* 19 (Warsaw & Crakow: Państwowe Wydawnictwo Naukowe).

Kolesnikov, Ch. M. 1982. Biogeokhimicheskoe izuchenie gidrokhimii i termiki melovykh limnicheskikh vodoyomov Mongolii. *In* G. G. Martinson (ed.), *Mezozoiskie ozernye basseiny Mongolii.* (Leningrad: Nauka).

Kurochkin, E. N. 1988. Melovye ptitsy Mongolii i ikh znachenie dla razrabotki filogenii klassa. *In* E. N. Kurochkin (ed.) *Iskopaemye reptilii i ptitsy Mongolii, Trudy, Sovmestnaya Sovetsko-Mongol'skaya Paleontologicheskaya Ekspeditsiya* 34.

Kurzanov, S. M. 1989. Dinozavry bez sensatsii. *Priroda* 9: 47–54.

Kurzanov, S. M., & Mikhailov, K. E. 1989. Dinosaur egg-shells from the Lower Cretaceous of Mongolia. *In* D. D. Gillette & M. G. Lockley (eds.), *Dinosaur Tracks and Traces*. (New York: Cambridge University Press).

Lefeld, J. 1971. Geology of the Djadokhta Formation at Bayn Dzak (Mongolia). *In* Z. Kielan-Jaworowska (ed.), *Results of the Polish-Mongolian Paleontological Expeditions* 3, *Palaeontologia Polonica* 25 (Warsaw & Crakow: Państwowe Wydawnictwo Naukowe).

Mikhailov, K. E. 1987. Some aspects of the structure of the shell of the egg. *Paleontological Journal* 21: 54–61.

1991a. Classification of fossil eggshells of amniotic vertebrates. *Acta Palaeontologica Polonica* 36: 193–238.

1991b. The microstructure of avian and dinosaurian eggshell: phylogenetic implications. *In* K. Campbell (ed.), *Papers in Avian Paleontology Honoring Pierce Brodkorb. Contributions in Science* (Los Angeles: Natural History Museum of Los Angeles County), pp. 361–73.

Osmólska, H. 1980. The Late Cretaceous vertebrate assemblages of the Gobi Desert, Mongolia. *Mémoirs de la Société Géologique de France, N.S.* 139: 145–50.

Sabath, K. 1991a. Upper Cretaceous amniotic eggs from the Gobi Desert. *Acta Palaeontologica Polonica* 36: 151–92.

1991b. A new look at the dinosaur nests: Mongolian perspective. *In* Z. Kielan-Jaworowska, N. Heintz, & H.A. Nakrem (eds.), *Fifth Symposium on Mesozoic Terrestrial Ecosystems and Biota, Extended Abstracts, Contributions from the Paleontological Museum, University of Oslo* 364.

Schwartz, L., Fehse, F., Muller, G., &ersson, F., & Sieck, F. 1961. Untersuchungen an Dinosaurier-Eischalen von Aix-en-Provence und der Mongolei (Shabarakh Usu). *Zeitschrift für wissenschaftliche Zoologie A* 165: 344–79.

Seymour, R. S. 1979. Dinosaur eggs: gas conductance through the shell, water loss during incubation and clutch size. *Paleobiology* 5: 1–11.

Shuvalov, V. F. 1975. Stratigrafiya mezozoya Tsentral'noi Mongolii. *In* N. S. Zaytsev, B. Lunsandansan, G. G. Martinson, et al. (eds.), *Stratigrafiya mezozoiskikh otlozhenii Mongolii, Trudy, Sovmestnaya Sovetsko-Mongol'skaya Paleontologicheskaya Ekspeditsiya* 13.

1982. Paleogeografiya i istoriya razvitiya ozernykh sistem Mongolii v jurskoe i melovoe vremya. *In* G. G. Martinson (ed.), *Mezozoyskie ozernye basseiny Mongolii* (Leningrad: Nauka).

Sochava, A. V. 1969. Dinosaur eggs from the Upper Cretaceous of the Gobi Desert. *Paleontological Journal* 4: 517–27.

1971. Two types of egg shells in Cenomanian dinosaurs. *Paleontological Journal*, 3: 353–61.

1972. The skeleton of an embryo in dinosaur egg. *Paleontological Journal* 4: 527–33.

1975. Stratigrafiya i litologiya verkhnemelovykh otlozhenii Mongolii. *In* N. S. Zaytsev, B. Lunsandansan, G. G. Martinson et al. (eds.), *Stratigrafiya mezozoiskikh otlozhenii Mongolii, Trudy, Sovmestnaya Sovetsko-Mongol'skaya Paleontologicheskaya Ekspeditsiya* 13.

Stankevitch, E. S. 1982. Ostrakody pozdnego mela i osobennosti ikh obitaniya. *In* G. G. Martinson (ed.), *Mezozoyskiye ozerniye basseyni Mongolii* (Leningrad: Nauka).

Straelen, V. Van 1925. The microstructure of the dinosaurian egg-shells from the Cretaceous beds of Mongolia. *American Museum Novitates* 173: 1–4.

Thulborn, A. 1991. The discovery of dinosaur eggs. *Modern Geology* 16: 113–26.

1992. Nest of the dinosaur *Protoceratops*. *Lethaia* 25: 145–9.

Young, C. C. 1954. Fossil reptilian eggs from Laiyang, Shantung, China. *Scientia Sinica* 3: 505–22.

1965. Fossil eggs from Nanhsiung, Kwangtung and Kanchou, Kiangsi. *Vertebrata PalAsiatica* 9: 141–70.

Zhao Z. 1975. The microstructure of the dinosaurian eggshells of Nanhsiung, Kwangtung (1). *Vertebrata PalAsiatica* 13: 105–17.

1978. A preliminary investigation on the thinning of the dinosaurian eggshells of Late Cretaceous and some related problems. *Vertebrata PalAsiatica* 16: 213–21

1979. Discovery of the dinosaurian eggshells from Alxa, Ningxia and its stratigraphical significance. *Vertebrata PalAsiatica* 17: 304–9.

1988. A new structural type of the dinosaur eggs from Anly county, Hubei Province. *Vertebrata PalAsiatica* 26: 107–15.

Zhao Z., & Ding S. R. 1976. Discovery of the dinosaurian eggshells f.om Alxa, Ningxia and its stratigraphical significance. *Vertebrata PalAsiatica* 14: 42–51.

Zhen S., Zhen B., Mateer, N. J., & Lucas, S. G. 1985. The Mesozoic reptiles of China. *In* S. G. Lucas, & N. J. Mateer (eds.), *Studies on Chinese Fossil Vertebrates, Bulletin of the Geological Institute of the University of Uppsala* N.S. 11: 133–50.

8 Comparative taphonomy of some dinosaur and extant bird colonial nesting grounds

JOHN R. HORNER

Abstract

The distribution and size ranges of skeletal remains on some dinosaur nesting horizons in Montana are examined and compared to extant colonial bird nesting grounds. Isolated skeletal elements and partial or complete skeletons on these horizons reflect the sizes of the young that inhabited the nesting grounds. No adult remains were found on either the dinosaur or bird nesting horizons. The hypsilophodontid *Orodromeus* apparently grew to half adult length before leaving the nesting area, whereas the hadrosaurid *Maiasaura* and the lambeosaurid *Hypacrosaurus* grew only to one-fourth adult length before leaving. One of the hypacrosaur nesting grounds may have been devistated by a volcanic event during the nesting season.

Introduction

Over the past decade, the Upper Cretaceous Two Medicine Formation of western Montana and southern Alberta has been the focus of an extensive study of dinosaur social behavior. Evidence that continues to be collected indicates that at least some dinosaurs nested in colonies (Horner, 1982, 1987). Data also indicate that the young of some species were altricial, whereas the young of others were precocial (Horner & Makela, 1979; Horner & Weishampel, 1988). Although numerous kinds of eggs have been found in the Two Medicine Formation, only four types have been found with embryonic remains, permitting their identification. The dinosaurs include the hypsilophodontid *Orodromeus makelai*, the hadrosaurid *Maiasaura peeblesorum*, a new species of *Hypacrosaurus* (see Horner & Currie, Chapter 21), and ?*Troodon* (species indeterminant).

The colonial birds used in this study include the American white pelican (*Pelecanus erythrorhynchos*), the double crested cormorant (*Phalacrocorax auritus*), and a seagull (*Larus argentatus*). The bird nesting sites are located on islands in Lake Bowdoin, within the Bowdoin National Wildlife Refuge in north central Montana.

The questions to be addressed include:

1. Are the juvenile remains found on the different nesting horizons similar in size?
2. Are adult remains likely to be found in the nesting areas?
3. Were any of the nesting sites destroyed by catastrophic events?
4. Is the enormous quantity of isolated baby bones found on the dinosaur nesting grounds analogous to that seen at colonial bird nesting grounds?
5. Can we determine how large the dinosaur babies were when they left their nesting areas?

Egg Mountain nesting horizons

The geology and distribution of bioclasts within the sediments of Egg Mountain, Montana, together with clutch distribution and egg orientation have been described in detail by Horner (1982, 1987). The distribution and taphonomy of skeletal elements and skeletons on the nesting surfaces have not been described. From 1982 to 1984, all fossil material was mapped using triradiate coordinates. Skeletal elements and skeletons of both *Orodromeus makelai* and *Troodon* sp. were found on the nesting horizons together with skeletal remains of lizards and mammals. Specimen mapping that was performed from 1986 to the present has been limited to egg clutches and skeletons.

The only fossil specimens found to have a preferred orientation or a predictable distribution were the eggs of *Orodromeus makelai*. The eggs of ?*Troodon* sp. were found in unpredictable locations. Skeletons and skeletal elements of all species appear to have been distributed randomly (Fig. 8.1). Although none of the skeletal elements show signs of predation (tooth marks), the skeletons are often found incomplete. Skeletons of *O*.

makelai commonly consist of the tail, pelvic region and legs, and are lacking all portions anterior to the posterior dorsals. Interestingly, all skeletons and skeletal elements of both dinosaur species represent juvenile individuals. No adult or subadult skeletal remains have been found on any of the three nesting horizons. The majority of specimens of *O. makelai* represent embryonic or small juvenile individuals. The femur length ranges from 4.5 to 14 cm. Adult individuals (determined from histological studies and isolated elements) are estimated to have a femur about 30 cm long.

Troodon specimens are very rare on Egg Mountain. Only one partial skeleton, with a femur length of 12.5 cm, has been found. Adult specimens, known from other areas, have femora up to 28 cm long. All isolated *Troodon* skeletal elements represent individuals with femora less than 10 cm long. The most common specimens of *Troodon* are shed teeth.

Other *Orodromeus* nesting sites

Egg clutches of *Orodromeus makelai* have been found at three additional sites besides Egg Mountain. These sites include Egg Island (Horner, 1987), the north end of Willow Creek Anticline (2 km north of Egg Mountain), and the Hall Ranch in northern Montana (Glacier County). At each of these sites the egg clutches appear to have been undisturbed. Abundant isolated skeletal remains of *O. makelai* are found strewn randomly over the nesting horizon. These elements fall in the same size range as those found at Egg Mountain. Shed *Troodon* teeth and a partial skeleton of *Saurornitholestes* sp. were found adjacent to an *Orodromeus* nest at the Hall Ranch Site.

Maiasaura nesting horizons

Three *Maiasaura peeblesorum* nesting horizons have been discovered, all in the area of the Willow Creek Anticline, Teton County, Montana. On the lower horizon, eight nest sites were found, two of which contained babies (Horner & Makela, 1979; Horner, 1982). Randomly strewn over the nesting horizon were the isolated skeletal remains of baby hadrosaurid dinosaurs presumed to be *Maiasaura peeblesorum* (Fig. 8.2). A single articulated partial skeleton was found adjacent to one of the nests containing young. The isolated skeleton was found 2 m from the nest. The specimen, with a femur length of 7 cm, was half the size of the individuals found in the nest. No eggshell fragments were found associated with the isolated baby. It is quite likely that the baby died shortly after hatching, and that it was removed from the nest by an adult. Isolated skeletal elements found on the nesting horizon represent individuals with femora ranging from 6 cm (embryonic) to 19 cm in length. Skeletal remains of the smaller size individuals far outnumber remains of larger individuals.

Adult specimens of *Maiasaura peeblesorum* have femora with lengths averaging about 110 cm.

Two upper *Maiasaura* nesting horizons have been found, each of which have also yielded both eggs and skeletal remains. The skeletal remains represent individuals with femora ranging from 5 to 10 cm. *Troodon* and unidentified theropod teeth are present but not abundant. On all three horizons, many of the isolated skeletal elements have desiccation cracks, indicating that they had lain on the surface for some time before being covered with sediment. Other bones appear to have suffered postmortem breakage, apparently from trampling.

Hypacrosaurus nesting horizons

Two very different nesting horizons containing the egg and baby remains of a new species of *Hypacrosaurus* (see Horner and Currie, Chapter 21) have been found in northern Montana. One horizon, located near Landslide Butte (Glacier County) can be followed for 3 km, and appears to consist of several hundred nests. Gilmore (1917) discovered and described this horizon as a marker bed, but thought that the eggshell fragments were fragments of unionid shells. Although the eggs are found in clusters, all of the skeletal elements are disarticulated and spread randomly over the surface (Fig. 8.3). A volcanic ash (bentonite) rests on the nesting horizon.

Femora from the nesting area range in length from 4.5 to 9 cm (adult femora are about 120 cm long). The absence of larger juveniles may show that the volcanic ash prematurely terminated the use of the nesting ground. Isolated *Troodon* and unidentified theropod teeth occur rarely on the nesting horizon.

A second, and much less extensive, nesting horizon was found on Blacktail Creek, Teton County. Although only two nests were found, juvenile skeletal elements were abundant in the vicinity, apparently concentrated by a small stream. One nest contained only a concentration of eggshell, whereas the other contained eggs with embryonic skeletal remains (femur length of 8 cm).

The fluvially concentrated elements, representing at least twelve individuals, have femora lengths ranging from 16 cm to 23 cm (Fig. 8.4). Some elements were found in the channel, whereas the majority were found in a crevasse splay deposit (Fig. 8.5). Although these specimens were transported to some degree by water, there is no indication of any abrasion. Very few elements were found articulated, and there is no osteological evidence of predation. Ten isolated *Troodon* teeth were, however, found concentrated with these lambeosaurid specimens, possibly indicating that some predation had occurred on the nesting horizon. The *Troodon* teeth also show no indication of abrasion. Desiccation cracks on the bones are rare and are found exclusively on the elements representing the largest individuals.

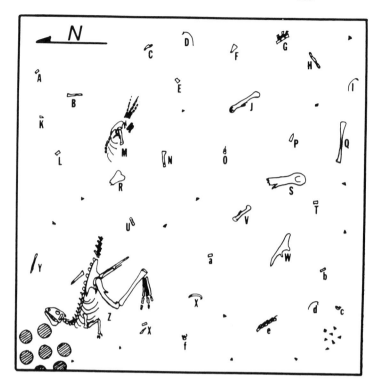

Figure 8.1. Map (in square meters) of a portion of Nesting Horizon B on Egg Mountain (see Horner, 1982) showing distribution of skeletal remains and eggs. See Appendix 1 for identification of lettered skeletal material. Δ are eggshell fragments.

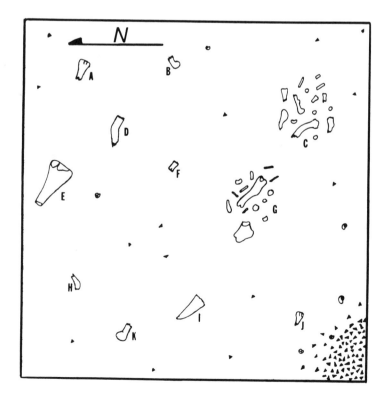

Figure 8.2. Map (in square meters) of lower *Maiasaura peeblesorum* nesting horizon, Willow Creek Anticline (see Horner, 1982) showing distribution of isolated skeletal elements. See Appendix 1 for identification of lettered skeletal material. @ are terrestrial gastropods. Δ are eggshell fragments.

Figure 8.3. Map (in square meters) of *Hypacrosaurus* nesting horizon at Landslide Butte (see Gilmore, 1917) showing egg and skeletal distribution at Site TM-035. See Appendix 1 for identification of lettered skeletal material. Δ are eggshell fragments. Cross-hatched areas represent crushed eggs.

Figure 8.4. Map (in square meters) of a portion of baby *Hypacrosaurus* bone bed on Blacktail Creek nesting horizon (TM-066) area showing distribution of skeletal elements along stream deposit. Cross-hatched area is edge of locality, dotted area represents streambed. Skeletal remains are identified in Appendix 8.1.

Bowdoin nesting horizons

Three islands in Bowdoin Lake are inhabited by colonial nesting birds during late spring and summer. The two largest islands are inhabited by pelicans, cormorants, and gulls, whereas the smallest island is inhabited exclusively by gulls (at least during the 1991 season). The largest of the islands, about 3,000 m², is barren in its central area, and is fringed by fireweed and small bushes. The other two islands are completely barren of vegetation. The pelicans and cormorants nest in colonies generally, but not exclusively, segregated from one another. The gulls nest on the fringes within the weeds and bushes. The cormorants construct sophisticated pillarlike nests made from twigs, whereas the pelicans have low, shallow nests made of soil but no plant debris. Pelican nests are preserved only because of the hardened remnants of their regurgitated food.

Throughout the nesting season the entire nesting surface of the islands is randomly scattered with the partial or complete carcasses of young and an abundance of isolated skeletal elements (Fig. 8.6). In no instance were any bodies or skeletal elements of adults found on the islands. Rare adult gull carcasses were found in the water offshore. A shallow excavation in the nesting area revealed isolated skeletal elements throughout the soil to a depth of about 12 cm. Because the islands are slightly raised, a few small streamlets (most likely produced during heavy rains) have washed over the surface and concentrated some small elements along their

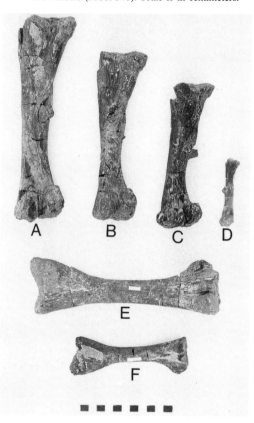

Figure 8.5. Femora and tibiae from the Blacktail Creek nesting horizon showing the size distribution. **A–C.** Femora (MOR 548). **D.** Femur (MOR 559). **E–F.** Tibiae (MOR 548). Scale is in centimeters.

Figure 8.6. Map (in square meters) of a portion of the largest island on Bowdoin Lake showing distribution of cormerant (CN) and pelican (PN) nests, and skeletal remains. Skeletal remains are identified in Appendix 8.1.

courses. Deposition of regurgitated and defecated materials, as well as nest debris mixed with wind blown dirt and alkali, form a shell protecting the island from excessive erosion. Most of the skeletal elements have desiccation cracks, and some (i.e., skulls) have postmortem breakage, apparently from trampling.

Lake Bowdoin is an alkaline lake and contains no fish. Both the pelicans and cormorants fly to other freshwater lakes and rivers to get both fish and crustaceans for feeding their respective young. Gull predation and scavenging is most active when many of the adult pelicans and cormorants have left the island to feed. The gulls occasionally kill hatchling or very young pelicans and cormorants, but were most often observed scavenging carcasses of individuals that apparently died for other reasons. During the nesting period when the pelican and cormorant chicks are unable to walk, one of the parents remains behind to guard the young.

Cormorant chicks fledge before the pelicans and leave the nesting area long before the pelican chicks can walk. When the pelican chicks attain mobility, they are kept in creches on the islands and cared for by a few adults. During this period the young are able to swim offshore, but always remain in a group. Pelican skeletal remains far exceed cormorant skeletal remains, most likely because their nesting period is much longer. Also, because pelicans reach a size almost equivalent to the adults before leaving the nesting area, some of the isolated skeletal elements are relatively large. Histolgical analyses verified that the larger elements were not those of adults.

Discussion

Except for nesting horizons, baby dinosaur remains are extremely rare in the fossil record, suggesting that most, if not all, baby dinosaur mortality occurred in the nesting areas. Baby remains in other areas were most likely transported by water or predators. For the hadrosaurid, lambeosaurid, and hypsilophodontid dinosaurs, there is very good evidence to indicate that the young were kept in their respective nesting areas for some time. The abundance of baby skeletal elements on the nesting horizons indicates relatively high mortality during nesting periods. The abundance of juvenile skeletal remains and an absence of adult remains on the dinosaur nesting horizons is analogous to that seen in the colonial nesting grounds of extant birds.

Although the babies of *Orodromeus* were apparently precocial (Horner & Weishampel, 1988), they remained on their island nesting grounds from the time they hatched until they reached about 1 m in length or about half adult size (Horner, 1984). Insufficient data preclude speculation concerning the amount of time, if any, *Troodon* young spent in their nesting areas. *Maiasaura* young were apparently altricial from the time they hatched until they were at least 1 m in length (Horner & Makela, 1979; Horner & Weisham-

pel, 1988). Isolated elements representing individuals that were 1.5 m in length indicate that they either remained altricial until they grew to that length, or that they left their nests and lived in groups within the nesting area. The hypacrosaurs, although none larger than hatchlings have been found in nests, apparently lived in the nesting areas from the time they hatched until the time they reached about 2 m in length. Hypacrosaur eggs and hatchlings were apparently larger than the eggs and hatchlings of *Maiasaura* (see Horner and Currie, Chapter 21).

This comparative taphonomic study suggests that the most logical places to find baby dinosaurs are on nesting horizons. Isolated baby skeletal elements that are not associated with nesting horizons (Carpenter, 1982; Fiorillo, 1987) can give us little or no behavioral information or nesting site location.

References

Carpenter, K. 1982. Baby dinosaurs from the Late Cretaceous Lance and Hell Creek formations and a description of a new species of theropod. *University of Wyoming, Contributions to Geology* 20: 123–34.

Fiorillo, A. R. 1987. Significance of juvenile dinosaurs from Careless Creek Quarry (Judith River Formation), Wheatland County, Montana. In P. J. Currie, & E. H. Koster (eds.), *Fourth Symposium on Mesozoic Terrestrial Ecosystems, Short Papers* (Drumheller, Canada: Tyrrell Museum of Palaeontology), pp. 88–94.

Gilmore, C. W. 1917. *Brachyceratops*, a ceratopsian dinosaur from the Two Medicine Formation of Montana with notes on associated fossil reptiles. *U.S. Geological Survey, Professional Paper* 103: 1–45.

Horner, J. R. 1982. Evidence of colonial nesting and "site fidelity" among ornithischian dinosaurs. *Nature* 297: 675–6.

   1984. The nesting behavior of dinosaurs. *Scientific American* 250: 130–37.

   1987. Ecologic and behavioral implications derived from a dinosaur nesting site. In S. J. Czerkas, & E. C. Olson (eds). *Dinosaurs Past and Present*, vol. 2 (Seattle: University of Washington Press), pp. 51–63.

Horner, J., & Makela, R. 1979. Nest of juveniles provides evidence of family structure among dinosaurs. *Nature* 282: 296–8.

Horner, J., & Weishampel, D. 1988. A comparative embryological study of two ornithischian dinosaurs. *Nature* 332: 256–7.

Appendix 8.1. List of specimens within boxes, Figures 8.1–8.4 and 8.6

Letter symbol	Specimen number	Material	Taxon

Figure 8.1. Egg Mountain: Locality TM-006 (B)

Letter symbol	Specimen number	Material	Taxon
A	EM-8-2-83	Vertebra	*Orodromeus*
B	EM-B-B	Tibia shaft	*Orodromeus*
C	MOR 406	Proximal ulna	*Orodromeus*
D	EM-7-9-84	Rib	*Orodromeus*
E	EM-7-8-84	Vertebra	*Orodromeus*
F	EM-B-F	Distal tibia	*Orodromeus*
G	EM-B-D	Articulated vertebra	*Orodromeus*
H	EM-7-17-84	Distal tibia	*Orodromeus*
I	EM-7-18-84	Rib	*Orodromeus*
J	EM-B-J	Femur	*Orodromeus*
K	EM-B-K	Tooth	*Troodon*
L	EM-7-26-84	Caudal vertebra	*Orodromeus*
M	MOR 331	Partial skeleton	*Orodromeus*
N	EM-8-11-82	Proximal metatarsal	*Orodromeus*
O	EM-7-22-83	Vertebra	Lizard
P	MOR 408	Ungual	*Orodromeus*
Q	MOR 404	Humerus	*Orodromeus*
R	EM-B-R	Distal tibia	*Orodromeus*
S	EM-B-S	Proximal femur	*Orodromeus*
T	EM-7-21-84	Vertebra	*Orodromeus*
U	EM-8-9-83	Metatarsal	*Orodromeus*
V	MOR 407	Femur	*Orodromeus*
W	MOR 401	ilium	*Orodromeus*
X		Unidentified fragments	
Δ=		Eggshell fragments	
Y	EM-B-Y	Proximal fibula	*Orodromeus*
Z	MOR 294	Articulated skeleton	*Orodromeus makelai*
a	EM-8-3-83	Vertebra	*Orodromeus*
b	EM-B-b	Vertebra	*Orodromeus*
c	MOR 318	Dentary	*Alphadon halleyi*
d	EM-B-d	Rib	*Orodromeus*
e	MOR 684	Articulated vertebrae	Lizard
f	EM-7-18-83	Vertebra	*Orodromeus*

Figure 8.2 *Maiasaura peeblesorum* lower nesting ground (Willow Creek Anticline)

Letter symbol	Specimen number	Material	Taxon
A	YPM-PU 22458	Proximal tibia	?*Maiasaura*
B	YPM-PU 23434	Distal femur	?*Maiasaura*
C	MOR 236	Partial skeleton	*Maiasaura peeblesorum*
D	YPM-PU 23439	Partial humerus	?*Orodromeus*
E	WL-8-6-81	Proximal tibia	?*Maiasaura*
F	MOR 279	Distal femur	?*Maiasaura*

Letter symbol	Specimen number	Material	Taxon
G	YPM-PU 23444	Skeletal elements	*Maiasaura*
H	YPM-PU 23438	Humerus	*Maiasaura*
I	MOR 259	Tooth	Theropod
J	WCA-WL-J	Proximal tibia	?*Maiasaura*
K	YPM-PU 22472	Distal femur	?*Maiasaura*
@			Terrestrial gastropods
Δ		eggshell fragments	

Figure 8.3 *Hypacrosaurus* nesting ground (Locality TM-035).

Letter symbol	Specimen number	Material	Taxon
A	MOR 434-A	Mid-shaft femur	?*Hypacrosaurus*
B	MOR 434-B	Proximal femur	*Hypacrosaurus*
C	MOR 434-C	Sacral vertebra	*Hypacrosaurus*
D	MOR 434-D	Proximal tibia	*Hypacrosaurus*
E	MOR 434-E	Proximal femur	*Hypacrosaurus*
F	MOR 434-F	Mid-shaft femur	*Hypacrosaurus*
G	MOR 434-G	Mid-shaft fibula	*Hypacrosaurus*
H	MOR 434-H	Proximal femur	*Hypacrosaurus*
I	MOR 434-I	Distal tibia	*Hypacrosaurus*
J	MOR 434-J	Distal femur	*Hypacrosaurus*
K	MOR 434-K	Distal tibia	*Hypacrosaurus*
L	MOR 434-L	Proximal dentary	*Hypacrosaurus*
M	MOR 434-M	Dorsal vertebra	*Hypacrosaurus*
N	MOR 434-N	Cervical vertebra	*Hypacrosaurus*
O	MOR 434-O	Associated caudal vertebrae (3)	*Hypacrosaurus*
P	MOR 434-P	Distal tibia	*Hypacrosaurus*
Q	MOR 434-Q	Distal tibia	?*Hypacrosaurus*
R	MOR 434-R	Distal femur	*Hypacrosaurus*
S	MOR 434-S	Mid-shaft femur	*Hypacrosaurus*
T	MOR 434-T	Mid-shaft femur	*Hypacrosaurus*
U	MOR 434-U	Mid-shaft ulna	*Hypacrosaurus*
V	MOR 434-V	Proximal ulna	*Hypacrosaurus*
W	MOR 434-W	Mid-shaft humerus	*Hypacrosaurus*
X	MOR 434-X	Caudal vertebra	*Hypacrosaurus*
Y	MOR 434-Y	Mid-shaft tibia	*Hypacrosaurus*
Z	MOR 434-Z	Caudal vertebra	*Hypacrosaurus*
a	MOR 434-a	Unknown mid-shafts	*Hypacrosaurus*
b	MOR 434-b	Caudal vertebrae (2)	*Hypacrosaurus*
c	MOR 434-c	Mid-shaft radius	*Hypacrosaurus*
d	MOR 434-d	Sacral vertebra	*Hypacrosaurus*
e	MOR 434-e	Cervical vertebra	*Hypacrosaurus*
f	MOR 434-f	Proximal ulna	*Hypacrosaurus*
g	MOR 434-g	Cervical vertebra	*Hypacrosaurus*
h	MOR 434-h	Caudal vertebra	*Hypacrosaurus*

Letter sym-bol	Specimen number	Material	Taxon

Figure 8.4. *Hypacrosaurus* nesting ground (Locality TM-066). All specimens catalogued MOR 548.

d	Dentary
f	Femur
fi	Fibula
h	Humerus
i	Ilium
j	Jugal
M	Maxilla
mt	Metatarsal
o	Postorbital
q	Quadrate
r	Radius
t	Tibia
v	Vertebra

Letter sym-bol	Specimen number	Material	Taxon

Figure 8.6 Pelican/cormorant nesting ground, Bowdoin Lake (large island).

Letter symbol	Material	Taxon
A	Rib fragment	?
B	Phalange	?
C	Ulna	Pelican
D	Femur	Pelican
E	Bone fragment	?
F	Rib fragment	?
G	Rib	Pelican
H	Femur	Pelican
I	Rib	Pelican
J	Pelvis	Pelican
K	Tarso–metatarsus	Pelican
L	Braincase	Cormorant
M	Ribs (2) and scapula	?
N	Tarso–metatarsus	Pelican
X	Mating crests	Pelican

9 Predation of dinosaur nests by terrestrial crocodilians

JAMES I. KIRKLAND

Abstract

During the Late Jurassic and Early Cretaceous a predatory role is postulated for small cursorial mesosuchian-grade crocodiles at dinosaur nesting sites. This theory is based primarily on the occurrence of small crocodile remains at a majority of the dinosaur nesting sites known from the Upper Jurassic and Lower Cretaceous of North America.

Introduction

Eggs are highly nutritious and today nests are exploited for food by a wide variety of organisms including raccoons, jackals, hyenas, black bears, pigs, primates, vultures, storks, crows, gulls, varanid lizards, large teiid lizards, colubroid snakes, and of course, man (Cott, 1961; Minton & Minton, 1973; Gans, 1974; Coombs, 1989). All of these animal and avian groups are aerial or terrestrial omnivores or carnivores with a mass of 1 to 500 kg.

Direct evidence for predation of eggs is rare in the fossil record. Small holes with characteristic flaking around the inner margin have been attributed to predators at Pleistocene and Eocene egg sites (Williams, 1981; Hirsch, personal communication). As yet, similar examples of this feature have not been reported from Mesozoic egg sites.

Determining which organisms may have been the major predators on dinosaur eggs is made difficult by the few known nesting sites and the practical limits of interpreting the behavior of extinct animals. Most of the animals and birds known to feed on eggs in modern environments did not exist during the Mesozoic. Varanid lizards are a notable exception. For extinct organisms, the association of potential egg predators in or near recognized dinosaur nesting sites provides one tentative line of evidence. This association, however, must be weighed against the possibility that the nesting organism brought it as a food item for the young (Hayward, Amlaner, & Young, 1989), or that the fossil material represents temporal or spatial mixing. It is also important to recognize that plausible egg predators might be found in association with nesting sites because they had been preying on the newly hatched young.

Although Upper Cretaceous dinosaur nesting sites are now becoming well known on several continents as indicated by papers in this volume, few dinosaur nesting sites have been recognized in the Lower Cretaceous (Jensen, 1970; Kurzanov & Mikhailov, 1989; Winkler & Murry, 1989) and Upper Jurassic (Hirsch, this volume; Hirsch, Young, & Armstrong, 1987; Scheetz, 1991; Young, 1991). I provide evidence to suggest that cursorial mesosuchian crocodiles (Fig. 9.1) were effective egg predators during the mid-Mesozoic prior to the advent of their closest modern counterparts, the varanid lizards. Potential Upper Cretaceous egg predators are reviewed, and nest raiding by varanid lizards is discussed. These discussions are followed by evidence to suggest that varanids were preceded by mesosuchian crocodiles as predators of dinosaur eggs.

Potential Upper Cretaceous egg predators

Mammals

Mammals have often been accused of raiding dinosaur nests for eggs (e.g., Wieland, 1925). Throughout the Mesozoic, the majority of mammals were extremely small (Lillegraven, 1979) and can be considered too small to have broken open a dinosaur egg, let alone to have dug into a dinosaur nest. Possible exceptions to this are the Lower Cretaceous triconodont, *Gobiconodon* (Jenkins & Schaff, 1988), and the stagodontid marsupials of the Upper Cretaceous (Cifelli & Eaton, 1987). These marsupials reached sizes comparable to *Gobiconodon* in the uppermost Cretaceous genus, *Didelphodon* (Clemens, 1979). Lofgren (1992) has indicated that the premolars of stagodontid marsupials were specialized for crushing and that these animals were quite capable of feeding on eggs. Although both *Gobiconodon* and *Didelphodon* were large enough (cat-size) and physically capable of eating dinosaur eggs, there is no evidence of this having taken place.

Pterosaurs and birds

Because most, if not all, dinosaurs, probably buried their eggs to at least some degree (Combs, 1989), raiders of dinosaur nests would need to have the ability to dig open the nest. Such an ability is unlikely for most pterosaurs or birds. A possible exception is the large pterosaur found in the vicinity of Horner's (1982, 1987) Upper Cretaceous dinosaur nesting sites (Padian, 1984; Padian & Smith, 1992). It is possible that such a large pterosaur could have uncovered a nest. However, a pterosaur the size of a small airplane would not go unnoticed by an adult dinosaur guarding the nest. Therefore, scavenging was likely, and abandoned nests would be much more likely targets for pterosaurs and birds.

Dinosaurs

Both oviraptorid and ornithomimid dinosaurs have been suggested as major dinosaur nest raiders during the Upper Cretaceous. Osborn (1924) proposed the name *Oviraptor philoceratops* (meaning "egg thief fond of ceratopsian eggs,") for the remains of an unusual edentulous theropod that was found 4 in. above a clutch of *Protoceratops* eggs. He suggested that the individual had been trapped by a sandstorm while raiding the nest, but he also cautioned that the occurrence may have been chance. Based on the great strength of their jaws, Barsbold (1983; Barsbold, Maryanska & Osmolska, 1990) has proposed that oviraptorids fed on bivalves. In contrast, Smith (1990) has reported that functionally *Oviraptor* was most likely a herbivore that fed on leaves.

Ornithomimids have been described as omnivores, feeding on plants, small animals, and eggs (e.g.,

Gregory, 1951; Russell, 1972; Osmolska, Roniewicz & Barsbold, 1972). It has also been suggested that they were selective feeders on plants (Osborn, 1917; Nicholls & Russell, 1985). Most recently, Barsbold and Osmolska (1990) proposed that ornithomimids were predominantly carnivores that fed on small prey, which were manipulated with the grasping forelimbs. However, opportunistic feeding on eggs is difficult to discount for either taxon. It is almost certain that if either dinosaur was able to secure a dinosaur egg, they would have had no trouble breaking it open by either their beak or by using their "hands" to strike the egg against the ground.

Another possible dinosaurian egg predator from Upper Cretaceous nesting sites is *Troodon* (Horner 1984, 1987). Their numerous, relatively small teeth, binocular vision, large brain, and specialized hands and feet suggest that they preyed on small vertebrates like lizards and mammals (Currie, 1989). Horner (1984, 1987) has suggested that *Troodon* remains in the ornithopod nesting sites in Montana represent individuals that preyed upon young dinosaurs. Opportunistic predation on eggs is certainly an additional role these sophisticated dinosaurs could have played within a nesting site.

In general, although small theropods may have been able to open the nests of other dinosaurs and manipulated the eggs in order to open them, they would not have been able to accomplish this unnoticed. Even small theropods would have had a difficult time approaching a nest unnoticed, because they would have presented a high profile silhouette. They would have been noticed by any adult dinosaur guarding the nest, and the small theropod would have been forced to flee if discovered. It is possible, however, that small theropod egg predators could have developed behavior aimed to draw the guarding adult dinosaurs away from the nest. Large theropods would have been immune to the defensive strategies of most adult herbivorous dinosaurs guarding the nests.

Figure 9.1. Skeletal reconstruction of a cursorial mesosuchian crocodile about 1 m long from the Upper Jurassic Morrison Formation in western Colorado. (Based on data of Clark, 1985.).

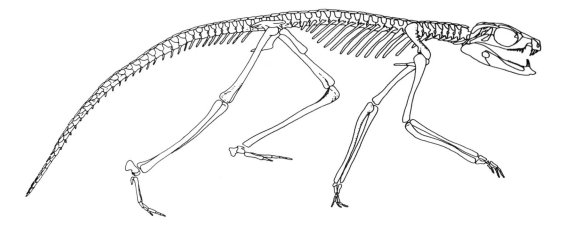

Teiid lizards

The large, extant heterodont teiid lizard, *Tupinambis* is known to prey on caiman eggs (Coombs, 1989). *Polyglyphanodon*, with a skull length of up to 8 cm, and *Chamops*, with a skull length of up to 5 cm, from the Upper Cretaceous of North America (Estes, 1983) could have preyed on smaller dinosaur eggs. Both dinosaur eggs (Jensen 1966) and *Polyglyphanodon* (Estes, 1983) are known from the Upper Cretaceous North Horn Formation in Utah. However, as yet there have been no studies showing a connection between the distribution of the dinosaur nesting sites and these large fossil teiid lizards. In addition, the transversely expanded posterior teeth of *Polyglyphanodon* are suggestive of a herbivorous diet (Carpenter, personal communication).

Varanid lizards

Most living varanid lizards will readily eat any egg encountered and are important predators of the nests of reptiles and mound-building birds (Cott 1961; Lincoln 1974; Auffenberg, 1981; Magnusson 1982). As a matter of fact, they may eat most of crocodile eggs laid during the year (Cott, 1961; Magnusson, 1982; Coombs, 1989). Magnusson (1982) observed that where crocodile nesting areas overlapped the preferred habitat range of *Varanus indicus*, nearly all the nests were raided. The exceptions were nests guarded by adult crocodiles. Cott (1961), however, reports that the large African monitor, *Varanus niloticus*, shows little fear of crocodiles. This species is reported to often work in pairs, with one distracting the adult crocodile, while the other quickly digs into the nest, to be joined subsequently by the first varanid (Cott, 1961). Varanids will also appear from any nearby cover to search the nesting grounds for eggs when the crocodiles leave the nest site.

Varanid lizards dig only with their front feet; first digging with one foot and then the other. An adult Komodo monitor (*Varanus komodoensis*) can dig a 1-m deep hole in unconsolidated earth in less than 1 hour. Auffenberg (1981) reports that, while young Komodo monitors are apparently prevented from feeding on sea-turtle eggs because of their depth of burial, specimens of 1.5–3 m regularly feed on these eggs.

In captivity, varanids swallow entire chicken eggs and crush them in their throats by strong flexion of their neck muscles (Auffenberg, 1981). Varanid lizards are also highly active carnivores and will readily eat newly hatched young if they are found (Cott, 1961). The Komodo monitor, when investigating mound building bird nests, will eat the adult birds, as well as dig into the nest for the eggs (Auffenberg, 1981). Although monitors will feed on adult sea turtles and their eggs, there is no evidence that they feed on baby sea turtles (Auffenberg, 1981). This may be due to the short time interval that newly hatched sea turtles are accessible on the beach.

Varanid lizards have a number of advantages in stealing eggs from nesting sites. They present a low silhouette in approaching and digging into a nest. They are able to move fast for short distances, with Komodo monitors of 0.5–1.2 m attaining speeds of 4.1–5.6 m/sec (Auffenberg, 1981). As a result, they are usually able to escape the adult crocodiles guarding the nests. Typically, the lizards are not pursued for great distances from the nest and, consequently, they often return to the nest once the guardian animals have relaxed (Cott, 1961).

Finally, not all varanids feed on eggs. Auffenberg (1988) found that eggs placed in the cages of captive *Varanus olivaceus* were often ignored. This Philippine varanid is a frugivore, which selectively feeds on the fruit it collects from the forest floor, although in captivity it will feed on fresh meat. In northern Australia, *Varanus indicus* readily eats eggs, while *Varanus gouldi* from the same area does not feed on eggs (Magnusson, 1982).

These examples indicate that broad generalizations on the feeding behavior of fossil varanids should be made with caution. Nevertheless, Horner (1987) has noted the close association of the earliest known large varanids with the nesting sites of the hypsilophodontid *Orodromeus* in the Upper Cretaceous of Montana. He speculated that this close association indicated predation of the *Orodromeus* eggs and that this predation behavior was established early in the history of varanid lizards (Carroll & DeBraga, 1992). It remains to be determined, however, how the 90 cm long varanid (Horner, personal communication) ate the 15 cm long hypsilophodontid eggs, because the eggs were larger than its head. Assuming that the varanid lizard did eat *Orodromeus* eggs, the small size of the Cretaceous lizards suggests that the eggs were probably not buried very deeply. Once hatched, precocial hatchling dinosaurs, such as *Orodromeus* (Horner & Weishampel, 1988), may have provided few predation opportunities within the nesting site, whereas altricial hatchling dinosaurs, such as *Maiasaura*, provided many more predation opportunities.

Mid-Mesozoic nesting sites and potential egg predators
Upper Jurassic

Dinosaur eggshell is known from the Upper Jurassic of Colorado, Utah, and Portugal (Hirsch 1989; Chapter 10). At present four possible dinosaur nesting sites are known, but only two of them have possible fossil egg predators in their vicinities. One of the localities is northwest of Delta, Colorado (Fig. 9.2) (Hirsch et al., 1987). It is approximately 38 m below the top of the Salt Wash Member of the Morrison Formation, making this the oldest known Morrison nesting site (Young, 1991). The Young egg site is a lenticular accumulation of very abundant eggshell fragments of one structural type, together with at least one crushed egg. Young (1991) interprets the site as the nest of a small theropod

Figure 9.2. Stratigraphic section of mid-Mesozoic rocks at the Fruita Paleontological Resource Area in western Colorado, showing the stratigraphic levels of the *Dryosaurus* nesting site and of the diverse microvertebrate faunas of Callison (1987). Wavy dashed lines indicate smectitic floodplain deposits and straight dashed lines indicate nonsmectitic floodplain deposits; other lithologic symbols are standard. On the locality map of Morrison nesting sites, FPA indicates Fruita Paleontological Area, S indicates the Scheetz (1991) nesting site, and Y indicates the Young (1991) nesting site.

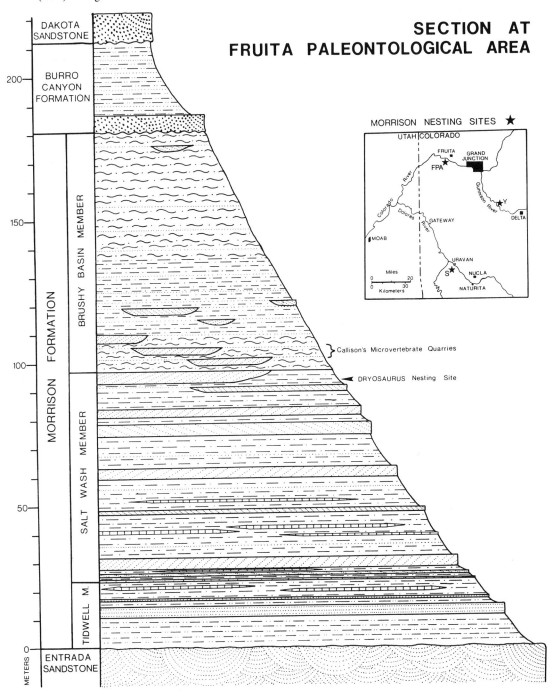

because of the abundance in the nest of diverse small fragmentary bones and teeth representing turtles, goniopholid crocodiles, sphenodontids, dinosaurs, mammals, and possibly birds. Hirsch (personal communication, 1992) has suggested that some of the associated micro fauna may have been washed into the site prior to burial. The density of eggshell is much greater than would be expected from one nest, and it has been suggested that the nesting site was occupied repeatedly (Hirsch, personal communication). Although the presence of at least one intact egg suggests minimal water disturbance, the presence of eggshell in the overlying sandstone demonstrates that at least some taphonomic modification of the nest had occurred (Young, 1991).

Chure, Madsen & Armstrong (1991) reported on the occurrence at the Young egg site of a small (5 cm long) adult mesosuchian crocodile skull with articulated jaws and some associated postcranial material. The skull has a broad skull table and short snout with a premaxillary–maxillary notch and heterodont dentition. An estimated size for the animal based on comparisons with similar taxa is about 50 cm long. The specimen shows a similarity to *Platyognathus* from the Upper Triassic of China and is the only articulated vertebrate from the Young egg site. Because the jaws are in occlusion with the skull, it was probably not transported over any great distance.

The third possible nesting site was discovered by me in June, 1991. The site is 3.6 m below the top of the Salt Wash Member of the Morrison Formation, southwest of Fruita, Colorado, in the Fruita Paleontological Resource Area (FPA) (Fig. 9.2). Three types of eggshell are present (Hirsch, Chapter 10), as well as over 200 *Dryosaurus* bones and fragments. At least two size classes are represented, including hatchling. At least two hatchlings are present, the distal end of their femora measuring 9–11 mm wide and the vertebrae centra 5–8 mm long. A minimum of two, and perhaps four, juveniles are also present. The distal end of the femora are 41–48 mm wide, and the centra 18–28 mm long. All of the centra lack fused neural spines. Galton (1981) indicates that adult *Dryosaurus altus* Marsh had a distal femur width of up to 98 mm and centra 40–53 mm long.

Two other taxa occur at the site. Turtles are represented by a few shell fragments, and cursorial mesosuchian crocodiles represented by several vertebrae and skull fragments. Owing to the fragmentary nature of the crocodiles, it is not certain if they represent the same taxa studied by Clark (1985) from the FPA (see below). The crocodilian material appears to represent a slightly larger animal. Clark (1985) reported a specimen about 60 cm long with vertebra-centra lengths of approximately 5–6 mm, whereas the crocodilian from the nesting site has vertebra-centra lengths of approximately 8-9 mm, suggestive of adults about 1 m long.

Fossils from this site are mostly permineralized with calcite and are generally encased in small carbonate nodules. These nodules are associated with root casts in a drab, gray mudstone indicating that the fossils were preserved by soil forming processes (Dodson et al., 1980; Retallack, 1990). Fragmentation of some of the bone occurred prior to burial, but there is no evidence of any extensive transport of the remains. This interpretation is based upon the lack of size sorting or selective preservation of the elements, and because partial associated hatchling *Dryosaurus* bones occur together in carbonate nodules. However, no nest has been delineated, and it is believed that the site is proximal to a nest or nests.

The basal part of the Brushy Basin Member of the Morrison Formation in the FPA, is known for its particularly rich and diverse Late Jurassic fauna (Callison, 1987). The fossiliferous interval is a few meters thick and begins approximately 10 m above the *Dryosaurus* nesting site (Fig. 9.2). The fauna are preserved in strata deposited under a variety of fluvial environments (Kirkland et al., 1990). The channel sandstones and abandoned channel ponds have produced a diverse assemblage of large dinosaurs and aquatic organisms. Floodplain environments have produced a wide variety of terrestrial vertebrates, including eight kinds of mammals, four kinds of lizards, three kinds of sphenodontids, a "snake" jaw, a small pterosaur, a mesosuchian crocodile, a small ("50" cm long) coelurosaur, and a small ("50" cm long) ornithischian, cf. "*Echinodon*" (Callison & Quimby, 1984; Callison, 1987). Small, isolated fragments of apparently transported eggshell have also been recovered from this level (Hirsch et al., 1987).

Among these taxa only the large sphenodontid, *Eilenodon robustus* (Rasmussen & Callison, 1981), and a cursorial mesosuchian crocodile (Clark, 1985; Benton & Clark, 1988) are about 1 m long as adults, and would appear to be in the size range to be effective egg predators. However, *Eilenodon* has transversely expanded acrodont teeth that are highly worn in adults, suggesting a herbivorous diet; it therefore may not be a good candidate for an egg predator.

The mesosuchian crocodile (Fig. 9.1) provides a very good candidate for an egg predator at the FPA. It is the most common small vertebrate at the FPA represented by as many as ten individuals (Clark, 1985, and subsequent collections). The crocodile has a broad skull table, a short snout with a heterodont dentition, a notch between the premaxilla and maxilla, and front lower jaws that may be edentulous. The animal is larger (skull up to 10 cm long) and differs in the palate from the specimen from the Young egg site (Chure et al., 1991; and personal communication). The FPA crocodile is the same size and is superficially similar to *Theriosuchus* from the Upper Jurassic Purbeck Formation of England. However, it is more primitive and is not closely related to *Theriosuchus* (Clark, 1985).

Not all nesting sites have possible egg predators associated with them. The largest such site was discovered in 1972 in the lower Brushy Basin Member of the Morrison Formation, southwest of Uravan, Colorado (Fig. 9.2). It was first reported because of the occurrence of juvenile *Dryosaurus* remains (Galton & Jensen, 1973). The site is currently under study by its discoverer, Rodney Scheetz (1991), who recognized among the more than 2,000 *Dryosaurus* bone fragments, a minimum of eight individuals ranging in size from 25 cm (embryonic or hatchling) to 165 cm long. An adult *Dryosaurus* would measure approximately 400 cm in length (Galton, 1981). In addition, eggshell of at least three morphologic types have been recognized (Scheetz 1991; Hirsch, Chapter 10).

Approximately 10 percent of the material from the Scheetz site represents other taxa. The specimens include sauropod and theropod tooth fragments, crocodile material, turtle shell, sphenodontid material, probable mammal material, and amiioid fish vertebrae. The crocodile material from the Scheetz locality includes a fragment of a scute and a vertebra 2 cm long (Scheetz, personal communication). Both specimens appear to be too large to represent a cursorial mesosuchian crocodilian and probably belong to an aquatic goniopholid crocodilian. Nearly all the specimens from this site result from surface collection and screen-washing sediment piles produced from road building (Scheetz, 1991). Therefore, taphonomic conclusions can not be made at this time.

Another possible nesting site has recently been discovered by Carpenter (personal communication; Hirsch, Chapter 10). More than 750 eggshell fragments have been found at Garden Park, near Canon City, Colorado. As yet no egg predator has been found; however, collection and analysis are still in progress.

In Europe, there is some dinosaur eggshell from the Upper Jurassic of Portugal, although not clearly from a specific dinosaur nest (Hirsch, 1989; personal communication). The cursorial mesosuchian crocodile, *Theriosuchus,* is known from the area (Brinkmann, 1989). It is the best nondinosaurian candidate for an egg predator, because lizards from this region are too small to have preyed on dinosaur eggs (Seiffert, 1973). Other small mesosuchian crocodiles from the Upper Jurassic of Europe (Wellnhofer, 1971) may also have been potential egg thieves.

Lower Cretaceous

Dinosaur nesting sites are very rare in the Lower Cretaceous. Sites are known from Utah and Texas in North America, and Mongolia in central Asia. In Utah, a number of sites with abundant dinosaur eggshell occur in the Aptian–Albian Cedar Mountain Formation (Jensen, 1970; Nelson & Crooks, 1987; Cifelli, personal communication). At present, no possible egg predators have been identified with certainty (DeCourten, 1991a, 1991b). A diverse fauna of small terrestrial vertebrates are associated with several new University of Oklahoma

sites, and may yield such a taxon in the future (Cifelli, personal communication).

The Proctor Lake dinosaur site of central Texas is probably a hypsilophodontid nesting site despite the absence of eggshell. It occurs in the lower part of the Aptian Twin Mountain Formation (Winkler et al., 1988, Winkler et al., 1989; Winkler & Murry, 1989). This site contains numerous adult hypsilophodontids and several concentrations of juveniles of the same size class. Taphonomical data indicate that the carcasses dried out undisturbed on a floodplain prior to burial (Winkler et al., 1989).

In contrast to the high diversity of other Lower Cretaceous vertebrate sites in central Texas, only three taxa are recognized at the Proctor Lake site (Winkler & Murry, 1989). These include a dromaeosaur (based on a few teeth), the abundant undescribed hypsilophodontids, and several small mesosuchian crocodilians. These crocodiles have been compared to *Theriosuchus*, and may have fed on the juvenile hypsilophodontids (Winkler et al., 1989). In addition, they may have also fed on hypsilophodontid eggs. Although lizards appear to have been diverse in the Lower Cretaceous of Texas (Winkler et al., 1990), they are too small to have been effective in raiding dinosaur nests.

Eggshells have been reported from a probable nest of a small theropod at Builjasutuin–Khuduk, Mongolia (Kurzanov & Mikhailov, 1989). This site occurs in the Lower Cretaceous Kkukhtekian (late Aptian–Albian) of Jerzykiewicz and Russell (1991). The oldest known varanid lizards occur at this locality, as do primitive ornithomimids and abundant *Psittacosaurus*.

Small cursorial mesosuchian crocodilians were apparently absent at this time in Asia, although they are known from North America, Europe, and Africa (Joffe, 1967; Buffetaut, 1982; Brinkmann, 1989, 1991). Instead, the first varanid lizards occured, suggesting possible competitive exclusion of the crocodiles.

Cursorial mesosuchian crocodiles as egg predators

Evidence for cursorial mesosuchian crocodiles as egg predators at dinosaur nesting sites is purely circumstantial. These small terrestrial crocodiles have been found closely associated at two of the three known possible Upper Jurassic dinosaur nesting areas. In addition, they are the only other common animal in a possible dinosaur nesting site in the Lower Cretaceous of central Texas.

A number of small cursorial mesosuchian crocodiles have been reported from Upper Jurassic; however, not all of these taxa are potential egg predators. *Hoplosuchus kayi*, from Dinosaur National Monument in northeastern Utah, is about 15 cm long, yet appears to be an adult (Gilmore, 1926). This small size would appear to exclude it from preying on dinosaur eggs. Another crocodilomorph, *Hallopus victor*, is known from near the Cope Quarry, high in the Morrison Formation near Canon City, Colorado (Carpenter, personal communication). Although the skull is unknown, this cursorial mesosuchian crocodile would appear to have been about 1 m long and may have been an effective predator of eggs (Walker, 1970). *Macelognathus vagans* Marsh, from Quarry Nine in the Upper Morrison at Como Bluff, Wyoming, is known only from the anterior half of a pair of dentaries approximately 10 cm long. The front of the jaws is spatulate and edentulous in the area of the long symphysis (Ostrom, 1971). The size of the jaws indicates that the animal was larger than known cursorial mesosuchians. *Macelognathus* may not be a mesosuchian, and the dentaries do not appear to be constructed so as to be able to prey on eggs.

The mesosuchian crocodile from the basal Brushy Basin Member of the Morrison Formation at Fruita, Colorado, resembles *Macelognathus* in that the very front of the jaws may also have been edentulous. However, the short edentulous section is followed by a double pair of enlarged caniniform teeth (Clark, 1985) (Fig. 9.1). The mesosuchian skull described by Chure et al. (1991; Madsen, personal communication) also has an enlarged pair of caniniform teeth on the lower jaw, as well as a pair of smaller teeth farther forward.

The two mesosuchian crocodiles found in association with the possible dinosaur nesting sites in western Colorado appear to be the best candidates for egg predators among the known mesosuchian crocodiles from the Morrison Formation. Both have a small, anteriorly projecting process at the suture of the lower jaws, which is bordered posteriorly by a transverse groove (Clark, 1985; Chure, personal communication). This structure might have been encased in keratin and could have been used to open eggs, but further study is needed.

The organization of the teeth in these crocodilians may also have been advantageous for feeding on eggs. Large teeth border the notch in the upper jaw for the reception of the two caniniform teeth of the lower jaw when the jaws are in occlusion. The posterior teeth are laterally compressed and rectangular in lateral view (Clark, 1985). During feeding, the large teeth of the upper jaw could brace the egg between them, while the caniniform teeth of the lower jaw could pierce a hole in it. A hypsilophodontid sized egg (estimated to be 6 cm in diameter) would not have fit in the mouth of a 1 m long crocodilian without extreme extension of the jaws. However, the egg could have been braced against one side of the mouth by a forefoot and worked open by the caniniform teeth.

The small mesosuchian crocodiles from the hypsilophodontid nesting site in Texas have been compared with *Theriosuchus* (Winkler et al., 1989). Another small, unnamed mesosuchian crocodilian from the Albian Paluxy Formation was illustrated by Langston

Figure 9.3. Restoration of a *Dryosaurus* nesting site being raided by cursorial mesosuchian crocodiles. (Based on information collected from the Fruita Paleontological Resource Area in western Colorado.)

(1974, Plate. 1, Figs. 1–3). It has a distinct notch in the upper jaw for the reception of a large dentary tooth. Benton and Clark (1988) suggest that this undescribed species is the sister group to the Eusuchia and thus, is not closely related to either *Theriosuchus* or to the more primitive Fruita crocodilian. It is difficult to support the independent development of nest-predation behavior in these distantly related taxa. However, such distantly related mammals as raccoons, skunks, and pigs will all eat eggs.

Theriosuchus is known from the Upper Jurassic and Lower Cretaceous of Europe (Owen, 1879; Joffe, 1967; Buffetaut, 1982; Brinkmann, 1989, 1991). It, and a number of other small cursorial crocodilians, lack the premaxillary–maxillary notch, but are similar in size to the mesosuchian crocodiles from western Colorado (Clark 1985). These other crocodiles have not yet been found in association with any potential dinosaur nesting sites and therefore lack evidence supportive for nest-predation behavior.

All of the mesosuchian crocodiles that are thought to have been dinosaur egg predators have a number of features in common. They are moderately brevirostrine, with a broad skull table and heterodont dentition. The vertebra are procoelous, and they have long slender limbs with forelimbs only slightly shorter than the hind limbs. No features specifically adapted for opening eggs have been positively identified, but these taxa have not been examined with this possible behavior in mind, so that functional studies are needed.

Morphologically, these small crocodiles have a number of features beneficial to an animal preying on dinosaur nesting sites. As a small quadruped, they would have presented a low silhouette in approaching and digging into the nest unnoticed. Their relatively long legs and erect stance would permit rapid escape if they were detected, as well as would allow them to return quickly to the nest. The crocodilians from the FPA have longer distal than proximal limb elements (Clark, 1985), indicating that they were runners. *Hallopus* has been reconstructed as a digitigrade runner (Walker, 1970), but data on the feet of the FPA crocodilian is not available, so it is reconstructed as plantigrade (Fig. 9.1). As in all modern crocodilians, *Hallopus* would have been capable of digging its own nest in addition to digging up the nests of dinosaurs.

It is very probable that, as active medium sized carnivores, these crocodilomorphs would also readily take hatchling dinosaurs as prey. However, because the articular surfaces of the hatchling *Dryosaurus* limb bones are well developed, the hatchlings were probably precocial at birth (see Horner & Weishampel, 1988). Therefore, there would have been few opportunities for the crocodiles to prey on the young dinosaurs within the nesting area. One complication of this hypothesis is the presence of juvenile and hatchling *Dryosaurus* together, suggesting that these animals stayed in the nesting area

for a prolonged period of time even though precocial by nature.

Conclusions

Although circumstantial, small, fleet-footed mesosuchian crocodiles may have raided dinosaur nesting sites during the Upper Jurassic and Lower Cretaceous (Fig. 9.3). Both dinosaur eggs and newly hatched young dinosaurs were items of prey. At present, there is no evidence to suggest that dinosaurs were egg predators during the mid-Mesozoic, although medium sized theropods would in all probability have been effective in this role.

Acknowledgments

A preliminary study of the baby *Dryosaurus* site at the Fruita Paleontological Resource Area was conducted by Randy Nydam, George Callison, Kay Fredette, Bonnie Carter, and me. Several participants of the Dinamation International Society's Dinosaur Discovery Expeditions Program also aided in this study. Preliminary examination of collected specimens benefited from the observations and comments by Rodney Scheetz, David Weishampel, and Dan Chure. Dan Chure and Dale Winkler are also gratefully acknowledged for making literature available, as well as for discussions about the matters discussed in this chapter. Unpublished information provided by Kenneth Carpenter and Richard Cifelli is also gratefully acknowledged. Thanks are due to Rick Adleman for Figure 9.1 and to Claudette Kennedy for Figure 9.3.

References

Auffenberg, W. 1981. *The Behavioral Ecology of the Komodo Monitor.* (Gainesville, Florida: University of Florida Press).

—— 1988. *Gray's Monitor Lizard.* (Gainesville, Florida: University of Florida Press).

Barsbold, R. 1983. Carnivorous dinosaurs from the Cretaceous of Mongolia. *The Joint Soviet-Mongolian Paleontological Expeditions Transactions* 19: 1–117.

Barsbold, R., Maryanska, T., & Osmolska, H. 1990. Oviraptorosauria. *In* D. B. Weishampel, P. Dodson, & H. Osmolska (eds.), *The Dinosauria.* (Berkeley: University of California Press).

Barsbold, R., & Osmólska, H. 1990. Ornithomimosauria. *In* D. B. Weishampel, P. Dodson, & H. Osmólska (eds.), *The Dinosauria.* (Berkeley: University of California Press).

Benton, M. J., & Clark, J. M. 1988. Archosaur phylogeny and the relationships of the Crocodylia. *In* M. J. Benton (ed.), *The Phylogeny and Classification of Tetrapods.* 1 (*Amphibians, Reptiles, Birds*). Systematics Association Special Volume 35A. (Oxford: Clarendon Press).

Brinkmann, W. 1989. Vorlaufige Mitteilung über die krokodilier-faunen aus dem Ober-Jura (Kimmeridgium) der Kohlegrube Guimarota, bei Leiria (Portugal) und der

Unter-Kreide (Barremium) von Una (Provinz Cuenca, Spanien). *Documenta Naturae* 56: 1–28.

1991. The crocodilians from Una (Early Cretaceous, Spain). *In* Z. Kielan-Jaworowska, N. Heintz, & H. A. Nakrem (eds.), *Fifth Symposium on Mesozoic Terrestrial Ecosystems and Biota; Extended Abstracts; Contributions from the Paleontological Museum* (Oslo: University of Oslo) 364: 11–12.

Buffetaut, E. 1982. Radiation évolutive, paléoécologie et biogéographie des crocodiliens mésosuchiens. *Mété Géologique de France N.S.* 60: 1–88.

Callison, G., 1987. Fruita: a place for wee fossils. *In* W. R. Averett (ed.), *Paleontology and Geology of the Dinosaur Triangle* (Grand Junction, Colorado: Colorado, Museum of Western Colorado).

Callison, G., & Quimby, H. M. 1984. Tiny dinosaurs: are they fully grown? *Journal of Vertebrate Paleontology* 3: 200–9.

Carroll, R. L., & DeBraga, M. 1992. Aigialosaurs: mid-Cretaceous varanoid lizards. *Journal of Vertebrate Paleontology* 12: 66–86.

Chure, D. J., Madsen, S. K., & Armstrong, H. J. 1991. An unusual mesosuchian crocodilian from the Salt Wash Member of the Morrison Formation, Upper Jurassic, Western Colorado. *Journal of Vertebrate Paleontology* 11 (supplement to No. 3): 22A–23A.

Cifelli, L., & Eaton, J. G. 1987. Marsupial from the earliest Late Cretaceous of Western U.S. *Nature* 325: 520–2.

Clark, J. M. 1985. *A New Crocodilian from the Upper Jurassic Morrison Formation of Western Colorado With a Discussion of the Relationships Within the "Mesosuchia."* Unpublished Master's Thesis. (Berkeley: University of California.

Clemens, W. A. 1979. Marsupialia. *In* JU. A. Lillegraven, Z. Kielan-Jaworowska, & W. A. Clemens (eds.), *Mesozoic Mammals; The First Two-Thirds of Mammalian History* (Berkeley: University of California Press).

Coombs, W. P., Jr. 1989. Modern analogs for dinosaur nesting and parental behavior. *In* J. O. Farlow (ed.), Paleobiology of the Dinosaurs, *Geological Society of America Special Paper* 238: 21–53.

Cott, H. B. 1961. Scientific results of an enquiry into the ecology and economic status of the Nile crocodile (*Crocodylus niloticus*) in Uganda and northern Rhodesia. *Transactions of the Zoological Society of London* 29: 211–356.

Currie, P. J. 1989. Theropod dinosaurs of the Cretaceous. *In* K. Padian, & D. J. Chure (eds.), *The Age of Dinosaurs, The Paleontological Society, Short Courses in Paleontology* 2: 113–20.

DeCourten, F. L. 1991a. The Long Walk Quarry and tracksite: unveiling the mysterious Early Cretaceous of the Dinosaur Triangle Region. *In* W. R. Averett (ed.), *Guidebook for Dinosaur Quarries and Tracksites Tour, Western Colorado and Eastern Utah* (Grand Junction, Colorado: Grand Junction Geological Society).

1991b. New data on Early Cretaceous dinosaurs from the Long Walk Quarry and tracksite, Emery County, Utah. *In* T. C. Chidsey, Jr. (ed.), *Geology of East-Central Utah, Utah Geological Association Publication* 19: 311–24.

Dodson, P., Behrensmeyer, A. K., Bakker, R. T., & McIntosh, J. S. 1980. Taphonomy and paleoecology of the dinosaur beds of the Jurassic Morrison Formation. *Paleobiology* 6: 208–32.

Estes, R. 1983. Sauria terrestria, Amphisbaenia. *Encyclopedia of Palaeoherpetology* Part 10 A. (Stuttgart: Gustav Fisher Verlag).

Galton, P. M. 1981. *Dryosaurus*, a hypsilophodontid dinosaur from the Upper Jurassic of North America and Africa: postcranial skeleton. *Paläontologische Zeitschrift* 55: 271–312.

Galton, P. M., & Jensen, J. A. 1973. Small bones of the hypsilophodontid dinosaur *Dryosaurus altus* from the Upper Jurassic of Colorado. *Great Basin Naturalist* 33: 129–32.

Gans, C. 1974. Analysis by dissection and observation: egg eating in snakes. *In* C. Gans *Biomechanics: An Approach to Vertebrate Biology.* (Philadelphia: J. B. Lippincott Company).

Gilmore, C. W. 1926. A new aetosaurian reptile from the Morrison Formation of Utah. *Annals Carnegie Museum* 16: 325–49.

Gregory, W. K., 1951. *Evolution Emerging, A Survey of Changing Patterns from Primeval Life to Man*, Vol. I. (New York: Macmillan Press).

Hayward, J. L., Amlaner, C. J., & Young, K. A. 1989. Turning eggs to fossils; a natural experiment in taphonomy. *Journal of Vertebrate Paleontology* 9: 196–200.

Hirsch, K. F. 1989. Interpretations of Cretaceous and pre-Cretaceous eggs and shell fragments. *In* D. D. Gillette, & M. G. Lockley (eds.), *Dinosaur Tracks and Traces.* (Cambridge: Cambridge University Press).

Hirsch, K. F., Young, R. G., & Armstrong, H. J. 1987. Eggshell fragments from the Jurassic Morrison Formation of Colorado. *In* W. R. Averett (ed.), *Paleontology and Geology of the Dinosaur Triangle* (Grand Junction, Colorado: Museum of Western Colorado).

Horner, J. R. 1982. Evidence of colonial nesting and "site fidelity" among ornithischian dinosaurs. *Nature* 297: 675–6.

1984. The nesting behavior of dinosaurs. *Scientific American* 250: 130–7.

1987. Ecologic and behavioral implications derived from a dinosaur nesting site. *In* S. J. Czerkas, & E. C. Olson (eds.), *Dinosaurs Past and Present*. Vol. II. (Seattle: University of Washington Press).

Horner, J. R., & Weishampel, D. B. 1988. A comparative embryological study of two ornithischian dinosaurs. *Nature* 332: 256–7.

Jenkins, F. A., Jr., & Schaff, C. R. 1988. The Early Cretaceous mammal *Gobiconodon* (Mammalia, Triconodonta) from the Cloverly Formation in Montana. *Journal of Vertebrate Paleontology* 8: 1–24.

Jensen, J. A. 1966. Dinosaur eggs from the Upper Cretaceous North Horn Formation of central Utah. *Brigham Young University Geology Studies* 13: 55–67.

1970. Fossil eggs in the Lower Cretaceous of Utah. *Brigham Young University Geology Studies* 17: 51–65.

Jerykiewicz, T., & Russell D. A. 1991. Late Mesozoic stratigraphy and vertebrates of the Gobi Basin. *Cretaceous Research* 12: 345–77.

Joffe, J. 1967. The "dwarf" crocodiles of the Purbeck Formation, Dorset, a reappraisal. *Palaeontology* 10: 629–39.

Kirkland, J. I., Mantzios, C., Rasmussen, T. E., & Callison, G. 1990. Taphonomy and environments; Fruita Paleontol. Resource Area, Upper Jurassic, Morrison Formation, W. Colorado. *Journal of Vertebrate Paleontology* 10 (Supplement to No. 3): 31A.

Kurzanov, S. M., & Mikhailov, K. E. 1989. Dinosaur eggshells from the Lower Cretaceous of Mongolia. *In* D. D. Gillette & M. G. Lockley (eds.) *Dinosaur Tracks and Traces.* (Cambridge: Cambridge University Press).

Langston, W., Jr. 1974. Nonmammalian Comanchean tetrapods. *Geoscience and Man* 8: 77–102.

Lillegraven, J. A. 1979. Introduction. *In* J. A. Lillegraven, Z. Kielan-Jaworowska, & W. A. Clemens (eds.), *Mesozoic Mammals; The First Two-Thirds of Mammalian History.* (Berkeley: University of California Press).

Lincoln, G. A. 1974. Predation of incubator birds (*Megaodius frecinet*) by Komodo Dragons (*Varanus komodoensis*). *Journal of Zoology (London)* 174: 419–28.

Lofgren, D. L. 1992. Upper premolar configuration of *Didelphodon vorax* (Mammalia, Marsupialia, Stagodontidae). *Journal of Paleontology* 66: 162–4.

Magnusson, W. E. 1982. Mortality of eggs of the crocodile *Crocodylus porosus* in Northern Australia. *Journal of Herpetology* 16: 121–30.

Minton, S. A., Jr., & Minton M. R. 1973. *Giant Reptiles.* (New York: Charles Scribner's Sons).

Nelson, M. E., & Crooks, D. M. 1987. Stratigraphy and paleontology of the Cedar Mountain Formation (Lower Cretaceous), eastern Emery County, Utah. *In* W. R. Averett (ed.), *Palaeontology and Geology of the Dinosaur Triangle* (Grand Junction, Colorado: Museum of Western Colorado).

Nicholls, E. L., & Russell, A. P. 1985. Structure and function of the pectoral girdle and forelimb of *Struthiomimus altus* (Theropoda: Ornithomimidae). *Paleontology* 28: 638–67.

Osborn, H. F. 1917. Skeletal adaptions of *Ornitholestes, Struthiomimus,* and *Tyrannosaurus. Bulletin, American Museum of Natural History* 35: 733–71.

1924. Three new Theropoda, *Protoceratops* zone central Mongolia. *American Museum of Natural History Novitates* 144: 1–12.

Osmolska, H., Roniewicz, E., & Barsbold, R. 1972. A new dinosaur, *Gallimimus bullatus* n. gen., n. sp. (Ornithomimidae) from the Upper Cretaceous of Mongolia. *Palaeontologica Polonica* 27: 103–43.

Ostrom, J. H. 1971. On the systematic position of *Macelognathus vagans. Postilla* 153: 1–10.

Owen, R. 1879. On the association of dwarf crocodiles (*Nannosuchus* and *Theriosuchus pusillus,* e.g.) with the diminutive mammals of the Purbeck Shales. *The Quarterly Journal of the Geological Society of London* 35: 148–55.

Padian, K. 1984. A large pterodactyloid pterosaur from the Two Medicine Formation (Campanian) of Montana. *Journal of Vertebrate Paleontology* 4: 516–24.

Padian, K., & Smith, M. 1992. New light on Late Cretaceous pterosaur material from Montana. *Journal of Vertebrate Paleontology* 12: 87–92.

Rasmussen, T. E., & Callison, G. 1981. A new herbivorous sphenodontid (Squamata: Reptilia) from the Jurassic of Colorado. *Journal of Paleontology* 55: 1109–16.

Retallack, G. J. 1990. *Soils of the Past; An introduction to Paleopedology.* (Boston: Unwin Hyman).

Russell, D. A. 1972. Ostrich dinosaurs from the Late Cretaceous of western Canada. *Canadian Journal of Earth Science* 9: 375–402.

Scheetz, R. D. 1991. Progress report of juvenile and embryonic *Dryosaurus* remains from the Upper Jurassic Morrison Formation of Colorado. *In* W. R. Averett (ed.), *Guidebook for Dinosaur Quarries and Tracksites Tour, Western Colorado and Eastern Utah* (Grand Junction, Colorado: Grand Junction Geological Society).

Seiffert, J. 1973. Upper Jurassic lizards from central Portugal. *Servicos Geologicos de Portugal Memoria* 22 (N.S.): 1–85.

Smith, D. K. 1990. Osteology of *Oviraptor philoceratops,* a possible herbivorous theropod from the Upper Cretaceous of Mongolia. *Journal of Vertebrate Paleontology* 10 (supplement to No. 3): 42A.

Walker, A. D. 1970. A revision of the Jurassic reptile *Hallopus victor* (Marsh), with remarks on the classification of crocodiles. *Philosophical Transactions of the Royal Society London* Ser. B 257: 323–72.

Wellnhofer, P. 1971. Die Atoposauridae (Crocodilia, Mesosuchia) der Oberjura-Plattenkalke Bayerns. *Palaeontographica Abt. A* 138: 133–65.

Wieland, G. 1925. Dinosaur extinction. *American Naturalist* 59: 557–65.

Williams, D. L. G. 1981. *Genyornis* eggshells (Dromornithidae; Aves) from the Late Pleistocene of South Australia. *Alcheringa* 5: 133–40.

Winkler, D. A., Jacobs, L. L., Branch, J. R., Murry, P. A., Downs, W. R., & Trudel, P. 1988. The Proctor Lake dinosaur locality, Lower Cretaceous of Texas. *Hunteria* 2(5): 1–8.

Winkler, D. A., & Murry, P. A. 1989. Paleoecology and hypsilophodontid behavior at the Proctor Lake dinosaur locality (Early Cretaceous), Texas. *In* J. O. Farlow (ed.), Paleobiology of the Dinosaurs, *Geological Society of America Special Paper* 238: 55–61.

Winkler, D. A., Murry, P. A., & Jacobs, L. L. 1989. Vertebrate paleontology of the Trinity Group, Lower Cretaceous of central Texas. *In* D. A. Winkler, P. A. Murry, & L. L. Jacobs (eds.), *Field Guide to the Vertebrate Paleontology of the Trinity Group, Lower Cretaceous of Central Texas* (Dallas: Southern Methodist University), pp. 1–22.

Winkler, D. A., Murry, P. A., & Jacobs, L. L. 1990. Early Cretaceous (Comanchean) vertebrates of central Texas. *Journal of Vertebrate Paleontology* 10: 95–116.

Young, R. G. 1991. A dinosaur nest in the Jurassic Morrison Formation, western Colorado. *In* W. R. Averett (ed.), *Guidebook for Dinosaur Quarries and Tracksites Tour, Western Colorado and Eastern Utah.* (Grand Junction, Colorado: Grand Junction Geological Society), pp. 1–15.

III Eggs

10 Upper Jurassic eggshells from the Western Interior of North America

KARL F. HIRSCH

Abstract

Eggs and eggshell fragments have been found at six localities in the Upper Jurassic Morrison Formation of Colorado and Utah. A parataxonomic family, Prismatoolithidae, is established for a prismatic dinosauroid basic type of eggshell. A new genus and species, *Prismatoolithus coloradensis*, is described for eggs and eggshells from two localities. In addition, a very thin eggshell is briefly described from four other sites. Their limited number and partial alteration prevents detailed description and naming. The localities, the diagenetic and pathological phenomena in the eggshells found at all sites, and the difficulties encountered in studying this material are discussed briefly.

Introduction

Eggs and eggshell fragments older than Cretaceous are relatively rare worldwide (Hirsch, 1989). There is a questionable egg from the Permian of Texas (Hirsch, 1979), and Triassic eggs from one locality in Argentina (Bonaparte & Vincent, 1979). Jurassic eggshells have been described from an egg clutch in South Africa (Kitching, 1979; Grine & Kitching, 1987), from an egg and eggshells in Portugal (Lapparent & Zbyszewski, 1957; Dantas, 1991), and from localities in Colorado and Utah (Hirsch, Young, & Armstrong, 1987; Hirsch, et al., 1989; Scheetz, 1991; Young, 1991; Kirkland, Chapter 9).

The eggs and eggshells in Colorado and Utah have been found in six localities at different stratigraphic levels in the Upper Jurassic Morrison Formation (Fig. 10.1). These are not as abundant as those from the Upper Cretaceous of the Western Interior. The Morrison specimens and localities are listed below.

Material and methods
Material

All eggshell is catalogued at the institution where located, and sections of the same material are in the Hirsch eggshell catalogue (HEC). The HEC collection will be deposited at the University of Colorado Museum.

There are two major types of Jurassic eggshells – one type with fragments about 0.8–1.0 mm in shell thickness and thin eggshells less than 0.5 mm in thickness. The thin fragments are small, varying from 5.0 × 4.0 mm to about 3.0 × 3.0 mm. The specimens from the Callison Locality were heavily coated with preservative which could not be completely removed and which interfered with some of the photography and examination.

Methods

The eggshell from each locality was ultrasonically cleaned, and, if feasible, pieces were sorted by their general morphology into type groups under the light microscope (LM). When possible, a series of specimens of each group type were studied using polarizing light microscopy (PLM) and scanning electron microscopy (SEM). Some thin sections were also examined under cathodoluminescence (CL) to detect recrystallization and diagenetic alterations of the shell structure. Complete eggs were checked by computerized axial tomography (CAT) for embryonic remains. The techniques used have been described in Hirsch & Packard (1987), Pooley (1979), Hirsch (1979, 1983), and Hirsch and Quinn (1990).

Eggshell fragments (N = 112) were studied in radial and tangential thin sections; 107 specimens were studied under the SEM, and 8 specimens with CL. In fourteen specimens, elements were analyzed with a microprobe.

Localities

Although the Morrison Formation in Colorado and Utah, is well known for its dinosaur fauna, in the past little attention had been paid to eggshells, perhaps due to lack of recognition. Recent discoveries raise the hope of more extensive finds.

1. The Cleveland–Lloyd Dinosaur Quarry, a dinosaur bone bed, yielded only one unexpected and interesting dinosaur egg (Hirsch et al. 1989; Madsen, 1991). The quarry is located in the Brushy Basin Member of the Morrison Formation in Emery County, Utah. The site was definitely not a nesting area, but a dinosaur cemetery (Stokes, 1985). The egg may have been retained in the body of the dying mother (Hirsch et al., 1989). The preservation of the egg, the pathological eggshell, and an embryo-like image within the egg support this assumption.

2. The Garden Park Area, near Canyon City, Colorado, is a classic locality for dinosaur bones. Several years ago Kenneth Carpenter (DMNH) found, when surveying for gastropods, a single, well-preserved eggshell on the surface of grayish mudstone in the Brushy Basin Member of the Morrison Formation. The locality is about 300 m north of the Garden Park monument. This shell fragment could have been close to the place of its origin, or a strong wind could have transported it to the place where it was found. Sporadic prospecting in the area produced only one thin eggshell fragment of a different morphotype. However, a new site south of the monument with abundant eggshell similar in structure to the first eggshell type, as well as a few thin eggshells, has been discovered recently.

3. The Callison Locality is located in the grayish mudstone of the lower part of the Brushy Basin Member within the Fruita Paleontological Area near Fruita, Colorado. This quarry is well known for its fossils of small vertebrates (Callison, 1987). Eggshells are rare in the fluviatile sediment. All are thin and weathered or altered by diagenesis. However, two variants of one structural morphotype were identified.

4. The Uravan Locality in Montrose County, Colorado, is situated in the lower part of the Brushy Basin Member of the Morrison Formation. Galton and Jensen (1973) described *Dryosaurus altus* juvenile remains from this site. In 1989 it yielded embryonic bones of *Dryosaurus*, eggshells of three different mor-

photypes, and other associated fauna (Scheetz, 1991).

5. The Kirkland Locality, 3.6 m below the top of the Salt Wash Member of the Morrison Formation, is located within the Fruita Paleontological Area near Fruita, Colorado. Besides hatchling and juvenile dinosaur bones, fourteen eggshell fragments of three different morphotypes have been found at this site.

6. The Young Locality in Delta County was discovered in 1986 by geologist Bob Young. This locality, situated in the upper Salt Wash Member, yielded innumerable eggshell fragments concentrated in a 4-m² area. Hirsch et al. (1987) published a preliminary description of the locality based on surface material collected at the site. The eggshell is of one morphotype, and the occurrence of the associated fauna is difficult to interpret. The locality probably represents a large and disturbed assemblage of more or less disintegrated eggs from a repeatedly used nesting site.

During preparation of the plaster casts collected here, a few faint indications of entire or partial eggs could be observed. However, only one very compressed, poorly preserved egg (Fig. 10.3A) could be saved. Other concentrations of eggshells (Fig. 10.3B) suggest that more whole or partial eggs did exist, but were lost during preparation of the friable mudstone. Because of this loss, the horizontal and vertical relationship of the eggs to each other could not be established. However, the vertical distribution of eggshell concentrations suggests multiple laying events. It is possible that this nesting site was situated at the edge of a river, which was weakened by erosion. The sediments of the site slumped and compacted, causing a disturbance of the original egg and eggshell distribution and a compaction of the still-preserved eggs.

Young (1991) has described the locality and the associated fauna recovered through screen-washing. He presented a scenario interpreting the features of the locality, types and distribution of the eggshell and microfauna. However, the examined eggshell material from this locality consists of a single structural morphotype (described below) contrary to Young (1991). Much of the eggshell from the Young Locality is of varying thickness, indicating that the eggshells were exposed to

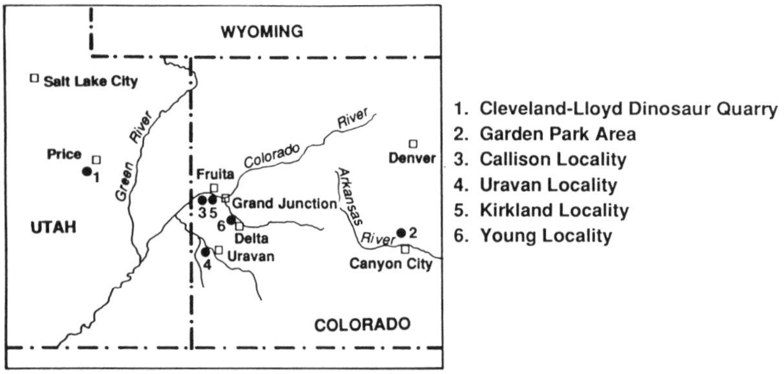

Figure 10.1. Morrison localities in Colorado and Utah.

1. Cleveland-Lloyd Dinosaur Quarry
2. Garden Park Area
3. Callison Locality
4. Uravan Locality
5. Kirkland Locality
6. Young Locality

weathering, dissolution, and alteration of the outer and/ or inner surfaces. This may have occurred before their final burial. Flood waters produced by heavy rains might be responsible for some of the associated fauna found within the nesting area.

The presence of one partially preserved egg and eggshell fragments representing disintegrated eggs suggests that the hatchlings left the nest soon after hatching. If the hatchlings had been raised in the nest, they would have trampled the eggs and egg fragments (Kurzanov & Mikhailov, 1989; Horner, 1987). As yet, however, no identifiable embryonic or hatchling remains have been found; thus, we cannot identify with any confidence the animal that laid the egg.

The spongy calcitic masses associated with some of the eggshell do not represent calcium reservoirs for embryonic bone as previously suggested by Young (1991). Embryos withdraw the needed calcium from the eggshell layer itself (Packard et al., 1984; Packard, et al., 1985; Simkiss, 1967), rather than from a calcium reservoir.

Classification of dinosaur eggshell
Brief history
The development of a system for classifying eggs and eggshell structure has gone through many, and often confusing, stages. However, during the past few years a system of structural and systematic classification has taken shape. Zhao and his colleagues (Zhao 1975; Zhao & Ding, 1976; Zhao 1979; Zhao & Li, 1988) have established a parataxonomic classification for eggs and eggshell. While useful, it is not enough by itself. Mikhailov (Mikhailov 1987a, 1987b; Kurzanov & Mikhailov, 1989; Mikhailov, 1992) has improved this system. He has treated the eggshells as a biocrystalline structure, as did Erben (1970). Mikhailov has established basic types of eggshell organization (geckonoid, testudoid, crocodiloid, ornithoid, dinosauroid) and eggshell morphotypes (dendrospherulitic, etc.).

Hirsch and Quinn (1990) described eggshells from the Upper Cretaceous of Montana, and slightly modified Mikhailov's (1992) morphotypes. These researchers did not, however, establish egg genera and species, nor did they assign the eggshells to existing egg families or to new families; they felt that this was premature.

Mikhailov (1991) has recently correlated structural morphotypes with the parataxonomic classification of Zhao. These two systems, combined in Figure 10.2, clarify the structural and systematic classification of eggs and eggshells. This is the system used in this chapter.

Examination of eggshells
When examining eggshells, it is necessary to distinguish between the general morphology (egg size and shape, shell thickness, sculpture of outer shell surface, and pore pattern) and the histostructure of the eggshell (Mikhailov, 1991). The histostructure, best studied in radial sections, is divided into (1) general histostructure,

which includes the microfeatures of the shell units (organic core, eisospherite, mammillae, wedges and prisms or continuous layer) and type of pore system, and (2) texture of eggshell, that is, the sequence and composition of horizontal ultrastructural zones.

Eggshell organization, morphotypes, and pore systems
Mikhailov (1991) has identified the basic eggshell types and has established, described, and illustrated the structural morphotypes. As an overview, a short description is given below of the basic types, morphotypes, and pore systems found in Jurassic eggshells. In addition, a new pore system is described in detail.

Dinosauroid prismatic basic type
The eggshell is composed of two distinct layers: (1) a continuous layer of interlocking shell units composed of prisms (zone of prisms) and (2) a thinner mammillary layer (zone of wedges). The wedges of the mammillae are radiating and slender. They change upward into a layer of vertical prisms. The extinction pattern seen in polarized light microscopy is columnar and sharp. The basic eggshell type is divided into two morphotypes based on the pore system (see below).

Ornithoid basic type with ratite morphotype
Two distinct shell layers are present and are separated by an abrupt change in structure. The eggshell structure is distinct only in the mammillary layer that comprises the inner 1/10 to 3/5 of the shell thickness. The remainder of the shell thickness is of the continuous layer.

Dinosauroid spherulitic basic type
Only one morphotype, tubospherulitic, has been found in Jurassic eggshells. The shell units are sharply separated from each other; a fanlike pattern can be traced from the nucleation point to the surface of the eggshell units. The shells have arched accretion lines and exhibit tubercle elevations on the shell surface.

Pore system
Of the pore systems described by Sochava (1969), Erben, Hoefs, & Wedepohl (1979), Hirsch and Quinn (1990), and Mikhailov (1991), tubocanaliculate and angusticanaliculate types have been found previously in the Colorado Jurassic eggshells. In addition, a new pore system, obliquicanaliculate, has been identified.

Obliquicanaliculate pore system
The pore canal has a large diameter of up to nearly one-fourth of the shell thickness. The diameter of the canal is constant throughout its length; however, some canals are wider and irregular in the middle of the shell layer (Figs. 10.3E & 10.7E). The canal extends obliquely 50°–65° relative to the shell surface from in-

terstices between the mammillae, and cuts through several shell units (Figs. 10.3E,F, 10.4B; & 10.7E).

The dinosauroid prismatic basic type and angusticanaliculate pore system has been described in protoceratopsid eggs (Mikhailov 1991; Sabath, 1991) and in hypsilophodontid eggs (Hirsch & Quinn, 1990). No family name for this type of structure has been established. For this reason, a new name is proposed below for the dinosauroid prismatic structure. It can be divided into two morphotypes with angusticanaliculate or obliquicanaliculate pore systems.

Description of eggs and eggshells

Eggshells from the Jurassic Morrison Formation have been known since 1987. All sites are still actively being worked and studied. The material from the Young locality, the pathologic egg from the Cleveland–Lloyd Quarry, and the shells from the newly discovered Garden Park site allow a detailed description and a systematic classification of a new eggshell type.

Prismatoolithidae fam. nov.
Diagnosis

Dinosauroid prismatic basic type; prismatic morphotype agusticanaliculate or obliquicanaliculate pore system; outer surface smooth to undulating; elongate ellipsoid or elongate ovoid eggs; medium shell thickness (0.8 to 1.0 mm).

Prismatoolithus gen. nov.
Diagnosis

Obliquiprismatic morphotype, obliquicanaliculate pore system. Outer shell surface smooth, probably with faint striations. Shell thickness 0.7–1.0 mm. Restored egg approximately 110 × 60 mm in size, of elliptical shape, one end slightly more pointed.

Prismatoolithus coloradensis sp. nov.
(Figs. 10.3A–G, 10.4A–H, & 10.7D–F)

Diagnosis
As for genus.

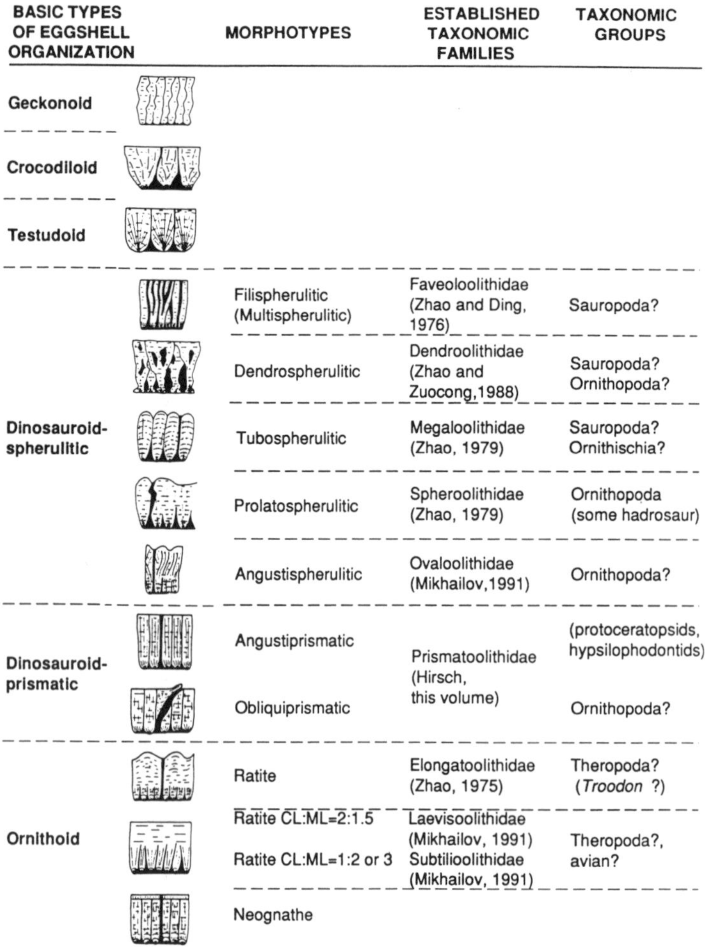

BASIC TYPES OF EGGSHELL ORGANIZATION		MORPHOTYPES	ESTABLISHED TAXONOMIC FAMILIES	TAXONOMIC GROUPS
Geckonoid				
Crocodiloid				
Testudoid				
Dinosauroid-spherulitic		Filispherulitic (Multispherulitic)	Faveoloolithidae (Zhao and Ding, 1976)	Sauropoda?
		Dendrospherulitic	Dendroolithidae (Zhao and Zuocong,1988)	Sauropoda? Ornithopoda?
		Tubospherulitic	Megaloolithidae (Zhao, 1979)	Sauropoda? Ornithischia?
		Prolatospherulitic	Spheroolithidae (Zhao, 1979)	Ornithopoda (some hadrosaur)
		Angustispherulitic	Ovaloolithidae (Mikhailov,1991)	Ornithopoda?
Dinosauroid-prismatic		Angustiprismatic	Prismatoolithidae (Hirsch, this volume)	(protoceratopsids, hypsilophodontids)
		Obliquiprismatic		Ornithopoda?
Ornithoid		Ratite	Elongatoolithidae (Zhao, 1975)	Theropoda (*Troodon* ?)
		Ratite CL:ML=2:1.5	Laevisoolithidae (Mikhailov, 1991)	Theropoda?, avian?
		Ratite CL:ML=1:2 or 3	Subtilioolithidae (Mikhailov, 1991)	
		Neognathe		

Figure 10.2. Schematic views of basic groups with their structural morphotype and corresponding taxonomic egg families.

Figure 10.3. A-G, *Prismatoolithus coloradensis* eggshells from the Young locality. **A.** *Prismatoolithus coloradensis* gen. et sp. nov, Holotype (MWC 122.3.I/HEC 457). Compressed egg, approximately 70% complete. **B.** Mudstone matrix from nest (MWC 122.2.3/HEC 418). Eggshell concentration and arcs of shell fragments (arrows) suggest disintegrated egg or large egg fragment. Bar = 1 cm. C–E, radial thin sections, PLM; outside of eggshell is at top. Bars = 100 μ. C-D, MWC 122.2/HEC 462-1. Note shell units above mammillae (small arrows), horizontal layering, columnar extinction pattern, and pore opening (large arrow). **C.** Normal light. **D.** polarized light. **E.** MWC 122.2/HEC 532-2, normal light. Note oblique pore canal originating between two mammilla (arrows) and cutting through shell units. Horizontal growth lines rise or dip near edge of pore canal. **F.** MWC 122.2/HEC 418-3, SEM view of freshly fractured radial view. Two pore canals (arrows), only partially visible, rise oblique to shell surface; their preservation suggests an originally fibrous wall. Note columnar structures in shell units above mammillae (small arrows). Bar = 100 μ. **G.** MWC 122.2/HEC 462-4, outer surface of eggshell. Note striations and pore opening. Bar = 100 μ.

Holotype

One compressed egg MWC 122.3.1/HEC 457 (Fig. 10.3A).

Type locality

Young Locality, Delta County, Colorado; locality MWC 73.00.86.

Age and formation

Upper Jurassic, Morrison Formation, upper part of Salt Wash Member.

Material

Four plaster jackets with eggs and eggshells (MWC 122.2/HEC 418, 489, 532, 462; MWC 122.3/ HEC 457; MWC 122.6; MWC 122.8) and thousands of eggshell fragments (MWC 122.1 /HEC 418). The largest cast yielded, among other eggshell concentrations, the one illustrated in Figure 10.3B (MWC 122.2.3/HEC 418). The small cast (MWC 122.3) yielded the egg MWC 122.3.1/HEC 457.

Preservation of material

Preservation of the eggshell material varies from a few well preserved shells to many eroded ones. A few diagenetically or pathologically altered eggshells were also found.

Description

About 70 percent of the badly crushed egg (Fig. 10.3A) is preserved; the limits of the egg were confirmed by tomography. The slightly eroded outer surface is smooth, although some shell fragments show faint striations (Fig. 10.3G). The single pore openings are irregularly spaced; occasionally they are arranged in groups (Fig. 10.4E). Radial thin sections show shell units with prismatic structure and a sharp columnar extinction pattern (Figs. 10.3C,D & 10.4C,D). The mammillae are composed of fine radiating crystals (Figs. 10.3C–E; & 10.4C,D).

The horizontal growth lines at the margin of pore canal walls rise or drop, indicating that the pore canals are part of the original growth structure of shell (Figs. 10.3E; 10.7E). The pore canals are almost the same diameter throughout their length, except occasionally near the middle. On the outer shell surface, the area surrounding the pore openings is depressed on one side of the opening and elevated on the other (Fig. 10.4E). The pore openings are arranged on the outer surface of the shell in an irregular pattern; sometimes three or more pore openings are clustered. Hirsch et al. (1987), in their report of Jurassic eggshells from Colorado, misinterpreted these pore canals as bore tubes of worms.

The relatively large pore canals and their preservation suggest that the canals may have had a fibrous wall (Fig. 10.3F). The elevation of the outer shell surface at the pore opening is mostly on the side where the

angle between axis of pore and shell surface is acute. The depression is on the side where the angle is obtuse.

P. coloradensis shell from other localities

A single egg (11584/BYU VP13019/HEC 464, 488) is known from the Cleveland–Lloyd Dinosaur Quarry. When found, some of the pathologic shell had separated from the primary shell, but lay immediately next to it in an undisturbed condition. The primary and the pathological shell are well preserved and complement the structural features of the egg from the Young Locality. The outer surface of the pathologic shell layer shows pore openings without depressions or elevations at the openings. It is smooth without any striations (Fig. 10.4H). Although the pathologic shell layer is prismatic in structure, the mammillary layer is dwarfed and is missing the typical mammilla structure (Fig. 10.4B-D). Pathologic eggshells are not always structurally similar to the primary shells (Hirsch, in preparation), and may also differ in shell thickness.

The inner surface (Fig.4F) of the pathological eggshell exhibits a reversed image of the outer surface sculpture of the primary layer (Fig 10.4E). The striations on the inner surface of the pathologic layer (Fig. 10.4G) are similar to the ones found on the primary shell (Fig. 10.4E). However, in places where the pathologic shell had probably been missing for a longer time, these striations were faint or absent due to erosion.

As expected, the eggshell shows no continuous pore canals through both layers. However, some of the pores of the primary shell penetrate a short distance into the pathological layer (Fig. 10.4G). At these places the depression and elevations at the pore opening on the outer surface of the primary shell are present in reverse on the inner surface of the pathological shell layer (Fig. 10.4F).

The Kirkland Locality (LACM uncatalogued/ HEC 629) has produced one *Prismatoolithus* eggshell. The Uravan Locality has produced one specimen of four eggshells stacked on top of each other (MWC 732.002.OO1/HEC 575). The first eggshell found in Garden Park (UCM 54471/HEC 268) and additional eggshells recently found in this area (DMNH 2734/HEC 647, 648), except for a few thin shell fragments, also belong to *P. coloradensis*. *Prismatoolithus*-like eggshells from sites in Portugal are being studied by colleagues in Madrid and Lisbon.

Description of thin eggshell

Thin eggshells are also known from the Morrison Formation. Such fragments from the Young Locality, however, are badly weathered or partially dissolved eggshells of *P. coloradensis*. Thin eggshells from the other localities are still scarce; about 110 fragments have been found, most of which are very small and poorly preserved. Thus, it is not possible to approximate the size and shape of an egg using the traditional technique of

Figure 10.4. A -H, *P. coloradensis* eggshell from Cleveland–Lloyd Dinosaur Quarry (BYU VP13019/HEC 464, 488).
A. Pathological double-shelled egg, broken open with two halves connected by eggshell. Bar = 1 cm. **B**. HEC 488-3,
SEM radial view. Note normal layer (NL) and pathologic layer (PL), pore canal (arrows) ends at margin of the
pathologic layer. Bar = 100 μ. C, D, HEC 488-1, PLM radial thin section. Note large mammillae in normal layer and
dwarfed mammillae in pathologic layer (arrows), horizontal growth lines, indications of shell units. Bar = 100 μ. **C**.
Normal light. **D**. Polarized light. Note columnar extinction pattern in both shell layers. **E**. HEC 464, SEM of outer
shell surface of normal eggshell. Note three pore openings with elevation of shell surface on one side of pore and
depressions on opposite side, and faint striations. Bar = 100 μ **F**. HEC 488-6, SEM view of inner shell surface of
pathological shell. Three pore openings indicate that the pore canal of normal shell penetrates partially into pathologic
shell. Note reversed depressions as elevations of normal eggshell; striations absent. Bar = 100 μ. G, H, HEC 488,
SEM view of pathologic eggshell layer. Note that the partial pore canal (arrow) is a continuation of pore canal of
normal shell layer. Bar = 100 μ. **G**. Inner shell surface (IS) and edge of shell. Note second pore opening and striations
that indicate that both shells were in close contact in this area. **H**. Outer shell surface (OS) and edge of shell.

Figure 10.5. Thin and very thin eggshell of ?prismatic and spherulitic morphotype. **A.** UCM 54 322/HEC 214, Radial view of thin shell fragment. Bars = 100 μ. A-B, PLM view. Note fine radiating structure in lower two-thirds and distinct horizontal growth lines in upper third of shell layer. A, normal light. **B.** polarized light. Extinction pattern sweeping in lower two-thirds and blocky, columnar in upper third. **C.** SEM view. Note radiating structure. **D.** UCM OS 1128/HEC 307, radial view of modern turtle eggshell (*Gopherus flavomarginatus*), polarized light. Note fine radiating structure through all of shell, arched growth lines in lower part, gradually changing into horizontal growth lines, typical for testudoid eggshell. Bar = 100 μ. E–G, MWC 731.002/HEC 596-1, radial view of two stacked, very thin eggshells, PLM. Similair ?prismatic morphotype structure as in A, B, shell units not as pronounced due to some diagenetic alterations (above arrow). Bar = 100 μ. **E.** Normal light. F, G, polarized light. **G.** Thin section rotated to show change in extinction. **H–J.** LACM uncatalogued, HEC 630-1, radial view of thin shell fragment of spherulitc morphotype, PLM. Above mammillae (arrows), note distinct shell units with radiating, fanlike structure ending in a node on shell surface. Bar = 100 μ. H, I normal light. J, polarized light, sweeping extinction pattern.

radial measurement (Sauer, 1968; Williams, 1981). The small size also precludes studying parts of the same fragment under the SEM and in thin sections to obtain images that complement one another. The preservation of the eggshells also makes it difficult to assign them definitely to a specific structural morphotype. Final description and classification must wait until more eggshell material is found, allowing the compositional study of the ultrastructural zones. For these reasons only the general histostructure of the eggshell will be described and illustrated here.

?Prismatic morphotype

Most of the thin eggshells belong to the ?prismatic morphotype. Ninety fragments of thin eggshell can be divided into two general groups by shell thickness and into several variants based on the ratio of mammillary layer to continuous layer: thin eggshell (0.30–

Figure 10.6. **A–C.** MWC 731.002/HEC 596-3, radial view of crocodilianlike eggshell, PLM. Bar = 100 μ. A, normal light. Due to some diagenetic alteration, structural features are faint, indications of large shell units composed of wedges radiating from basal plate groups (arrows), horizontal growth lines, secondary deposit (small arrows). B, C, polarized light. Note distinct, typical extinction pattern of crude wedges. **D–E.** MWC 122.2/HEC 532-4, two radial views of pathologic shell fragment, normal light. Probable double-layered eggshell with both shell layers very irregular. Bar = 100 μ. **F.** From upper left to lower right: LACM 115728, 115731a,b, 120501/HEC 210, 211a,b, 212, ?pseudo eggs. Scale in millimeters. **G.** Radial thin sections of ?pseudo eggs shown in F. Note lack of eggshelllike structures. Bar = 100 μ.

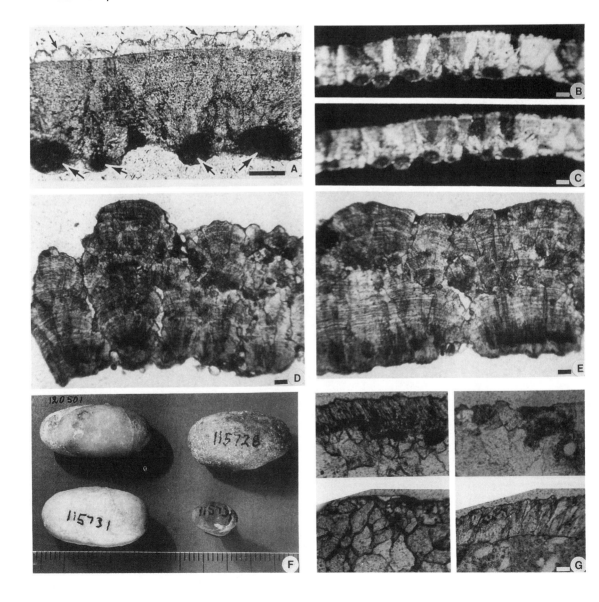

Figure 10.7 A–H, diagenetic alterations of eggshell structure observed in radial thin sections. Bars = 100 μ. A–C, UCM 54321/HEC 339-2; three views of a 3 mm long shell fragment from the Callison Locality; outer shell surface contaminated with secondary deposits, arrows point to outer border of eggshell. **A.** Little diagenetic alteration. Note fine radiating structure in lower two-thirds and horizontal layering in outer third of shell. **B.** Shell with pore canal and alteration in outer part of shell. **C.** Heavily altered shell. A, B, normal light microscopy; C, polarized light microscopy. **D.** LACM uncatalogued/HEC 629, *P. coloradensis* eggshell from Kirkland Locality, normal light

0.46 mm; ratio 1:1 to 2:1) and very thin eggshell (0.20–0.24 mm; ratio about 1:1). The mammillae in both groups are composed of fine radiating crystals (Fig. 10.5A–C,E). The continuous layer shows pronounced horizontal growth lines (Fig. 10.5A,B,E) and in some cases indications of shell units or columns. Under crossed nicols, the extinction pattern is sweeping in the mammillae and blocky or irregular in the continuous layer (Fig. 10.5B,F,G). An abrupt change of structure between the two layers, as is found in the ratite morphotype from the Cretaceous of Montana (Hirsch & Quinn, 1990), could not be observed.

The majority of the eggshell fragments of this morphotype are diagenetically altered to varying degrees (Fig. 10.7A–C). Thus, it is difficult to observe and interpret the original structure of the shell layer. The pore canals are narrow, more or less straight, and of same diameter (angusticanaliculate).

Most of the thin eggshells (0.30-0.46 mm) are from the Callison Locality (UCM 54322/HEC 213, 214; UCM 54321/HEC 259, 339) (Fig. 10.5A–C) and may represent two different variants of the ?prismatic morphotype. Hirsch et al. (1987) wrongly interpreted these fragments as avianlike (prismatic) and turtlelike (Fig. 10.5D).

All but one of the very thin eggshell fragments from the Uravan Locality (MWC 731.002/HEC 576, 596) (Fig. 10.5E–G) represent another variant of the ?prismatic morphotype based on shell thickness and the 1:1 ratio between the layers.

Indeterminate morphotype (prismatic or ratite)

Eleven thin shell fragments (0.26–0.38 mm) with a ratio of 1:4 between mammillary and continuous layer, are known from the Kirkland Locality (LACM uncatalogued/HEC 628). Four shell fragments from the Uravan Locality (MWC 731.002/HEC 576, 577) have about the same shell thickness and ratio. These fragments show very faint horizontal striations and a structural change between the mammillary and continuous eggshell layers. However, additional specimens are needed before more detailed descriptions can be given.

Dinosauroid spherulitic basic type

Two tiny fragments from the Kirkland Locality (Fig.5H–J) (LACM uncatalogued/HEC 630) are of the dinosauroid spherulitic basic type, possibly the tubospherulitic morphotype. The fragments are about 0.34 mm thick. The single shell layer is composed of distinct, separated shell units with a radiating (fanlike) crystal pattern (Fig. 10.5H). These shell units end in nodes on the outer surface. The pore canals rise between the shell units.

?Crocodiloid basic type

Three shell fragments from the Uravan Locality (MWC 731.002/HEC 596-3) are about 0.31 mm thick. They show large shell units with indications of wedge-like structures similar to those in crocodilian eggs (Fig. 10.6A–C). In addition, the extinction pattern is also similar to that seen in crocodilian eggs. However, before a final identification is made, the specimens must be studied in more detail.

Questionable eggshells

A few very thin fragments of even thickness and with eggshell-like curvature are problematic. If they are eggshells, diagenesis has so altered them that no structural eggshell features are preserved. Four specimens (LACM 115728, 11573a,b, 120501 /HEC 210, 211 a,b, 212) (Fig. 10.6D) ranging in size from 23 × 11 mm to 10 × 7 mm are known from the Callison Locality. They do not show any structural features on their outer surface that resemble known eggshell structure (Fig. 10.6E).

One specimen from the Uravan locality (MWC 731.002/HEC 586) is 40 × 30 mm in size and is egg-shaped. It is partially covered with a thin shelllike outer

Caption to Fig. 10.7 (*cont.*)

microscopy. Note distinct horizontal growth lines and abrupt change to large areas replaced by silica. **E.** UCM 54471/HEC 268, *P. coloradensis* eggshell with pore canal from Garden Park location. Large areas within shell irregularly replaced by silica (arrows). **F.** UUVP 11584/BYU VPI 3019/HEC 464-3, multilayered *P. coloradensis* eggshell from Cleveland–Lloyd Dinosaur Quarry, normal light microscopy. Note horizontal growth lines and fan-shaped area of herringbone pattern in the primary shell indicating recrystallization in only this specific area. **G–H,** MWC 122.2/HEC 462-5, radial views of eggshell from Young Locality, **G.** Polarized light microscopy. Eggshell units, with growth lines in lower two-thirds, terminate in nodes, irregular structure in upper third of shell layer. **H.** Cathodoluminescence analysis shows diagenetic changes (arrows) in mammillae, outer shell part, and horizontal growth lines through bright colored area. Luminescence of growth lines indicates replaced network of organic matter; mammillae, rich in organic matter and of different crystalline structure, are very often diagenetically changed. Luminescence of outer layer indicates probable dissolution of normal smooth outer shell surface and build-up of nodes through secondary deposits.

layer. Some eggshelllike fragments are found in the outer part of the matrix filling the specimen. The outer layer and the fragments resemble the shell layer found in modern sea turtle eggs, which have a very pliable and poorly organized calcareous layer of aragonitic, spherulitic shell units. Although elemental analysis of the fossil specimen showed the units to be calcium, no radiating structures could be found in thin sections nor under the SEM.

Pathologic eggshell

Pathological eggshell is relatively rare in the Western Interior compared to other localities in the world. It is more abundant in the tubospherulitic morphotype of the dinosauroid spherulitic basic type found in France, India, and Argentina. Two pathologic eggshells were found at the Young Locality (MWC 122.2/ HEC 418-R2, 532-4). One specimen (Fig. 10.6D,E) shows a peculiar arrangement of growth units indicating a very irregularly structured pathological shell layer deposited on a primary layer, which is also somewhat irregular. The other specimen has several extra growth units nucleating within the middle of the single shell layer.

Diagenetically altered eggshell

Although recrystallization is common in eggshells throughout the world, completely altered eggshell structures are more rare and seem to be more abundant in some localities than in others.

Among the Jurassic eggshells are several very unusual diagenetic alterations. One eggshell fragment 3 mm long and 0.20 mm thick demonstrates different stages of alteration (Fig. 10.7A–C). The shell varies from small changes in structure to partial replacement and dissolution of the shell. Other fragments (Fig. 10.7D,E) show complete replacement by silica in the middle of the shell, whereas the outer and/or inner parts still retain a normal shell structure.

In the pathologic eggshell from Utah (Fig. 10.7F), a fan-shaped, herringbone pattern in the primary shell can be observed under a pore of the pathological layer. The herringbone pattern must be due to recrystallization because the eggshell on both sides of the fan is normal in structure. This demonstrates that the herringbone pattern is predominantly a diagenetic feature, and is not reliable for identification of fossil eggshell structure as Erben (1970) had suggested. In another piece of eggshell, cathodoluminescence showed secondary calcite replacement of what were probably dissolved sections of eggshell. The replacement even formed nodes on the outer surface of the shell (Fig. 10.7G,H).

Conclusion

Until recently, Jurassic eggshell localities were almost nonexistent, with just one report of half an egg from the Jurassic of Portugal (Lapparent & Zbyszewski, 1957). In recent years, however, eggs and eggshells have been discovered at six different locations in the Morrison Formation of the Western Interior, thus opening a challenging new area of study. All but one of these sites have produced only a small amount of eggshells to date. An egg and an immense amount of eggshell have been found at the Young Locality, and a pathological egg has been found at the Cleveland–Lloyd Quarry. These specimens have allowed us to divide the dinosauroid prismatic basic type into two morphotypes, to describe a new pore system, and to establish a parataxonomic family, as well as a new genus and species of egg.

The current challenge is presented by the thin (0.25–0.5 mm) and very thin (<0.25 mm) eggshells found almost exclusively at three of the sites (Uravan, Kirkland, and Callison localities). The study of these shells is difficult due to the small size of the shell fragments, the unusual diagenetic alterations of structure, and poor shell preservation. Most of the thin and very thin eggshells probably belong to variants of the prismatic structural morphotype. Three small fragments seem to be of a crocodiloid basic type, and two other very small fragments seem to be of a dinosauroid spherulitic basic type. Eleven of the fragments are still indeterminate as to type.

Some of the thin eggshell may not be dinosaur, but could be sphenodontid. The modern *Sphenodon* (tuatara) lays pliable eggs with a somewhat loose calcitic structure interwoven by fibers of the shell membrane (Packard, Hirsch, & Meyer-Rochow, 1982) and with a very poor chance for fossilization. However, it is possible that the Jurassic sphenodontids showed as much variation in egg structure as do modern turtles, having both pliable and rigid shells. At present no final conclusion can be reached on the taxonomy of the thin and very thin shell material. More study and especially more material are needed.

The structural and systematic classification of dinosaur eggshell has been established mostly on Upper Cretaceous and a few Lower Cretaceous specimens. It is, therefore, not surprising that slightly different or altogether new morphotypes have been found within Jurassic eggshells.

The Uravan, the Kirkland, and the Garden Park localities, which are still being worked intensively, may prove to be nesting sites. The Young Locality, with an immense amount of eggshell material, shows there are, as would be expected, nesting sites in Jurassic sediments.

Acknowledgments

Special thanks are due to all the people and institutions who furnished the specimens I used for this study. I am indebted to Judith Harris, Hans-Peter Schultze, Kenneth Carpenter, Richard Stucky, and Arthur Binkley for reading and discussing the manuscript. My thanks are also extended to John

Drexler and Bob McGrew for elemental analyses and assistance with the SEM and CL, and to Ed Hendrick for scanning the holotype egg. The study was supported in part by the University of Colorado in Boulder, the Denver Museum of Natural History, and by a generous donation from the Greater Denver Area Gem and Mineral Council.

References

Bonaparte, J. F., & M. Vincent. 1979. El hallazgo del primer nido de dinosaurios triásicos (Saurischia Prosauropoda), Triásicos superior de Patagonia, Argentina. *Ameghiana* 16: 173–82.

Callison, G. 1987. Fruita: A place for wee fossils. *In* W. Averett, (ed.), *Paleontology and Geology of the Dinosaur Triangle Guidebook.* (Grand Junction: Museum of Western Colorado), pp. 91–6.

Dantas, P. 1991. Dinossaurios de Portugal. *Gaia* 2: 17–26.

Erben, H. K. 1970. Ultrastrukturen und Mineralisation rezenter und fossiler Eischalen bei Voegeln und Reptilien. *Biomineralisation* 1: 1-66.

Erben, H., Hoefs, J., and Wedepohl, K. H. 1979. Paleobiological and isotopic studies of eggshells from a declining dinosaur species. *Paleobiology* 5: 380–414.

Galton, P. M., & Jensen, J.A. 1973. Small bones of the hypsilophodontid dinosaur *Dryosaurus altus* from the Upper Jurassic of Colorado. *Great Basin Naturalist* 33: 129–32.

Grine, F. E. & Kitching, J. W. 1987. Scanning electron microscopy of early dinosaur egg shell structure: a comparison with other rigid sauropsid eggs. *Scanning Microscopy* 1: 615–30.

Hirsch, K. F. 1979. The oldest vertebrate egg? *Journal of Paleontology* 53:1068–84.

1983. Contemparary and fossil chelonian eggshells. *Copeia* 1983: 382–97.

1989. Interpretations of Cretaceous and pre-Cretaceous eggs and shell fragments. *In* D. Gilette, & M. Lockley, (eds.), *Dinosaur Tracks and Traces.* (New York: Cambridge University Press), pp. 89–97.

Hirsch, K. F., and Packard, M. J. 1987. Review of fossil eggs and their shell structure. *Scanning Microscopy* 1: 383–400.

Hirsch, K. F., & Quinn, B. 1990. Eggs and eggshell fragments from the Upper Cretaceous Two Medicine Formation of Montana. *Journal of Vertebrate Paleontology* 10: 491–511.

Hirsch, K. F., Stadtman, K. L., Miller, W., & Madsen, J. 1989. Upper Jurassic dinosaur egg from Utah. *Science* 243: 1711–13.

Hirsch, K. F., Young, R.G., & Armstrong, H.J. 1987. Eggshell fragments from the Jurassic Morrison Formation of Colorado. *In* W. Averett, (ed.), *Paleontology and Geology of the Dinosaur Triangle, Guidebook.* (Grand Junction: Museum of Western Colorado), pp 79–84.

Horner, J. R. 1987. Ecologic and behavioral implications derived from a dinosaur nesting site. *In* S. Czerkas, & E. Olsen, (eds.), *Dinosaurs Past and Present*, Vol 2, (Seattle: University of Washington Press), pp. 51–63.

Kitching, J. W. 1979. Preliminary report on a clutch of six dinosaurian eggs from the Upper Triassic Elliot Formation, northern Orange Free State. *Paleontologica Africana* 22: 72–7.

Kurzanov, S. M., & Mikhailov, K. E. 1989. Dinosaur eggshells from the Lower Cretaceous of Mongolia. *In* D. Gilette, & M. Lockley, (eds.), *Dinosaur Tracks and Traces.* (New York: Cambridge University Press), pp. 109–13.

Lapparent, A. F., & Zbyszewski, G. 1957. Les Dinosauriens du Portugal. *Mémoires du Service géologique de Portugal* 2: 1–63.

Madsen, J.H. 1991. Egg-siting (or) first Jurassic dinosaur egg from Emery County, Utah. *In* W. Averett (ed.), *Dinosaur Quarries and Tracksites, Western Colorado and Eastern Utah, Guidebook.* (Grand Junction: Museum of Western Colorado), pp. 55–6.

Mikhailov, K. E. 1987a. Some aspects of the structure of the shell of the egg. *Paleontological Journal* 3: 54–61.

1987b. The principle structure of the avian eggshell: data of SEM studies. *Acta Zoologica Cracoviensia* 30: 53–70.

1991. Classification of fossil eggshells of amniote vertebrates. *Acta Palaeontologica Polonica* 36 (2): 193–238.

1992. The microstructure of avian and dinosaurian eggshells – phylogenetical implications. *In* K. Campbell (ed), *Papers in Avian Paleontology Honoring Pierce Brodkorb, Contributions in Science* (Los Angeles: Natural History Museum of Los Angeles County), pp. 361–73.

Packard, M. J., Hirsch, K. F., & Meyer-Rochow, V. B. 1982. Structure of the shell from eggs of the tuatara, *Sphenodon punctatus. Journal of Morphology* 174: 197–205.

Packard, M. J., Packard, G. C., Miller, J. D., Jones, M. E., & Gutzke, W. H. 1985. Calcium mobilization, water balance, and growth in embryos of the agamid lizard *Amphibolurus barbatus. Journal of Experimental Zoology* 235: 349–57.

Packard, M. J., Short, T. M., Packard, G. C., & Gorell, T. A. 1984. Sources of calcium for embryonic development in eggs of the snapping turtle *Chelydra serpentina. Journal of Experimental Zoology* 230: 81–7.

Pooley, A. S. 1979. Ultrastructural relationships of minerals and organic matter in avian eggshells. *Scanning Electron Microscopy* 11: 475–82.

Sabath, K. 1991. Upper Cretaceous amniotic eggs from the Gobi Desert. *Acta Palaeontologica Polonica* 36: 151–92.

Sauer, E. G. F. 1968. Calculations of *Struthios* egg sizes from measurements of shell fragments and their correlation with phylogenetic aspects. *Cimbebasia Ser. A*, 1: 27–55.

Scheetz, R. D. 1991. Progress report of juvenile and embryonic *Dryosaurus* remains from the Upper Jurassic Morrison Formation of Colorado. *In* W. Averett (ed.), *Dinosaur Quarries and Tracksites, Western Colorado and Eastern Utah, Guidebook.* (Grand Junction: Museum of Western Colorado), pp. 27–9.

Simkiss, K. 1967. *Calcium in Reproductive Physiology.* (London: Chapman and Hall LD.).

Sochava, A. V. 1969. Dinosaur eggs from the Upper Cretaceous of the Gobi Desert. *Paleontological Journal* 4:
517–27.

Stokes, W. L. 1985. *The Cleveland–Lloyd Dinosaur Quarry
– Window to the past.* (Washington, D.C.: U.S. Gov-
ernment Printing Office).

Williams, D. L. G. 1981. *Genyornis* eggshell (Dromornithidae; Aves) from the Late Pleistocene of South Australia. *Alcheringa* 5: 133–40.

Young, R. G. 1991. A dinosaur nest in the Jurassic Morrison
Formation, Western Colorado. *In* W. Averett (ed.).
*Dinosaur Quarries and Tracksites, Western Colorado
and Eastern Utah, Guidebook.* (Grand Junction: Museum of Western Colorado), pp. 1–15.

Zhao Z. 1975. The microstructure of dinosaurian eggshells of
Nanhsiung, Kwangtung and Guangdong. *Vertebrata
PalAsiatica* 13: 105–17.

1979. The advancement of researches on the dinosaurian
eggs in China. *Mesozoic and Cenozoic Red Beds of
South China.* (Beijing: Science Publishing Co.), pp.
330–40.

Zhao Z. & Ding S. 1976. Discovery of the dinosaurian eggshells from Alxa, Ningxia and its stratigraphic significance. *Vertebrata PalAsiatica* 14: 42–4

Zhao Z. & Li Z. 1988. A new structural type of the dinosaur
eggs from Anlu County, Hubei Province. *Vertebrata
PalAsiatica* 26: 107–15.

11 Review of French dinosaur eggshells: morphology, structure, mineral, and organic composition

MONIQUE VIANEY-LIAUD,
PASCALE MALLAN,
OLIVIER BUSCAIL,
AND CLAUDINE MONTGELARD

Abstract

A review of the Upper Cretaceous dinosaur eggshells from southern France is presented. It does not include egg-laying and nesting data. After a brief historical survey, the egg-shell bearing localities are placed in geographical and stratigraphical context. The morphology and the structure of new material from the Aix-en-Provence Basin and Languedoc localities are described. A parataxonomy (egg genus: nomen + oolithus; egg species: nomen) is used and three new egg genera and six egg species are named. The effects of weathering and diagenesis are clearly distinguished from the diagnostic characters. The parataxonomy provides a useful tool for paleobiological and evolutionary studies of dinosaur eggs. For comparison, the egg morphotypes of Erben, Kérourio, and Penner are discussed. The stratigraphical implications of the egg taxa are also examined. A short review of the geochemical analysis of dinosaur eggshell is given and its paleobiological significance is considered. Finally, after a review of the literature on the organic components of eggshells, we present our method of extraction and identification of this organic matrix.

Introduction

A detailed knowledge of dinosaur eggs should be based upon the study of complete eggs so that their size, shell thickness, morphology, ultrastructure, and mode of nesting can be determined. Furthermore, the only reliable taxonomic identification of these eggs is based on the association of eggs and embryonic bones. Such an ideal situation is exceptional, however. Usually, only eggshell fragments are found, although sometimes crushed or solitary eggs may be found and, on occasion, nests lacking baby dinosaur bones. Unless previously identified with skeletal remains, eggshell fragments are not sufficient to establish dinosaur egg types (Hirsch & Packard, 1987; Hirsch, 1989). In many localities, egg-shell fragments are more common than whole eggs. In France, the main regions that have yielded fragments and sometimes, fortunately, complete eggs and nests, are the Aix-en-Provence Basin and localities at Languedoc and Corbières in southern France.

Dinosaur egg localities in France

Aix-en-Provence Basin

Brief historical survey

The most famous French localities are found in the Aix-en-Provence Basin, which has been prospected for over a century. The initial work was done by Matheron (1869), and later by Van Straelen and Denayer (1923). Dughi and Sirugue regularly worked the basin during the 1950s (see Chapter 5) and recognized at least ten types of eggshell and possibly five others (Dughi & Sirugue, 1976). Unfortunately, they did not describe in detail nor figured these different types. The main part of their collection was under study by P. Kérourio, who has only sporadically published on the eggshells (Kérourio, 1981, 1982, 1987). During the 1970s, Erben started a more complete study of the Aix Basin's egg-shell fragments (Erben, 1970; Erben & Newesely, 1972; Erben, Hoefs, & Wedepohl, 1979). He combined microstructural and chemical analyses with a statistical approach. Another approach, by M.-F. Voss-Foucard (1968), examined the organic matrix of the dinosaur eggshells of the Aix Basin. During the last decade, research on eggs has shifted to the mass extinctions at the K/T boundary (Penner, 1983, 1985; Morin, 1989; Bochérens et al., 1988; Mallan, 1990).

Localities

Because of recent plundering of fossil-bearing sites by commercial fossil collectors not guided by scientific interest, we are giving the various localities only by schematic stratigraphical section (Fig. 11.1). The geographic coordinates for the localities are available to the scientific community in the locality files of our laboratory. Because various authors have sometimes given different names for the same eggshell locality, and other times they have given the same name to different localities, we have attempted to standardize the locality names.

In the Aix region, the Arc Basin is the most fossiliferous (see Durand, 1989). The eggshell-bearing levels range from the Fuvelian to the Begudo-Rognacian. The continental stages Fuvelian + Begudian have been correlated with the Campanian, and the Rognacian with the Maastrichtian (Babinot & Durand, 1980, Babinot and Freytet, 1983). This correlation is supported by paleomagnetic data (Galbrun, 1989; Hansen, Gwodz, & Rasmussen, 1989; Krumsiek & Hahn, 1989; Westphal, 1989).

Most of the sites are in Rognacian strata. The lower Rognacian sediments are mainly fluvial deposits composed of red, silty clays with red sandstone layers or lenses. The Middle Rognacian is represented by the Rognac Limestone that is a "white-grey lacustrine, algal, calcareous mass, with interbedded black shales reflecting swamping and emergence of the open lake landscape" (Sittler, 1989). The Rognac Limestone is overlain by about 150 m of upper Rognacian red, sandy marls and clays, with sandstone or pebble lenses. This is capped by the La Galante Conglomerate.

The K/T boundary was initially placed at some calcareous lenses about 10 m below the La Galante Conglomerate (Dughi & Sirugue, 1957). Erben, et al. (1983), however, placed the K/T boundary lower, within the Rognac Limestone, even though dinosaurs have long been reported from the overlaying strata. More recently, Kérourio (1982) has indicated that he had found eggshells between the calcareous lenses and the La Galante Conglomerate. As a result, the K/T boundary is now placed at the base of the La Galante Conglomerate.

In the Arc Basin, dinosaur eggshell remains extend from the Fuvelian to the Upper Rognacian, about a few meters below the La Galante Conglomerate. The localities considered here are shown on a composite stratigraphic section (Fig. 11.1).

Upper Campanian. Clos-la-Neuve (Fig. 11.1A). North of road D6 toward Trets, we have collected thin eggshell fragments in red, sandy marls intercalated with sandstones. Kérourio (1982) has studied at least two Fuvelian sites from which he described a new ornithoid eggshell type.

Figure 11.1. Stratigraphic sections and fossiliferous localities from the Aix-en-Provence Basin. **A.** Composite section made near "Montagne du Cengle," Les Bréguières, and Clos-la-Neuve. **B.** Section from Roquehautes, south of the "Crête du Marbre." C. Section from Roquehautes-Grand Creux, north of the "Crête du Marbre."

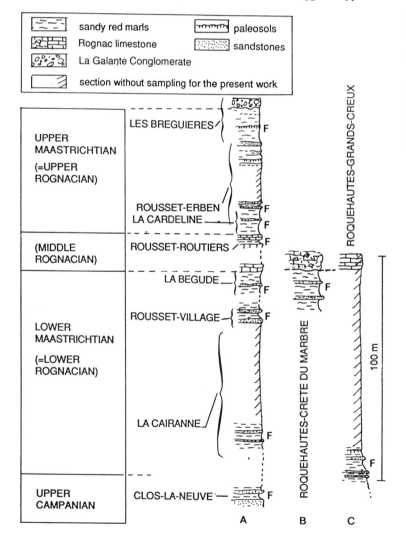

Lower Rognacian. 1. La Cairanne (Fig. 11.1A) (Penner, 1983; Williams, Seymour, & Kérourio, 1984; Morin, 1989; Mallan 1990). This site, 3 km east of Rousset, is situated about 50 m below the Rognac Limestone. The eggshell fragments have been collected from a single level. Some of the fragments were concentrated and probably represent isolated eggs.

2. Roquehautes (Fig. 11.1B) (Erben, 1970; Erben & Newesely, 1972; Erben et al., 1979; Kérourio 1981; Penner, 1983, 1985; Morin, 1989; Williams et al. 1984). This, the most famous site, is located at the foot of Sainte Victoire Mountain. The 100 m of red, silty clays and sandstones have yielded numerous eggshell fragments, eggs, and nests. We have collected eggshell fragments from red, sandy marls south to the Crête du Marbre (Roquehautes). We have also collected fragments farther north at the base of the section (Iatsoura, Cojan, & Renard, 1991) here named Roquehautes-Grand Creux (Fig. 11.1C). A paleomagnetic study of this site has been done by Hansen et al. (1989).

3. Rousset-Village (Morin, 1989; Mallan 1990). The village of Rousset is situated on sandy clays and pink sandstone just below the Rognac Limestone. We have collected some eggshell fragments west of the village.

4. La Bégude. Guided by F. Sirugue, we collected eggshell fragments in the upper part of the lower Rognacian, near La Bégude and south to RN7, under the first Rognac Limestone bar. This locality, named "La Bégude" by Sirugue, is different from the locality "La Bégude" in Iatsoura et al. (1991). The latter is the same one studied by Erben et al. (1979) under the name of "Rousset," and is Upper Rognacian in age. To avoid confusion, we propose to use "La Bégude" for the Middle Rognacian locality. This locality is the one named Rousset-Horizon A by Erben et al. (1979).

Middle Rognacian. 1. Rousset-Les Routiers. In the grey clays just below the last Rognac Limestone bar, we have collected very rare eggshell fragments.

2. Maupague. Near Maupague Hamlet, there are exposures of Rognac Limestone where Kérourio has found spherical eggs (sample D) that he described as type 4 (Williams et al., 1984).

Upper Rognacian. 1. Rousset-Erben (Fig. 11.1A). Named "Rousset" by Erben et al. (1979), Penner (1983) has proposed the name "Rousset-Erben" for this classical locality. Nevertheless, Iatsoura et al. (1991) used "La Bégude" for this locality. It is situated above the Rognac Limestone, north of RN7. The eggshell fragments studied were collected 20 m above the level of the road. Five samples were collected from a 50-m² area (ERR1,2,4,A,B) at the same level. A single egg (ERR3) was collected 1 m below the five samples, and one egg (ERRC), 5 m above.

2. La Cardeline. This locality occurs north of RN7, not far from Châteauneuf le Rouge. The sediments are red, silty clays, like at Rousset-Erben, with sandy and conglomeratic layers and lenses. They clearly overlie the Rognac Limestone that are exposed below, south to the road. One of the samples was collected 1.5 m above the level of the road, and the others 3 m higher.

3. Les Bréguières. This locality was called "the Carriere profile" by Hansen et al. (1989) and was studied by Iatsoura et al. (1991). "The section is 34 meters thick and consists of red silty-clayey sediments with slight color changes through the section" (Hansen et al. 1989). It is overlain by La Galante Conglomerate. The eggshell fragments were collected 2 m above the base of the section. They were not common but were locally concentrated, and appear to correspond to isolated, crushed eggs.

Bas Languedoc (Hérault) localities

1. Montpellier Basins. Here, the Upper Cretaceous series is reduced to only a few meters of discontinuous exposures north and east of Montpellier. At least three localities have yielded eggshells (Mattauer & Thaler, 1961). Clapiers, near Montpellier, is a conglomeratic sandstone that has yielded about ten badly preserved eggs. Two eggs have been collected in sandstones near Argelliers. South of this locality, between Montarnaud and Argelliers, nests of several eggs have been discovered in the yellow Rognacian sandstones. More recently, Crochet (1984) reported some eggshell fragments in small exposures of yellow clays, near the Pic Saint Loup structure, north of Montpellier.

2. Villeveyrac Basin (Hérault) (Fig. 11.2). The Upper Cretaceous series is thicker here than near Montpellier, reaching more than 750 m (Mattauer & Thaler, 1961). B. Sigé has found various eggshell fragments in a site (Villeveyrac–Valmagne) that has also yielded the lower Rognacian charophytes *Septorella brachycera* and *Peckichara sertulata* (Freytet, 1981).

3. Saint-Chinian (Hérault). Eggshells have been found near Saint-Chinian in the main fluviatile series, considered upper Campanian in age (Alabouvette, et al., 1982).

Corbières (Aude) localities

1. Albas. The stratigraphy is a sequence of a hard conglomerate, overlain by reddish marls (Albas 1) and siltstones with abundant eggshell fragments. The series is capped by grey, cross-bedded, hard sandstones and conglomerates (Erben et al., 1979).

2. Northeast of the Pyrénées. See Chapter 5 this volume.

Description of eggshells

As noted by Hirsch and Packard (1987), the study of eggshells is still in its infancy, even though the first discoveries of dinosaur eggs occurred during the nineteenth century! Some authors have attempted to identify who laid the eggs even though the identity of these eggs is unknown. The taxonomic identifications are sometimes based on bones associated with the same deposit (see Chapter 7). Many others, however, have been more cautious (e.g., Dughi & Sirugue, 1958, p. 198; Erben, 1970; Sochava, 1971; Williams et al. 1984; Penner, 1985). Dughi and Sirugue, unfortunately, never systematized their observations.

The only reliable taxonomic assignment of an eggshell type to a dinosaur genus or species occurs when the egg contains a dinosaur baby. Typological classification of eggshells without embryonic bones has been attempted, but proposed systems are not really practical: Erben (1970) designates types A, B, and C; Penner (1983) types 1, 2, and 3; and Williams et al. (1984), types 1, 2, 3-1, 3-2, and 4 (see Introduction to this volume).

We propose treating eggshells (complete or fragments) as fossils, describing the various types from the different localities, and, because we do not know the species they represent, to use a parataxonomy for them. For example, Dughi and Sirugue (1962) use the parageneric term "*Ornitholithus*" for Tertiary avian eggs. Use of a parataxonomic scheme allows similarities and relationships to be established (and with baby remains, real taxonomic identifications!). As Hirsch (1989) has noted, "Evolution studies [of eggs] should wait until sufficient descriptive literature is available and a universal classification exists."

Schleich and Kestle (1988), in their reptile eggshell SEM atlas, proposed following the recommendations for zoological nomenclature. They suggest using the organ genus term "*-ovum*" preceded by the name of the reptile (e.g., *Gekkodinovum textilis*). For fossil material, they proposed, for example, *Sauropsidarum ovum* or *Testudinarum ovum*. But with dinosaur eggshells, we often do not know if they are ornithischian or saurischian and, sometimes we can not be certain if they are dinosaurian. The Schleich and Kestle taxonomy, therefore, is too general and not very helpful.

For practical reasons, it is better to use a neutral parataxonomy like that introduced by Zhao (1975, 1979; Zhao et al., 1991). In this classification, the paragenus is based on a name (author, region, etc.) with the suffix *-oolithus*, followed by a paraspecies name or term. Zhao placed the different paragenera within four families, based on morphological and microstructural similarities of the eggshells. He established three families for Chinese eggs and one (Megaloolithidae) for the French and Spanish material.

Above the parataxonomic generic and specific level, at least five structural morphotypes have been distinguished among dinosauroid and ornithoid eggshells: prolatospherulic, tubospherulitic, ?tubospherulitic, prismatic, and ratite (Hirsch & Quinn 1990, Fig. 2; modified from Mikhailov 1987 a, b, 1991, and Kurzanov & Mikhailov, 1989).

Our eggshell species from France are placed among these structural types using the characters established after Hirsch and Packard (1987) and Mikhailov (1991):

1. Egg size
2. Egg shape
3. Sculpture of the outer surface
4. Pore pattern and pore diameters
5. Shell thickness
6. Microstructure and unit arrangement (size and shape of the units)
7. Texture of eggshells
8. Mode of nesting

For isolated fragments, only points 3 through 7 are given. The more precise information we can determine,

Bauxite

Jurassique

Fuvelian (F)
+ Upper Campanian
or Lower Maastrichtian

Rognacian

Paleocene

Lower Eocene

Upper Eocene
+Oligocene

Alluvial deposits

Villeveyrac

★ Eggshells Locality

Figure 11.2. Villeveyrac–Valmagne eggshell locality shown on a schematic geological map. (Based on a geological map of France, 1/50,000.)

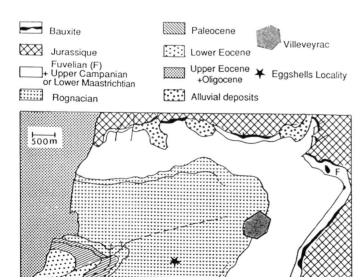

500m

the more credible the paleophysiological and paleobiological interpretations will be.

Because much of our work is in preliminary stages, we present results mainly for eggshell fragments collected from 1988 to 1991. Part of the samples were also used in the analysis of the organic matrix. Thickness measurements were made with calipers. Radial thin section (about 25 μm thick) were prepared, and some fragments were studied with the scanning electron microscope. All of the material studied is kept in the collections of the Laboratory of Paleontology at the University Montpellier II.

Systematics—tubospherulitic structural morphotype

The eggshells described in the following text belong to the dinosauroid basic organizational group and are of the tubospherulitic structural morphotype.

Megaloolithus oogen. nov.

Diagnosis

Tuberculate tubospherulitic dinosauroid eggshell, with fan-shaped units; tubocanaliculate pore system; pore diameters (at mid-height of the eggshell) greater than 50 μm.

Megaloolithus mammilare oospec. nov.

Type locality

Rousset-Erben (ERR), near La Bégude (Aix-en-Provence Basin), above the Rognac Limestone.

Age

Upper Rognacian.

Other localities

La Cardeline (CRD); Les Bréguières (LBR) (Upper Maastrichtian, Aix-en-Provence Basin).

Diagnosis

Short fan-shaped units; thickness in the main part of the eggshells from 1.2 to 2.1 mm; average node diameter about 1 mm; pore diameters range from 75 to 120 μm.

Description

The preservation of the Rousset-Erben (ERR) eggshells is generally bad. The surfaces are weathered or encrusted. On the external surface, the diameter of the nodes varies from 0.3 to 1.2 mm, with an average of about 1 mm (Fig. 11.3A). Some fragments show a dominance of small nodes (about 0.5 mm in diameter). But because they are widely separated from one another, this small size may be due to weathering (Fig. 11.3B). The pores are difficult to see on the outer surface because of weathering and encrustations. The size of the few pores observed is about 75 to 120 m. The inner surface of the shell is frequently very worn.

Thin sections show divergent crystal margins, units margins, centers of the spherulites, and curved growth lines. The units are fan-shaped, but some show parallel margins (Fig. 11.4B, C). Frequently, black minerals outline all these margins. Some specimens show a very marked herring-bone pattern within the units, indicating calcite recrystallization (Fig. 11.4). The range of thickness for most of the eggshell fragments (Fig. 11.5) is 1.2 to 2.1 mm; very few are thinner.

In sample ERR-C, 55 fragments were collected from a single crushed egg. Six fragments, under 1 mm thick, appear to be completely worn and weathered. Some very rare fragments, mainly in samples ERR-1 and ERR-C, are a little thicker than the others, averaging about 2.15 mm. They may represent another eggshell type. Most of the eggshell fragments have short fan-shaped units, but the thicker ones, which have the same node size, have fans that appear taller (Fig. 11.6 D, E) and are similar to those in *Megaloolithus siruguei* nov. from La Bégude or *M. petralta* nov. from Roquehautes-Grand Creux-D (see below).

The preservation of the eggshell fragments from La Cardeline (CRD1, samples A, B, C; CRD2, samples A, B, C, D, E, and F) is variable, but better than from Rousset-Erben. The external surface is not weathered, just slightly encrusted between the well-defined nodes. On the other hand, the internal surface is rather worn. In the thin section, the crystal margins, the unit margins, the centers of the spherulites, and the growth lines are well defined. Some specimens, however, are more weathered than the others, especially specimens CRD2-A and F (Figs. 11.6 and 11.7). These show an enlargement of the pore canals and of the unit margins at the base of the mammillary margins. These enlargements are filled with calcite crystals that are enhanced by black mineral spots. The herringbone pattern occurs in some parts of CRD2-A, indicating the beginning of recrystallization of the primary calcite units. Some specimens, such as La Cardeline 2B and 1C, do not show any traces of recrystallization, or mineralization.

The shell thickness in CRD1 (n 375) and CRD2 (n 428) form a roughly unimodal frequency histogram (Fig. 11.5). However this distribution is wide, showing a possible mixing of several eggs of the same or possibly of two oospecies. Some of the thickest fragments are pathological, with two layers of eggshell (ovum in ovo, e.g., CRD1-D, Fig. 11.8A,B). Another pathological morphotype shows the addition of an incomplete second layer that lacks the mammillary zone (Fig. 11.8C,D). Some thick eggshell fragments may represent a distinctive eggshell type similar to those from Rousset–Erben (Fig. 11.6).

For all eggshells the units are fan shaped. One node corresponds to single or double-fused or triple-fused mammillae. In the thinnest shells, the fans appear wider. The outer surface is covered with nodes (0.3 to

Figure 11.3. Surfaces of *Megaloolithus mammilare*. **A**. ERRC-3, from Rousset-Erben (Aix Basin). **B**. ERRC-4, from Rousset–Erben (Aix Basin). **C**. ERRC-3, from Rousset–Erben (Aix Basin), showing two pore openings. **D**. CRD2-D, from La Cardeline 2 (Aix Basin), showing a pore opening. **E**. CRD2-Did, less magnified outer surface. **F**. CRD2-E, from La Cardeline 2 (Aix Basin) showing the weathered bases of the units (inner surface). **G**. CRD2-E, enlarged base of a spherulite. **H**. CRD1-b, from La Cardeline 1 (Aix Basin), shape of nodes on outer surface; *Megaloolithus* aff. *petralta* or aff. *siruguei*. **I, J**. CRD2-A and CRD1-a, from La Cardeline 2 and 1 (Aix Basin), shape of nodes on outer surface.

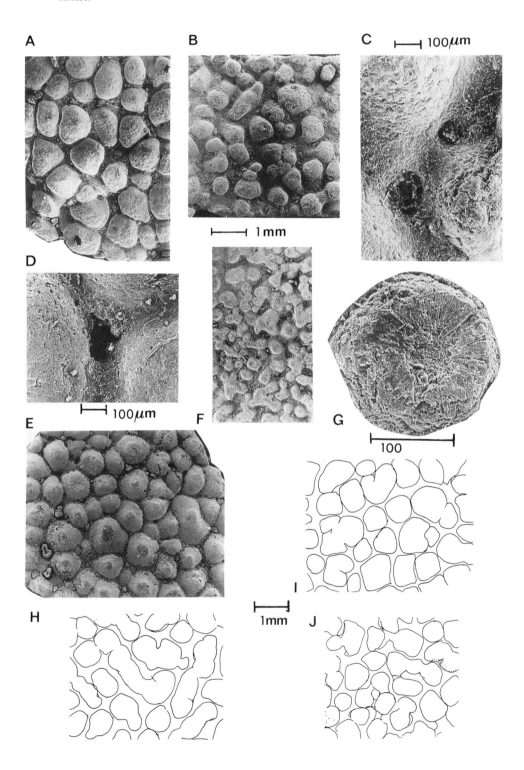

Figure 11.4. Radial thin sections of *Megaloolithus mammilare* from Rousset-Erben (Aix Basin). **A**. ERR1-A, showing the fan-shaped units and herringbone pattern. **B**. Same specimen as A enlarged, showing weathered mammilary zone. **C**. ERRB-A, with a herringbone pattern in the unit and with spherulite better preserved. **D**. ERRA-1, showing mineralization along the units concentrated at surface of the nodes; numerous recrystallization encrusting the mammilary zone. Bars = 1 mm.

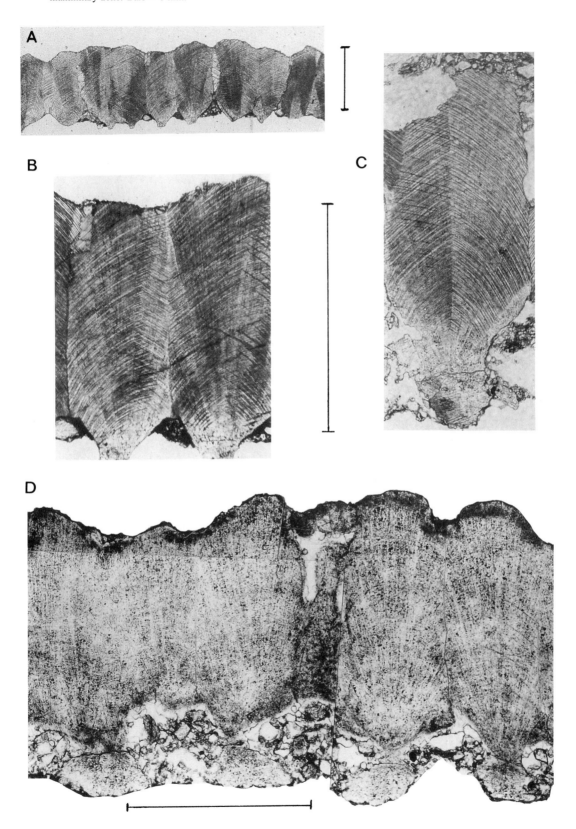

1.2 mm diameter) isolated or linked into irregular crests. Their size and shape seem to be the same regardless of shell thickness (Fig. 11.3E–J). The pores are difficult to locate and we cannot determine the pore pattern. The pores we can find have an opening diameter of about 100 to 150 μm.

Megaloolithus aff. *mammilare* oospec. nov.

Five samples from Les Bréguières (LBR) have been studied (LBR1 to LBR5), each containing a variable number of fragments (from 20 to 378). Their preservation is better than the specimens from Rousset–Erben, with no sign of recrystallization. The surfaces are somewhat weathered or encrusted, mainly on the internal side. The pores are filled with matrix, and their di-

ameter varies from between 65 to 150 μm. The nodes look like those of the Rousset–Erben eggshells (Fig. 11.9). The units are fan shaped; growth lines increase in undulations from the base to the top of the units (Fig. 11.7F,G, H).

Sample LBR5 (a single crushed egg) shows considerable variation in shell thickness (1.21 to 1.84 mm). This thickness, however, is less than that of the scattered fragments from the same level (compare Figs. 11.5 and 11.10). The average thickness of the eggshell is less than those from Rousset–Erben, because they come from a single egg. The significance of this difference is difficult to determine because the variation in thickness of three different samples (LBR2, 3, and 4) is quite small (0.9, 0.95, and 1 mm).

Figure 11.5. Frequency histograms of the thickness of *Megaloolithus mammilare* eggshells from three samples from Rousset–Erben and two samples from La Cardeline (Aix Basin).

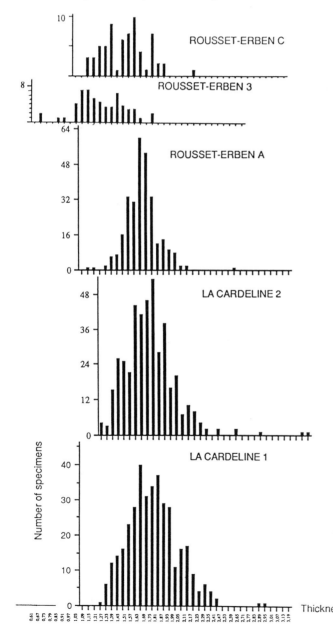

Figure 11.6. Radial thin sections of *Megaloolithus* aff. *siruguei* or aff. *petralta*. **A**. CRD2-A, from La Cardeline 2A (Aix Basin) showing a herringbone pattern, mainly on the left, divergent units, pores filled with recrystallized calcite, and black mineralization along the unit margins. **B**. CRD1-C, from La Cardeline 1C (Aix Basin) showing the fine preservation of the crystals, with their margins and growth lines. **C**. CRDA-a, from La Cardeline A (Aix Basin), showing the strong mineralization both on the outer and inner part of the eggshell. **D**. ERR1-1, from Rousset–Erben 1 (Aix Basin), showing the elongated fan-shaped units and mineralization along the margins. **E**. Same for ERR1-C. Bars = 1 mm.

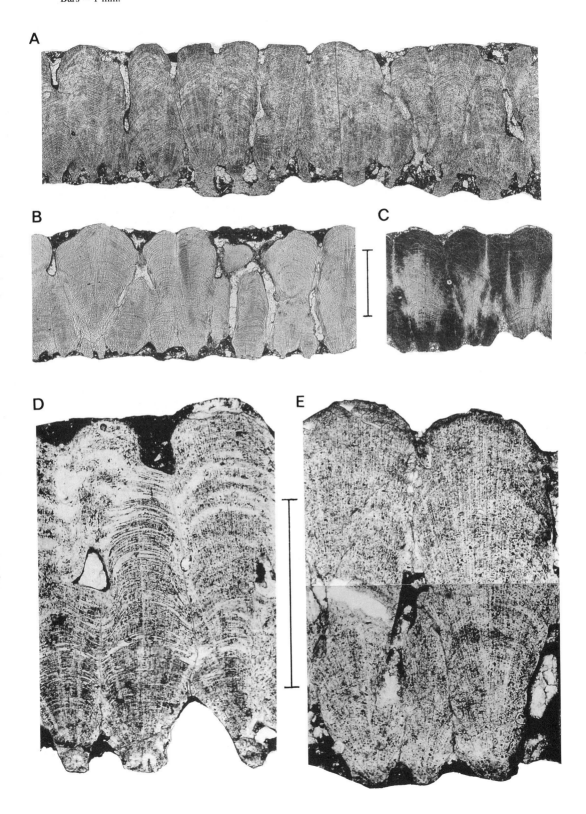

Figure 11.7. *Megaloolithus mammilare*. **A**. CRD2-F, from La Cardeline 2 (Aix Basin). **B**. CRD1-b, from La Cardeline 1 (Aix Basin). **C**, **D**. CRD2-B, from La Cardeline 2 (Aix Basin). **E**. LBEA-6, from La Bégude (Aix Basin). **F**. LBR2-A, from Les Bréguières (Aix Basin). **G**, **H**. LBR3-A, from Les Bréguières (Aix Basin). Bars = 1 mm.

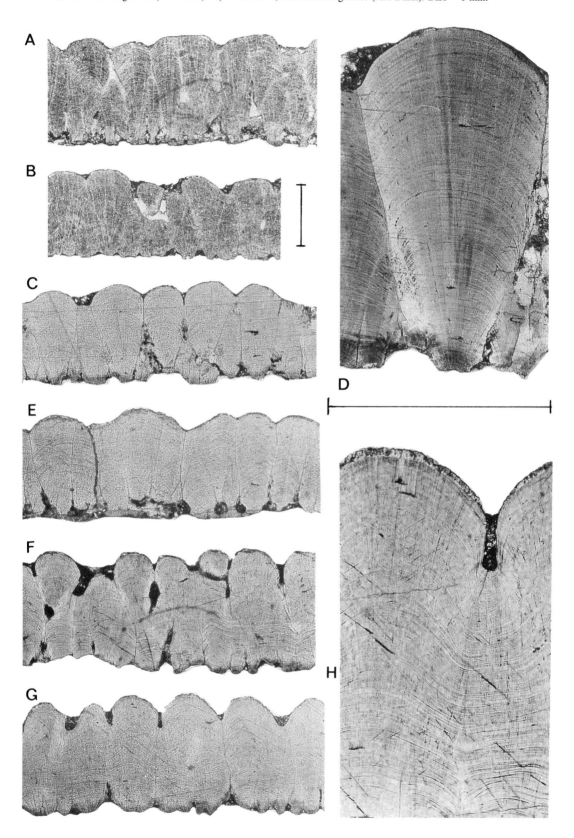

Figure 11.8. Pathological eggshells of *Megaloolithus mammilare* from La Cardeline (Aix Basin). **A**. CRD1-d, two layered eggshell with numerous recrystallizations between the two layers; **B**. Same eggshell, enlarged and polarized, to show numerous extra-spherulites between the two layers. **C**. CRD2-C, enlarged to show contact between the first eggshell layer and beginning of the incomplete second one, which lacks spherulitic structures. **D**. Same as C, less magnified. Bars = 1 mm.

Rare eggshell fragments from La Bégude (LBE-A-7, LBE-A-6) can be referred to *M.* aff. *mammilare* (Fig. 11.7E).

Discussion

Zhao (1979) established the family Megaloolith-idae for the large eggs from the Upper Cretaceous of France. However, he did not assign a genus or species to this family, thus making the parataxon invalid as noted by Mikhailov (1991). To correct this error of Zhao, we propose the name *Megaloolithus*.

The egg species *Megaloolithus mammilare* is identical to that studied by Erben et al. (1979) from Horizons B, C, and D at Rousset-Erben (near Rousset, in Erben et al. 1979). This species is the same as type 3 of Williams et al. (1984) and of Penner (1983, 1985).

Megaloolithus siruguei oospec. nov.

Type locality

La Bégude (LBE), near Rousset Horizon A in Erben et al. (1979), under the Rognac Limestone.

Age

''Middle'' Rognacian

Other localities

Rousset-Routiers (RSR), marls within the Rognac Limestone; Rousset-Village (RSV), sandstone under the Rognac Limestone; Roquehautes South, below the ''Crête du Marbre'' (RHM); rare fragments at Rousset-Erben and at La Cardeline; Maupague (Williams et al., 1984); Villeveyrac Basin (VIO); Roquefumade (Corbières).

Diagnosis

Probably spherical eggs; shell units taller than in *M. mammilare*; thickness range 0.65 to 0.70 mm; pore diameter range 50 to 80 μm.

Description

The La Bégude (LBE) material (Figs. 11.11 and 11.12) shows wear of some nodes due to weathering of the outer surface. Nevertheless, the nodes are well preserved. Their diameters range from 0.4 to 1.1 mm, with most about 0.65 to 0.70 mm. The inner surfaces are more weathered and worn. The surfaces and pore canals are encrusted with recrystallized calcite. The preservation of the units is good, the growth lines are well defined, and the herringbone pattern is very rare and faint. In thin section, the margins of the units appear nearly parallel. The growth lines curve down steeply near the unit margins. Pore canals are straight along these margins, but there is some enlargement due to calcite recrystallization. The pore diameters are about 50 to 80 μm. Some small extra spherulites are present along the unit margins and sometimes along certain growth lines.

The thickness frequency histogram for sample LBE-A is wide (Fig. 11.13A). Most of the eggshell fragments are between 1.84 and 3.18 mm thick. The thinner specimens are *Megaloolithus* cf. *mammilare*.

The material from Rousset-Les Routiers (RSR), which is close to road RN7, has yielded only two fragments. Their microstructure is the same as the samples from La Bégude (Fig. 11.12D).

The Rousset-Village material (RSV) is not abundant and is very encrusted. The thickness of the ten fragments ranges from 2.33 to 2.68 mm. Their morphology is like those from La Bégude, except there are more numerous, tiny extra-spherulites all along the growth lines (Fig. 11.14).

Figure 11.9. *Megaloolithus* aff. *mammilare*, from Les Bréguières (Aix Basin). LBR5-a. **A**. Shape of nodes and pore openings on outer surface. **B**. Inner surface. **C**. Outer surface. Bar = 1 mm.

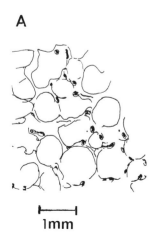

A

1mm

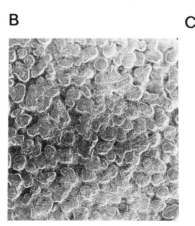

B

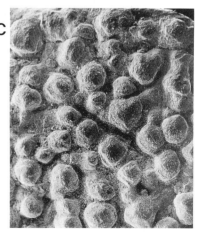

C

The small sample from Roquehautes, south of the "Crête du Marbre" (RHM), includes two egg species. The most abundant fragments are referred to *M. siruguei*. They show the same thickness and the same microstructure as those from Rousset-Village, especially with their numerous, tiny extraspherulites.

The material from the Villeveyrac-Valmagne locality (VIO) contains at least four, and maybe more, morphotypes that can be attributed to the egg genera defined in the Aix Basin. Among the 200 specimens, only four can be attributed to *M. siruguei*. The outer surface of this material is covered with nodes of the same size and arrangement as those from the La Bégude eggshell. The thickness is also the same; the units are of an elongated fan shaped.

Specimens from Roquefumade (Aude Valley, Corbières) have been studied by Beetschen (1986), who described ornithoid and testudoid eggshell fragments. The ornithoid fragment has been tentatively assigned to the hadrosaurs. The elongated units, the small nodes, and the thickness of the testudoid type from Roquefumade identifies it as *Megaloolithus siruguei*. Numerous eggshell fragments and eggs have been collected in various localities in this region. They are still under study.

Discussion

This oospecies, *Megaloolithus siruguei*, is type 4 of Williams et al. (1984) and was studied by Erben from Horizon A. Because Kérourio has found spherical eggs of type 4 in the Maupague Locality, we can infer that

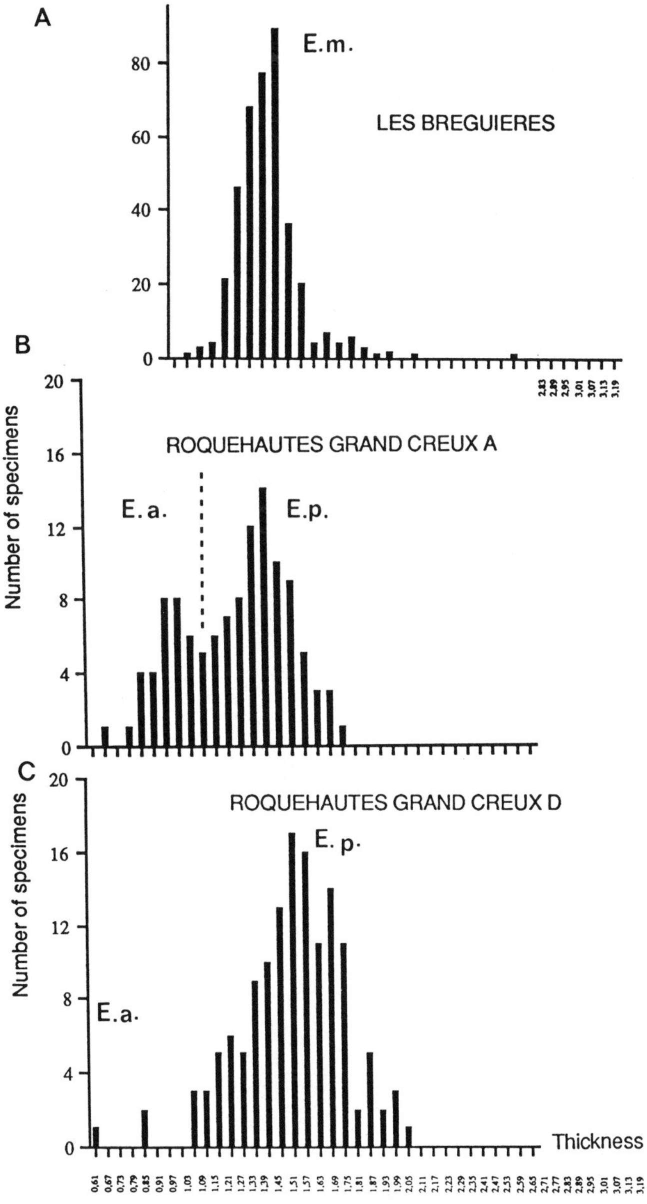

Figure 11.10. Frequency histograms of eggshell thickness. **A**. *Megaloolithus* aff. *mammilare* (M. m.), from Les Bréguières (Aix Basin). **B, C**. *Megaloolithus aureliensis* (M. a.) and *M. petralta* (M. p.), from Roquehautes-Grand Creux (Aix Basin).

Figure 11.11. *Megaloolithus siruguei* from La Bégude (Aix Basin). **A**. LBEA-1, outer surface. **B**. LBEA-2, inner surface. **C**. LBEA-5, radial thin section showing growth lines and units and small extra spherulites along unit margins (arrows). **D**. LBEA-5, another unit with a pore canal. Bars = 1 mm.

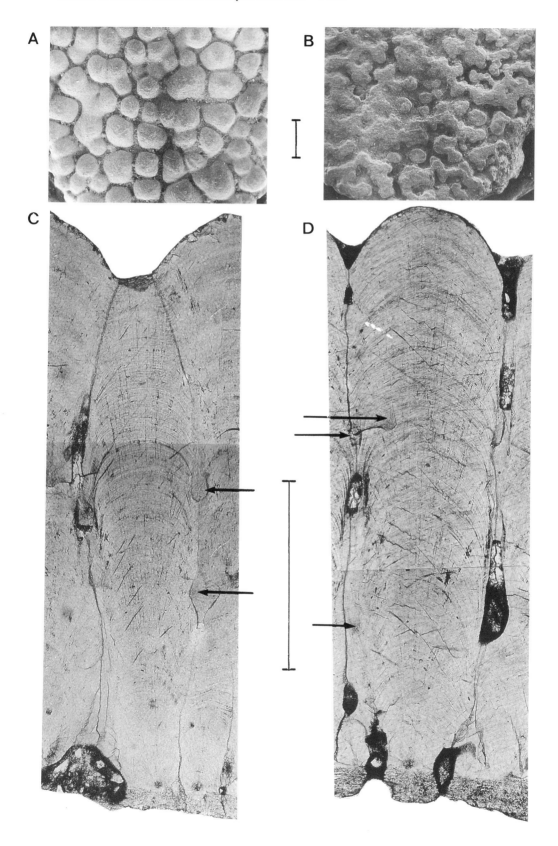

Figure 11.12. Radial thin sections of *Megaloolithus siruguei*. **A.** LBEA-5 from La Bégude (Aix Basin); **B.** Same as A, enlarged, showing the down turning of the growth lines at a pore canal margin. **C.** Same as A, enlarged and polarized, showing the base of two units and recrystallization of the inner layer under the spherulite centers. **D.** RSR-1, from Rousset–Les Routiers, showing same pattern as for LBEA-5. Bars = 1 mm.

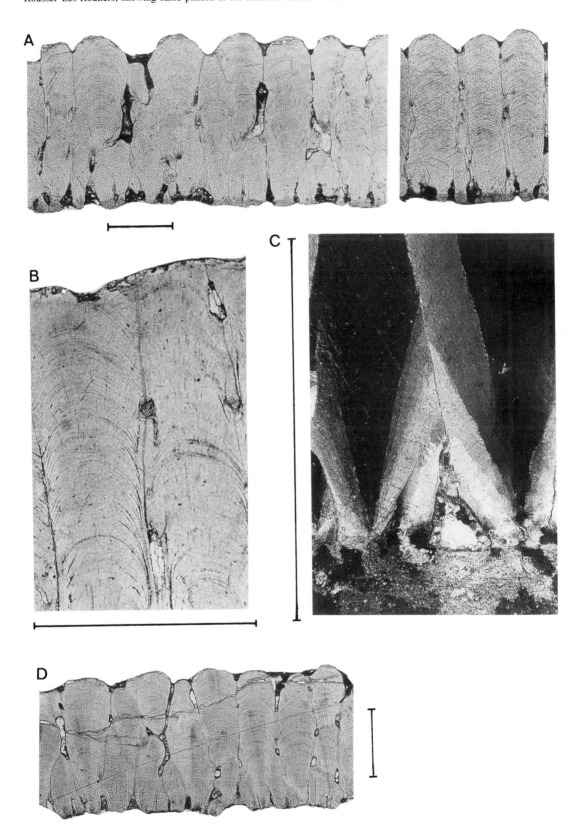

Megaloolithus siruguei from La Bégude was probably a spherical egg.

Fragments with numerous, tiny extra-spherulites, occur in at least two places, Rousset-Village and Roquehautes-Crête du Marbre. These localities represent two different environments, fluviatile and lacustrine (Iatsoura et al., 1991). The extra-spherulites do not have the same significance as the rare occurrence of "ovum in ovo" structures observed in eggshells from La Cardeline. They appear to be a normal microstructure of eggshell fragments of *M. siruguei* from the level just below the Rognac Limestone. Either they are a peculiar character of this egg species, or they have an unknown pathologic significance.

Erben et al. (1979) thought there was a decrease in thickness of dinosaur eggshells from La Bégude to the top of the Rousset-Erben Locality. As suggested by Penner (1985), this is not a decrease of eggshell thickness within one dinosaur lineage, since it occurs in a succession of different dinosaur types.

Megaloolithus aureliensis oospec. nov.
Type locality

Clos-La-Neuve (CLN) (near Trets, Bouches du Rhône).

Age

Begudian (= Campanian).

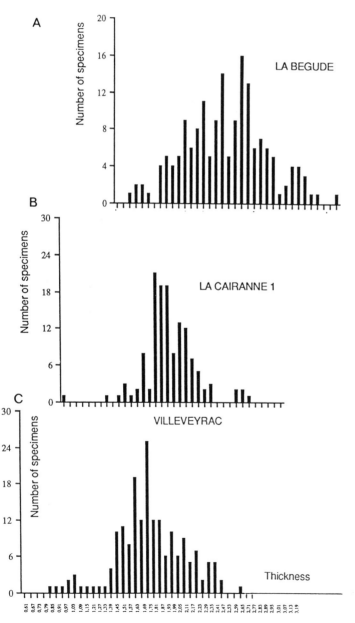

Figure 11.13. Frequency histograms of the eggshell thickness. **A.** *Megaloolithus siruguei* from La Bégude (Aix Basin). **B.** *Cairanoolithus dughii* from La Cairanne (Aix Basin). **C.** *Megaloolithus* aff. *aureliensis* (M. a.), *M. siruguei* (M. s.), *?Cairanoolithus* + *?Dughioolithus* (C + D), from Villeveyrac–Valmagne (Languedoc).

Figure 11.14. Radial thin sections of *Megaloolithus siruguei*, from Rousset-Village (A, B, C) and Roquehautes-Crête du Marbre (D) (Aix Basin). **A**. RSV3-9. **B**. RSV3-8. **C**. RSV3-9, enlarged and polarized, showing numerous extra-spherulites along a unit. **D**. RHM-K, enlarged, showing numerous, tiny extra spherulites. Bars = 1 mm.

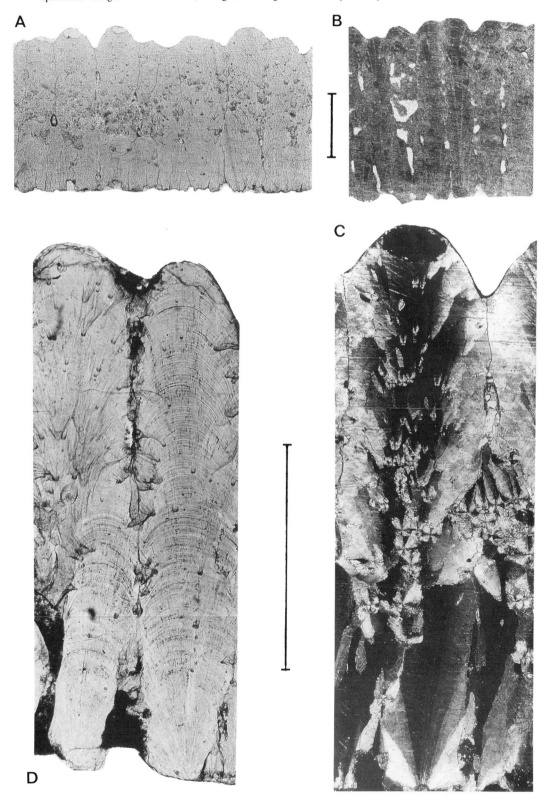

Other localities

Roquehautes–Grand Creux-A (RHG-A), 10 m above the base of the section, just above the third sandstone layer; Roquehautes-Grand Creux-B and D (RHG-B; RHG-D); Villeveyrac Basin (VIO); and Saint-Chinian (SCH).

Diagnosis

Probably ovoid-shaped eggs; short fan-shaped units; thickness ranges from 0.82 to 1.52 mm; average node diameter about 0.5 mm; pore diameters range from 100 to 120 μm.

Description

The fragments from Clos–La-Neuve (CLN) (Fig. 11.15A,B,C) are badly preserved because of erosion. In thin section, the surfaces are worn and encrusted. The pore canals have been enlarged. Because the pores are completely filled with recrystallized calcite, it is difficult to recognize their pattern, but they are probably tubo-canaliculate. The shell units are fan shaped. Growth laminations and the margins of the crystals are accented by black mineralization (Fig. 11.15A).

In specimens in which the outer surface is not much worn, the nodes are close to one another. Their size varies from 0.3 to 0.7 mm in diameter, with most of them being about 0.5 mm (Fig. 11.15B). When surface wear is greater, the nodes are farther apart and the pore openings are well developed at the surface (Fig. 11.15C); their diameter varies from 100 to 120 m. The shell thickness of the few specimens collected varies from 0.82 to 1.52 mm.

Roquehautes–Grand Creux-A (RHG-A) material (Fig. 11.15D,E,F) is not well preserved. The inner and outer surfaces are worn, and recrystallization and mineralization frequent. It is therefore difficult to recognize the finest microstructure. For RGH-A, the frequency histogram of shell thickness is bimodal. The fragments were collected from one egg and its immediate surroundings. It is possible that originally there were two or more eggs. The thickest fragment is similar to those from Rousset-Erben A and C, while the thinnest fragment is clearly thinner than any from those two sites. A thin section of one of the thinnest fragments (RHG-A-2) shows that the inner surface is much weathered.

The nodes are well defined on the outer surface. These nodes are smaller than in *Megaloolithus mammilare*. Most of them are about 0.5 mm in diameter, similar to those at Clos-La-Neuve. They are sometimes linked in more or less long chains. The pores are located at the base of the nodes, and their external opening is about 70 to 100 μm in diameter. In the thickest specimens, which all come from a single egg, the outer surface appears more weathered than the inner one. The nodes are faint and sometimes are almost gone. On the inner surface, the base of the spherulites is frequently recrystallized. It is difficult to decide whether these thick specimens can be attributed to *M. aureliensis* or to another species.

The few badly preserved fragments from Villeveyrac Basin (VIO) (Fig. 11.19 A,B) are thin (0.9 to 1.1 mm). Their outer surface shows nodes, the diameter of which varies between 0.3 to 0.72 mm. The pore opening diameter varies from 70 to 160 μm. In thin section, the units are not as fan shaped as in *M. mammilare*; they are more like those of *M. aureliensis*. It is difficult to compare these two species because in Clos–La-Neuve and Roquehautes–Grand Creux-A, B and D, the preservation is bad and the growth lines are not clear. In the VIO specimens, we can see that the growth lines are less undulating than in *M. mammilare*. A distinctive feature of these shells is the occurrence of small extra-spherulites on the outer half of the eggshell (Fig. 11.19A,B).

The eggs from Saint-Chinian (SCH) (Fig. 11.19E) are ovoid in shape. On the basis of the best preserved partial egg (SCH-5), the measurements are 18.5 cm high and 14.5 cm wide (Fig. 11.19E). The eggs are imbedded in a coarse sandstone, and some of the outer surfaces are visible. The nodes are isolated or arranged in more or less long rows. Their size varies from 0.3 to 0.7 mm in diameter.

On a surface etched with 5% acetic acid to remove matrix, the pores have diameters from 80 to 140 μm. In thin section (Fig. 11.19C, D), the shell units appear weathered on their tops and bottoms. They are fan shaped, and each shows a herringbone pattern that indicates calcite recrystallization. The eggs are more like *M. aureliensis* than *M. mammilare* because of the size of nodes and shell thickness (1.12 to 1.32 mm).

Megaloolithus petralta oospec. nov.

Type locality

Roquehautes–Grand Creux-D (RHG-D), 3 m above the fourth sandstone layer.

Age

?Lower Rognacian.

Other localities

?Roquehautes-Grand Creux (RHG-A), Roquehautes–Grand Creux-B and C (RHG-B and RHG-C), 50 cm under the fourth sandstone layer.

Diagnosis

Eggshell thickness similar to that of *M. mammilare*, but units less wide (about 0.5 mm).

Description

Thin and thick fragments of *Megaloolithus* (Fig. 11.16) occur together in sample RHG-D. There are only a few thin fragments that can be referred to *M. aureliensis*. The others are thicker than at RHG-A (mode 1.52 mm instead of 1.40 mm). The nodes are about 0.4

Figure 11.15. *Megaloolithus aureliensis* from Clos–La–Neuve (A, B, C) and Roquehautes-Grand Creux-A (D, E, F) (Aix Basin). **A**. CLN-1, radial thin section. **B**. CLN-2, pattern of nodes on outer surface. **C**. CLN-3, pattern of nodes on outer surface. **D**. RHG-A-2, enlarged, showing the base of unit with worn spherulite. **E**. Same as D, less enlarged, showing the general shape of the units. **F**. Node pattern on outer surface.

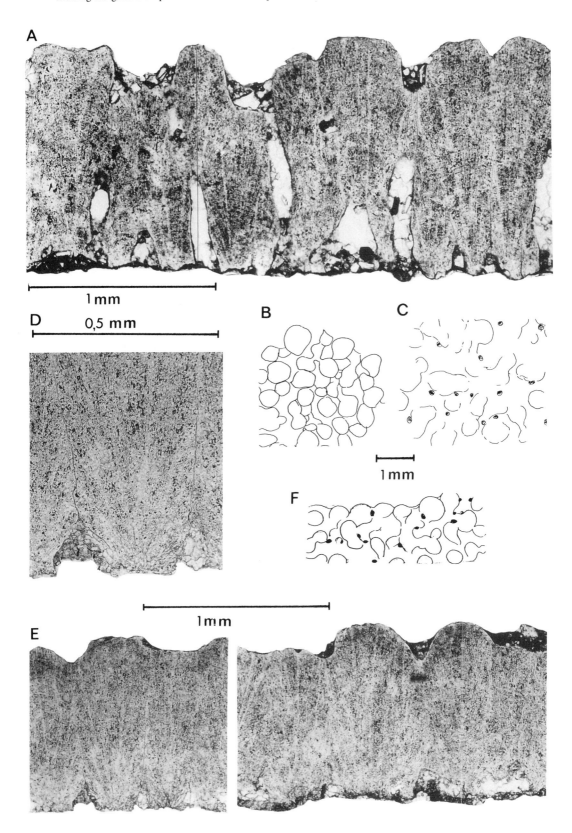

Figure 11.16. *Megaloolithus*, from Roquehautes–Grand Creux (Aix Basin). **A**. *M. petralta* (RHG-C-1) radial thin section. **B**. Same as A, pattern of nodes on outer surface. **C**. *M. petralta* or *M. aureliensis* (RHG-D-2), a weathered outer surface with remains of nodes and some pore openings. **D**. Same as C, radial thin section, showing strong weathering of outer surface. Bars = 1 mm.

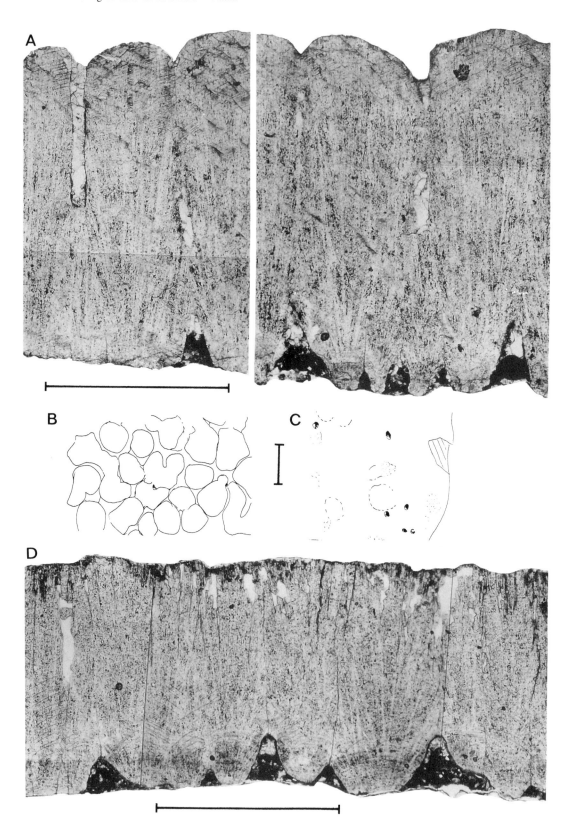

to 0.6 mm in diameter. In all cases, the shell units are fan-shaped, but those of the thickest fragments from RHG-D and RHG-C are more elongated than in *M. aureliensis*. They are similar to *M. siruguei*, but the shells are not as thick. We have chosen to refer this morphotype to a new egg species, even though it is difficult to differentiate fragments of intermediate thickness. A detailed analysis of more complete material from the whole Roquehautes section could help solve this problem.

Systematics—?Tubospherulitic structural morphotype

This structural morphotype differs from the typical tubospherulitic by the less arched accretion lines and faint tuberculous elevations on the shell surface.

Cairanoolithus oogen. nov.

Diagnosis

Faint tuberculate to nearly smooth dinosauroid eggshell, with nearly columnar units that are not completely interlocking; growth lines are horizontal in the interlocking shell units, but more or less arched in the single units; narrow pore diameters, about 25 μm.

Cairanoolithus dughii oospec. nov.

Type locality

La Cairanne (LCA), east of Rousset, about 50 m under the Rognac Limestone.

Age

Lower Rognacian.

Other localities

Roquehautes-Grand Creux (RHG), about 100 m under the Rognac Limestone; and Villeveyrac Basin (VIO).

Description

The La Cairanne (LCA) material (Fig. 11.17) is weathered on the external and inner surfaces and encrusted. Because the nodes are not very distinct, the external surface frequently appears smooth. The pores are fairly distinct. On the internal surface, the bases of the small spherulites of the mammillary zone are linked and form a network; only a few of them are isolated. Secondary calcite sometimes covers the surfaces and fills the pores and the pore canals. The canals are narrower in diameter than in *Megaloolithus*. They are straight along the unit margins and are about 25 μm wide; they are about 50 μm wide at their opening. In thin section, the growth lines and the top of the nodes, are not much curved. Yet they gently but clearly turn down at the margins of the shell units. The units are more columnar than in *Megaloolithus siruguei*, with a rather constant diameter from above the mammillary zone to the external surface. The faint nodes correspond to a single, dou-

ble fused or triple fused mammillae. The eggshell fragments appear less thick than *M. siruguei* and thicker than *M. mammilare*. Most of the fragments are 1.57 to 2.41 mm thick (Fig. 11.13B).

A great part of the Villeveyrac material can be referred to *Cairanoolithus*, although to an undetermined species. The outer surface is sometimes it is smooth because of wear or sometimes covered with faint nodes. The pore pattern is the same as those from La Cairanne. The pore canals are narrow, about 25 μm in diameter, and between 40 to 75 μm at their opening. The growth lines are straight to slightly undulating and turn down at the pore canal. The units are not so regularly columnar as those from La Cairanne (Fig. 11.20A,B,C).

Discussion

Cairanoolithus dughii corresponds to Type 1 of Williams et al. (1984) and Penner (1983, 1985). It is not completely smooth as they described it, however.

Dughioolithus oogen. nov.

Diagnosis

Dinosauroid eggshell with faint nodes; nearly columnar units that are not completely interlocking; growth lines slightly undulating; pores diameters from 50 to 100 μm and ?tubocanaliculate.

Dughioolithus roussetensis oospec. nov.

Type locality

Rousset-Village (RSV), just under the Rognac Limestone.

Age

"Middle" Rognacian.

Other localities

Roquehautes-Crête du Marbre; Villeveyrac Basin; Argelliers–Montarnaud.

Diagnosis

Shell less thick than *Cairanoolithus dughii* (1.49 to 1.77 mm).

Description

The rare fragments from Rousset-Village (RSV) (Fig. 11.18) were collected at the surface of the sandstones. They are much weathered, especially on their inner surface where the spherulites are completely worn. The outer and inner surfaces of the shell are generally encrusted, and the pore canals are filled with calcite recrystallizations. The columnar shell units are shorter than in *Cairanoolithus*. The growth lines are slightly undulating, producing faint nodes on the outer surface. The growth lines continue from one unit to the other, sometimes slightly turning down at the margins of the pore. The nodes are sometimes completely, sometimes partly, surrounded by a chain of pores in an irregular

Figure 11.17. *Cairanoolithus dughii* from La Cairanne (Aix Basin). **A.** LCA1-A, radial thin section showing different levels of interlocking of columnar units; (a) from end of mammilary zone, (b) from the midheight of eggshell, (c) not interlocking. **B.** LCA1-A, magnification of a pore canal zone. **C.** LCA1-B, same, for another fragment. **D.** Pore pattern on a weathered outer surface. Bars = 1 mm.

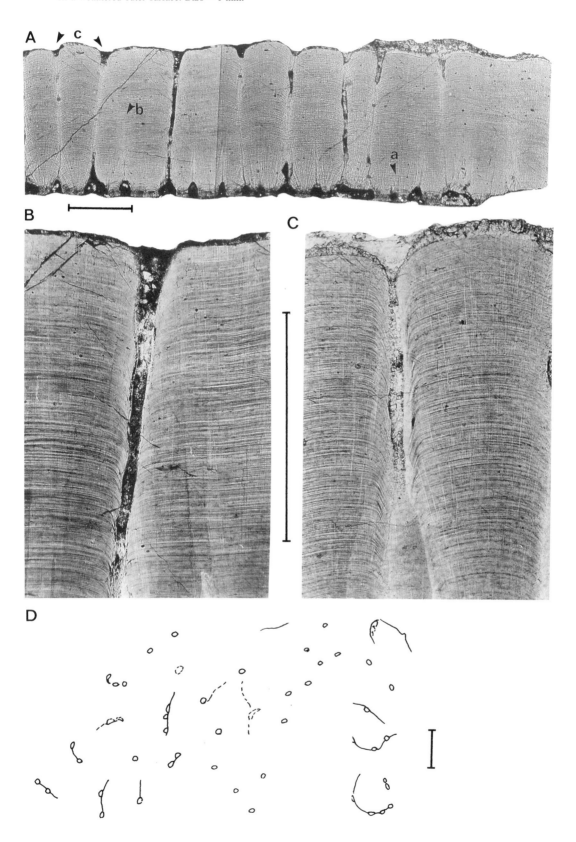

Figure 11.18. *Dughioolithus roussetensis* from Rousset-Village (Aix Basin). **A**. RSV3-11, radial thin section showing the pore canals, the more or less short columnar units, and the nearly straight growth lines with slight undulations producing faint nodes on the surface. **B**. RSV3-12, radial thin section thinner than RSV3-11, and badly preserved. **C**. RSV3-14, outer surface showing the pore pattern and some faint nodes. Bars = 1 mm.

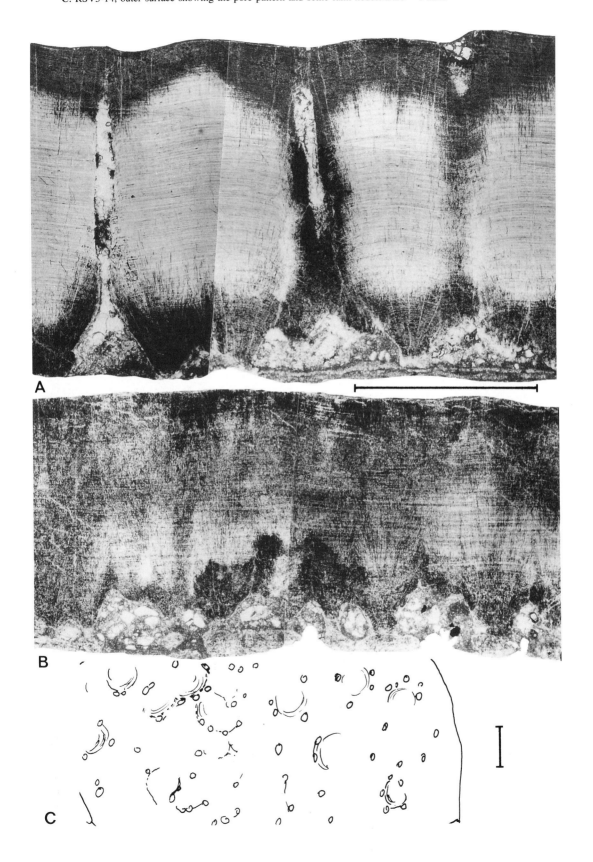

Figure 11.19. **A**. *Megaloolithus* aff. *aureliensis* from Villeveyrac-Valmagne (Languedoc), V10-3, radial thin section showing the tiny extra spherulites in the upper half of the eggshell. **B**. Same as A. **C**. *M*. aff. *aureliensis* from Saint-Chinian (Languedoc), SCH-5, showing general shape of the units. **D**. Same as C. **E**. Same as C, shape of the egg (d) diameter; (h) height). **F**. ?*Dughioolithus* oospecies indet., from Montarnaud–Argelliers, MAO-3, radial thin section. **G**. Same as F, shape of the egg (d) diameter, (h) height). **H**. Same as F, radial thin section. Bars = 1 mm.

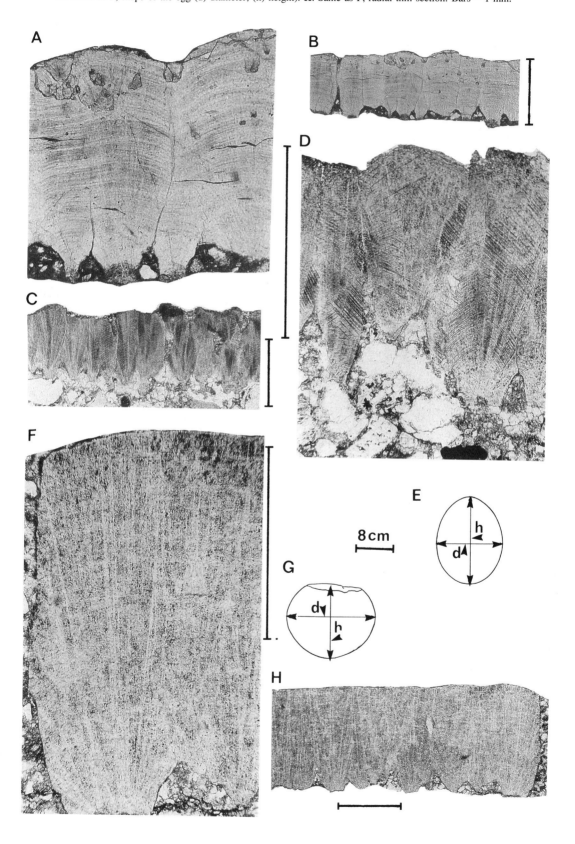

Figure 11.20. **A.** *Cairanoolithus* oospecies indet., from Villeveyrac-Valmagne (Languedoc), V10-2, radial thin section. **B.** Same as A, Radial thin section showing pore canal. **C.** Outer surface showing pore pattern and some faint nodes. **D, E.** ?*Dughioolithus* oospecies indet., from Villeveyrac-Valmagne (Languedoc), V10-4. **F, G.** V10-1,. Bars = 1 mm.

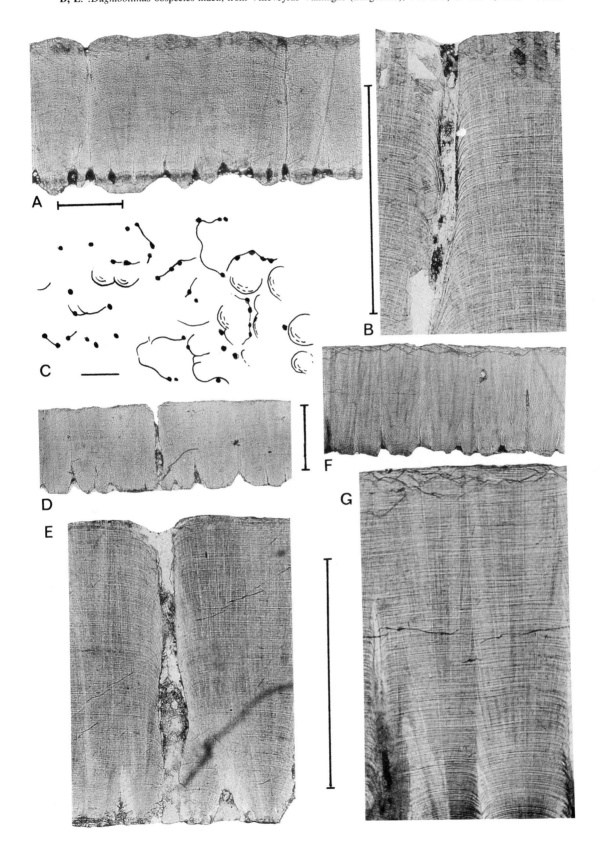

pattern. The pores are wider than in *Cairanoolithus*, being 50 to 100 μm in diameter. They are tubocanaliculate rather than angusticanaliculate. The thickness of the ten specimens collected varies from 1.49 to 1.77 mm.

The Villeveyrac Basin material (Fig. 11.20D,F,G) includes some eggshell fragments that are quite smooth, but it appears that the outer surface is worn. On a few specimens we can identify faint nodes. On the surface the diameter of the pore openings varies between 100 and 150 μm. The pore canals are 70 m in diameter. The shell units have parallel margins, making them columnar. The growth lines are sometimes very slightly undulating, but generally straight. Near the pores, the lines turn downward (Fig. 11.20E).

The best preserved egg from Argelliers-Montarnaud (MAO) is the one described by Mattauer and Thaler (1961) (Fig. 11.19G,F,H). The general shape is spherical (19 cm in diameter). The base and the top are flattened, and the top is broken about 14.5 cm above the base. The outer surface is irregular with some faint nodes. The pore openings are scattered or in rather long sinuous rows. Their diameter varies from between 60 to 140 μm. The inner surface is very weathered and the spherulites are partly worn. Although the preservation is better than at Saint-Chinian (no herringbone pattern is present), it is not perfect. Mineralization is extensive along the crystal margins, and the growth lines are hardly recognizable. The general shape of the units is similar to that of *Dughioolithus* from Villeveyrac–Valmagne.

Some conclusions about French dinosaur eggs

In the Aix Basin, we have defined six egg species among the eggshell fragments collected (Fig. 11.21). The main diagnostic characters are shell thickness, diameter of nodes, size of pore canals, and shape of the shell units. Examination of numerous eggshell fragments and radial thin sections shows that preservation of the microstructures is variable. Recrystallization results in herringbone patterns and mineralization produces black dots along the shell unit margins; neither of these change the basic shape of the units. On certain fragments, the effect of weathering and erosion is noticeable. A weathered tuberculate fragment cannot be confused with an originally smooth eggshell, because faint nodes are often present and because the growth lines arch into the units. Shell thickness measurements must be taken at face value and must not be exaggerated. In a sample some shell fragments always remain undeterminable.

The new egg species from Aix Basin is also present in four other basins from Languedoc to Corbières. The occurrence of stratigraphically limited egg species allows us to correlate findings from other basins (Fig. 11.22). These correlations agree with previous stratigraphical data for Saint-Chinian, Villeveyrac-Valmagne and Roquefumade. The small Upper Cretaceous exposures at Argelliers-Montarnaud can be tentatively referred to the lower-middle Rognacian.

At present it is too premature to discuss the taxonomic identity of these egg species. At present the egg species laid by *Hypselosaurus* remains unknown. When the whole collection of eggs and eggshell fragments from the Upper Cretaceous of southern France is described, we may go further with this taxonomic discussion.

We have studied eggshell fragments of the dinosauroid tubospherulitic type from the Upper Cretaceous of India which show affinities with European eggshells (Vianey-Liaud, Jain, & Sahni, 1987). The morphology of the eggshells from Jabalpur is close to that of the egg genus *Megaloolithus* (see also Chapter 13). The thickness of the shell units is similar to *M. petralta*. A few eggshells from Takli resemble *M. aureliensis*.

To establish the relationship between dinosaur eggshells and the extinction of the dinosaurs at the end of the Cretaceous, Erben et al. (1979) focused on supposedly pathological eggshell structures. The "ovum in ovo" phenomenon, generating bi- or multilayered eggshells, was supposed to illustrate one of these relationships. But these multilayered eggshells are neither frequent in the Aix Basin nor localized at particular stratigraphic levels in the Rognacian series (Kérourio, 1981).

Figure 11.21. Schematic organization for units of newly described egg species. **A.** *Megaloolithus mammilare.* **B.** *M. aureliensis.* **C.** *M. siruguei.* **D.** *M. petralta.* **E.** *Cairanoolithus dughii.* **F, G.** *Dughioolithus roussetensis*, showing pore canal (p).

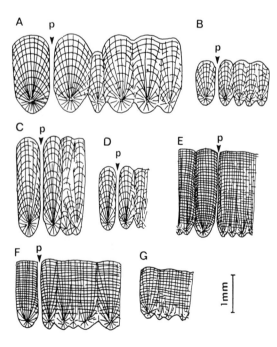

The occurrence of numerous, tiny extra-spherulites is common along the growth lines of the shell units in *Megaloolithus siruguei*. These microstructures occur in specimens from the Arc Basin under the Rognac Limestone (Rousset-Village) and from the Roquehautes-Crête du Marbre section. In the Arc Basin, they occur far from the two different proposed K/T boundaries (within the Rognac Limestone or under the La Galante Conglomerate). Therefore, these extra-spherulite pathologies do not seem to be linked with the extinction of the dinosaurs. Their cause remains unknown.

Erben et al. (1979) thought they had observed another pathological tendency in the Rousset section. They reported a reduction of shell thickness in shells from Horizon A (La Bégude) to Horizon C (top of Rousset-Erben). But Penner (1985) and we have determined that there are different eggshell types in this section. There is no indication that these eggshells belong to a single evolutionary lineage that shows a reduction in shell thickness through time. Moreover, there are thin "testudoid" dinosaur eggshells in the Upper Campanian, such as *Megaloolithus aureliensis*, well below the K/T boundary.

Chemical analysis of eggshells
Calcite components

Isotopic and chemical studies of dinosaur eggshells from southern France was begun by Folinsbee et al. (1970). Erben et al. (1979) made an extensive analysis of eggshells to correlate chemical changes in alleged pathologic eggshells with possible environmental changes. The stable isotopes $\delta^{18}O$ and $\delta^{13}C$ and the trace element ratios of magnesium and strontium were measured. The analysis was continued by Morin (1989) and Iatsoura et al. (1991). However, there are two problems with these studies:

1. The effect of diagenesis on the calcite of the eggshell was not taken into account. We have seen a considerable range of preservation, from very poor to very good. It is important that only unrecrystallized and unmineralized fragments be used for thin-section and for X-ray studies. Furthermore, the isotopic composition of the egg-bearing limestones differs substantially from that of the eggshells from the same level (Folinsbee et al. 1970; Erben et al. 1979).

2. The possible influence of the various chemical components of the ancient environment must also be taken into account. For example, it is well established that there is a relationship between the $\delta^{13}C$ ratio of the

Figure 11.22. Stratigraphical distribution of described egg species.

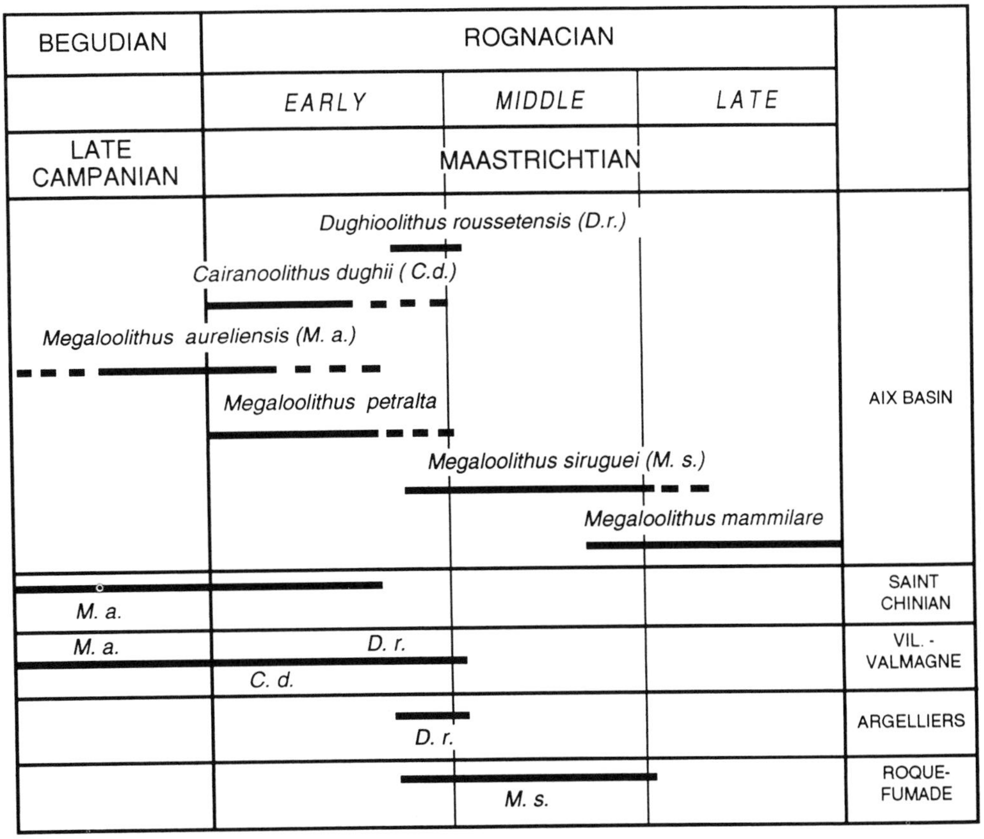

organic matrix in vertebrates bones and the diet of the animal (De Niro & Epstein, 1978; Bocherens et al. 1988). The effect of diet does not appear so clear for the mineralized component of the animal's bones or for eggshells. Erben et al. (1979, p. 405) are very cautious and conclude that observations on chicken eggshells "perhaps indicate that both food and dissolved carbonate carbon participate in the formation of eggshells." They were not very optimistic for the $\delta^{18}O$ ratio of the eggshell carbonates: "the result is not yet very promising in our endeavor to get information on the climate and environment of extinct reptiles from oxygen isotopes in eggshells" (Erben et al. 1979, p. 403).

Morin (1989), on the other hand, analyzing Penner's collection of thin sections from La Cairanne, Rousset-Erben and Roquehautes–Grand Creux, showed there is a clear correlation between $\delta^{18}O$ and $\delta^{13}C$ (Fig. 11.23). It appears that $\delta^{18}O$ is controlled by the isotopic composition and concentration by evaporation of drinking water. Moreover, strontium and magnesium concentrations are related to the chemical composition of drinking water (Folinsbee et al., 1970; Renard, 1985).

Iatsoura et al. (1991, p. 1344) conclude that

dinosaur eggshell geochemistry is clearly constrained by drinking water and food composition. Therefore, environmental changes can be deduced from geochemical composition of the eggs. In the Aix-en-Provence basin, a contamination by Sr from the nearby Trias below the Calcaires de Rognac Formation and a modification of fluvial influxes above is recognized. Characterization of climatic evolution is more complex because external temperature does not directly interfere with the shell's isotopic composition at the time of carbonate precipitation. As for Cretaceous-Tertiary boundary, no important modification of

Figure 11.23. Correlation between the ^{18}O and ^{13}C ratios of eggshell carbonates from Aix Basin.

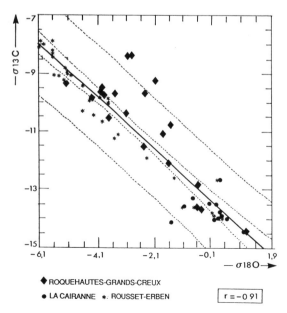

♦ ROQUEHAUTES-GRANDS-CREUX

● LA CAIRANNE ∗. ROUSSET-ERBEN   r = −0.91

shell chemical composition can signify the proximity of a crisis.

We would like to note that this conclusion did not consider the different eggshell types analyzed. The authors assumed that the different dinosaurs had eaten the same food and had drunk the same water, and that they had used the chemical elements in the same metabolic way to build their eggshells. This may have been true, but it needs demonstration such as by chemical testing of the different eggshell types found at the same stratigraphic level. This is all the more important because several egg species succeed one another in the Roquehautes-Grand Creux section, where temporal stability of Strontium and increase of $\delta^{18}O$ have been determined.

The geochemical analysis of the stable isotopes and trace elements of eggshell carbonates could give interesting information on the paleoenvironment of the egg-layers (even though this may not give any indication of the possible reasons for the K/T extinction). Furthermore, eggshells must be treated as fossils (i.e., remains of biological objects) and not as anonymous pieces of carbonate.

Organic matrix

Another approach, complementing the microstructural study of dinosaur eggshells, is the biochemical analysis of the organic matrix preserved in fossil material (Voss-Foucart, 1968; Kolesnikov & Sochava, 1972; Mallan, 1990; Marin & Dauphin 1991). Such studies deal with the amino acid composition of the organic material recovered after removal of the eggshell component. This is usually done with ethylene dinitrilotetracetic acid at neutral pH (EDTA, 0.5M, pH 7.5).

The amino acid composition of the soluble protein matrix of dinosaur eggshells is presented in Table 11.1. This material was obtained from eggshells found in two Cretaceous deposits in the Gobi Desert (Kolesnikov & Sochava, 1972) and southeastern France (Voss-Foucart, 1968; Mallan, 1990; Marin & Dauphin, 1991). Comparison of the amino acid distributions shows that significant differences exist. Notable differences are even seen in eggshells from the same localities, as is the case for extracts of eggshell from La Cairanne and Rousset-Erben specimens (Fig. 11.24). These samples from the Aix-en-Provence Basın have been analyzed separately by Mallan (1990) and by Marin and Dauphin (1991). The amino acid content of the La Cairanne and Rousset-Erben specimens studied by Mallan are much more comparable to the distributions observed in eggshells from the Gobi Desert (Fig. 11.25).

It is noteworthy that all dinosaur eggshell amino acid composition shows the same global pattern. These amino acids can be classified according to their observed percentages: high values (>8%) for aspartic acid, glutamic acid, glycine, serine, and alanine; intermediate

Table 11.1. Amino acid composition of dinosaur eggshell soluble matrix (in %)

Amino acid	Gobi	CAIR-1	CAIR-2	ROUS-1	ROUS-2	RH	FA	VF-1	VF-2
ASP	11,3	14,0	12,2	8,2	1,9	7,6	14,2	10,5	13,3
THR	5,8	8,4	8,1	4,4	7,9	16,8	7	6,9	5,5
SER	9,4	9,0	4,2	12,5	11	7,5	4,6	6,6	11,9
GLU	13,2	11,4	20	12,0	15,1	7,6	16,5	11,4	13,3
PRO	4,7	5,8	4,2	5,2	8,1	3,8	11,7	6,4	–
GLY	13,2	10,9	25,5	18,2	18,2	25	13,8	10,3	19,7
ALA	8,6	11,8	16,7	8,3	22,3	21,1	15,2	10	9
CYS	1,2		0,8	–	0,0	–	–	–	–
VAL	8,7	6,0	4,8	4,8	7,2	4,3	6,1	7	5,6
MET	0,7	1,3	–	1,5	–	–	–	0,9	–
ILE	0,0	3,4	1,7	2,9	4,6	4,1	3,1	4	3,6
LEU	4,3	5,6	1,7	6,1	4,6	4,1	3,1	7,6	7
TYR	3,1	1,7	0,4	2,7	0,2	0,2	tr	3,7	2,6
PHE	2,7	3,1	tr	3,0	0,5	0,3	1,7	4	2,6
HIS	1,0	0,8	1,8	1,9	0,5	0,5	0,5	1,2	–
LYS	6,5	3,7	tr	4,2	0,3	0,8	0,3	5,5	5,6
ARG	0,1	2,4	0,5	4,1	0,9	0,5	5,5	3,8	–

Data sources: Kolesnikov and Sochava (1972) for Gobi; Mallan (1990) for La Cairanne (CAIR-1) and Rousset-Erben (ROUS-1); Marin and Dauphin (1991) for La Cairanne (CAIR-2), Rousset-Erben (ROUS-2), Roquehautes (RH) and Fox Amphoux (FA); Voss-Foucart (1968) for White span (VF-1) and brown pellicle (VF-2) from a Rognacian deposit in Provence.

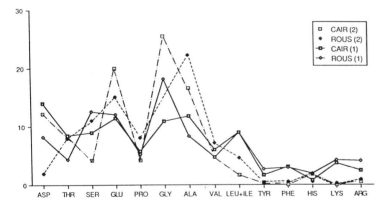

Figure 11.24. Comparison of amino acid content (%) of dinosaur eggshells from La Cairanne and Rousset-Erben. (Data from Mallan, 1990, CAIR-1 and ROUS-1; and Marin & Dauphin, 1991, CAIR-2 and ROUS-2.)

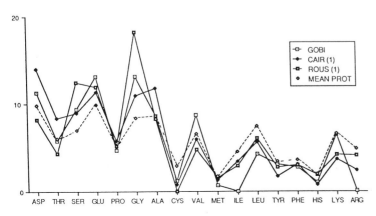

Figure 11.25. Amino acid content (%) of dinosaur eggshells from the Gobi Desert (from Kolesnikov & Sochava, 1972) and southern France: CAIR-1 and ROUS-1 (from Mallan, 1990) compared with the composition of a "standard protein" (from Dayhoff et al., 1978).

values (4–8%) for threonine, proline, valine, leucine, ly-sine, and arginine; and low values (<4%) for cysteine, methionine, isoleucine, tyrosine, phenylalanine, and his-tidine. These amino acid distributions have been com-pared with those obtained for modern hen and turtle eggshells (Krampitz, Erben, & Kriesten, 1972; Salevsky & Leach, 1980; Mallan, 1990). In Table 11.2 and Figure 11.26, the three amino acid distributions available for hen eggshells are similar and show the same general pattern as previously described for the dinosaur egg-shells.

This similarity between recent and fossil amino acid profiles has led some authors to conclude that there exist "paleoproteins" in dinosaur eggshells. Krampitz et al.(1972) suggest that reptile and bird shell proteins should show differences in their amino acid composi-tions. However, it is difficult to compare precisely these different distributions for two reasons.

First, the general pattern observed in fossil and recent eggshell amino acid distributions is quite similar to the profile of a "standard protein" (Figs. 11.25 and 11.26). The composition of this standard protein is the average of the amino acid content of 314 proteins as calculated by Dayhoff, Hunt, and Hurst-Calderone (1978). Actually, the amino acid analysis of the eggshell matrix was performed on samples composed of a mix-

Table 11.2. *Amino acid composition of turtle and chicken eggshell matrix compared to standard protein (in %)*

Amino acid	Turtle	Hen (1)	Hen (2)	Hen (3)	Mean prot.
ASP	10,5	8,9	8,0	8,2	9,8
THR	8,3	6,2	5,2	6,1	6,1
SER	10,0	9,4	7,38,5		7,0
GLU	11,7	11,3	12,6	11,4	9,9
PRO	6,0	6,0	11,3	8,4	5,2
GLY	7,7	13,3	16,0	13,2	8,4
ALA	7,1	10,6	7,7	6,9	8,6
CYS	1,9	0	1,8	4,5	2,9
VAL	5,6	6,6	7,5	6,2	6,6
MET	0,9	1,3	1,5	2,4	1,7
ILE	3,5	3,3	3,2	2,8	4,5
LEU	6,1	6,0	5,6	6,2	7,4
TYR	3,5	1,9	1,3	1,6	3,4
PHE	4,9	3,0	1,6	2,3	3,6
HIS	2,2	2,3	3,2	2,9	2,0
LYS	7,0	4,0	2,8	3,4	6,6
ARG	3,2	5,8	3,5	5,7	4,9

Data sources: Mallan (1990) for Hermann's turtle and hen (1); Krampitz et al. (1972) for hen (2); Salevsky and Leach (1980) for hen (3); Dayhoff et al. (1978) for an "average protein."

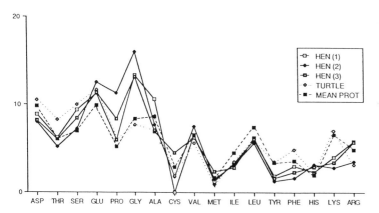

Figure 11.26. Amino acid composition (%) of modern turtle and chicken egg matrix compared with the distribution of a "standard protein" (= "average protein"). Abbreviations are as in Table 11.2.

ture of unknown proteins. As a result, the amino acid composition gives only a little information compared to the amino acid content from bone extract. In bone, organic components correspond essentially to the collagen whose composition is very specific (one-third glycine and traces of hydroxyproline and hydroxylysine).

Second, when comparing the amino acid of dinosaur eggshells from many sources, the possibility of contamination by bacteria or fungi from the surrounding sediment remains a possibility. For this reason, it is necessary to use other techniques, such as electrophoresis, to analyze fossil eggshell extracts.

Preliminary studies have been performed in our laboratory using electrophoresis on SDS-polyacrylamide slab gels in a gradient followed by protein silver staining. Electrophoresis has been carried out on eggshell extracts, as well as on organic material recovered from the sediment surrounding the eggshells. The electrophoretic patterns observed show several bands in common between the eggshell and the sediment. There are also discrete bands peculiar to the eggshells samples. Hence, it seems that the original organic matrix remains in the fossil dinosaur eggshells (Fig. 11.27).

Finally, progress in the study of fossil proteins requires a better knowledge of the protein composition

Figure 11.27. Electrophoretic pattern of soluble extracts of eggshells from Rousset–Erben (well No. 2) and La Cairanne (well No. 4), also of eggshells and their surrounding sediments (Rousset: well No. 3; La Cairanne: well No. 5). Electrophoretic markers (collagen: MW = 120,000; albumin: MW = 70,000; actin: MW = 42,000) are shown in well No. 1. Electrophoresis was performed on a vertical polyacrylamide slab gel with a gradient between 4.4% and 17.5%. The eggshell specific bands are marked with a star.

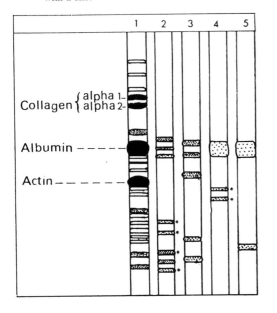

of the eggshells of present-day animals. Yet little is known about chicken eggshell composition, and until now, few proteins have been identified. These proteins include carbonic anhydrase (Krampitz, Engels, & Helfgen, 1974), chonchiolin (Krampitz & Graser, 1988), and collagen (Schleich & Kestle, 1988), although the presence of the latter protein has been controversial (Baker & Balch, 1962). The study of homologous proteins and the use of sensitive techniques such as immunological methods could improve the analysis of fossil eggshell proteins and its application to the study of dinosaur eggs. We are now working in that direction.

Conclusions

We have reviewed our present knowledge of dinosaurs eggs from southern France. To assess the different paleobiological studies on a concrete and practical, systematic basis, we have used a parataxonomic classification. Among the diagnostic criteria used in this classification, those observed in radial thin section are the most reliable.

In the previous ten to fifteen undescribed egg types of Dughi and Sirugue, there are at least seven egg species. One is the dinosauroid prismatic type of Kérourio (1982) from the Campanian and Upper Maastrichtian that resembles the eggs of the hypsilophodontid *Orodromeus*, described by Hirsch and Quinn (1990). The six other egg species belong to the dinosauroid organizational group. The four species of *Megaloolithus* nov. are referred to the dinosauroid tubospherulitic structural morphotype. The egg species of *Cairanoolithus* and *Dughioolithus* are questionably attributed to the same structural morphotype.

The diversity of eggshell structures found in dinosaur remains from different basins of southern France is greater than that discussed here. Several localities rich in eggshell are now under study. The stratigraphic position for the egg species is now known. The eggs characterize three periods in southern France: (1) Upper Campanian + Lower Rognacian/Lower Maastrichtian, (2) Middle Rognacian/Middle Maastrichtian, and (3) Upper Rognacian/Upper Maastrichtian (Fig. 11.22).

The eggs found in France show a close affinity with those from the Indian Intertrappean. Some of the Indian eggshells can be referred to *Megaloolithus*, but with the present state of study, it is not possible to draw conclusions as to the specific level. The extraction of organic material seems to be possible and may provide support for further paleobiological studies.

Acknowledgments

We are particularly indebted to J. Guiraud, who made, in a short time, the fine thin sections used in our studies. We would like to thank F. Sirugue for driving us around in the Aix-en-Provence Basin, B. Sigé and B. Marandat in the Villeveyrac Basin, and the colleagues of the Laboratory of Paleontology that accompanied us. We are grateful to J. Galtier and B. Meyer-Berthaud who gave us their time and the use of their cameras and microscopes. We also wish to thank J. Martin, who printed the photos.

References

Alabouvette, B., Arthaud, F., Bambier, A., Freytet, P., & Paloc, H. 1982. Notice explicative de la feuille de Saint-Chinian à 1/50,000. *Bureau de Recherche Géologique et Minière*.

Babinot, J. F., & Durand, J. P. 1980. Valdonien, Fuvélien, Bégudien, Rognacien, Vitrollien. Les étages français et leurs stratotypes. *Mémoires du Bureau de Recherche Géologique et Minière* 109: 171–92.

Babinot, J. F., & Freytet, P. 1983. Le Sénonien supérieur continental de la France méridionale et de l'Espagne septentrionale: état des connaissances biostratigraphiques. *Géologie Méditerraneenne* 10: 245–68.

Baker, J. R., & Balch, D. A. 1962. A study of the organic material of hen's egg shell. *Biochemistry Journal* 82: 352–61.

Beetschen, J. C. 1986. Sur les niveaux à coquilles d'oeufs de Dinosauriens de la région de Rennes-le-Château (Aude). *Actes du Colloque Les Dinosaures de la Chine à la France 1986*: 113–26 (Toulouse: Muséum de Toulouse).

Bocherens, H., Fizet, M., Cuif, J.-P., Jaeger, J.-J., Michard, J. G., & Mariotti, A. 1988. Premières mesures d'abondances isotopiques naturelles en ^{13}C et en ^{15}N de la matière organique fossile de Dinosaure. Application à l'étude du régime alimentaire du genre *Anatosaurus* (Ornithischia, Hadrosauridae). *Compte rendu hebdomadaire des séances de l'Académie des Sciences* 306: 1521–5.

Crochet, J. Y. 1984. Géologie et paléontologie de la partie septentrionale du fossé Oligocene des Matelles (Hérault, Sud de la France). *Géologie de la France* 1-2: 91–104.

Dayhoff, M. O., Hunt, L. T., & Hurst-Calderone, S. 1978. Composition of proteins. *In* M. O. Dayhoff, (ed.), *Atlas of Proteins Sequence and Structure* 5: 363–73.

De Niro, M. J., & Epstein, S. 1978. Influence of diet and the distribution of carbon isotopes in animals. *Geochemica Cosmochemica Acta* 42: 495.

Dughi, R., & Sirugue, F. 1957. La limite supérieure des gisements d'oeufs de Dinosauriens dans le bassin d'Aix-en-Provence. *Compte rendu hebdomadaire des séances de l'Académie des Sciences* 245: 907–9.

———1958. Sur les oeufs de dinosaures du bassin fluviolacustre de Basse-Provence. *83rd Congrès des Sociétés Savantes, section des Sciences*, pp. 183–205.

———1962. Distribution verticale des oeufs d'oiseaux fossiles de l'Eocène de Basse Provence. *Bulletin de la Société Géologique de France* 4: 69–78.

———1976. L'extinction des dinosaures à la lumière des gisements d'oeufs du Crétacé terminal du Sud de la France, principalement dans le bassin d'Aix-en-Provence. *Paléobiologie Continentale* 7: 1–39.

Durand, J.-P. 1989. Le synclinal de l'Arc et la limite Crétacé–Paleocene. *In La limite Crétacé–Tertiaire dans le bassin de l'Arc (Sud-Est, France). Cahiers de la réserve géologique de Provence* 1: 2–10.

Erben, H. K. 1970. Ultrastrukturen und Mineralisation rezenter und fossiler Eischalen bei Voegeln und Reptilien. *Biomineralisation* 1: 1–66.

Erben, H. K., Ashraf, A. R., Krumsiek, K., & Thein, J. 1983. Some dinosaurs survived the Cretaceous "final event." *Terra Cognita* 3: 211–12.

Erben, H. K., Hoefs, J., & Wedepohl, K. H. 1979. Paleobiological, and isotopic studies of eggshells from a declining dinosaur species. *Paleobiology* 4: 380–414.

Erben, H. K., & Newesely, H. 1972. Kristalline Bausteine und Mineralbestand von Kalkigen Eischalen. *Biomineralisation* 6: 32–48.

Folinsbee, R. E., Fritz, P., Krouse, H. R., & Robblu, A. R. 1970. Carbon, and oxygen-18 in dinosaur, crocodile, and bird eggshells indicate environmental conditions. *Science* 168: 1553.

Freytet, P. 1981. Crétacé supérieur, Paléocène, Eocène. *In* Notice de la Carte Géologique de la France au 1/50,000 Pézenas. *Bureau de Recherche Géologique et Minière*, pp. 13–14.

Galbrun, B. 1989. Résultats magnétostratigraphiques à la limite Rognacien–Vitrollien: Précisions sur la limite Crétacé–Tertiaire dans le bassin d'Aix-en-Provence. *In* La limite Crétacé–Tertiaire dans le bassin de l'Arc (Sud-Est, France). *Cahiers de la réserve géologique de Provence* 1: 34–7.

Hansen, H. J., Gwodz, R., & Rasmussen, K. L. 1989. The continental Cretaceous–Tertiary boundary in the Aix-en-Provence region, South France). *In* La limite Crétacé–Tertiaire dans le bassin de l'Arc (Sud-Est, France). *Cahiers de la réserve géologique de Provence* 1: 43–50.

Hirsch, K. F. 1989. Interpretations of Cretaceous and pre-Cretaceous eggs and shell fragments. *In* D. D. Gillette & M. G. Lockey (eds.), *Dinosaur Tracks and Traces*, (New York: Cambridge University Press), pp. 89–97.

Hirsch, K. F., & Packard, M. J. 1987. Review of fossil eggs and their shell structure. *Scanning Microscopy* 1: 383–400.

Hirsch, K. F., & Quinn, B. 1990. Eggs and eggshell fragments from the Upper Cretaceous Two Medicine Formation of Montana. *Journal of Vertebrate Paleontology* 10: 491–511.

Iatsoura, A., Cojan, I., & Renard, M. 1991. Variations de température exprimées par la géochimie des coquilles d'oeufs de dinosaures (Maastrichtien; Bassin d'Aix-en-Provence, France). *Compte rendu hebdomadaire des séances de l'Académie des Sciences* 312: 1343–9.

Kérourio, P. 1981. La distribution des coquilles d'oeufs de dinosauriens multistratifiés dans le Maestrichtien continental du Sud de la France. *Geobios* 14: 533–6.

———1982. Un nouveau type de coquille d'oeuf présumé Dinosaurien dans le Campanien et le Maestrichtien continental de Provence. *Paleovertebrata* (Montpellier) 12: 141–7.

———1987. Les nids de dinosaures en Provence. *La Recherche* 18: 256–7.

Kolesnikov, C. M., & Sochava, A. V. 1972. A paleobiochemical study of Cretaceous dinosaur eggshell from the Gobi. *Paleontological Journal* 2: 235–45.

Kurzanov, S. M., & Mikhailov, K. E. 1989. Dinosaur eggshells from the lower Cretaceous of Mongolia. *In* D. D. Gillette & M. G. Lockley (eds.), *Dinosaur Tracks and Traces*, (New York: Cambridge University Press), pp. 109–113.

Krampitz, G., Erben, H. K., & Kriesten, K. 1972. Uber Aminosäurenzusammensetzung und Struktur von Eischalen. *Biomineralisation* 4: 87-99.

Krampitz, G., Engels J., & Helfgen, I. 1974. Uber das Vorkommen von Carboanhydratase in der Eischale des Huhnes. *Experientia* 30: 228–9.

Krampitz, G., & Graser, G. 1988. Molecular mechanisms of biomineralisation in the formation of calcified shells. *Angewandte Chemie Internazional* 27: 1145–56.

Krumsiek, K., & Hahn, G. G. 1989. Magnetostratigraphy near the Cretaceous/Tertiary boundary near Aix-en-Provence (Southern France). *In* La limite Crétacé–Tertiaire dans le bassin de l'Arc (Sud-Est, France). *Cahiers de la réserve géologique de Provence* 1: 2–10.

Mallan, P. 1990. Etude des coquilles d'oeufs de dinosaures du bassin d'Aix-en-Provence (France): microstructure et matière organique. *Mémoire de D.E.A., Université de Montpellier* II: 1–37 (Unpublished dissertation.)

Marin, F., & Dauphin, Y. 1991. Composition de la phase protéique soluble des coquilles d'oeufs de dinosaures du Rognacien (Crétacé) du Sud Est de la France. *Neues Jahrbuch für Geologie und Paläontologie* 4: 243–55.

Matheron, P. 1869. Notice sur les reptiles fossiles des dépôts fluvio-lacustres cretaces du bassin à lignite de Fuveau. *Mémoires de l'Académie Impériale des Sciences, Belles-Lettres et Arts de Marseille 1869*: 345–79.

Mattauer, M., & Thaler, L. 1961. Découverte d'oeufs et d'os de Dinosaures dans le Crétacé terminal des environs de Montpellier (Hérault). *Comptes rendus sommaires des séances de la Société Géologique de France* 1: 7–8.

Mikhailov, K. E. 1987a. The principal structure of avian eggshell: data of SEM studies. *Acta Zoologica Cracoviensis* 30: 53–70.

—— 1987b. Some aspects of the structure of the shell of the egg. *Paleontological Journal* 21: 54–61.

—— 1991. Classification of fossil eggshells of amniotic vertebrates. *Acta Paleontologica Polonica* 36: 193–238.

Morin, N. 1989. Géochimie des coquilles d'oeuf de Dinosaures du Sud de la france. *Mémoire de D.E.A., Université de Paris* VI: 1–30 (Unpublished dissertation.)

Penner, M. M. 1983. Contribution à l'étude de la microstructure des coquilles d'oeufs de Dinosaures du Crétacé supérieur dans le bassin d'Aix-en-Provence (France): Application Biostratigraphique. Doctoral thesis, University of Paris VI. *Mémoires des Sciences de la Terre* 83: 1–234.

1985. The problem of dinosaur extinction. Contribution of the study of terminal Cretaceous eggshells from Southeast France. *Geobios* 18: 665–9.

Renard, M. 1985. Geochimie des carbonates pelagiques: Mise en evidence des fluctuations des eaux oceaniques depuis 140 ma. Essai de chimiostratigraphie. *Document du Bureau de Recherche Géologique et Minière* 85: 1–650.

Salevsky, E., & Leach, R. M. 1980. Studies on the organic components of shell gland fluid and the hen's egg shell. *Poultry Science* 59: 438–43.

Schleich, H. H., & Kestle W. 1988. *Reptile Egg-Shells SEM Atlas*. (Stuttgart: Gustav Fischer).

Sittler, C. 1989. The fluvio-lacustrine mineralogical environment at the Cretaceous-Tertiary boundary in the Aix-en-Provence basin (SE France). *In* La limite Crétacé-Tertiaire dans le bassin de l'Arc (Sud-Est, France). *Cahiers de la réserve géologique de Provence* 1: 11–15.

Sochava, A. V. 1971. Two types of eggshells in Senonian Dinosaurs. *Paleontological Journal* 5: 353–61.

Straelen, V., Van & Denaeyer, M. E. 1923. Sur les oeufs fossiles du Crétacé supérieur de Rognac en Provence. *Bulletin de l'Académie Royale des Sciences de Belgique* 9: 14–26.

Vianey-Liaud, M., Jain, S. L., & Sahni, A. 1987. Dinosaur eggshells (Saurischia) from the Late Cretaceous Intertrappean and Lameta formations (Deccan, India). *Journal of Vertebrate Paleontology* 7: 408–24.

Voss-Foucart, M. F. 1968. Paléoprotéines des coquilles fossiles d'oeufs de dinosauriens du Crétacé Supérieur de Provence. *Comparative Biochemistry and Physiology* 24: 31–6.

Westphal, M. 1989. Magnétostratigraphie du Crétacé supérieur du Bassin d'Aix-en-Provence. Etude préliminaire. *In* La limite Crjétacé–Tertiaire dans le bassin de l'Arc (Sud-Est, France). *Cahiers de la réserve géologique de Provence* 1: 28–33.

Williams, D. L. G., Seymour, R. S., & Kérourio, P. 1984. Structure of fossil dinosaur eggshell from the Aix Basin, France. *Paleogeography, Paleoclimatology, Paleoecology* 45: 23–37.

Zhao Z. 1975. The microstructure of the dinosaurian eggshells of Nanhsiung, Kwangtung. *Vertebrata PalAsiatica* 13: 105–17.

—— 1979. The advancement of researches on the dinosaurian eggs in China. *South China Mesozoic and Cenozoic "Red Formation."* (Beijing: Science Publishing Co.), pp. 329–40.

Zhao Z., Ye J., Li H., Zhao Z., & Yan Z. 1991. Extinction of the dinosaurs across the Cretaceous–Tertiary boundary in Nanxiong Basin, Guandong Province. *Vertebrata PalAsiatica* 29: 1–20.

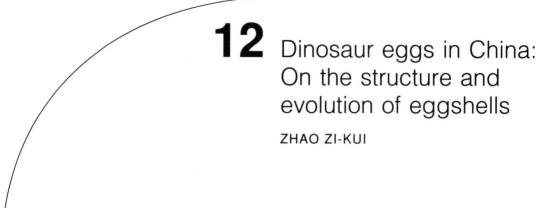

12 Dinosaur eggs in China: On the structure and evolution of eggshells

ZHAO ZI-KUI

Abstract

Since 1951, several hundred more or less complete eggs and many eggshell fragments have been discovered in the vast area of China. These discoveries are examined for clues to understanding the evolution and distribution of dinosaur eggs during the Cretaceous period and the extinction of the dinosaurs at the end of the Cretaceous. The Cretaceous dinosaur egg horizons and egg faunas of China, the evolution of the dinosaur eggshell, and the possible cause of the extinction of the dinosaurs are discussed.

The known Cretaceous dinosaur eggs of China are classified according to their macromorphological and histostructural characters into five families: Dictyoolithidae, Faveoloolithidae, Dendroolithidae, Spheroolithidae, and Elongatoolithidae. The stratigraphy of the eggshell-bearing Cretaceous beds and egg faunas known from three representative basins are briefly discussed.

Since the evolution of the dinosaur eggshell structure was directional, proceeding along definite courses during the Cretaceous period; there is no doubt that the calcareous eggshell allowed the dinosaurs to make quick use of available ecological and evolutionary opportunities. It is suggested that changes in eggshell structure near the end of the long dinosaur history reveal that these animals were fighting a losing battle against a drying environment.

Introduction

Dinosaur eggs are a special group of vertebrate fossils. The opportunities for eggs to be preserved as fossils are usually far fewer than those for bones. As a result, fossilized dinosaur eggs have always been regarded as rare items. Although the study of dinosaur eggs has been carried on for more than a hundred years, it is still in its early stage. The earliest reported fossil eggshells come from southern France (see Buffetaut and Le Loeuff, Chapter 2). Matheron (1869) believed that they might be dinosaurian, but the paleontological value of these specimens was virtually ignored at that time.

Dinosaur eggs were first discovered in the Mongolian Gobi area by the Central Asiatic Expedition of the American Museum of Natural History in the 1920s, and created a sensation (van Straelen, 1925; Andrews, 1932). Since then, our knowledge of dinosaur eggshells has increased considerably.

On a global scale, dinosaur eggs and eggshell fragments have been reported from China and Mongolia (Andrews, 1932; Sochava, 1969; Kurzanov & Mikhailov, 1989; Mikhailov, 1991a; Mikhailov, Sabath & Kurzanov, Chapter 7), Kazakhstan (Bazhanov, 1961), Uzbekistan and Kirghizia (Nesov & Kaznishkin, 1986), Romania (Grigorescu et al., 1990, Chapter 6), India (Sahni & Gupta, 1982; Sahni, 1989; Mohabey, 1983, 1984a; Jain, 1989; Sahni et al., Chapter 13), Japan (Azuma, personal communication, 1989), Peru (Sige, 1968), Brazil (Price, 1951), the United States (Jepsen, 1931; Jensen, 1966, 1970; Horner, 1982; Hirsch & Packard, 1987; Hirsch & Quinn, 1990; Hirsch, Chapter 10), France, and Spain (Lapparent, 1947; Dughi & Sirugue, 1964, 1966; Erben, 1970; Erben, Hoefs, & Wedepohl, 1979; Williams, Seymour, & Kerourio, 1984; Cousin et al., Chapter 5; Vianey-Liaud et al., Chapter 11) (see also Carpenter and Alf, Chapter 1). Most of these eggs are Cretaceous in age.

Dinosaur eggs in China first received scientific attention when Chow (1951) reported on fossil eggshells and dinosaur bones in the Upper Cretaceous Wangshih Series of Laiyang County, Shandong Province. During the 1950–51 field season, the famous fossil skeleton of the hadrosaur *Tsintaosaurus spinorhinus* and many complete eggs and eggshell fragments were collected and described respectively by Young (1954, 1958, 1959) and Chow (1954). By the beginning of the 1960s, the discovery of the rich collection of dinosaur eggs in Nanxiong Basin, Guangdong Province, stirred up new interest (Zhang & Tong, 1963; Young, 1965; Zheng et al., 1973). It also stimulated a more intense search for fossil eggs in other regions.

In the last two decades, with the growth in our knowledge of the stratigraphical distribution of dinosaur eggshells and the advancement and availability of modern techniques, new types of eggshell structure have been found in many new localities (Zhao & Jiang, 1974; Zhao, 1975, 1978, 1979a, 1979b; Zhao & Ding, 1976; Zeng & Zhang, 1979; Zhao & Li, 1988; Mateer, 1989). New techniques have opened new areas for detailed studies of these egg specimens, the nesting habits of dinosaurs, and the possible causes for the extinction of the dinosaurs. However, a systematic study of Chinese dinosaur eggs has so far not been made, and a good number of specimens remain uninvestigated. I provide a general review of the fossil eggshell record from China and discuss their classification, evolutionary trend of eggs, and the role of eggs in the extinction of the dinosaurs at the end of Cretaceous.

Cretaceous dinosaur egg horizons and egg faunas of China

China has extensive Mesozoic continental deposits known as "Redbeds" because of their color. These deposits are extremely rich in dinosaur eggs. Since 1951, localities of Cretaceous dinosaur eggs have been successively discovered in many different regions (Fig. 12.1): from Shandong Peninsula in the east to Xinjiang Uygur Autonomous Region in the west and from the Gobi Desert in Inner Mongolia Autonomous Region in the north, to Guangdong Province in the south. However, only in Shandong, Henan, and Guangdong provinces do the dinosaur eggs occur with an appreciable frequency.

Identification and classification

Considerable differences in size and shape occur in Chinese dinosaur eggs, as well as in the structure of the eggshell and in the arrangement of the eggs within the clutch (Chow, 1951, 1954; Young, 1954, 1959, 1965; Zhao & Jiang, 1974; Zhao, 1975, 1979a, 1979b; Zhao & Ding, 1976; Zhao & Li, 1988). Differences also occur in the stratigraphical distribution of the egg types. Because of these differences, the first step in the study of the biology and stratigraphical distribution of the dinosaur eggs is the classification of the eggs into an orderly system (i.e. naming, describing, and classifying all known dinosaur eggs).

Because dinosaur eggshells are independent of dinosaurian skeletons, they are never structurally connected with the latter, making it difficult to assign a certain egg to a certain dinosaur taxon. The exception occurs when the eggs contain embryonic skeletons or the nest contains eggshells and hatchlings. Such embryos and hatchlings have been found in the Upper Cretaceous (Campanian) Two Medicine Formation of western Montana, and have been identified as hypsilophodontid and hadrosaurid (Horner, 1982, 1984, 1987; Horner & Gorman, 1988; Horner & Weishampel, 1988). This discovery allowed Hirsch and Quinn (1990) to identify specific eggshell structural morphotypes with known dinosaur taxa. Such discoveries are rare, and some paleontologists have attempted to identify eggs on the basis of dinosaur bones found in the same strata or deposit. Probably this method is not accurate because fossil eggs, in general, are very rarely associated with bones. When such an association is found, there is no sound reason to assume that the eggs were laid by the animal whose bones are present.

An additional problem in the identification of di-

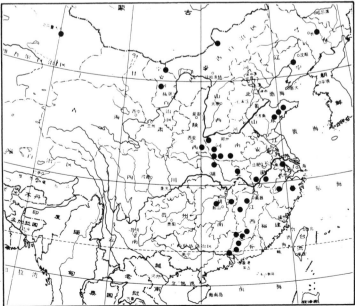

Figure 12.1. Distribution of dinosaur eggs in China.

nosaur eggs is that our studies of fossil eggs are based on modern eggshells. Most reptiles today produce parchment-covered eggs bearing little structural resemblance to the calcareous eggshells of the dinosaurs. This difference makes comparison and taxonomic identification of dinosaur eggs difficult (Zhao & Jiang, 1974).

In spite of these problems, there has been much speculation about the identity of the egg layers. For example, the eggshells from the Djadochta beds of Shabarakh Usu, Mongolia, have been assigned to the genus *Protoceratops*, while the eggshells from Erlian (Iren) Dabasu beds, Inner Mongolia, China, have been assigned to the hadrosaurs (van Straelen, 1925). Chow (1951, 1954) and Young (1954) identified as hadrosaur two types of eggshell from Laiyang County, Shandong Province. Sochava (1969, 1971) divided the eggshells from the Mongolian Gobi Desert into three types based on the pattern of their pore canals. One of these types – the angusticaniculate eggshells with a ribbed outer surface (ornithoid) formerly thought to belong to *Protoceratops* – probably belongs to the hadrosaurs.

Erben Hoefs, and Wedepohl (1979) classified eggshells from Mongolia, France, and Spain into four types, which more or less agreed with Sochava's classification based on eggshell microstructure and pore canals. Some of the eggshells from France and Spain, previously attributed to the sauropod *Hypselosaurus*, have tentatively been assigned to the genus *Megalosaurus*, a theropod (Voss-Foucart, 1968). Williams et al. (1984) distinguished at least four types of dinosaur eggshells from southern France alone, whereas Penner (1985) recognized three types, and Dughi and Sirugue (1958, 1976) nine types. In China, dinosaur eggs have

been classified on the basis of similarities or differences in their macromorphological and histostructural characters. This scheme is similar to that used in the classification of fossil spore pollen.

I have previously suggested, based on my study of the dinosaur eggs from Shandong and Guangdong provinces, a general scheme for the classification of all known dinosaur eggs (Zhao 1975, 1979a). This classification consists of defining groups or categories of dinosaur eggs on a hierarchical scale. The result is that all the dinosaur eggs can be classified in ascending rank, from the "species" to the "family" or "order."

According to my observations, the structural characters may be somewhat arbitrarily arranged under three headings:
1. Macromorphological characters
    a. Shape and size of the egg
    b. Sculpturing of the outer shell surface
    c. Thickness of the eggshell
2. Histostructural characters
    a. Basic structural units and pore canal including size, shape, and arrangement of the mammilla, cone, columna, and type of pore canal system
    b. Texture of eggshell including the sequence and composition of horizontal ultrastructural zones
3. Ethological characters of the nest

All of these characters have received increasing attention as taxonomic criteria for dinosaur eggshells, particularly in the pattern of the basic structural unit and the pore canal (e.g., Mikhailov et al., Chapter 7; Hirsch, Chapter 10; Vianey-Liaud et al., Chapter 11).

The structure of the modern avian eggshell has been studied extensively and has been used as the standard for comparison (Fig. 12.2; Schmidt, 1965; Erben, 1970; Hirsch, 1979; Hirsch & Quinn, 1990; Mikhailov, 1991a). Under the scanning electron microscope (SEM) and polarizing light microscope (PLM), the single structural unit of the dinosaur eggshell is more or less similar to that of Aves. The basic structural units in different dinosaur eggshells show striking variations in size, shape, and arrangement. Therefore, these characters are important in the diagnosis and separation of the various dinosaur eggshells.

In addition, the texture of eggshell has also been used in structural classification (Erben, 1970; Hirsch & Quinn, 1990; Mikhailov, 1991a, 1991b). However, such features may be characteristic of the whole "family" or a series of "families" and thus may be more important in understanding categories at higher taxonomic levels.

Ethological characters are also important in egg taxonomy because it is well known that the pattern of egg arrangement within the clutch differs from group to group. For example, the elongated egg group contains several different morphotypes of eggshells. Yet these elongated eggs are always arranged in two or more circular layers within the clutch. This suggests that the

Figure 12.2. Schematic diagram and terminology of eggshell structure based on the avian eggshell. (After Erben, 1970 and Hirsch & Quinn, 1990, slightly modified.)

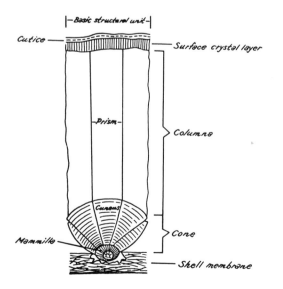

nesting behavior of the different animals is much the same and that they should be regarded as being systematically very close. In China, at least three different egg arrangement patterns have been found.

In egg taxonomy, the delimitation and ranking of the different groups of dinosaur eggs is often subjective. Nevertheless, they appear to have a biological and structural correlation with some objective criteria. One major problem is that it is often not possible to assign most dinosaur eggs to known dinosaur taxa. Therefore, it is necessary to provide labels for the different dinosaur eggshells to facilitate communication among paleontologists. Every taxonomic rank should be assigned a name derived from characteristics of the fossil eggshells. I have suggested using a binary nomenclature for the different forms of fossil eggshell under the rules of zoological nomenclature (Zhao, 1975, 1979a). Such names are not definitions, nor are they descriptions. They are merely handles by which the form of the eggshell is known. To avoid possible misunderstanding or confusion, I have suggested that generic name of the dinosaur eggshell should have the uniform ending *-oolithus* (Zhao, 1975, 1979a).

As has been mentioned, there have been other attempts at dinosaur egg classification, with a consequent lack of agreement about the classification and morphotypes of the specimens and about terminology. Like any scheme of classification based upon a single character or a set of closely related characters, my classification and nomenclature for the dinosaur eggs is not perfect, but it does give a useful arrangement for the study of groups of dinosaur eggs, as we shall discuss later.

Dinosaur egg horizons and egg fauna

One of the major advances in eggshell paleontology in China during the past two decades is the discovery of abundant Cretaceous eggshells and the recognition of thick sequences of Cretaceous strata in several provinces. In the middle and southern regions of China, there are many intermontane basins with Cretaceous and Tertiary clastic sediments of considerable thickness. Because the sediments in these basins are almost universally dark red or purplish red in color, they are called the Redbeds of South China by Chinese geologists. Owing to the scantiness of fossils, opinions differed for a long time about the age, correlation, and subdivisions of these beds.

By the beginning of 1960s, dinosaur eggs were discovered in Nanxiong Basin in the northern part of Guangdong Province. Field work has gradually extended to many other basins with similar redbeds, resulting in a large amount of information on the stratigraphy and in large collections of dinosaur eggshells.

The discovery of Cretaceous dinosaur eggs, especially those from Lower Cretaceous strata, is of considerable interest, not only to geologists using them to date and subdivide the redbeds, but also to paleontologists to gain a general view of the evolution and distribution of dinosaur eggs. Because most of the information on specimens collected during the past two decades has not yet been published or is still in preparation, only a short account of the stratigraphy of the eggshell-bearing Cretaceous beds and egg faunas from three representative basins is given below.

Dinosaur eggs from Nanxiong Basin

Nanxiong Basin is an intermontane basin situated mostly in northern Guangdong Province, which encompasses Nanxiong and Shixing counties, and in the western part of Xinfeng County in Jiangxi Province. The basin is elongated with its axis oriented northeast – southwest. Upper Cretaceous dinosaur eggs and Paleocene mammals were discovered in 1961. Inspired by this discovery, further investigations were carried out, and a stratigraphical sequence based on fossil vertebrates was established. Within the basin the topography is a hilly terrain, with rolling hills less than 50 m high. The Upper Cretaceous and Tertiary sediments attain a maximum thickness of over 7,000 m. The exposed strata can be divided into three groups and five formations (Zhao et al., 1991).

The Upper Cretaceous Nanxiong Group, formerly called the Nanxiong Formation, consists of the Yuanpu and Pingling formations. The Yuanpu Formation comprises, from bottom to top, a sedimentary cycle of fanglomerate, fluviolacustrine, and lacustrine facies. The Pingling Formation consists of fluvial deposits. The overlying early Tertiary Loufuzhai Group is darker colored and finer grained than the Nanxiong Group, and is made up of the Shanghu, Nonshan, and Gucheng formations. The Danxia Group is a suite of coarse, sandy conglomerate along the north margin and at both ends of the basin. This unit overlaps or intertongues with the Nanxiong and Luofuzhai groups.

The fossils from Nanxiong Group include nearly 300 more or less well-preserved eggs and more than 40,000 eggshell fragments collected by the IVPP and other institutions. In addition, several eggs preserving embryos have been found recently in the Pingling Formation. This discovery will provide a wealth of information concerning the dinosaur that laid the eggs.

Dinosaur eggshells are restricted to the uppermost and middle parts of the Yuanpu Formation, and throughout the Pingling Formation. The egg fauna consists of the following taxa:

Yuanpu Formation

Elongatoolithidae Zhao, 1975

    Macroolithus yaotunensis Zhao, 1975

    Macroolithus rugustus Young, 1965

    Macroolithus sp. nov.

    Elongatoolithus andrewsi Zhao, 1975

    Elongatoolithus elongatus Young, 1954

    Elongatoolithus sp. nov.

Nanshiungoolithus chuetienensis Zhao, 1975
Spheroolithidae Zhao 1979
    Ovaloolithus cf. *O. chinkangkouensis* Zhao
    and Jiang, 1974
Pingling Formation
    Elongatoolithidae Zhao, 1975
        Macroolithus yaotunensis Zhao, 1975
        Macroolithus rugustus, Young, 1965
        Macroolithus sp. nov.
        Elongatoolithus andrewsi Zhao, 1975
        Elongatoolithus elongatus Young, 1954
        Elongatoolithus sp. nov.
        Nanshiungoolithus chuetienensis Zhao, 1975
    Spheroolithidae Zhao, 1975
        Ovaloolithus cf. *O. chinkangkouensis* Zhao
        and Jiang, 1974
        Ovaloolithus cf. *O. laminadermus* Zhao and
        Jiang, 1974
        Ovaloolithus sp. nov.
        Shixingoolithus erbeni Zhao et al., 1991
    Fam. indet.
        Stromatoolithus pinglingensis Zhao et al.,
        1991

Besides the egg fauna, the Nanxiong Group also contains four dinosaurian species, – *Tarbosaurus* sp., *Nanshiungosaurus brevispinus*, *Nemegtosaurus* sp., and *Microhadrosaurus nanshiungensis* (Dong, 1979) – and the large turtle, *Nanshiungchelys wuchingensis* (Ye, 1966). This assemblage indicates an Upper Cretaceous age for the Nanxiong Group. This is supported by recent magnetostratigraphy and K/Ar dates that indicate that the Pingling Formation and the upper part of the Yuanpu Formation are Maastrichtian in age (Zhao et al., 1991).

Dinosaur eggs from the Laiyang area

In the Laiyang-Zhucheng Basin of Shandong Province, there occurs a thick series of redbeds named the Wangshih Series that unconformably overlies the Lower Cretaceous Chinshan Formation. The geology of the Laiyang area was studied by Tan (1923) and Wang (1930). The Wangshih Series is subdividable into three lithologic units: (1) a lower *Cyrena* shaley sandstone, (2) a middle clayey silt with lenses or sheets of sandstone and conglomerate, and (3) an upper sandstone intercalated with conglomerate.

More than 100 completely preserved dinosaur eggs and many eggshell fragments have been discovered in Laiyang County. All the fossils, including dinosaur bones and eggs, come from the middle and upper parts of the Wangshih Series. The egg fauna includes:

Upper part of Wangshih Series
    Spheroolithidae Zhao, 1979a
        Paraspheroolithus irenensis Zhao and Jiang,
        1974
        Ovaloolithus chinkangkouensis Zhao and

        Jiang, 1974
        Ovaloolithus tristriatus Zhao, 1979a
        Ovaloolithus mixtistriatus Zhao, 1979a
        Ovaloolithus monostriatus Zhao, 1979a
        Ovaloolithus laminadermus Zhao and Jiang,
        1974
    Elongatoolithidae Zhao, 1975
        Elongatoolithus andrewsi Zhao, 1975
        Elongatoolithus elongatus Young, 1954
Middle part of Wangshih Series
    Elongatoolithidae Zhao, 1975
        Elongatoolithus andrewsi Zhao, 1975
    Spheroolithidae Zhao, 1979a
        Spheroolithus chiangchintingensis Zhao and
        Jiang, 1974
        Paraspheroolithus irenensis Zhao and Jiang,
        1974
        ?*Spheroolithus megadermus* Young, 1959

Other fossils associated with the eggs are mostly dinosaurs, especially hadrosaurs. The hadrosaurs indicate that the Wangshih Series possibly represents only a part of the Upper Cretaceous (Young, 1958). Recently, a fossil skeleton of a protoceratopsid dinosaur was found in the upper part of the Wangshih Series in Jiaoxian County adjacent to Zhucheng County (Hu & Cheng, 1988). This dinosaur suggests that these beds may correlate with the Djadokhta Formation of Mongolia, believed to be Santonian-Campanian age (Savage & Russell, 1983).

Based on the stratigraphical distribution of the dinosaur eggshells, *Ovaloolithus* seems to be restricted to the upper part of the Wangshih Series, and *Spheroolithus* to the middle part. *Paraspheroolithus* is most predominant in the middle part. The eggshell structure of *Spheroolithus* is similar to that of prolatocanaliculate eggshells described by Sochava (1969) from the Turonian and Cenomanian of the Gobi Desert. In consideration of this similarity, it is safer to regard the age of the Wangshih Series as early-Middle Upper Cretaceous.

Dinosaur eggs from Xixia Basin

The Xixia Basin, in Xixia and Neixiang counties, Henan Province, is situated on the eastern flank of the Qinling Range. It is an elongated basin with its axis oriented northwest–southeast. The redbeds unconformably overlying Proterozoic rocks are extensively developed in the basin and can be divided into three formations: the Sigou, Majiacun, and Gaogou formations. The dinosaur eggs from these formations (Fig. 12.3) are mostly new forms and their eggshell structures are more primitive than those from Nanxiong and Laiyang basins. The egg-bearing formations in the Xixia Basin are considered to be Lower Cretaceous based on spore–pollen assemblages (Song, 1965). Whether they represent a part or the entire Lower Cretaceous is a question impossible to settle at present.

The egg fauna mainly includes the following taxa:

> Elongatoolithidae Zhao 1975
>   *?Elongatoolithus* sp.
> Spheroolithidae Zhao, 1979a
>   *Paraspheroolithus* sp. nov.
>   *Spheroolithus* sp. nov.
> Dendroolithidae gen. et sp. nov.
> Faveoloolithidae gen. et sp. nov.
> Dictyoolithidae Zhao
>   *Dictyoolithus hongpoensis* Zhao
>   *Dictyoolithus neixiangensis* Zhao

Because collecting in this basin is still in progress, only a brief diagnosis of *Dictyoolithus* is given here to make the new name available for use.

Systematics

Dictyoolithidae fam. nov.
Diagnosis. Same as for the genus.

Dictyoolithus gen. nov.
Diagnosis. Radial view of eggshell with reticulated organization framed by irregular basic structural units.

Dictyoolithus hongpoensis sp. nov.
Holotype. Two broken eggs. Field No. 79001.
Locality and horizon. Hongpo, Chishuigou, Xixia County, Henan Province. Lower Cretaceous, Sigou Formation.
Diagnosis. Eggs nearly oval. Eggshell about 2.5–2.8 mm thick. Outer surface smooth with grainy appearance or with very low, sculptured nodes. Basic structural units vary in length and width; separated at their origin forming interstices. Shell layer of five or more superimposed shell units with a reticulate organization (Fig. 12.4). Shape of pores irregular.

Dictyoolithus neixiangensis sp. nov.
Holotype. Three eggs. Field No. 79007.
Locality and horizon. Shibangou, Chimei, Neixiang County, Henan Province. Lower Cretaceous, Gaogou Formation.
Diagnosis. Eggs spheroidal, with longitudinal diameter about 120 mm. Eggshell 1.5–1.7 mm thick. Outer surface and shell microstructure similar to *Dictyoolithus hongpoensis*, but with only two or three superimposed shell units (Fig. 12.5).

Evolution of the dinosaur eggshell
The eggshell taxa known from China include over thirty species allocated to five families and eleven genera. The egg assemblages span most of the Cretaceous, thus giving us a general view of the evolution of dinosaur egg faunas in Asia. Because most of the fossils

have not yet been studied in detail or adequately described, evolutionary inferences are of very preliminary nature and subject to modification.

Structural characteristics of the eggshell
Cretaceous dinosaur eggs found in China fall into two categories on the basis of the structural characteristics of the eggshell. One category contains the Elongatoolithidae, of which eggs from Mongolia were thought to belong to *Protoceratops*. These were later were called angusticanaliculate by Sochava (1969, 1971). The other category includes the Spheroolithidae,

Figure 12.3. Stratigraphic distribution of dinosaur eggshell from Xixia Basin. **1**. *Dictyoolithus hongpoensis*. **2**. *Spheroolithus* n. sp., *Paraspheroolithus* n. sp., Dendroolithidae n. gen., n.sp. Faveoloolithidae n. gen., n. sp. **3**. *Spheroolithus* n. sp., *Paraspheroolithus* n. sp. and *Dictyoolithus neixiangensis* n. sp.

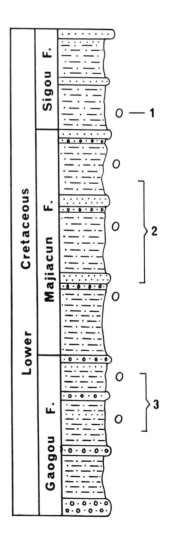

Dendroolithidae, Faveoloolithidae, and Dictyoolithidae.

Members of the Elongatoolithidae, *Elongatoolithus*, *Macroolithus*, and *Nanshiungoolithus* are major components in all the Upper Cretaceous egg faunas, irrespective of their stratigraphic level. The eggs are elongated and somewhat obtuse at one end and slightly pointed at the other. The outer surface of the eggshell is covered by a discontinuous chain of subcircular or elongate nodes separated by deep channels. The egg clutches are characterized by a very regular circular arrangement of the eggs. Radial views of the eggshell show elements and layers characteristic of avian eggshell. The pore canals are narrow and more or less straight, having regular and almost subparallel walls.

The basic structural units are relatively small and closely crowded together. The cone is similar to that of the avian egg. A radial view of *Macroolithus rugustus* (Fig. 12.6A) reveals the junctions between the cunei radiating outward from the mammilla. The zones of tabular structure in the cunei are visible. The columns in the columnar layer are laterally interlocking, with some tendency to fuse. The herringbone pattern of the columnar layer is less developed. Instead, the columnar layer shows an undulating sublayering, roughly parallel to the outer surface of the eggshell (Fig. 12.6B). The Elongatoolithidae made its first appearance in early Late Cretaceous, perhaps in the Early Cretaceous, and was very abundant and diversified by the late Late Cretaceous.

In the second category, represented by the Spher-oolithidae, Dendroolithidae, Faveoloolithidae, and Dictyoolithidae, the egg shape is more or less spherical. Based on microstructural characteristics of the eggshell, four structural levels can be recognized.

Avianlike shell structure

This eggshell type is represented by *Ovaloolithus* and *Shixingoolithus* of the family Spheroolithidae. These eggs differ in that *Ovaloolithus* is an oval shape, while *Shixingoolithus* is nearly spherical. In radial views the difference between them is even more pronounced. The eggshell structure is tighter and less porous in *Ovaloolithus*. The shell units are tall and slender, and originate from relatively low mammillae on the shell membrane. The units are tightly packed and interlocked forming a continuous layer in which only the slender prisms (columns) with well-marked horizontal sublayering (growth layers?) can be recognized (Fig. 12.7A). The cone is thin, one-ninth or less of the eggshell thickness. The columns can be divided into inner and outer zones. The prisms of the inner zone are arranged in a columnar manner, and those of outer zone radiate from the inner zone to the surface. Hence, they show an interlacing appearance (Fig. 12.7B). The outer zone probably corresponds to the "corymbs" described by Schmidt (1968) in avian eggshells. The pore canals are almost straight.

Shixingoolithus also has an avianlike structure (Fig. 12.8). The thickness of the cone layer is 0.6 mm, approximately one-fourth that of the eggshell. The crystallites of adjacent shell units are interlocking, with some fusion. The subvertical structural margins are slightly divergent and faintly developed. The herringbone pattern of the columnar layer is usually well developed. No horizontal sublayering occurs. The pore canals are narrow but irregular.

Figure 12.4. *Dictyoolithus hongpoensis*, radial thin section, PLM. Bar = 100 μm.

Figure 12.5. *Dictyoolithus neixiangensis*, radial thin section, PLM. Bar = 100 μm.

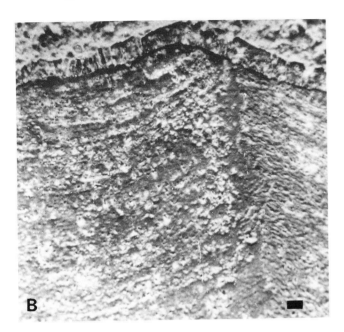

Figure 12.6. *Macroolithus rugustus*; Upper Cretaceous, Nanxiong Basin (No. 72092), SEM. **A**. Cone in lateral view. Bar = 5 μm. **B**. Outer part of columnar layer; note growth lines in the columnar layer. Bar = 100 μm.

Figure 12.7. *Ovaloolithus chinkangkouensis*, Upper Cretaceous, Upper Wangshih Series, Laiyang, Shandong Province (V735), SEM. **A**. Radial view of eggshell. Note slender shell units (U) and pore canal (arrow). Bar = 100μm. **B**. External zone of columnar layer. Bar = 5 μm

Primitive avianlike eggshell structure

This type of egg is recorded from Shandong, Henan, Hubei, Xinjiang, and Inner Mongolia. *Spheroolithus* and *Paraspheroolithus* of the Spheroolithidae, and Dendroolithidae and the Mongolian ornithoid called prolatocanaliculate by Sochava (1969, 1971) may be referable to this type. Incomplete structural units are the most characteristic feature of this type of eggshell.

In *Spheroolithus* (Fig. 12.9), the cone is apparently irregular in shape, and fairly isolated at its origin, so noticeable spaces remain between the cones. The cunei of each cone are not compressed together, and intercuneate spaces occur between the cunei. The columns in the columnar layer are laterally interlocking with some fusion. The subvertical structural margins are slightly divergent. The pore canals are irregular in shape, are variable in diameter along their length, and are contorted.

In Dendroolithidae, the eggshell is also composed of basic structural units of irregular shape (Fig. 12.10). These incomplete structural units in most cases do not abut tightly, but are laterally interlocking with some fusion near the outer surface of the eggshell. The subvertical structural margins are slightly divergent. The pore canals, originating in intercone spaces, are expanded in diameter and branched in their middle part to form a labyrinth. Most of them terminate immediately beneath the outer surface of the eggshell, and only a

Figure 12.8. *Shixingoolithus erbeni*, Upper Cretaceous Pingling Formation, Nanxiong Basin (No. 901). Radial thin section. Bar = 100 μm.

small number of them perforate the outer surface.

The Dendroolithidae are present in Hubei and Henan provinces and possibly also in Inner Mongolia (Zhao & Li, 1988); they have also been recently discovered in the Mongolian Gobi area (Mikhailov, 1991a).

Honeycomblike eggshell structure

This type of eggshell includes several genera of Faveoloolithidae that are in the process of being described. The Faveoloolithidae, established by Zhao and Ding (1976), were first known from the Albian–Cenomanian of the Gobi Desert, Mongolia (Sochava, 1969, 1971), and were assigned the name multicanaliculate. The discovery of these eggs in the Cretaceous of China is important for correlation and for an understanding of the phylogeny and provenance of the groups concerned.

Youngoolithus is the most noteworthy of the Faveoloolithidae. Its basic structural units are relatively small and rather irregular in shape. The individual units have varying lengths and widths and are fairly isolated at their origin. Radial views of the eggshell reveal that they usually consist of two or three superimposed shell units (Fig. 12.11). The pore canals are more or less straight, extremely numerous, and closely spaced, looking as a whole like a honeycomb.

Reticulated-like eggshell structure

This type of egg, represented by the Dictyoolithidae (Figs. 12.4 & 12.5), has only been discovered in the Lower Cretaceous of Xixia Basin, Henan Province. As has been shown above, the eggshell is composed of five or six superimposed, irregular structural units. The radial views of the eggshell show a reticulated-like organization.

Hypothesis for shell formation in dinosaur eggs

The Elongatoolithidae, Spheroolithidae, and Dendroolithidae all have an avianlike shell structure. The crystalline layer is composed of a single layer of well-defined, tightly packed, and interlocking shell units. Therefore, it is reasonable to assume that formation of these and avian eggshells might have occurred by fundamentally similar mechanisms.

The physiological mechanisms producing avian eggshells are well known (Simkiss 1968; Simkiss & Taylor, 1971; Erben, 1970; Board, 1982). The initial step of forming the eggshell in birds is the production

Figure 12.9. *Spheroolithus chiangchintingensis*, Upper Cretaceous, Middle Wangshih Series, Laiyang, Shandong Province (G5547). Radial view of eggshell. Note shell units with incomplete cones (U) and irregular pore canal (arrow). Bar = 100 μm.

of small, organic cores on the outer surface of the shell membrane. These cores serve as a nucleus for the deposition of crystalline calcite. The primary direction of calcite growth is both upwards and lateral because of the presence of the underlying shell membrane. Further calcification produces the cone layer of the shell. Eventually the calcite crystals fuse to form the columnar layer or the spherulitic units, which grow outward. As a result, there is a clear demarcation between the fibers of shell membrane and the crystalline layer of the eggshell as illustrated in Figure 12.2. The formation of the shell membrane and crystalline layer occurs sequentially (Fujii, 1974), with the shell membrane formed before the initiation of calcification.

In view of the comparable shell structures in bird, Elongatoolithidae, Spheroolithidae, and Dendroolithidae

eggs, it is reasonable to assume that these dinosaur eggs were formed by a mechanism similar to that in living birds. The eggshell structures in the Faveoloolithidae and Dictyoolithidae differ from those of birds, Elongatoolithidae, Spheroolithidae, and Dendroolithidae. Therefore, the mechanism and process of calcareous eggshell formation might have been different from those observed in extant birds.

A clue to interpreting the formation of these dinosaur eggs may be derived from the formation of eggs in the tuatara, *Sphenodon punctatus*, presented by Packard et al. (1988). The shell membrane and the calcareous columns in tuatara eggs form more or less simultaneously, rather than sequentially; that is, membrane formation may slightly precede crystal formation, thereby providing active sites for nucleation of the crystalline

Figure 12.10. Dendroolithidae, Cretaceous, Xichuan Basin, Henan Province (No. 75030). Radial view. Note incomplete shell units (U). SEM. Bar = 100 μm.

shell units during the initial stage of the eggshell formation. Afterward, the growth of both components is closely matched for a time. In the Faveoloolithidae and Dictyoolithidae, egg development might begin this way and then new components are formed over the first one. Additional shell units and shell membranes are repeatedly formed until growth of the crystalline shell units exceeds that of the shell membrane. The result is an eggshell in which the shell units are framed into a reticulated- or honeycomblike structure. This hypothetical description is depicted in Figure 12.12.

Trends in dinosaur eggshell evolution

The geological history of dinosaur eggs is comparatively restricted, being represented mostly from the Cretaceous. Their scarcity, or even absence, in earlier geological times has been explained by Sochava (1969, 1971), who suggested that the eggs may have had parchment shells as in many extant reptiles (except the crocodile, some turtles, and the gecko). The chance of fossilization for parchment eggshells is slim, and the same may have been true for the majority of pre-Cretaceous dinosaur eggs. This suggestion is supported by the apparent primitive structural characteristics of the

Dictyoolithidae discussed above. The eggs of the extant *Sphenodon* (Packard, Hirsch, & Meyer-Rochow, 1982; Packard et al., 1988) might provide the best analogy.

The development of a rigid calcareous eggshell is an adaptive achievement to satisfy certain functional needs, such as to maintain moisture needed by the embryo and to protect it against mechanical injury. It is possible, therefore, that the later development of the calcareous eggshell in the evolution of dinosaurs may have been caused by changes in the paleoecological conditions. When the calcareous eggshell developed is not certain, but recent discoveries in the Jurassic sediments of South Africa and North America (Grine, 1987; Grine & Kitching, 1987; Hirsch, Young, & Armstrong, 1987; Hirsch et al., 1989; Hirsch, 1989; Chapter 10) indicate that the rigid calcareous eggshell first appeared earlier than previously had been thought.

From the discussions, the evolution of dinosaur eggshell structure during the Cretaceous was apparently directed along certain definite courses from the Early Cretaceous to the end of the Late Cretaceous. The Elongatoolithidae developed an avianlike eggshell structure as shown by the basic structural units that changed very little throughout the Cretaceous. This family appeared

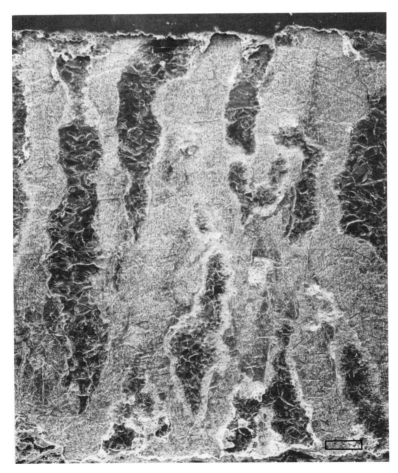

Figure 12.11. *Youngoolithus xiaguanensis*, Cretaceous, Xiaguan Basin, Henan Province (No. 75029). Radial view; SEM. Bar = 100 μm.

in the early Late Cretaceous – perhaps in the Early Cretaceous – and continued until the extinction of the dinosaurs at the end of the Cretaceous. During the late Late Cretaceous, this group was among the most common type of dinosaur egg. The elongatoolithid eggs were the most successful of the dinosaur eggs, if geographical and stratigraphical distribution is the criterion for success.

The other line of dinosaur eggshell evolution involves the Spheroolithidae, Faveoloolithidae, and Dictyoolithidae. Eggshells of this category evolved during the Cretaceous along three definite courses that led to a considerable variation in shell structure. Each of the three evolutionary trends may be an adaptation to the selective pressure imposed by environmental changes. A summary of the three evolutionary courses in this category is given in Figure 12.13.

The first trend is toward the avianlike eggshell structure that made the eggshell more effective in preventing dehydration of the egg. This evolutionary

branch is represented by the Dendroolithidae and Spheroolithidae. The Dendroolithidae and certain genera of Spheroolithidae appeared in Early Cretaceous, continued through early Late Cretaceous, and were replaced by *Shixingoolithus* and *Ovaloolithus*, which had more advanced avianlike eggshell structure, in the late Late Cretaceous.

The second trend seems to have evolved toward the honeycomblike structure, which was a new pattern. This evolutionary line is seen in members of Faveoloolithidae. It starts from an ancestral type toward the primitive form, *Youngoolithus*, in the Early Cretaceous, and was replaced in the early Late Cretaceous by *Faveoloolithus*.

The third trend, even more aberrant than those of Faveoloolithidae, is represented by the members of Dictyoolithidae in Early Cretaceous. They are characterized by a reticulated-like organization formed by irregular shell units.

Figure 12.12. Schematic diagram of shell formation in the Dictyoolithidae and Faveoloolithidae.

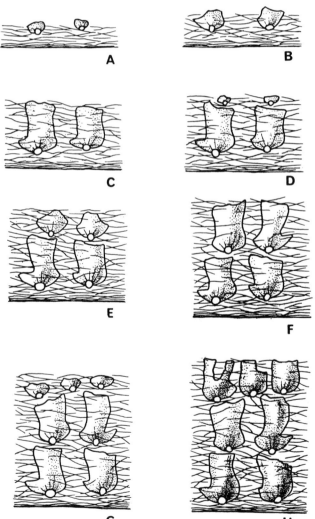

This account, brief as it is, may give some idea of the rather involved history of the dinosaur eggshells during the Cretaceous period. It is summarized in Figure 12.14.

Cause of the extinction of the dinosaurs

It is reasonable to assume that improvements in the dinosaur eggshell structure during the Cretaceous period opened new areas for the evolution of dinosaurs. These changes provided the mechanism by which these animals could liberate themselves completely from any dependence upon water during their life history. Thus, the dinosaurs could quickly make use of the available ecological and evolutionary opportunities. However, they all vanished from the earth in the global events at the end of the Cretaceous. The cause poses a difficult question for which no ready answer is apparent. Many now believe that the extinction was caused by the direct or indirect effects of the collision of a large meteoritic body with the earth, as first proposed by Alvarez et al. (1980).

The extinction of the dinosaurs appears to have been a sudden global phenomenon. However, we do not have an adequate global record for the extinction of di-nosaurs. The best record is from the Western Interior of North America. Another area that holds promise for clues to dinosaur extinction is the fluviolacustrine section in the Nanxiong Basin of Guangdong Province. As has been mentioned above, a sequence of vertebrate faunas there spans the Cretaceous–Paleocene boundary. Sedimentation was cyclic and continuous, and much of the physical environment can be reconstructed. My own research and that carried out by our research group in the Nanxiong Basin provide some direct evidence for interpreting the dinosaur extinction (Zhao, 1978, 1990; Zhao et al., 1991).

As indicated above, dinosaur eggs occur with an appreciable frequency in the Nanxiong Basin. At least twelve species of dinosaur eggshell have been identified from the Upper Cretaceous Pingling Formation. Their stratigraphic distribution shows that they disappear within the early–middle span of geomagnetic chron 29R. Thus, the number of species and number of dinosaur eggshells found are reduced abruptly and disappear during the last 200,000 to 300,000 years of the Cretaceous. Only one "species," *Macroolithus yaotunensis*, ranges up to the Cretaceous–Paleocene boundary, where it becomes extinct.

Figure 12.13. Diagram showing evolutionary trends of the eggshell structure in spheroid eggs. **A.** Dendroolithidae. **B, C.** Spheroolithidae. **D.** Faveoloolithidae. **E, F.** Dictyoolithidae.

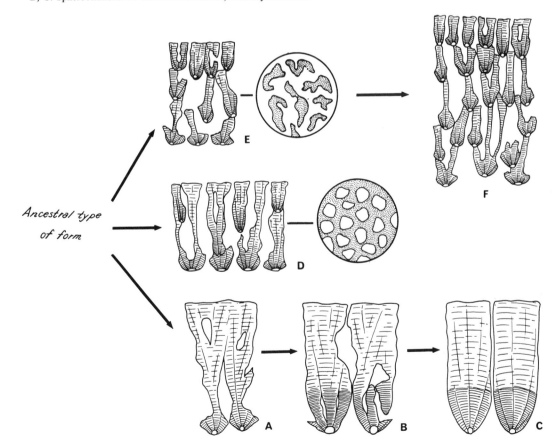

Quite unexpectedly in our study, many eggshells revealed abnormalities. Two types of pathological development were reported by Zhao (1975, 1978, 1990) and Zhao et al. (1991) and occur almost within the horizons containing the last dinosaur eggshells. One pathology is an anomalous variation in eggshell thickness. A tendency toward a reduction in the average values of shell thickness is revealed from the Yuanpu Formation up to the Pingling Formation. The other pathology affects shell structure, especially in the formation of bi- or multilayered cones (Fig. 12.15). This pathological development is different from the teratological cases called "ovum in ovo" in living birds (Romanoff & Romanoff, 1949) and "multilayered eggshells" in living and fossil turtles and dinosaurs (Erben 1970, 1972; Erben et al., 1979; Ewert, Firth, & Nelson, 1984; Mohabey 1984b; Hirsch 1989, 1990; Hirsch et al., 1989).

The abnormal phenomena of ovum in ovo and multilayered eggshells may be caused by unintended retention of egg in the oviduct beyond the time of normal oviposition. The physiological mechanisms producing ovum in ovo are caused by repetitive eggshell formation in the oviduct (Sturkie, 1965). This occurs in birds, not because of a deficiency in vasotocin, but because the egg is forced by antiperistaltic movements back to the isthmus where another layer of shell is laid down. In reptiles retention of eggs occurs due to abnormal bio-logical or environmental conditions. Multilayered eggshells result without antiperistaltic movements because the formation of the shell takes place in one long section of the oviduct that is homologous to the isthmus in birds. As a result, several eggs may receive multiple layers at one time (Aitken & Solomon, 1976; Ewert et al., 1984; Hirsch et al., 1989).

The microstructural characteristics of the eggshell in Figure 12.15 suggests that the eggshell with a bi- or multilayered cone was apparently not caused by the antiperistaltic movements of the oviduct or by the retention of eggs beyond the time of normal oviposition. Instead, the formation of this pathological eggshell may have been correlated with changes in the protein profiles of the organic matrix in the calcareous eggshell. Although the overall mechanism of eggshell formation is not fully understood on a molecular basis, it is quite clear that the organic matrix plays a key role during eggshell calcification. According to Krampitz and Witt (1979) and Krampitz (1982), the protein profiles in an avian eggshell can be influenced by environmental toxins, such as Cd, Pb, Hg, Be, etc., and changes in the protein profile may in turn lead to a pathological formation of eggshell.

A study was made of the trace elements and the isotope ratios of the dinosaur eggshells from horizons yielding a large number of pathological eggshells. The

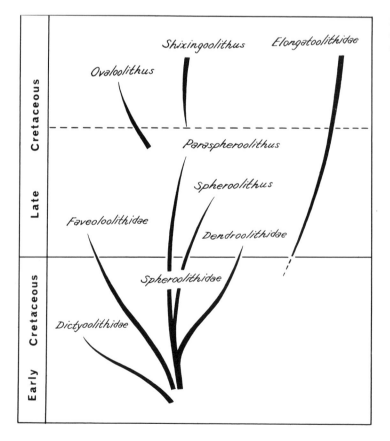

Figure 12.14. Evolution and relationships of the major groups of dinosaur eggs.

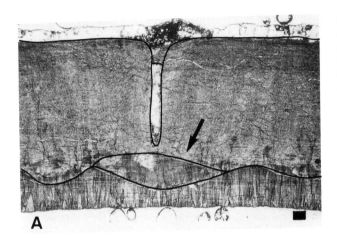

Figure 12.15. Radial thin section of pathologic eggshell, Maastrichtian, Nanxiong Basin. **A.** *Macroolithus* sp., eggshell with double-layered cone (arrow) (CGD112). Bar = 100 μm. **B.** Detail of A showing double cone, PLM. Bar = 50 μm. **C.** *Macroolithus yaotunensis*; note multilayered cone (CGD115), PLM. Bar = 100 μm.

results show that the concentration is very high for certain trace elements such as Co, Cr, Cu, Mn, Ni, Pb, Sr, V, and Zn, and the $\delta^{18}O$ values are also anomalous (Zhao et al., 1991). Therefore, it is reasonable to assume that the formation of pathological dinosaur eggshells was caused by the presence of trace elements, which changed the shell protein profile.

It is also well known that trace elements play an important role in the normal biological processes of the embryo, although most are required in minute quantity only. According to Romanoff and Romanoff (1949), the hatchability of eggs may be impaired by a deficiency of some essential mineral elements, or by an excess of toxic elements transmitted to the egg from a bird's diet.

The high level of trace elements in some dinosaur eggs and the deleterious effect that these elements can play in avian eggs suggests a new hypothesis for the extinction of the last dinosaurs. Near the end of the Cretaceous, there was an extremely dry climate in southern China. This resulted in an increase in evaporation causing an enrichment of trace elements in soil and water. These trace elements became concentrated in the dinosaurian body through the food chain. As a result, the elements interfered with the formation of eggshell and the mineral composition of the egg. Under these circumstances, the hatchability of the dinosaur eggs may have been considerably impaired and the number of progeny decreased until, finally, the population collapsed completely. This entire process of extinction, from the initial ecological changes up to the disappearance of dinosaurs, was rather short, lasting for only 200,000 to 300,000 years.

Conclusions

For the past several decades, numerous specimens of Cretaceous dinosaur eggs have been found in different parts of China. The different structural patterns of these dinosaur eggshells and their geographical and stratigraphical distribution provide important information about the evolutionary trends in the eggs. Furthermore, the eggs may provide a possible clue to explain the extinction of the dinosaurs at the end of the Cretaceous.

Eggshells are independent of dinosaurian skeletons and never structurally connected with the latter. Therefore, it is difficult to assign an egg to a certain dinosaur taxon, except for eggs containing embryonic skeletons or eggshells from nests with the remains of hatchlings.

Through careful study of shell structure, all dinosaur eggshells can be classified in a hierarchy of ascending rank from the lower taxonomic level to the higher. On this basis, all the dinosaur eggs of China can be divided into five groups: Dictyoolithidae, Faveoloolithidae, Dendroolithidae, Spheroolithidae, and Elongatoolithidae.

Although the mineral component of dinosaur eggshells is calcite, several lines of evidence indicate that eggshell formation in the five groups may occur by two different mechanisms. The process of eggshell formation in Elongatoolithidae, Spheroolithidae, and Dendroolithidae may have been fundamentally similar to that of birds, with the formation of the shell membrane and calcareous layer occurring sequentially. In contrast, in the eggshells of Dictyoolithidae and Faveoloolithidae, the shell membranes and calcareous shell units formed simultaneously.

From the Early Cretaceous to the close of the Late Cretaceous the evolution of the dinosaur eggshell followed two general lines. Elongatoolithidae, which have an avianlike eggshell structure, changed very little throughout the Cretaceous. The other line consists of Dictyoolithidae, Faveoloolithidae, Dendroolithidae, and Spheroolithidae. In general, it is possible to trace the rise and the development of these taxa along three evolutionary trends. The Dendroolithidae and Spheroolithidae evolved towards an avianlike eggshell structure, and the Dictyoolithidae and Faveoloolithidae evolved toward a reticulated-like pattern and a honeycomblike pattern, respectively.

Changes in the eggshell of these dinosaurs near the end of the Cretaceous suggest that they were fighting a losing battle against a drying land environment. The different species of dinosaur eggshell near the Cretaceous–Tertiary boundary in South China display an abnormal eggshell structure and anomalous amounts of trace elements and isotopes. These facts suggest pollution by trace elements: a distinct change in climatic conditions interfered with the formation of normal eggshell, resulting in a considerable reduction in egg hatchability. This may be a contributing cause for the extinction of the dinosaurs.

Acknowledgments

I would like to thank those who helped with advice, criticism and reading of the manuscript. The reviewers, Yang Zun-yi, Sun Ai-lin, Qiu Zhan-xiang, and Karl F. Hirsch, provided helpful comments on an earlier version of this manuscript. I am also grateful to Shen Wen-long, who drew the sketches, and to Zhao Chun, who typed the manuscript. The study was supported by funds from IVPP, and by BSFPP, Academia Sinica.

References

Aitken, R. N. C. & Solomon, S. E. 1976. Observations on the ultrastructure of the oviduct of the Costa Rican green turtle (*Chelonia Mydas* L.). *Journal of Experimental Marine Biology* 21: 75–90.

Alvarez, L. W., Alvarez, W., Asaro, F., & Michael, H. 1980. Extraterrestrial cause for the Cretaceous Tertiary extinction. *Science* 280: 1095–1108.

Andrews, R. C. 1932. *The New Conquest of Asia.* (New York: American Museum of Natural History).

Bazhanov, V. S. 1961. First discovery of dinosaur eggshells in USSR. *Trudy of Institute of Zoology, AN., KSSR* 15: 177–81.

Board, R. G. 1982. Properties of avian egg shells and their adaptive value. *Biological Reviews of the Cambridge Philosophical Society* 57: 1–28.

Chow, M. C. 1951. Notes on the Late Cretaceous dinosaurian remains and the fossil eggs from Laiyang, Shantung. *Bulletin of the Geological Society of China* 31: 89–96.

1954. Additional notes on the microstructure of the supposed dinosaurian eggshells from Laiyang, Shantung. *Scientia Sinica* 3: 523–6.

Dong Z. M 1979. The Cretaceous dinosaurs in South China. *In* IVPP and NGPI (eds.), *Mesozoic and Cenozoic Redbeds in Southern China.* (Beijing: Science Press), pp. 342–50.

Dughi, R., & Sirugue, F. 1958. Observations sur les oeufs de dinosaures du Bassin d'Aix-en-Provence: les oeufs a coquilles bistratifiées. *Comptes rendus des seances de l'Académie des Sciences* 246: 2271–4.

1964. Sur la structure des coquilles des oeufs des sauropsides vivants ou fossiles; le genre *Psammornis* Andrews. *Bulletin de la Société géologique de France* 7: 240–52.

1966. Sur la fossilisation des oeufs de dinosaures. *Comptes rendus des séances de l'Académie des Sciences* 262: 2330–2.

1976. L'extinction des dinosaures a la lumière des gisements d'oeufs du Crétacé terminal du sud de la France, principalement dans le Bassin d'Aix-en Provence. *Paleobiologie Continentale, Montpellier* 7: 1–39.

Erben, H. K. 1970. Ultrastrukturen und Mineralisation rezenter und fossiler Eischalen bei Vogeln und Reptilien. *Biomineralisation* 1: 1–66.

1972. Ultrastrukturen und Dicke der Wand pathologischer Eischalen. *Abhandlungen Mathematisch-Naturwissenschaftliche Klasse, Akademie der Wissenschaften und der Literatur* 6: 191–216.

Erben, H. K., Hoefs, J., & Wedepohl, K. H. 1979. Paleobiological and isotopic studies of eggshells from a declining dinosaur species. *Paleobiology* 5: 380–414.

Ewert, M. A., Firth, S. J., & Nelson, C. E. 1984. Normal and multiple eggshells in batagurine turtles and their implications for dinosaurs and other reptiles. *Canadian Journal of Zoology* 62: 1834–41.

Fujii, S. 1974. Further morphological studies on the formation and structure of hen's eggshell by scanning electron microscopy. *Journal of the Faculty of Fisheries and Animal Husbandry* (Hiroshima University), 13: 29–56.

Grigorescu, D., Seclamen, M., Norman, D. B. & Weishampel, D. B. 1990. Dinosaur eggs from Romania. *Nature* 346: 417.

Grine, F. E. 1987. Early Jurassic Prosauropod eggs and the evolution of sauropsid egg shell structure. *Journal of Vertebrate Paleontology* 7, supplement to No. 3: 17.

Grine, F. E., & Kitching, J. W. 1987. Scanning electron microscopy of early dinosaur egg shell structure: A comparison with other rigid sauropsid eggs. *Scanning Microscopy* 1: 615–30.

Hirsch, K. F. 1979. The oldest vertebrate egg? *Journal of Paleontology* 53: 1068–84.

1989. Interpretations of Cretaceous and pre-Cretaceous eggs and shell fragments. *In* D. D. Gillette & M. G. Lockley (eds.), *Dinosaur Tracks and Traces.* (New York: Cambridge University Press), pp. 89–97.

1990. Pathological eggshell—fossil and modern. Paleopathology Proceedings (Texas Memorial Museum), 1989 Annual Meeting of the Society of Vertebrate Paleontology.

Hirsch, K. F., and Packard, M. J. 1987. Review of fossil eggs and their shell structure. *Scanning Microscopy* 1: 383–400.

Hirsch, K. F., and Quinn, B. 1990. Eggs and eggshell fragments from the Upper Cretaceous Two Medicine Formation of Montana. *Journal of Vertebrate Paleontology* 10: 491–511.

Hirsch, K. F., Stadtman, K. L., Miller, W. E., & Madsen, J. H., Jr. 1989. Upper Jurassic dinosaur egg from Utah. *Science* 243: 1711–13.

Hirsch, K. F., Young, R. G., and Armstrong, H. J. 1987. Eggshell fragments from the Jurassic Morrison Formation of Colorado. *In* W. R. Averett (ed.), *Dinosaur Triangle Guidebook* (Grand Junction: Museum of Western Colorado), pp. 78–84.

Horner, J. R. 1982. Evidence of colonial nesting and ''site fidelity'' among ornithischian dinosaurs. *Nature* 297: 675–6.

1984. The nesting behavior of dinosaurs. *Scientific American* 250: 130–7.

1987. Ecologic behavioral implications derived from a dinosaur nesting site. *In* S. J. Czerkas & E. C. Olsen (eds.), *Dinosaurs Past and Present*, Vol. II. (Los Angeles: Natural History Museum of Los Angeles County), pp. 51–63.

Horner, J. R. & Gorman, J. 1988. *Digging Dinosaurs.* (New York: Workman Publishing).

Horner, J. R. & Weishampel, D. B. 1988. A comparative embryological study of two ornithischian dinosaurs. *Nature* 332: 256–7.

Hu C. Z. & Cheng Z. W. 1988. New progress in the restudy on *Shantungosaurus giganteus. Bulletin of the Chinese Academy of Geological Sciences* 1988: 251–8.

Jain, S. L. 1989. Recent dinosaur discoveries in India, including eggshells, nest and coprolites. *In* D. D. Gillette & M. C. Lockley (eds.), *Dinosaur Tracks and Traces* (New York: Cambridge University Press), pp. 99–108.

Jensen, J. A. 1966. Dinosaur eggs from the Upper Cretaceous North Horn Formation of Central Utah. *Brigham Young University Geology Studies* 13: 55–67.

1970. Fossil eggs in the Lower Cretaceous of Utah. *Brigham Young University Geology Studies* 17: 51–65.

Jepsen, G. L. 1931. Dinosaur eggshell fragments from Montana. *Science* 73: 12–13.

Krampitz, G. P. 1982. Structure of the organic matrix in mollusc shell and avian eggshells. *In* G. H. Nancollas (ed.), *Biological Mineralization and Demineralization* (Stuttgart: Springer-Verlag), 219–32.

Krampitz, G. P., & Witt, W. 1979. Biochemical aspects of biomineralization. *In* F. L. Boschke (ed.), *Topics in Current Chemistry*, 78: 57–144.

Kurzanov, S. M., and Mikhailov, K. E. 1989. Dinosaur eggshells from the Lower Cretaceous of Mongolia. *In* D.

D. Gillete & M. G. Lockley (eds.), *Dinosaur Tracks and Traces* (New York: Cambridge University Press), pp. 109–13.

Lapparent, A. F. de. 1947. Les dinosauriens du Crétacé Superieur de Midi de la France. *Memoire, Société Geologique de France* 56: 1–54.

Mateer, N. J. 1989. Upper Cretaceous reptilian eggs from Zhejiang Province, China. *In* D. D. Gillette & M. C. Lockley (eds.), *Dinosaur Tracks and Traces* (New Yrok: Cambridge University Press), pp. 115–18.

Matheron, M. P. 1868. Reptiles fossiles des depots fluviolacustres Crétacés du Bassin a Lignite de Fuveau. *Memoires de l'Académie Impériale des Sciences, Belles Lettres et Arts de Marseille* 19: 345–79.

Mikhailov, K. E. 1991a. Classification of fossil eggshells of amniotic vertebrates. *Acta Palaeontologica Polonica* 36: 193–238.

1991b. The microstructure of avian and dinosaurian eggshell: phylogenetic implications. *In* K. Campbell (ed.), *Papers in Avian Paleontology Honoring Pierce Brodkorb*. Contributions in Science, Natural History Museum, Los Angeles County, pp. 361–73.

Mohabey, D. M. 1983. Note on the occurrence of dinosaurian fossil eggs from Infratrappean limestone in Kheda District, Gujarat. *Current Sciences* 52: 1194.

1984a. The study of dinosaurian eggs from Infratrappean limestone in Kheda District, Gujarat. *Journal of the Geological Society of India* 25: 329–37.

1984b. Pathologic dinosaurian egg shells from Kheda District, Gujarat. *Current Sciences* 53: 701–3.

Nesov, L. A., & Kaznishkin, M. N. 1986. Discovery of dinosaur eggshells from Lower and Upper Cretaceous of USSR. *Biology* 9: 35–49.

Packard, M. J., Hirsch, K. F., and Meyer-Rochow, V. B. 1982. Structure of the shell from eggs of the tuatara, *Sphenodon punctatus*. *Journal of Morphology* 174: 197–205.

Packard, M. J., Michael, B. T., Kenneth, N. G., and Marta, V. 1988. Aspects of shell formation in eggs of the tuatara, *Sphenodon punctatus*. *Journal of Morphology* 197: 147–57.

Penner, M. M. 1985. The problem of dinosaur extinction. Contribution of the study of terminal Cretaceous eggshells from Southeast France. *Geobios, Lyon* 18: 665–70.

Price, L. I. 1951. Um ovo de dinossaurio na Formacao Bauru, do Cretacico do Estado de Minas Gerais. *Diviao de Geologia e Mineralogia, Notas Preliminares e Estudos*. 53: 1–5.

Romanoff, A. L., & Romanoff, A. J. 1949. *The Avian Egg*. (New York: John Wiley & Sons, Inc.).

Sahni, A. 1989. Paleoecology and paleoenvironments of the Late Cretaceous dinosaur eggshell sites from Peninsular India. *In* D. D. Gillette & M. C. Lockley, (eds.), *Dinosaur Tracks and Traces* (New York: Cambridge University Press), pp. 179–85.

Sahni, A., & Gupta, V. J. 1982. Cretaceous eggs shell fragments from Lameta Formation, Jabalpur, India. *Bulletin, Indian Geologists' Association* 15: 85–8.

Savage, D. E., & Russell, D. E. 1983. *Mammalian Paleofaunas of the World*. (London: Addison-Wesley Publishing).

Schmidt, W. J. 1965. Die Eisosphäriten (Basalkalotten) der Schwanen-Eischale. *Zeitschrift für Zellforschung* 67: 151–64.

1968. Die Büschelsphäriten Corymben auf der Eischale von Lappentauchern. *Zeitschrift für Zellforschung* 88: 408–14.

Sige, B. 1968. Dents de micromammiferes et fragments de coquilles d'oeufs de dinosauriens dans la faune de vertebres du Crétacé supérieur de Laguna Umayo (Andes peruviennes). *Comptes rendus des seances de l'Academie des Sciences* 267: 1495–8.

Simkiss, K. 1968. The structure and formation of the shell and shell membranes. *In* T. C. Carter (ed.), *Egg Quality/A Study of the Hen's Egg. British Egg Marketing Board Symposium* No. 4: 3–25.

Simkiss, K. & T. G. Taylor 1971. Shell formation. *In* D. J. Bell & B. M. Freeman (eds.), *Physiology and Biochemistry of the Domestic Fowl* (London: Academic Press), pp. 1331–43.

Sochava, A. V. 1969. Dinosaur eggs from the Upper Cretaceous of the Gobi Desert. *Paleontological Journal* 4: 517–27.

1971. Two types of egg shells in Cenomanian dinosaurs. *Paleontological Journal* 3: 353–61.

Song Z. C. 1965. *Palynological Analysis*. (Beijing: Science Press).

Sturkie, P. D. 1965. *Avian Physiology*, 2nd ed. (Ithaca: Cornell University Press).

Tan, H. C. 1923. New research on the Mesozoic and Early Tertiary geology in Shantung. *Bulletin of the Geological Survey of China* 5: 95–135.

Straelen, V. van 1925. The microstructure of the dinosaurian eggshells from the Cretaceous beds of Mongolia. *American Museum Novitates* 173: 1–4.

Voss-Foucart, M. F. 1968. Paleoproteines des coquilles fossiles d'oeufs de dinosauriens de Crétacé Supérieur de Provence. *Comparative Biochemistry and Physiology* 24: 31–6.

Wang H. C. 1930. The geology in Eastern Shantung. *Bulletin of the Geological Society of China* 9: 79–91.

Williams, D. L. G., Seymour, R. S., & Kerourio, P. 1984. Structure of fossil dinosaur eggshell from the Aix Basin, France. *Palaeogeography Palaeoclimatology, Palaeoecology* 45: 23–37.

Ye X. K. 1966. A new Cretaceous turtle of Nanxiong, northern Guangdong. *Vertebrata PalAsiatica* 10: 191–200.

Young, C. C. 1954. Fossil reptilian eggs from Laiyang, Shantung, China. *Scientia Sinica* 3: 505–22.

1958. The dinosaurian remains of Laiyag, Shantung. *Palaeontologica Sinica* (Ser. C) 16: 1–138.

1959. On a new fossil egg from Laiyang, Shantung. *Vertebrata PalAsiatica* 3: 34–5.

1965. Fossil eggs from Nanshiung, Kwangtung and Kanchou, Kiangsi. *Vertebrata PalAsiatica* 9: 141–89.

Zeng D. M., & Zhang J. J. 1979. On the dinosaur eggs from the western Dongting Basin, Hunan. *Vertebrata PalAsiatica* 17: 131–6.

Zhang Y. P., & Tong Y. S. 1963. Subdivision of ''redbeds'' of Nanxiong Basin, Guangdong. *Vertebrata PalAsiatica* 7: 249–60.

Zhao Z. K. 1975. The microstructures of the dinosaurian eggshells of Nanxiong Basin, Guangdong Province.

(I) On the classification of dinosaur eggs. *Vertebrata PalAsiatica* 13: 105–17.

1978. A preliminary ivestigation on the thinning of the dinosaurian eggshells of the Late Cretaceous and some related problems. *Vertebrata PalAsiatica* 16: 213–21.

1979a. The advancement of research on the dinosaurian eggs in China. *In* IVPP and NGPI (eds.), *Mesozoic and Cenozoic Redbeds in Southern China.* (Beijing: Science Press), pp. 330–40.

1979b. Discovery of the dinosaurian eggs and footprint from Neixiang County, Henan Province. *Vertebrata PalAsiatica* 17: 304–9.

1990. Mass extinction across the Cretaceous-Tertiary boundary at Nanxiong Basin, Guangdong Province. *A Monthly Journal of Science* 35: 380.

Zhao Z. K., & Ding S. R. 1976. Discovery of the dinosaur eggs from Alashanzuoqi and its stratigraphical meaning. *Vertebrata PalAsiatica* 14: 42–4.

Zhao Z. K., & Jiang Y. K. 1974. Microscopic studies on the dinosaurian eggshells from Laiyang, Shandong Province. *Scientia Sinica* 17: 73–83.

Zhao Z. K. and Li Z. C. 1988. A new structural type of the dinosaur eggs from Anlu County, Hubei Province. *Vertebrata PalAsiatica* 26: 107–15.

Zhao Z. K., Ye J., Li H. M., Zhao Z. H., & Yan Z. 1991. Extinction of the dinosaurs across the Cretaceous-Tertiary boundary in Nanxiong basin, Guangdong Province. *Vertebrata PalAsiatica* 29: 1–20.

Zheng J. J., Tang Y. J., Qui Z. X., & Ye X. K. 1973. Notes on the Upper Cretaceous-Lower Tertiary of the Nanxiong Basin, N. Guangdong. *Vertebrata PalAsiatica* 11: 18–28.

13 Upper Cretaceous dinosaur eggs and nesting sites from the Deccan volcano–sedimentary province of peninsular India

ASHOK SAHNI, S. K. TANDON,
ASIT JOLLY, SUNIL BAJPAI,
ANIL SOOD, AND S. SRINIVASAN

Abstract

A fairly extensive record of dinosaur eggshell fragments and nesting sites is now known from the Upper Cretaceous (Maastrichtian) sedimentary rocks associated with the Deccan Volcanics of peninsular India. The specimens occur beneath the basal-most basaltic flows or within the thin sedimentary horizons between flows. The fossiliferous area encompasses more than 1,000 km from east to west. Since the early 1980s, nesting sites have been identified in nine areas at Kheda, Dohad, the Hathni River, and Jabalpur in western and central peninsular India. All the nesting sites recorded so far occur in a hard, sandy carbonate. Some of the nests have been attributed to titanosaurid sauropods. Five eggshell types are recognized: (?)Titanosaurid Type I, (?)Titanosaurid Type II, (?)Titanosaurid Type III, Ornithoid type and an indeterminate dinosaurian type.

Remapping of Matley's original localities around Jabalpur (central India) was conducted, along with a detailed analysis of the sedimentary facies. Paleogeomorphic surfaces of the upper and lower contacts of the nest-bearing horizon show the preservation of a palustrine flat with some relief and characteristic lithofacies associations. Pedogenic modification has resulted in the formation of calcimorphic paleosols (calcretes) indicating a semi-arid climate. Most of the eggs in nests are apparently unhatched, although in none have embryonic remains been found preserved. Host-rock characteristics indicate stressed and hostile conditions during incubation. Extensive nesting sites with morphologically identical eggs suggest colonial nesting. The widespread distribution of nesting sites in similar lithofacies across central and western India demonstrates strong nesting site selectivity.

Introduction

Dinosaur nesting and eggshell sites from the Upper Cretaceous sedimentary sequences of the Indian subcontinent represent one the most extensive fossil hatcheries in the world (Fig. 13.1). The terms "nesting site" or "nests" are used wherever it has been possible to demonstrate the association of at least two or more eggs in close juxtaposition showing the original pattern of egg laying. Eggshell fragments are usually found as isolated specimens in screen-washed material along with microvertebrates and other microfossils. In most nesting sites, eggshell fragments are also common and probably represent the remnants of eggs displaced from their nests and later crushed and broken.

Nesting sites provide data concerning the in situ distribution of individual nests, burial and preservational conditions, paleoenvironments, and paleoclimate. While a majority of the nests belong to titanosaurid sauropods, some extremely thin eggshell fragments of ornithoid (sensu Sochava 1969, i.e., avianlike) are also known. The nesting sites extend locally over several square kilometers and are associated with a characteristic lithology, usually a sandy carbonate; they reflect paleogeomorphic surfaces contemporaneous with the eruptions of the Deccan volcanics.

Stratigraphic framework

The eggshells and dinosaurian skeletal material occur just below the Cretaceous–Tertiary Boundary on the Indian landmass. The nests and eggshell fragments occur in strata associated with the basal basaltic flows. These sediments either underlie the flows (infratrappeans = Lameta Beds), or are intercalated within them (intertrappeans, Fig. 13.2). The relationship between the infratrappean and the intertrappean sediments has long been debated and is crucial to understanding the stratigraphic framework in which the nests and eggshells are found. As used here, the terms "infratrappean" and "intertrappean" describe the physical position of the beds in relation to the lava flows (Fig. 13.3) regardless of chronology.

Contrary to the traditional viewpoint, which envisaged a time separation of up to 40 Ma between the infratrappean and the intertrappean sequences (Krishnan, 1968, Alexander, 1981), several criteria now demonstrate that the sedimentary rocks are associated with

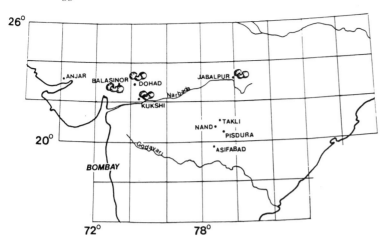

Figure 13.1. Dinosaur eggshell and nest localities of central and western peninsular India. Index map shows sauropod nesting sites (indicated by symbol showing three eggs).

Figure 13.2. Stratigraphic sections (from east to west) giving details of the egg-bearing lithologies.

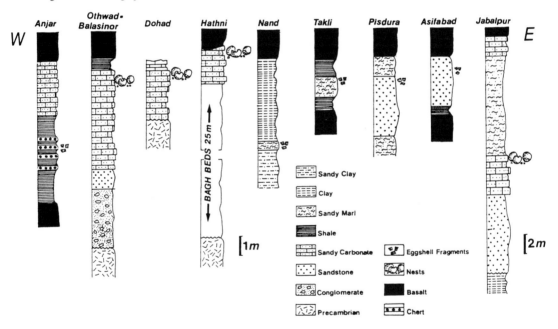

Figure 13.3. Schematic representation of possible relationships between infra-/inter-/supratrappean sequences, Deccan Basalt flows, and basement rocks in peninsular India.

the basal flows. For example, Chiplonkar (1986) traced an intertrappean bed in the Saurashtra region until it assumed an infratrappean position. Furthermore, micropaleontological studies of widespread infratrappean and intertrappean strata have shown that the biotic assemblages recovered by screen washing are essentially identical in composition (Prasad, 1989, Sahni, 1984). Other evidence that the Lameta Beds (Green Sandstone) are contemporaneous with volcanic activity elsewhere in the region is the presence of green, interstitial, multimineralogical clays dominated by montmorillonite and attributed to contemporaneous basaltic degradation (R. P. Kohli, personal communication).

Localities
Kachchh

The westernmost site (Figs. 13.1, 13.2, Table 13.1) yielding eggshells is located in the Anjar area (Bajpai 1990; Bajpai et al, 1990). Both sauropod and ornithoid eggshell morphotypes are present and are very well preserved in a poorly indurated argillaceous sediment having intercalations of chert. The eggshells between basalt flows 3 and 4 (Ghevariya, 1988) occur with other fossils, such as mollusks, microvertebrates, and ostracodes. Fragmentary theropod teeth have also been recovered from the same bed, while in an overlying bed, long-bone fragments and vertebrae occur. At the same stratigraphic level, Ghevariya and Srikarni (1990) have reported a unique association (at least for Indian localities) of large dinosaurian skeletal elements (fragmentary limb bones, scapula, etc.) with the eggs. Ornithischian eggs and egg clutches have also been reported, but the morphological details have as yet not been described.

Kheda

The Kheda District nesting sites (Figs. 13.1, 13.2, Table 13.1) in the Lameta Beds represent the most extensively mapped regional areas. Over 1,000 complete

eggs from over fifteen sites, including Balasinor and Othwad, have been discovered (Mohabey 1983; Srivastava et al., 1986; Mathur & Srivastava, 1987; Mohabey, 1990). The stratigraphy of the egg-bearing horizons has been studied (Srivastava et al., 1986) and a large-scale map has been published showing the distribution of the clutches in a 60 km^2 area at Rahioli. The Lameta Beds rest on a basement of Godhara Granitoids or Aravalli metamorphics, both Precambrian. The sauropod nests are found in a sandy carbonate which often forms thin caps on granitic hills that possibly represent paleotopographic "highs." The carbonate exhibits several features, such as prismatic fabric, color mottling, development of silcrete, etc., that have been interpreted as calcrete. In places, the calcrete contains randomly distributed large sand- to pebble-sized clasts of chert, jasper and quartz (Mohabey, 1990).

Dohad

Approximately 100 km east of the Kheda localities, near Dohad (Fig. 13.1, Table 13.1), Ganapathi (1986) noticed the presence of dinosaur eggs in Lameta Beds underlying the basal basaltic flows (Fig. 13.2). Mohabey and Mathur (1989) found several new localities (in a 30 km^2 area) where the Lameta Beds unconformably overlie the Aravalli phyllites and quartzites. Here, the carbonates are arenaceous in the lower part and become finer grained and less arenaceous upward to the level where the dinosaur eggs have been discovered. The eggs from the localities of Paori, Dholidhanti, Mirakheri, and Waniawao are spherical with diameters ranging from between 12–20 cm, and with a nodose surface ornamentation.

Hathni River (near Kukshi) and Bagh Caves

To the south and southeast of Dohad (Fig. 13.1, Table 13.1), sauropod eggshells and nesting sites have been found in the Hathni River section (Sahni, 1990b) and near the ancient "Buddhist" Bagh Caves about 50

Table 13.1 Distribution of dinosaur eggs and nests, Lameta Formation, India

					Localities				
Eggshell Types	Kachchh Anjar(I)	Kheda Balasinor(L)	Panchmahals Dohad (L)	Dhar Hathni River (L)	Chandrapur Pisdura (L)	Nagpur Takli (I)	Nagpur Nand (L)	Jabalpur Jabalpur (L)	Adilabad Asifabad(I)
(?)TST-1	N		N	N				N	
(?)TST-II		N	F					N	
(?)TST-III	F	N			F	F			
Orn.	F								
Indet.							F		F

Notes: N, nests; F, fragments; I, intertrappean beds; L, Lameta Beds; TST, Titanosaurid; Orn, Ornithoid; Indet., indeterminant

km east of the Hathni River (S. Bajpai & S. Srinivasan, unpublished field report, March, 1991). These discoveries are among some of the most interesting because the eggs occur in sandy carbonates which lay between the Bagh Beds containing abundant shallow marine benthonic invertebrates (Singh & Srivastava, 1981) and the Deccan Basalts (Fig. 13.2). The Hathni and Bagh Cave lithologies are indistinguishable in megascopic and petrographic sections from the egg-bearing horizons of Kheda and Jabalpur (see below). These occurrences demonstrate the uniformity of the Lameta nest bearing carbonates irrespective of the basement rocks. Furthermore, stratigraphic data suggest that the Hathni and Bagh Cave carbonates must be younger than the underlying Cenomanian–Turonian Bagh Beds (Chiplonker, 1986). A Campanian–Maastrichtian age is suggested by ostracod assemblages (Jain, 1975).

Jabalpur

Localities here (Figs. 13.1, 13.2, 13.4, Table 13.1) have been the subject of much attention (Chanda, 1967; Singh, 1981; Brookfield & Sahni, 1987; Tandon et al., 1990; Jolly, Bajpai & Srinivasan, 1990; Sahni & Tripathi, 1990). The nesting sites in the Lower Limestones will be dealt with later, along with taphonomy and sedimentology.

Pisdura and Nand

South of Jabalpur (Fig. 13.1, Table 13.1), eggshell fragments are now known from Pisdura (one of the original dinosaur localities of Huene & Matley, 1933), as well as from the Nand area southwest of Nagpur (Mohabey & Udhoji, 1990). At Pisdura, the eggshells were recovered from a "white sandstone" (= Lameta Formation) that underlies the basalts. A microvertebrate assemblage from the site is dominated by fishes similar to the ones described from the green, sandy, and pebbly marls of the Lower Limestone at Bara Simla Hill, Jabalpur (Jain & Sahni, 1983; Sahni, 1984; Sahni & Tripathi, 1990) and from several intertrappean horizons (Prasad, 1989).

At Nand, the eggshell fragments are comparatively thinner than at other localities (specimens are less than 0.4 mm thick). The shells occur in the Lameta limestones (calcretes?) underlying the Deccan basalts (Fig. 13.2). They, in turn, rest unconformably on the Precambrian metamorphics (Sakoli Group) in the east and over Gondwana sediments in the south and southwest (Udhoji, Mohabey, & Verma, 1990). The Lameta Beds attain a maximum thickness of 20 m and are divisible into two distinct units: (1) a lower unit of nodular limestone (calcrete?) that grades laterally into an argillaceous equivalent and (2) an upper unit that consists of red and green mottled or carbonaceous clays.

Nagpur (Takli) and Asifabad

These localities are spatially distant from each other (Figs. 13.1, 13.2, Table 13.1), but share a number of similar features. Both are intertrappean thin sedimentary horizons yielding abundant microvertebrates, including eggshell fragments. The Takli intertrappean at Nagpur typifies the mode of occurrence, biotic assemblages, stratigraphy and paleoenvironmental conditions for the Deccan intertrappean (Sahni, Rana, & Prasad, 1984). At present, the only way to demonstrate that the lowermost intertrappean horizons are coeval with one another is by the fossil assemblages (microvertebrates, ostracodes, charophytes, mollusks, etc.) from widely separated areas (Prasad, 1989). At Nagpur, paleomagnetic measurements suggest that the Takli intertrappean lies in the 29 R chron (Courtillot et al., 1986) and is therefore close to the Cretaceous–Tertiary Boundary.

Two distinct eggshell morphotypes are present in the Takli intertrappean. One is a thin eggshell recently shown to have lacertilian affinities (Hirsch & Packard, 1987; Bajpai, 1990), but which was earlier listed as uncertain (Sahni et al., 1984). The other is a thick eggshell belonging to dinosaurs (Vianey-Liaud, Jain, & Sahni, 1987) and is much rarer. The eggshells at Takli are associated with a few fragmentary theropod teeth (Lydekker, 1890; Rana,1984; Vianey-Liaud et al., 1987).

The stratigraphy of the Asifabad localities has been described by Prasad, Sahni, & Gupta, 1986). Eggshell fragments are more common than at Takli, and they appear to belong to thin, sauropod morphotypes.

Age of the eggshell localities

Research during the last decade strongly suggests that sedimentary rocks associated with the basal Deccan basaltic flows in areas fringing the volcanic province to the northwest, east and south, are Maastrichtian in age (Sahni & Bajpai, 1988 and references therein). This viewpoint is gaining acceptance (Mohabey & Udhoji, 1990; Courtillot et al., 1986), but differing opinions persist (Bande & Chandra, 1990).

Figure 13.4. Indexed sketch map showing the sauropod nesting sites at Jabalpur. Bara Simla Hill and Pat Baba Mandir (a), Chui Hill (b), Lameta Ghat (c).

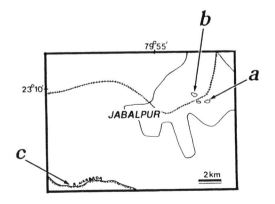

The most extensive nesting sites are located in the Lameta Beds, not only from the type section, but also from across peninsular India (Sahni, 1990b). The Lameta is highly variable in thickness, ranging from a few meters to about 75 m at Jhiraghat, west of Jabalpur (Lunkad, 1990). Initially the age of the Lameta Beds was based on large dinosaur skeletal remains described by Matley (1921) and Huene and Matley (1933) from the Jabalpur and Pisdura localities. These dinosaurs were found to be comparable to those from Madagascar and Argentina. Buffetaut (1987) has recently shown that the Turonian age previously assigned to the Lameta Beds was based upon misconceptions about the Upper Cretaceous stratigraphy of Madagascar and South America. There were also erroneous interpretations based on the supposed presence of allosaurids, in the primitive evolutionary grade of the coelosaurs, and the absence of the highly specialized nodosaurids.

Of the three dinosaur-yielding localities known 70 years ago, Kallamedu (Ariyalur Formation) was considered to be the youngest based on marine benthonic invertebrates; Pisdura, lowermost Senonian in age; and Bara Simla Hill at Jabalpur, Turonian in age (Huene & Matley, 1933). Screen-washing of sandstones at Pisdura and green, sandy, and pebbly marls associated with the Lower Limestone at Bara Simla Hill, Jabalpur, has brought to light diverse Maastrichtian microvertebrate assemblages with numerous eggshell fragments (Jain & Sahni, 1983, 1985; Tripathi, 1986; Sahni & Tripathi, 1990). The microfossils (including sauropod eggshell fragments) are characteristic of latest Cretaceous freshwater biotas known from other Laurasian localities. Ostracodes and charophytes (somewhat eroded) have affinities with those from the intertrappeans (S. B. Bhatia et al., 1990), which in turn bear a close resemblance to those known from the Nemegt Basin of Mongolia and adjacent basins of China (Sahni & Tripathi, 1990). These data clearly suggest that the Lameta Beds are of Maastrichtian age.

A regional correlation has been established based on dinosaurian assemblages, sauropod nesting sites, and the characteristic sandy calcrete facies. This correlation indicates that the localities at Kheda, Dohad, Hathni River, Bagh (although lacking dinosaurian skeletal assemblage), and the Pisdura-Nand area are all Maastrichtian (Srivastava et al., 1986; Mohabey & Udhoji, 1990).

The age of the Deccan intertrappeans has long been debated as to whether they are largely Cretaceous or Paleocene-Eocene (Sahni et al, 1988). The controversy originated from the pioneering work of Hislop (1860), and later expanded by Sahni (1940) based on paleobotanical data, and by Hora (1938) on the fish fauna. These studies indicated the intertrappean sequences were Eocene in age (the Paleocene as the earliest period of the Tertiary was not then accepted in India). This viewpoint was further enlarged through the work of the Paleobotanical Institute at Lucknow (Prakash, 1960; Bande & Prakash, 1982; Bande & Chandra, 1990).

We believe that the intertrappeans in which the eggshells occur may be terminal Cretaceous in age. This age is based upon the presence of dinosaurian remains in several intertrappean beds (e.g., at Kachchh, Jabalpur, Takli, and Asifabad). The palynomorph *Aquilapollenites bengalensis* from the Jabalpur Intertrappean at Padwar and Ranipur supports this interpretation (Prakash, Singh, & Sahni, 1990; Mathur, 1990; Venkatachala and Kar, personal communication), as do paleomagnetostratigraphy and Ar/Ar dating (Courtillot et al., 1986; Duncan & Pyle, 1988; Bakshi, 1987).

Eggshell structure and taxonomic relationships

Although dinosaur eggs and eggshell fragments are now known from several infra- and intertrappean localities, no serious attempt has been made to classify the various morphotypes. Admittedly, the lack of embryonic remains or hatchlings from Indian eggshell sites poses serious constraints on any classification based solely on morphological characteristics (see Vianey-Liaud et al., Chapter 11; Zhao, Chapter 12). This problem is further compounded by the possibility of environmentally induced variability in the eggshell, as well as that induced by diagenetic alteration (see Dauphin, 1990; Zhao, Chapter 12).

Numerous attempts to classify dinosaur eggshells have been made during the past three decades (e.g., Young, 1959; Sochava, 1969, 1971; Erben, 1970; Dughi & Sirugue, 1966; Zhao, 1979, Chapter 12; Erben, Hoefs, & Wedepohl, 1979; Williams, Seymour, & Kerourio, 1984; Kurzanov & Mikhailov, 1989; Hirsch & Quinn, 1990; Penner in Vianey-Liaud et al., 1987; Vianey-Liaud et al., Chapter 11). In a pioneer study using polarizing microscopy, Sochava (1969) distinguished three groups of dinosaur eggshells from the Gobi Desert of Mongolia, based mainly on the structure of the pore canals (angusticanaliculate, prolatocanaliculate, and multicanaliculate). Later, she distinguished two taxonomic categories of dinosaur eggshells: "Testudoid" (turtlelike, single-layered) and "Ornithoid" (avianlike, two-layered) (Sochava, 1971).

Erben (1970), who pioneered the use of the scanning electron microscopy (SEM) in the study of eggshells, examined European material. Later, Erben et al. (1979) classified the Cretaceous dinosaur eggshells into the three types recognized by Sochava (1969), and named a fourth type, "tubocanaliculate," for the French and Spanish eggshells.

Recently, Mikhailov (1991) has established four basic eggshell types based on biocrystalline ultrastructure: testudoid, crocodiloid, dinosauroid, and ornithoid.

Of these, the dinosauroid type includes two main groups, spherulitic and prismatic. These are subdivided into several morphotypes based mainly on the nature of shell limits and the structure of pore canals (see Hirsch & Quinn, 1990; Mikhailov et al., Chapter 7). These recent classifications are most useful because they use an integrated scheme to categorize the various known eggshells. However, its utility is limited by uncertainties regarding the morphological variation within a given taxon. Furthermore, the precise correlation of eggshell morphotypes may not be possible when morphologically intermediate varieties are encountered.

All of these points, especially variability, are taken into account in categorizing the Indian eggshells. Each proposed morphotype is based on a combination of parameters listed below. Furthermore, names used to denote the eggshells sometimes only serve to indicate their possible attribution to one or more of the dinosaur taxa known from skeletal remains. These bones are not necessarily in direct association with the eggs.

The parameters used by us include:

1. Size and shape of the egg. Little variability is visible in the spherical shape of the Indian sauropod eggs. The size usually ranges between 14 and 20 cm in diameter. There is some overlap in size ranges among morphotypes.
2. Shell thickness. The range of eggshell thickness is fairly constant for each of the categories. Nevertheless, given the possibility of variation due to diagenesis, this feature is used only as a subsidiary characteristic, except for some extremely thin eggshells.
3. External sculpture. The majority of the eggs have nodose ornamentation and the nodes may vary slightly in size. Nodes may be isolated from each other (each node corresponds to a single shell unit) or may be coalesced to form multiple nodes (representing several compressed shell units).
4. Shell unit. The nature of the shell unit, particularly its relationship with adjacent units, is one important parameter for distinguishing the Indian sauropod eggshells. Several patterns of shell units are recognized:
   a. Discrete, slender units of nearly constant width and not fused to adjacent units.
   b. Shell units of variable shape and width and not fused to adjacent units.
   c. Shell units with a tendency to fuse with adjacent units, except near the base of the eggshell.
5. Pore pattern. Where visible, the pores are subrounded or elliptical openings between nodes. In radial section, pore canals are often enlarged due to diagenetic alteration, but their general configuration allows them to be classified under known categories (e.g., Erben et al., 1979).
6. Growth striations. Growth striations exhibit a dis-

tinctive pattern in each of the three types of sauropod eggs:
   a. Highly convex growth striations either terminating at the lateral boundaries of the individual shell unit, or continuing into adjacent units with an intervening concavity.
   b. Growth striations that are convex near the base and becoming undulatory or almost straight near the surface where they follow the contour of the outer shell surface.
7. Mammillae. The distribution or spacing of mammillae is fairly uniform in certain cases, giving rise to shell units of nearly constant width. In other cases, the intermammillary spacing is highly variable, causing differences in the shape and size of shell units.

Classification of Indian dinosaur eggshells

Five morphotypes of dinosaur eggshells are recognized in India. These are (?)Titanosaurid Type I, (?)Titanosaurid Type II, (?)Titanosaurid Type III, Ornithoid type, and an indeterminate dinosaurian type.

Family Megaloothidae Zhao 1979
(?) Titanosaurid Type I [(?)TST-I, Figs. 13.5–13.7a–d]

Diagnosis. Eggs spherical, usually 14–18 cm in diameter; eggshell thick, commonly 2–3.5 mm; single-layered, spherulitic; shell units long, compressed, nearly cylindrical in shape and not fused to adjacent units; shell unit boundaries essentially parallel above small mammillae caps; growth striations highly arched throughout shell thickness and restricted to individual units; pore canals long, narrow, of tubocanaliculate type; extinction pattern sweeping; outer surface with circular or subcircular, well separated nodes, each corresponding to a single shell unit; node size constant; subcircular pores; inner surface with subcircular mammillae, 0.2–0.5 mm in diameter, tightly packed, generally distinct from each other.

Discussion. A large proportion of the Lameta eggs can be assigned to the morphotype (?)TST-I. Originally reported from the Kheda area of Gujarat (Kheda Type "B" of Srivastava et al., 1986), eggs of this type have since been recognized at the Jabalpur nesting sites (Tripathi, 1986; Sahni & Bajpai, 1988), Paori, Dholidhanti and Mirakeri in the Panchamahals District, Gujarat (Mohabey & Mathur, 1989), and more recently from the Hathni River Section in central India (Vertebrate Palaeontology Laboratory, Chandigarh, work in progress).

Morphologically, (?)TST-I corresponds to the recently proposed "tubospherulitic" morphotype of the "dinosauroid–spherulitic" group (Mikhailov, 1991). The close similarity of (?)TST-I with dinosaur eggshells long known from the Upper Cretaceous of Aix-en-

Provence, France (probably Penner's Type-I, in Vianey-Liaud et al,. 1987; see also Vianey-Liaud et al., Chapter 11) suggests inclusion of the Indian material in the family Megaloolithidae, a parataxon erected by Zhao (1979) for the Provence eggs. The eggs from Aix-en-Provence are usually referred to *Hypselosaurus* on the basis of associated bones in the same strata (see Buffetaut and Le Loewf, Chapter 2; Cousin et al., Chapter 5; Vianey-Liaud et al., Chapter 11). Sauropods from the Indian eggshell sites are represented by the family Titanosauridae, to which the eggs can be safely attributed. Without associated embryonic remains or hatchlings, it is not possible to attribute these eggs to any one of the three genera (*Titanosaurus, Antarctosaurus,* and *Laplatosaurus*) known from the Lameta Beds of peninsular India (Huene & Matley, 1933).

(?)Titanosaurid Type II(?)TST-II, Figs. 13.5, 13.7e–g, 13.8a–f.

Diagnosis. Eggs spherical, with diameters comparable to (?)TST-I; shell thickness between 1–1.5 mm; single-layered spherulitic; shell units compressed, of variable width and shape, usually conical and/or cylindrical, not fused together; growth striations arched throughout shell thickness becoming concave near the lateral boundaries; striations continue into adjacent shell units; pore canals long, narrow vertical or sub-vertical, of tubocanaliculate type; extinction pattern sweeping; outer surface nodose with well-separated nodes each corresponding to a single shell unit; nodes rarely coalescing; inner surface with mammillae distinctly smaller than (?)TST-I (diameter 0.1–0.5 mm); mammillae tightly packed; pores circular to elongate in tangential sections usually shared among three to four mammillae.

Discussion. (?)TST-II is distinguishable from (?)TST-I in being thinner and in having variable shaped shell units. Several smaller units occur of varying height ranging from one-third to one-half of the shell thickness. This feature may be related to the irregular spacing of the nucleation centers and the differential rate of spherulite growth which results in the arrested development of some units at the expense of others (Fig. 13.7g).

Tripathi (1986) discovered four nests of (?)TST-II containing up to seven eggs at Bara Simla Hill, Jabalpur. This eggshell type has also been reported from Waniawao in district Panchamahals, Gujarat (Mohabey & Mathur, 1989), the Hathni River Section, and a locality close to the Bagh Caves (Vertebrate Paleontology Laboratory, Chandigarh, work in progress).

Like (?)TST-I, this eggshell type also has a tubospherulitic structure and is referable to the family Megaloolithidae (Zhao, 1979). A "juvenile sauropod" from a nest that may be (?)TST-II eggs (Mohabey, 1984) needs to be re-examined in view of the serious doubts raised by Jain (1989) on its identification.

Family unknown

(?) Titanosaurid Type-III [(?)TST-III, Figs. 13.5, 13.8g–h, 13.9]

Diagnosis. Eggs spherical, 14–20 cm in diameter; shell thickness 1.0–1.5 mm, single layered, spherulitic; shell units variable in their relationship with each other; units may be not fused or only partially fused; where corresponding to a single node, individual units are fan shaped or conical, and relatively widely separated in the inner part near the mammillae caps; where forming part of a multiple node, units much compressed with lateral boundaries traceable only in the inner third to half of shell thickness; growth striations convex towards the base, becoming progressively undulatory upward where they continue across adjacent units following contour of shell surface; pore canals narrow, curved, often incomplete (?tubocanaliculate); extinction pattern sweeping; outer surface with coalescing nodes, some nodes solitary; node size variable so radial sections show variability in nodal relief; pores on outer surface subcircular to elliptical, often with overhanging nodes; inner surface with isolated or coalescing mammillae of variable diameter (0.15–0.30 mm).

Figure 13.5. Schematic representation of the five types of dinosaur eggshells recognized in the Indian Upper Cretaceous strata.

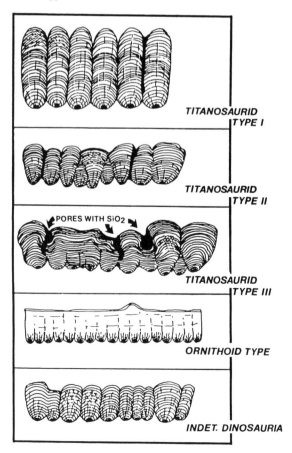

TITANOSAURID TYPE I

TITANOSAURID TYPE II

◄ PORES WITH SiO2 ►

TITANOSAURID TYPE III

ORNITHOID TYPE

INDET. DINOSAURIA

Figure 13.6. **a.** Three eggs from a nest at Pat Baba Mandir. **b.** Enlarged egg (28 cm maximum diameter) showing in situ fragmentation. c–h. (?)Titanosaurid Type-I. **c.** Outer surface, SEM, Kheda. **d.** Radial view, SEM, Kheda; note discrete shell units. **e.** Radial view; SEM; polished surface, Jabalpur. **f.** Radial thin section, not polarized, Jabalpur; note highly convex growth striations terminating at the lateral boundaries of individual shell units. **g.** Radial thin section, polarized, Hathni River; note sweeping extinction pattern. **h.** Radial view; SEM, weathered surface, Hathni River; note eroded pore canals (PC) along boundaries of shell units. Bar = 1,000 μm for c–h.

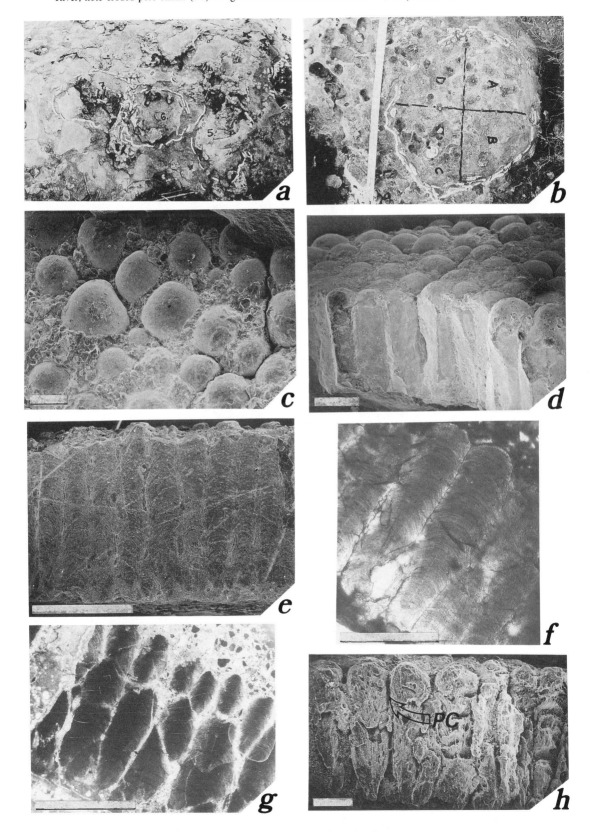

Figure 13.7. a–d, (?)Titanosaurid Type-I. e–g, (?)Titanosaurid Type-II. **a.** Tangential thin section of outer surface, not polarized, Hathni River; note spherolitic structure and pores (p). **b.** Radial view, SEM, polished surface, Jabalpur; note silicified areas (seen in relief). **c.** Inner surface, SEM, Kheda; note recrystallized mammillae. **d.** Enlarged mammillary cone, SEM, Hathni river; note radiating crystallites. **e.** Outer surface, SEM, Jabalpur; note variable node size.
f. Tangential thin section through mid-thickness of eggshell, polarized, Jabalpur; note uniaxial cross corresponding to each unit. **g.** Radial thin section, not polarized, Jabalpur; note variation in shape of shell units. Bar = 1,000 μm for a–c and e–g; = 500 μm for d.

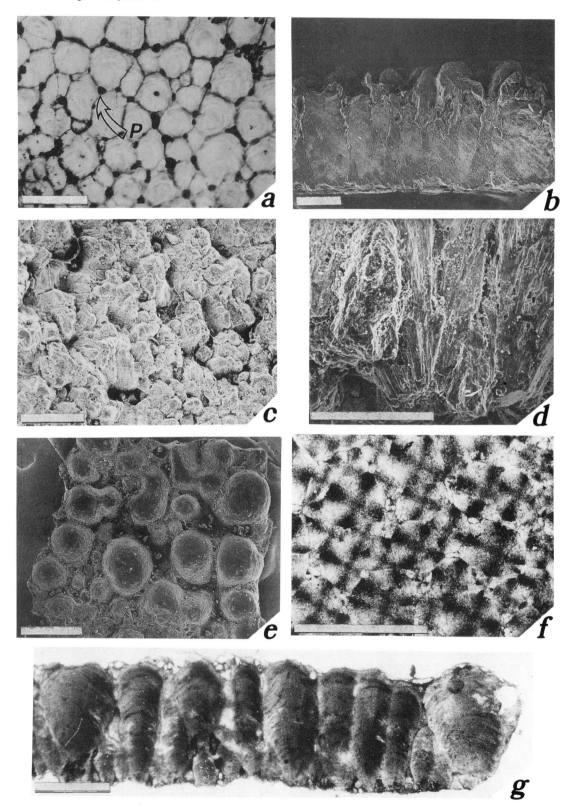

Figure 13.8. a–f, (?)Titanosaurid Type-II. g–h, (?)Titanosaurid Type-III. **a**. Tangential thin section of inner surface, not polarized, Jabalpur; note subcircular and elliptical pores. **b**. Single mammilla, SEM, Jabalpur; note central core (CC). **c**. Radial thin section, not polarized, Hathni River; note that the eggshell is split within the rock matrix along the boundaries of shell units. **d**. Radial thin section, not polarized, Jabalpur; note pronounced growth striations, faint radiating structure, and pore canal (PC). **e**. Radial thin section, not polarized, Jabalpur; note growth striations. **f**. Radial thin section, not polarized, Jabalpur. **g**. Outer surface, SEM, Pisdura; note prominent pores (P). **h**. Radial view, SEM, polished surface, Kheda; note discrete as well as fused shell units. Bar = 1,000 μm for a,c–h; = 100 μm for b.

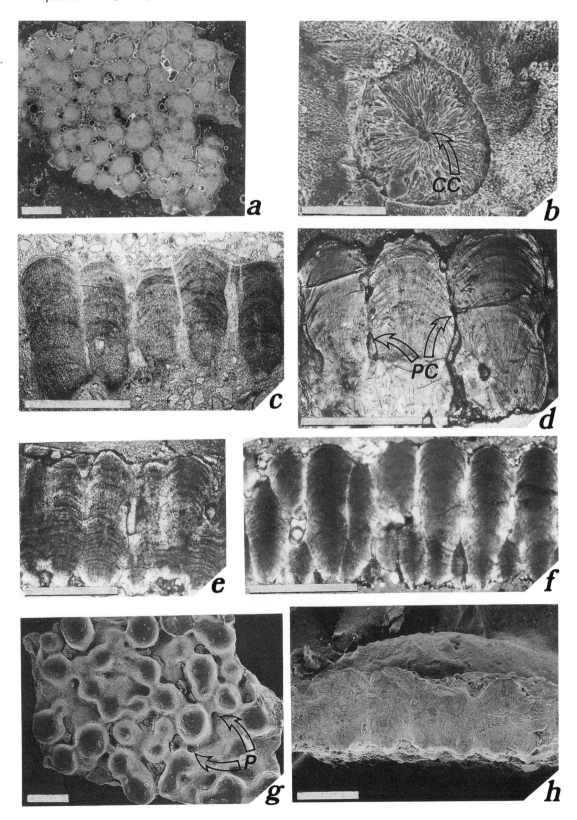

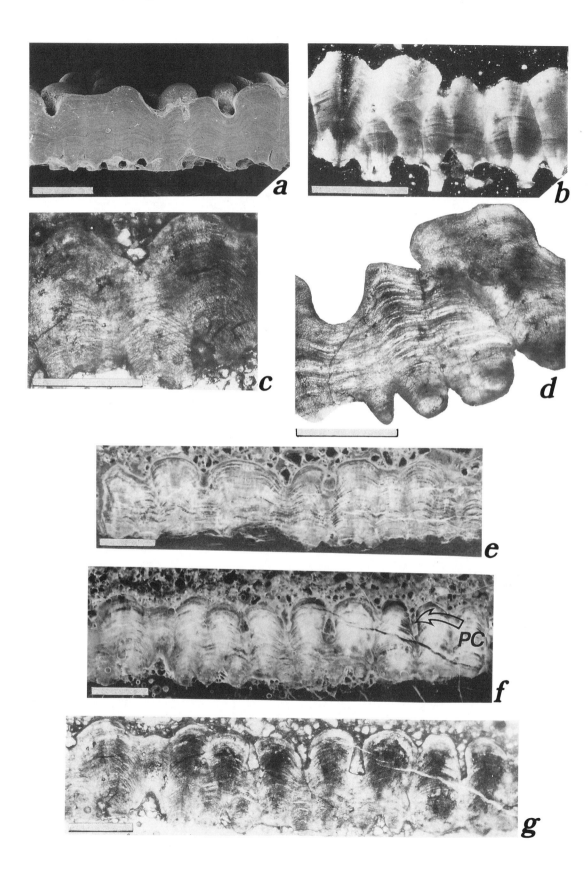

Discussion. Eggs of this type were originally reported as "Kheda A" (Srivastava et al., 1986) from the Balasinor Quarry in Kheda, Gujarat. Fragmentary eggshells from the Lameta Beds at Pisdura (Jain & Sahni, 1985), as well as from the Takli intertrappeans at Nagpur and the Kachchh intertrappeans (Vianey-Liaud et al., 1987; Bajpai et al., 1990), are also probably referable to this type. Additional material needs to be examined to confirm this association, however.

In size, shape, and external sculpturing of the eggs, (?)TST-III is not significantly different from other titanosaurid eggs [(?)TST-I and (?)TST-II)]. In its structural configuration, however, (?)TST-III is distinct and is characterized by shell units that grade laterally from unfused to partially fused, and near horizontal growth striations that follow the contour of the outer shell surface.

(?)TST-III cannot be correlated with any of the morphotypes recognized by Mikhailov (1991). It has similarities to both the tubospherulitic and the prolatospherulitic morphotypes, the latter being spherical with nodes on the outer surface and are identified as hadrosaur (Mikhailov, 1991; K. Hirsch, personal communication, 1992). As of now (?)TST-III is referable to the tubospherulitic structural morphotype as a possible variant.

Family Unknown
Ornithoid Type (Figs. 13.5, 13.10).

Diagnosis. Eggshell double-layered, thickness 0.4–0.5 mm, mammillary layer thick (1/3 to slightly less than 1/2 of shell thickness); mammillary cones distinct, radiating and tightly packed; continuous layer with no distinct structural pattern barring some faint columnlike structures; rare horizontal layering (?growth increments) widely spaced; no abrupt change from mammillary to continuous layer; extinction pattern columnar, better defined in mammillary layer; outer surface with irregularly spaced tubercles independent of shell units; inner surface with tightly packed mammillae (diameter, 0.03–0.05mm); pores not visible on the outer surface or in radial section.

Discussion. These eggshell fragments have only recently been reported from the intertrappean sequences near Anjar Kachchh (Bajpai et al., 1990). They represent the only record of two-layered eggshells from the Upper Cretaceous rocks of India. Complete "ornithischian eggs and egg clutches" have been discovered at the same locality (Anjar) by Ghevariya and Srikarni (1990). Mohabey (1990) has also mentioned the presence of "ornithischian eggs" from the Lameta Beds of the Pan-

chamahals district. However, because no illustrations or descriptions have been published, it is not possible to make any valid reference to these reports.

The Anjar eggshells can be classified in the Ratite morphotype of the ornithoid group (Hirsch & Quinn, 1990; Mikhailov, 1991). However, they cannot be assigned to any of the three ornithoid families because of minor differences. The family Elongatoolithidae differs from the Anjar eggshells in being thicker (approximately by a factor of 4) and in the details of surface sculpture. The other two families, Laevisoolithidae and Subtiliotlithidae are more similar, but the former has a smooth eggshell surface, while the latter has a continuous to mammillary layer ratio of 1:2 to 1:3, unlike the Anjar ornithoid eggshells with a ratio of 1:3 to 1:2.

The recently described ornithoid type eggshells from Montana (?*Troodon*, Hirsch & Quinn, 1990) are also easily distinguishable from the Anjar eggshells in being considerably thicker (average 1–2 mm vs. 0.03–0.05 mm), in having a much thinner mammillary layer (1/10 to 1/12 of shell thickness vs. 1/3 to 1/2), and in a pronounced change from the mammillary to the continuous layer. The Anjar eggshells are more comparable to the avianlike eggshells from the Upper Cretaceous of Montana, but the latter differ in being smooth on the outer surface (Hirsch & Quinn, 1990).

Eggshells of ornithoid type have long been known from Kazakhstan, Mongolia, and China (Erben et al., 1979). The two new ornithoid families (Mikhailov, 1991) with which the Anjar eggshells are comparable, Laevisoolithidae and Subtiliotlithidae, are known from the Nemegt Formation (Mongolia).

The taxonomic affinities of the ornithoid type eggshells have yet to be established. Some have suggested that these double-layered eggshells are ornithischian (Sochava, 1971; Erben et al., 1979), while others have suggested theropod (Kurzanov & Mikhailov, 1989; Mikhailov, 1991).

Family unknown
Indeterminate Dinosauria (Fig. 13.5)

Diagnosis. Eggshells extremely thin (0.1–0.4 mm); single-layered spherulitic, shell units not fused, usually conical in shape; growth striations arched, continuing into adjacent units; outer surface with discrete nodes; mammillae well developed, subcircular in shape, occasionally overlapping.

Discussion. Limited morphological details are available for these extremely thin eggshells, which were

---

Figure 13.9. (*facing*) a–g, (?)Titanosaurid Type-III. **a.** Radial view, SEM, polished surface; Kachchh; note several shell units corresponding to a single multiple-node and undulating growth striations following the contour of the shell surface. **b.** Radial thin section, polarized, Pisdura; note sweeping extinction pattern and widely separated mammillae. **c.** Radial thin section; not polarized, Kheda; note distinct herringbone pattern. **d.** Radial thin section,; not polarized, Kheda; note fused shell units and pronounced growth striations. **e.** Radial thin section, not polarized, Kheda; note discrete as well as fused shell units. **f.** Radial thin section, not polarized, Kheda; note enlarged pore canals (PC). **g.** Radial thin section, polarized, Kheda; note sweeping extinction pattern. Bar = 1,000 μm.

Figure 13.10. a–g, Ornithoid type. **a**. Outer and radial view, SEM, weathered surface; Kachchh, note irregularly spaced tubercles. **b**. Radial view, SEM, weathered surface, Kachchh; note mammillary layer and a secondary deposit on the outer surface. **c**. Radial view, SEM, weathered surface, Kachchh; **d**. Radial thin section, not polarized, Kachchh; note well defined mammillary layer (ML), faint columnlike structures in the continuous layer, and a secondary deposit on the outer surface. **e**. Radial thin section showing enlarged mammillary layer, not polarized, Kachchh; note small mammillae caps to the right. **f**. Inner surface. **g**. Radial thin section, polarized, Kachchh; note columnar extinction. Bar = 500 μm for a,c,d; = 100 μm for b,e–g.

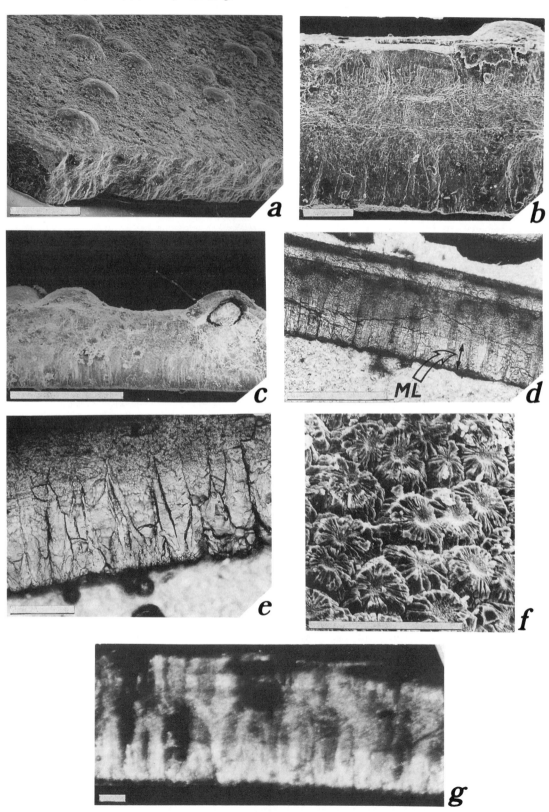

described from the Asifabad intertrappean sequences (Sahni et al., 1984). Recently, Mohabey (1990) has recorded their occurrence in the Lameta Beds of the Nand area in Maharashtra, but detailed descriptions and illustrations have yet to be published. These eggshell fragments may be sauropod on the basis of their gross structural similarity with titanosaurid eggshell types described above. Eggshells of comparable thickness are also known from Aix-en-Provence, France (Kérourio, 1982; Vianey-Liaud et al., Chapter 11).

Sedimentology of the nest-bearing sandy calcretes

The best-studied sections of the Lameta Beds are exposed in Jabalpur City, including the type section at Lameta Ghat along the banks of the Narbada River (Fig. 13.4). Major localities occur within 2–3 km of each other, and in the Jabalpur Cantonment sections, lithological units are often physically traceable.

Matley's (1921) original descriptions of the Lameta Beds are based on a detailed geological map (8 in. to 1 mile scale with a 5-ft contour interval) that subdivides the sequence into five units: Upper Sands, Upper Limestone, Mottled Nodular Beds, Lower Limestone, Green Sandstone.

Remapping of some segments on a scale of 1:300 using a plane table and microptic alidade reveals that the two uppermost units of Matley (1921) – the Upper Sands and the Upper Limestone – are not traceable laterally. In fact, the Upper Sands occurs only in a single section (western face of Bara Simla Hill). Hence a modified subdivision of the Lameta sequence at Jabalpur is suggested: Upper Calcretized Gritty Sandstone, Mottled Nodular Beds, Lower Limestone, Green Sandstone.

Dinosaur remains, fragmentary eggshells, and sauropod nests mostly occur in the Lower Limestone unit and its facies variants. Lithologically, the Lower Limestone can be described as a concretionary, brecciated, siliceous, and rarely laminated carbonate. Distinct color mottling and abundant terrigenous fine to coarse sand and pebble-sized clasts of quartz, jasper, and chert are among its other general features.

Several distinct subfacies are found associated within the Lower Limestone: green sandy and pebbly carbonate; sandy, brecciated, nodular carbonate; grey, silty, finely autobrecciated carbonate; sandy, brecciated carbonate; strongly brecciated carbonate; prismatic, brecciated carbonate; calcretised, clast-bearing, sandy carbonate; sandy carbonate with dominant (?)algal silicifications and moderate prismatic shrinkage; brecciated, pisolitic carbonate; and conglomeratic calcrete. These discrete subunits may have a gradational or sharp relationship with each other. However, the sauropod nests are usually found in a sandy, nodular carbonate, with buff and purple color mottling and shrinkage cracks occasionally filled with silica. At Bara Simla Hill, the nest bearing carbonates grade laterally into a green, sandy,

and pebbly marl which has yielded dinosaur skeletal material, including a few articulated skeletons (Chatterjee, 1992).

At the type section of the Lameta Beds at Lameta Ghat, the nest-bearing limestone is an impure carbonate which shows desiccation cracks, shrinkage features, and a marked prismatic-brecciated structure. Pebble-sized detritus including chert, jasper, and vein quartz is both present as thin discontinuous sheets and is dispersed chaotically within the carbonate. In places, the clasts fill cracks up to 15 cm wide giving rise to "fissure fill diamictite" (Tandon & Friend, 1989). Silcrete layers a few millimeters to a few centimeters thick are also commonly interspersed in the carbonates.

Pedogenic modification in the nest-bearing carbonate is inferred from several features. These include breccias exhibiting multiple episodes of fracturing and recementation, truncated paleosol profiles, and bioturbation due to animal or plant-root activity. There also occurs diffuse and irregular gray colored mottling within the buff colored limestone, shrinkage resulting in the development of a prismatic fabric, as well as resulting in spindle-shaped extension fractures that were later filled with silica. In addition, shrinkage result in the formation of circumgranular cracks and "peds." Soil microrelief, and pseudo-anticlines also occur (Reeves 1976; Goldbery 1982; Bowles 1979; Fitz-Patrick 1984).

Paleogeomorphology of the nesting horizon

Besides exhibiting rapid lateral facies variation, the sauropod nest-bearing carbonate unit shows markedly undulatory upper and lower contacts. Matley's (1921) map of the Lameta Beds of the Jabalpur Cantonment shows that several such undulations mark the bounding surfaces of the Lower Limestone. Furthermore, these undulations (reflected as contour intersections with the upper and lower contacts) are spatially variable, with the contacts being inclined or horizontal. From the southeast to west of Bara Simla Hill, the lower and upper contacts fall by 3 m over a distance of 100 m (Fig. 13.11). In the Bara Simla Hill and Pat Baba Mandir sections, the Lower Limestone also shows highly variable thickness (between 3 and 12.5 m).

Using Matley's (1921) map a data set was generated using each contour intersection point for the upper and lower contacts and perspective block diagrams were then constructed. The surfaces generated show an undulatory character with some well marked depressions (Fig. 13.12). The "relief highs" and "relief lows" have been interpreted to represent shoulders, ramps, and flats on a paleogeomorphic surface (Mittal, 1990).

The various subfacies observed at Bara Simla Hill and Pat Baba Mandir reflect the paleorelief. Predictably, the "relief highs" are associated with the nest-bearing nodular, brecciated carbonates, while the "relief lows" show green, sandy, and pebbly marl. The marl is interpreted as a palustrine facies with low pedogenic modification, while the topographically higher sandy carbonate represents emergent areas of the palustrine flat (Freytet, 1973; Freytet & Plaziat, 1982; Platt, 1989). Similarly, at Chui Hill and Lameta Ghat, the egg-bearing facies represent either subaerially exposed palustrine flats or a proximal fan surface. In all three areas, sediments show modification through pedogenesis, a well-developed calcrete profile, development of silcrete laminations, and a dispersed, widespread terrigenous influx in the form of sand grains and pebble-sized clasts.

Paleoenvironment of the nesting site

The Lameta sequences at and around Jabalpur are characteristic of fluvial and semi-arid, pedogenically modified fan–palustrine flat systems. The Lower Limestone represents a palustrine–pedostratigraphic unit separating events and units lying above or below it. Figure 13.13 is a simplified reconstruction of the various environments/events reflected in the Lameta sequence at the localities of Jabalpur Cantonment. The nest-bearing sandy carbonates at Kheda, Hathni River/Bagh Caves, and Nand have also been interpreted as pedogenic calcrete horizons (Mohabey, 1990; Mohabey & Udhoji,

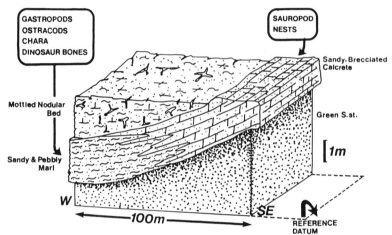

Figure 13.11. Upper and lower contacts of the Lower Limestone west to southeast of Bara Simla Hill where the contacts rise by as much 3 m over a distance of 100 m with respect to a reference datum surface.

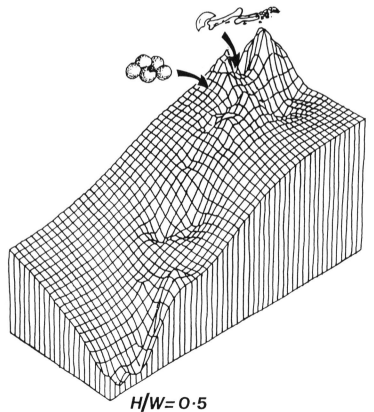

Figure 13.12. Perspective block diagram constructed for the upper contact of the Lower Limestone in a segment of Bara Simla Hill. Positions of the nest-bearing sandy calcrete and the dinosaur skeletal element bearing sandy and pebbly marl indicated.

1990). Aridity or semi-aridity, with intermittent wet periods, is reflected across a major portion of the Indian subcontinent. This possibly represents a widespread regional paleoclimatic condition.

The most useful indicator of palustrine conditions in the Lower Limestone is the recovery of microfossils (Tripathi, 1986; Sahni, 1984; 1985; 1989; Sahni & Tripathi, 1990). The green, sandy, pebbly marl at Bara Simla Hill is well known for its dinosaurian skeletal elements (Huene & Matley, 1933; Chatterjee, 1992). The horizon has also produced freshwater aquatic pulmonates, including *Paludina*, *Physa*, and *Lymnaea*; the ostracod taxa *Cyprinotus*, *Paracypretta*, *Eucandona*, and *Darwinula*; and charophytes, including *Platychara*, *Microchara* and *Peckichara*; the fish *Lepisosteus*, *Pycnodus*, *Phareodus*, *Apateodus*, and *Stephanodus*; as well as at least one species of frog referable to the family Pelobatidae.

The evidence of aquatic microfossils from the green, sandy, pebbly marl, a facies of the Lower Limestone, supports a palustrine interpretation. The marl represents "relief lows" on a reconstructed paleogeomorphic surface, and in addition, it characterizes a palustrine carbonate facies that formed in a small, shallow body of water that probably had a rapidly fluctuating shoreline due to intermittently wet and dry conditions.

Taphonomy of the nesting horizon

Although dinosaur skeletal remains have been known from the Lameta Beds for over a century, the remains of dinosaur nesting sites have been rigorously documented for only the past 10 years. Nevertheless, the sediment–fossil relationship is known only for the eggs and nests. Similar information for the dinosaur skeletal material is woefully inadequate. Furthermore, only

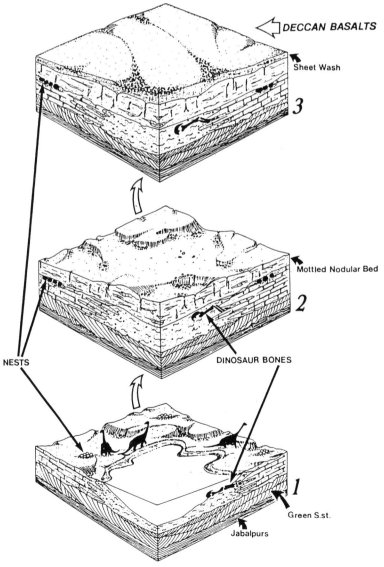

Figure 13.13. Reconstruction of the paleotopographic surface showing a palustrine flat with the relative position of the nests, sandy carbonate and green sandy and pebbly marl. Diagrams 2 and 3 show the burial of the nest-bearing surface, first by the Mottled Nodular Bed and subsequently by a sheet flood event marked by the Upper Calcretised Gritty Sandstone. The first trappean flows arrived in the area after the development of the Upper Calcretised Gritty Sandstones.

rarely do nests and dinosaur skeletal material occur in the same horizon.

Bara Simla Hill in Jabalpur Cantonment is one such example. Here the green, sandy, and pebbly marl facies of the nest-bearing Lower Limestone has produced fragmentary, partially articulated or articulated dinosaur skeletal material as well as dinosaur eggs. Much of this skeletal material recovered during recent excavations (e.g,. Chatterjee, 1992) is still in the process of being documented. We have attempted to relocate previously reported dinosaur bone sites and to determine the orientation, concentration, state of preservation, etc. of the bones but our success has been very poor.

Two major nesting sites have been identified at Jabalpur, both near Bara Simla Hill. One site occurs in a saddle between Bara Simla Hill and the ridge on which the Pat Baba Mandir is situated. The second site is located immediately adjacent to the temple itself. A nest bearing three eggs along with scores of eggshell fragments has been reported in the Chui Hill Section. The Lower Limestone on the right bank of the Narbada River at the Lameta-type locality has also yielded nests and eggshell fragments.

The Bara Simla Hill and Pat Baba Mandir sites have yielded as many as fifteen nests, with the number of eggs per nest varying from two to seven (thirteen eggs in one nest has been reported from Khempur,

Gujarat, by Mohabey, 1990). At Pat Baba Mandir, nine separate nests have been recorded in an area 100 × 50 m. All the nests or isolated eggs were found in a lithofacies which is representative of an "emergent" palustrine flat. In no instance have any dinosaur cranial or postcranial skeletal elements been found associated with nests at Bara Simla Hill, Pat Baba Mandir, Chui Hill, or Lameta Ghat. Such elements have, however, been reported from the palustrine facies of Bara Simla Hill situated in a "relief low" on the reconstructed paleogeomorphic surface (Figs. 13.12 and 13.13).

The nests and isolated eggs represent autochthonous remains, and the skeletal elements are thought to have accumulated in shallow ephemeral ponds or small lakes after having undergone limited transport from the site of death and/or initial accumulation. The most likely mechanism for transport of the skeletal elements is sheet flooding as recorded in the sandy calcrete facies.

Details of some of the nests at the Bara Simla/ Pat Baba Mandir nesting sites, as well as those at Rahioli in Gujarat, are illustrated in Figure 13.14. Most eggs are arranged in roughly circular groups. Some isolated eggs, as well as linearly arranged eggs, have also been recorded. Single nests have been observed at a new quarry face at Chui Hill, but it is likely that sampling, as well as observation, are incomplete.

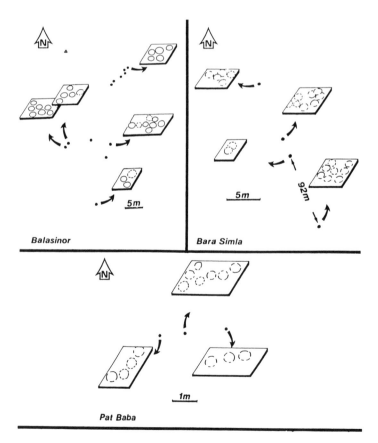

Figure 13.14. Distribution of nests at three nesting sites. Balasinor (modified after Srivastava et al., 1986), Bara Simla (modified after Tripathi, 1986), and Pat Baba showing detail of three of the nests.

Burial conditions

At Bara Simla Hill/Pat Baba Mandir, Chui Hill, and Lameta Ghat, the nests and eggs are often associated with sediments having a moderately higher sand and pebble content than that the horizon as a whole. This sandy and pebbly sediment is believed to have originated from sheet floods. The Lameta sauropods apparently made their nests at sites on relatively high areas fringing small, shallow, ephemeral lakes or ponds. Partly or wholly buried eggs/nests in a wetted (through sheet wash or palustrine high-stands) carbonate-rich, sandy soil would have had a particularly high fossilization potential. Later desiccation of the subaerial surface (due to generally arid climate) and the resultant extremely rapid cementation because of the high carbonate content of the water and soil would entomb a large proportion of the eggs permanently.

Breakage

No undamaged egg has been documented from the Jabalpur area. All eggs exhibit in situ fragmentation through intrastratal reworking even though they still retain some manifestation of their original sphericity and size (Figs. 13.6a,b, & 13.15).

Eggshell fragments (maximum dimension, 4 cm) are usually found scattered within nests and between adjacent nests. Isolated fragments are very rare or absent in areas where solitary nests have not been recovered. Breakage of the eggshell usually occurred prior to burial, however; several instances of in situ fragmentation were observed where two or more fragments have matching edges. These fragments are angular to subangular in shape and exhibit minor displacement relative to one another.

Some of the eggs appear to have abnormally large diameters. For example, near the Pat Baba Mandir, an isolated egg had a diameter of 28 cm (Fig. 13.15) compared to the normal range of 14–18 cm. Such phenomena may be the result of calcrete expansion processes (Watt, 1977; Saigal & Walton, 1988; Braithwaite, 1989). Repeated, alternating drying and wetting of the sediment causes calcite crystals to grow, resulting in grain breakage and reorientation (Buczynski & Chafetz, 1987). In addition, wetting followed by desiccation and shrinkage may have enhanced fragmentation of the eggs. Measurements (n = 30) taken on shrinkage cracks near the eggshells demonstrate a 10–27 percent range of contraction. Several eggs show an apparent decrease in diameter due to fragmentation or inward collapse of the shell.

The eggs and shell fragments from Bara Simla Hill–Pat Baba Mandir commonly show secondary replacement by silica. Complete silicification of individual eggshell fragments is rarely observed. Instead, silicification is usually random, affecting most parts of a shell unit while the adjacent units may be minimally or not affected at all. Silicification of the eggshell progresses inward from the inner and outer surfaces and is rela-

tively deeper in the internodal areas along pore canals (Fig. 13.7b).

The silicified eggshells occur in the Lower Limestone where various forms of silica occur. This includes thin, spindlelike, tapering silica bands with complex internal morphology, spherulites of chalcedony in the ground mass as well as within the cracks, and silica rims around quartz clasts (Kumar, 1990).

The nests, eggs, and eggshell fragments reported from Bara Simla Hill–Pat Baba Mandir and other Jabalpur localities occur on an exhumed paleogeomorphic surface. As a result, many fossils have been subjected to considerable weathering.

Fairly extensive sauropod nesting sites are also known from other locations in central and western India along a more or less east–west trending axis (Fig. 13.1). The eggs and the lithology of the nest-bearing strata closely resemble those from Jabalpur.

Paleophysiology of the Indian eggs

Inferences about the physiological processes in fossil eggshells are based upon comparison of features with those observed in extant avian and reptilian species (Erben, 1970; Erben et al., 1979). Features such as shell thickness, pore distribution and density, (?)pathologic thinning or thickening of eggshell units, and resorption of mammillae have been emphasized in such studies.

Pathological thickening (multilayering) or thinning of dinosaur eggshells from India has not been demonstrated, contrary to Mohabey (1984). Mohabey claimed that the double layered eggshell from Kheda was due to the "ovum-in-ovo" condition reported by Erben et al. (1979). However, we have discovered that improperly made radial sections of thick eggshells result in what is apparently a two-layered eggshell. Without any well-illustrated description and conclusions based

Figure 13.15. Detail of diametrically "enlarged" egg showing fragmentation due to brecciation and intrastratal reworking of host rock.

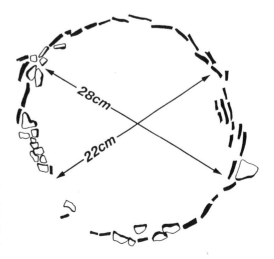

only on a few fragments (Mohabey, 1984, 1990), we suggest that any inference of a pathologic condition await more work.

Resorption craters may be present in the mammillary surface of the Lameta eggs. This feature has been cited as indicating actively growing embryos. We suggest that given the extensive secondary silicification we have discussed and the weathering observed in the eggshells, the presence of possible resorption craters should be considered in conjunction with other lines of evidence (e.g., embryonic or juvenile remains in the nest) that the embryos had begun development.

The porosity of eggshells may reflect the microenvironment of the dinosaur nest, such as nest humidity. Williams et al., (1984) suggested that the high values of water vapor conductance of the Aix-en-Provence sauropod eggs (Types 2–4) indicated that they were incubated under conditions of high humidity, probably underground. Our calculated conductance values for (?)TST-I (from Hathni River) and (?)TST-II (from Bara Simla Hill) range from between 2,650 mg/(day.torr) and 3,490 mg/(day.torr). These values are comparable to those reported by Williams et al. (1984), which range between 2,893 mg/(day.torr) and 4,500 mg/(day.torr). Comparative values for the Gobi sauropod eggs are much higher, 17,000 mg/(day.torr) (Seymour, 1980).

We feel that the implication for relative nest humidity in the three different localities has yet to be fully understood and must, in any case, be interpreted in conjunction with data for the lithofacies containing the eggs. The Lameta dinosaur eggs were apparently laid on wet, sandy-limey soils and were later subjected to prolonged intervals of aridity that resulted in a considerable loss of water.

Despite inherent taphonomical, preservational, and other such biases that color research on the Indian dinosaur eggs and nests, some aspects of the nesting behavior of the dinosaurs can still be inferred. At the Pat Baba Mandir/Bara Simla Hill nesting site, distances between adjacent nests (InterNest Distance, IND) are highly variable. Over small areas, the IND varies from between less than 1 m to over 10 m. This highly variable IND precludes the possibility of any relationship between IND and the size of the parent. Such a relationship was suggested for an Upper Cretaceous hadrosaur nesting site by Horner (1982). Nevertheless, the presence of several nests of sauropod eggs on the same paleogeomorphic surface is strongly indicative of colonial nesting.

Despite the many nesting sites from widely separated localities, no dinosaur skeletal remains have yet been recovered from them. The absence of embryos, hatchlings, and juveniles is particularly intriguing. Unfortunately, this lack makes it difficult to attribute the nests to any particular dinosaur species represented by skeletal remains. It is possible, however, that the absence of juveniles indicates that they did not inhabit the nests for long after hatching. Such an inference, however, is in our opin-

ion speculative as yet. The possibility that the eggs were entombed in the calcrete during the very early stages of incubation cannot be entirely ruled out given the lithofacies characteristics of the Lower Limestone.

Multiple layers of nests have been reported by Tripathi (1986) at Bara Simla Hill. Similarly, Mohabey (1990) has also reported at least two layers containing nests at Khempur (Gujarat). Both of these occurrences have not been illustrated in section and, because the egg-bearing horizon reflects the paleotopography, it is very likely that they represent laterally separated nests at different topographic levels on the same time surface (see another discussion by Cousin et al. in Chapter 5). Hermann H. Schleich (personal communication) has noted that sea turtles may utilize a nesting site repeatedly over years, resulting in different generations of nests on the same surface. Such a situation could very easily occur in a fossil nesting site, but would probably be impossible to recognize. However, the Bara Simla Hill–Pat Baba nests may be a part of a single nesting event. They have similar preservational characteristics, and it is very likely that they were buried and preserved together. Later or earlier generations of nests, if present, have apparently not been preserved. Hence, at present there is no direct evidence for site fidelity among the Lameta sauropods.

The repeated occurrence of nesting sites in similar lithofacies across the subcontinent suggests that the Lameta sauropods actively selected subaerial surfaces fringing small, shallow, ephemeral ponds or lakes. A semi-arid climate and intermittent wet events causing sheet floods or palustrine high stands further characterize the nesting sites. This evidence is further supported in the recent study by Sarkar, Bhattacharya, and Mohabey (1991). Their analysis of ^{18}O and ^{13}C in Indian sauropod eggs have yielded values which indicate that the dinosaurs consumed C3-type plants and ingested water from rivers and small evaporative pools.

Discussion and concluding remarks

The Upper Cretaceous (Maastrichtian) Lameta Beds preserve what are possibly the most geographically widespread dinosaur nesting sites in the world. These occurrences are supplemented by several coeval sites yielding eggshell fragments from the peninsular intertrappean sequences. Integrating the analysis of sedimentary facies, eggshell ultrastructure, and taphonomy of the nesting sites at Jabalpur, Balasinor, and the Hathni River has provided unique insights into the paleoecology of the nesting sites. Recently excavated dinosaur skeletal material from deposits contemporary with the nests may soon provide the taxonomic affinities of the eggs.

Despite rather large collections of eggs and eggshell fragments (over 1,000 complete eggs, Mohabey, 1990), little information is available on nest microenvironments, and the environmental or species specific physiological adaptations of the eggshell micro- and ultrastructure characters. Some information about eggshell water conductance was obtained from shell porosity. This conductance is believed to reflect the nest's microenvironment. The results, however, are based on a small sample size, so the sample needs to be enlarged before any serious inferences can be drawn.

The complete absence of microvertebrates and invertebrates in direct association with the nests may be a general character of the sites. In addition, the nest-bearing, sandy carbonates are pollen depauperate even though Maastrichtian microfloras have been recovered from the overlying beds at Jabalpur (B. S. Venkatachala, personal communication). Future work at the nesting sites must include a systematic and thorough search for pollen.

Despite obvious gaps in our present state of knowledge, some definite conclusions and broad inferences can be drawn about the eggs and nests from India:

1. Dinosaur eggshell fragments and nesting sites are found exclusively associated with the Deccan Volcano–sedimentary Province of peninsular India, which for a major part, represents Maastrichtian nonmarine strata. The nesting sites always occur underneath the basal-most basaltic flow in any area.

2. At present it is difficult to assign the eggshells to one or more dinosaur genera. Instead, the material is classified into five types based on morphological characteristics. The superficial, microstructural, and ultrastructural resemblance of the eggs to material known from southern France, along with the presence of sauropod skeletal material in the Lameta Beds, suggests that three of the eggshell types (TST-I, TST-II and TST-III) were laid by titanosaurid sauropods. A fourth type may also be sauropod, whereas the fifth type exhibits an ornithoid type of structure.

Eggshells recovered from the nesting sites are exclusively titanosaurid. However, the amount of morphological variability in the five eggshell types may not be an accurate measure of the taxonomic diversity of the dinosaur population and needs to be supplemented by other evidence.

3. A paleogeomorphic surface was reconstructed from Matley's (1921) map (8 in. = 1 mile, and with a 5-ft contour interval) of the Lameta Beds around Jabalpur. Superimposition of the lithofacies and fossils of the Lower Limestone has allowed the reconstruction of the Bara Simla Hill–Pat Baba Mandir nesting sites. The scenario from such a reconstruction suggests that the sauropods established their nests on wet surfaces in emergent areas of a palustrine flat system. Pedogenic modification resulted in the formation of calcimorphic paleosols.

The lithofacies indicates a semi-arid climate punctuated by several wet episodes that were followed by desiccation of the surface. Repeated wetting and drying of the soil resulted in widespread intrastratal reworking of the nest-bearing carbonates masking the original character of the soil sediment. In addition, frag-

mentation often accompanied by diametric enlargement or compaction of eggs occurred within nests. Tubular structures thought to be related to plant activity are referred to as rhizoconcretionary carbonate structures (Tandon et al., 1990).

The complete lack of embryos, hatchlings, or juveniles associated with the nesting sites may be attributable to several factors: early, deep burial of the eggs by sediments; rapid water loss from the eggs due to the high porosity of eggshell in a dry environment; or pedogenic modification of the soil with several episodes of brecciation and intrastratal reworking causing fragmentation of the egg or embryonic bones. One or more of these factors may have been partly responsible for inhibiting embryonic development or causing destruction of the entombed embryonic material.

4. The rather extensive nesting sites containing nests with morphologically identical eggs provide strong evidence for colonial nesting by the Lameta sauropods. Furthermore, the wide geographic distribution of the nesting sites in lithologically similar horizons indicates site selectivity by these dinosaurs.

Acknowledgements

We would like to thank Prof. V. K. Verma, Dr. Vinay Jhingran, Ravinder P. Kohli, Subodh Mittal, Shailaja Kumar, and Tanu Prakash who substantially helped as members of the 1989 field trip to Jabalpur. Asit Jolly, Sunil Bajpai, and S. Srinivasan acknowledge funding from the Council of Scientific and Industrial Research, New Delhi. We thank Karl Hirsch for examining the eggshell material from Anjar and for his comments on the manuscript. Walter Coombs, Jr., Herman Schleich, and B. P. Radhakrishna helped improve the manuscript through their review and comments. This chapter represents part of the work done under IGCP 245.

References

Alexander, P. O. 1981. Age and duration of Deccan volcanism: K-Ar evidence. *Memoir, Geological Society of India* 3: 244–58.

Bajpai, S. 1990. Geology and paleontology of some Late Cretaceous and Middle Eocene sequences of Kachchh, Gujarat, western India. Unpublished Ph.D. Thesis, Chandigarh: Panjab University.

Bajpai, S., Sahni, A., Jolly, A., & Srinivasan, S. 1990. Kachchh intertrappean biotas: affinities and correlation. *Cretaceous event stratigraphy and the correlation of the Indian nonmarine strata. Seminar cum Workshop I.G.C.P. 216 and 245*: 101–5. (Chandigarh: Panjab University).

Bakshi, A. K. 1987. Critical evaluation of the age of the Deccan Traps, India: Implications for flood basalt volcanism and faunal extinctions. *Geology* 15: 147–50.

Bande, M. B., & Chandra, S. 1990. Early Tertiary vegetational reconstructions around Nagpur-Chindwara and Mandla, central India. *Palaeobotanist* 38: 196–208.

Bande, M. B., & Prakash, U. 1982. Palaeoclimate and palaeogeography of central India during the Early Tertiary. *Geophytology* 12: 152–65.

Bhatia, S. B., Prasad, G. V. R., & Rana, R. S. 1990. Deccan volcanism; a Late Cretaceous event: conclusive evidence of ostracodes. *Cretaceous event stratigraphy and the correlation of the Indian nonmarine strata. Seminar cum workshop IGCP 216 and 245*: 47–49. (Chandigarh: Panjab University).

Bowles, J. E. 1979. *Physical and Geochemical Properties of Soils*, (New York: McGraw-Hill Book Company).

Braithwaite, C. R. J. 1989. Displacive calcite and grain breakage in sandstones. *Journal of Sedimentary Petrology* 59: 258–66.

Brookfield, M. E., & Sahni, A. 1987. Palaeoenvironments of the Lameta Beds (Late Cretaceous) at Jabalpur, Madhya Pradesh, India: soils and biotas of a semi-arid alluvial plain. *Cretaceous Research* 8: 1–14.

Buczynski, C., & Chafetz, H. S. 1987. Siliciclastic grain-breakage and displacement due to carbonate crystal growth: an example from the Lenders Formation (Permian) of North Central Texas, U.S.A. *Sedimentology* 34: 837–43.

Buffetaut, E. 1987. On the age of the dinosaur fauna from the Lameta Formation (Upper Cretaceous) of central India. *Newsletters in Stratigraphy* 18: 1–6.

Chanda, S. 1967. Petrogenesis of the calcareous constituents of the Lameta Group around Jabalpur, M.P. India. *Journal of Sedimentary Petrology* 37: 425–37.

Chatterjee, S. 1992. A kinematic model for the evolution of the Indian plate since the Late Jurassic. *In* S. Chatterjee & N. Hotton (eds.) *New Concepts in Global Tectonics*. (Lubbock: Texas Tech University Press).

Chiplonkar, G.W. 1986. Age of the Deccan Trap. *Publication of the Centre of Advanced Study in Geology, Panjab University* 2: 1–22.

Courtillot, V., Besse, J., Vandamme, D., Montigny, R. J., Jaeger, J. J. & Cappetta, H. 1986. Deccan flood basalts at the Cretaceous-Tertiary Boundary? *Earth and Planetary Science Letters* 80: 361–74.

Dauphin, Y. 1990. Comparative microstructural studies of eggshells. 1. Dinosaurs of Southern France. *Revue de Paleobiologie* 9: 121–37.

Dughi, R., & Sirugue, F. 1966. Sur la fossilisation des oeufs de Dinosaures. *Comptes rendus des séances de l'Académie des Sciences* 262: 2330–2.

Duncan, R. A., & Pyle, D. G. 1988. Catastrophic eruption of the Deccan Flood Basalts, western India. *Nature* 33: 841–3.

Erben, H. K. 1970. Ultrastruktur und mineralisation rezenter und fossiler Eischalen bei Vogeln und Reptilien. *Biomineralisation* 1: 1–66.

Erben, H. K., Hoefs, J., & Wedepohl, H. K. 1979. Paleobiological and isotopic studies of egg shells from a declining dinosaur species. *Paleobiology* 5: 380–414.

FitzPatrick, E. A. 1984. *Micromorphology of Soils* (London: Chapman and Hall Ltd.).

Freytet, P. 1973. Petrography and paleoenvironment of continental carbonate deposits with particular reference to Upper Cretaceous and Lower Eocene of Languedoc (Southern France). *Sedimentary Geology* 10: 25–60.

Freytet, P., & Plaziat, J. C. 1982. Continental carbonate sedimentation and pedogenesis – Late Cretaceous and Early Tertiary of Southern France. *In* B. H. Purser (ed.) *Contributions to Sedimentology 12*. [Stuttgart: E. Schwerzerbart'sche Verlagsbuchhandlung (Nagele u. Obermiller)].

Ganapathi, S. 1986. *Occurrence of molluscan fossils from the intertrappean beds of Dohad area and its depositional environments*, 27–28. Abstract Volume, Indian Association of Sedimentologists, Wadia Institute of Himalayan Geology, Dehradun.

Ghevariya, Z. G., 1988. Intertrappean dinosaurian fossils from Anjar area, Kachchh district, Gujarat. *Current Science* 57: 248–51.

Ghevariya, Z. G., & Srikarni, C. 1990. Anjar Formation, its fossils and their bearing on the extinction of dinosaurs. *Cretaceous event stratigraphy and the correlation of the Indian nonmarine strata. Seminar cum Workshop I.G.C.P. 216 and 245*: 106–109. (Chandigarh: Panjab University).

Goldbery, R. 1982. Structural analysis of soil microrelief in palaeosol of the Lower Jurassic "Laterite Derivate Facies" (Mishor and Ardon Formations) Makhtesh Ramon, Israel. *Sedimentary Geology* 31: 119–40.

Hirsch, K. F., & Packard, M. J. 1987. Review of fossil eggs and their shell structure. *Scanning Microscopy* 1: 383–400.

Hirsch, K. F., & Quinn, B. 1990. Eggs and eggshell fragments from the Upper Cretaceous Two Medicine Formation of Montana. *Journal of Vertebrate Paleontology* 10: 491–511.

Hislop, S., 1860. Geology and fossils of Nagpur. *Quarterly Journal of Geological Society of London* 16: 154–89.

Hora, S. L. 1938. On some fossil fish scales from the intertrappean beds at Deothan and Kheri, Central Provinces. *Records Geological Survey of India* 73: 267–94.

Horner, J. R. 1982. Evidence of colonial nesting and "site fidelity" among ornithischian dinosaurs. *Nature* 297: 675–6.

Huene, F. V., & Matley, C. A. 1933. The Cretaceous Saurischia and Ornithischia of the Central Province of India. *Memoir Geological Survey of India, Palaeontographica Indica* 21: 1–72.

Jain S. L. 1989. Recent dinosaur discoveries in India including eggshells, nests and coprolites. *In* D. D. Gillette & M. G. Lockley (eds.) *Dinosaur Tracks and Traces* (Cambridge, Cambridge University Press).

Jain, S. L. & Sahni, A. 1983. Some Upper Cretaceous vertebrates from central India and their palaeogeographic implications. *Cretaceous of India*. (Lucknow: Indian Association of Palynostratigraphers).

1985. Dinosaurian eggshell fragments from the Lameta Formation at Pisdura, Chandrapur District, Maharashtra. *Geoscience Journal* 2: 211–20.

Jain, S. P. 1975. Ostracoda from the Bagh Beds (Upper Cretaceous) of Madhya Pradesh. *Geophytology* 5: 188–212.

Jolly, A., Bajpai, S., & Srinivasan, S. 1990. Indian sauropod nesting sites (Maastrichtian, Lameta Formation): a preliminary assessment of the taphonomic factors at Jabalpur, India. *Cretaceous event stratigraphy and the correlation of the Indian nonmarine strata. Seminar cum Workshop IGCP 216 and 245*: 78–81 (Chandigarh: Panjab University).

Kérourio, P. 1982. Un nouveau type de coquille d'oeuf présumé dinosaurien dans le Campanien et le Maestrichtien continental de Provence. *Palaeovertebrata* 12: 141–7.

Krishnan, M. S. 1968. *Geology of India and Burma*. (Madras: Higginbothams Pvt. Ltd.).

Kumar, S. 1990. Facies, petrography and genesis of Lower Limestone unit, Lameta Group, Jabalpur area. Unpublished Masters of Science Thesis, Delhi: University of Delhi.

Kurzanov, S. M., & Mikhailov, K. E. 1989. Dinosaur eggshells from the Lower Cretaceous of Mongolia. *In* D. D. Gillette & M. G. Lockley (eds.), *Dinosaur Tracks and Traces*. (Cambridge: Cambridge University Press).

Lunkad, S. K. 1990. Stratigraphy, lithology and petrographic characterization of Late Cretaceous infratrappean (Lameta Group) sediments of Jhiraghat area in East Narmada valley. *Cretaceous event stratigraphy and the correlation of the Indian nonmarine strata. Seminar cum Workshop IGCP 216 and 245*: 90–3. (Chandigarh: Panjab University).

Lydekker, R. 1890. Note on certain vertebrate remains from the Nagpur District: Part of the chelonian plastron from Pisdura. *Records Geological Survey of India* 25: 329–37.

Mathur, U. B. 1990. Nonmarine Cretaceous of Rajasthan and Gujarat, western India. *Cretaceous event stratigraphy and the correlation of the Indian nonmarine strata. Seminar cum Workshop IGCP 216 and 245*: 84–5 (Chandigarh: Panjab University).

Mathur, U. B., & Srivastava, S. 1987. Dinosaur teeth from Lameta Group (Upper Cretaceous) of Kheda District, Gujarat. *Journal of the Geological Society of India* 29: 554–66.

Matley, C. A. 1921. On the stratigraphy, fossils and geological relationship of the Lameta beds of Jubbulpore. *Records Geological Survey of India* 53: 142–64.

Mittal, S. K. 1990. Facies analysis of Lameta beds, Bara Simla Hill, Jabalpur. Unpublished Masters of Science Thesis, Delhi: University of Delhi.

Mikhailov, K. E. 1991. Classification of fossil eggshells of amniotic vertebrates. *Acta Palaeontologica Polonica* 36: 193–238.

Mohabey, D. M. 1983. Note on the occurrence of dinosaurian fossil eggs from infratrappean limestone in Kheda District, Gujarat. *Current Science* 52: 1194.

1984. Pathologic dinosaurian egg shells from Kheda District, Gujarat. *Current Science* 53: 701–3.

1990. Dinosaur eggs from Lameta Formation of western and Central India: their occurrence and nesting behavior. *Cretaceous event stratigraphy and the correlation of the Indian nonmarine strata. Seminar cum Workshop IGCP 216 and 245*: 86–9 (Chandigarh: Panjab University).

Mohabey, D. M., & Mathur, U. B. 1989. Upper Cretaceous dinosaur eggs from new localities of Gujarat, India. *Journal of the Geological Society of India* 33: 32–7.

Mohabey, D. M., & Udhoji, S. G. 1990. Fossil occurrences and sedimentation of Lameta Formation of Nand area, Maharashtra – palaeoenvironmental, palaeoecological and taphonomical implications. *Cretaceous event stra-

tigraphy and the correlation of the Indian nonmarine strata. Seminar cum Workshop IGCP 216 and 245*: 75–7. (Chandigarh: Panjab University).

Platt, N. H. 1989. Lacustrine carbonates and pedogensis: sedimentology and origin of palustrine deposits from the Early Cretaceous Rupelo Formation, W. Cameros Basin, N. Spain. *Sedimentology* 36: 665–84.

Prakash, T., Singh R. Y., & Sahni, A. 1990. Palynofloral assemblage from the Padwar Deccan intertrappeans (Jabalpur), M.P. *Cretaceous event stratigraphy and the correlation of the Indian nonmarine strata. Seminar cum Workshop IGCP 216 and 245*: 68–9 (Chandigarh: Panjab University).

Prakash, U. 1960. A Survey of the Deccan intertrappean flora of India. *Journal of Paleontology* 34: 1027–40.

Prasad, G. V. R. 1989. Vertebrate fauna from the infra- and intertrappean beds of Andhra Pradesh: age implications. *Journal of the Geological Society of India* 34: 161–73.

Prasad, G. V. R., Sahni, A., & Gupta, V.J. 1986. Fossil assemblages from infra- and intertrappean beds of Asifabad, Andhra Pradesh and their geological implications. *Geoscience Journal* 7: 163–80.

Rana, R. S. 1984. Microvertebrate paleontology and biostratigraphy of the infra- and intertrappean beds of Nagpur, Maharashtra. Unpublished Ph.D. Thesis, Chandigarh: Panjab University.

Reeves, C.C. Jr. 1976. *Caliche origin, Classification, Morphology and Uses*. (Texas: Estacadeo Books).

Sahni, A. 1984. Cretaceous – Paleocene terrestrial faunas of India: lack of endemism and drifting of the Indian Plate. *Science* 226: 441–3.

1985. Upper Cretaceous – Paleocene palaeobiostratigraphy of India based on terrestrial vertebrate fauna. *Mémoires de la Société Géologique de France* 147: 125–37.

1989. Paleoecology of the Late Cretaceous dinosaur eggshell sites from peninsular India, *In* D. D. Gillette & M. G. Lockley (eds.), *Dinosaur Tracks and Traces*. (Cambridge: Cambridge University Press).

1990. Recent advances and shortcomings in understanding nonmarine Cretaceous basins of India. *Cretaceous event stratigraphy and the correlation of the Indian nonmarine strata. Seminar cum Workshop IGCP 216 and 245*: 18–21. (Chandigarh: Panjab University).

Sahni, A. & Bajpai, S. 1988. Cretaceous–Tertiary boundary events: the fossil vertebrate, paleomagnetic and radiometric evidence from peninsular India. *Journal of the Geological Society of India* 32: 382–96.

Sahni, A., & Tripathi, A. 1990. Age implications of the Jabalpur Lameta Formation and intertrappean biotas. *Cretaceous event stratigraphy and the correlation of the Indian nonmarine strata. Seminar cum Workshop IGCP 216 and 245*: 35–7. (Chandigarh: Panjab University).

Sahni, A., Rana, R. S., Kumar, K., & Loyal, R. S. 1984. New stratigraphic nomenclature for the infra- and intertrappean beds of Nagpur. *Geoscience Journal* 5: 55–8.

Sahni, A., Rana, R. S. & Prasad, G. V. R. 1984. SEM studies of thin eggshell fragments from the intertrappeans (Cretaceous-Tertiary transition) of Nagpur and Asifabad, Peninsular India. *Journal of the Palaeontological Society of India* 29: 26–33.

Sahni, A., Jaeger J. J., Rana, R. S., & Prasad, G. V. R. 1988. Takli Formation and coeval Deccan intertrappeans. *In* H. K. Maheshwar (ed.), *Palaeocene of India*. (Lucknow: Indian Association of Palynostratigraphers).

Sahni, B. 1940. The Deccan Traps: an episode of the Tertiary era. *General Presidential Address, 27th Indian Science Congress, Madras*.

Saigal, G. C., & Walton, E. K. 1988. On the occurrence of displacive calcite in Lower Old Red Sandstone of Carnoustie, Eastern Scotland. *Journal of Sedimentary Petrology* 58: 131–5

Sarkar, A., Bhattacharya, S. K., & Mohabey, D. M. 1991. Stable isotope analyses of dinosaur eggshells, paleoenvironmental implications. *Geology* 19: 1068–71.

Seymour, R. S. 1980. Dinosaur eggs: the relationship between gas conductance through the shell, water loss during incubation and clutch size. *Mémoires de la Société Géologique de France* 130: 177–84.

Singh, I. B. 1981. Palaeoenvironment and palaeogeography of the Lameta Group sediments (Late Cretaceous) in Jabalpur Area, India. *Journal of the Palaeontological Society of India* 26: 38–53.

Singh, S. K., & Srivastava, H. K. 1981. Lithostratigraphy of Bagh Beds and its correlation with the Lameta Beds. *Journal of the Palaeontological Society of India* 26: 77–85.

Sochava, A. V. 1969. Dinosaur eggs from the Upper Cretaceous of the Gobi Desert. *Paleontological Journal* 4: 517–27.

1971. Two types of eggshells in Senonian dinosaurs. *Palaeontological Journal* 3: 353–61.

Srivastava, S., Mohabey, M. B., Sahni, A., & Pant, S. C. 1986. Upper Cretaceous dinosaur egg clutches from Kheda District, Gujarat, India: their distribution, shell ultrastructure and palaeoecology. *Palaeontographica, Abt. A* 193: 219–33.

Tandon, S. K., & Friend, P. F. 1989. Near surface shrinkage and carbonate replacement processes, Arran Cornstone Formation, Scotland. *Sedimentology* 36: 1113–26.

Tandon, S. K., Verma, V. K., Jhingran, V., Sood, A., Kumar, S. Kohli, R. P., & Mittal, S. 1990. The Lameta beds of Jabalpur, central India: Deposits of fluvial and pedogenically modified semi-arid fan-palustrine flat systems. *Cretaceous event stratigraphy and the correlation of the Indian nonmarine strata. Seminar cum Workshop IGCP 216 and 245*: 27–30. (Chandigarh: Panjab University).

Tripathi, A. 1986. Biostratigraphy, palaeoecology and dinosaur eggshell ultrastructure of the Lameta Formation of Jabalpur, Madhya Pradesh. Unpublished M. Phil. Thesis, Chandigarh: Panjab University.

Udhoji, S. G., Mohabey, D. M., & Verma, K. K. 1990. Palaeontological studies of Lameta Formation of Nand area and their bearing on K-T boundary problem. *Cretaceous event stratigraphy and the correlation of the nonmarine strata. Seminar cum work shop IGCP 216 and 245*: 73–4. Chandigarh: Panjab University.

Vianey-Liaud, M., Jain, S. L. & Sahni, A. 1987. Dinosaur eggshells (Saurischia) from the Late Cretaceous intertrappean and Lameta Formation (Deccan, India) *Journal of Vertebrate Paleontology* 7: 408–24.

Watt, N. L. 1977. Pseudo-anticlines and other structures in some calcretes of Botswana and South Africa. *East Surface Process* 2: 63–74.

Williams, D. L. G., Seymour, R. S., & Kerourio, P. 1984. Structure of dinosaur eggshell from the Aix Basin, France. *Palaeogeography, Palaeoclimatology, Palaeoecology* 45: 23–37.

Young, C. C. 1959. New fossil eggs from Liyang Shantung. *Vertebrata PalAsiatica* 3: 34–5.

Zhao Z. 1979. Discovery of the dinosaurian eggs and foot prints from Neixiang County. Henan Province. *Vertebrata PalAsiatica* 17: 304–9.

IV Dinosaur babies

14 Life history syndromes, heterochrony, and the evolution of Dinosauria

DAVID B. WEISHAMPEL AND
JOHN R. HORNER

Abstract

Life history syndromes and heterochrony are important to our understanding of the evolutionary dynamics in many extant and extinct organisms, including Dinosauria. We review general issues of both of these aspects of growth and development. This review is followed by a discussion of the ways that skeletal (ontogenetic, histologic, etc.) and paleoecologic data can identify life history syndromes and heterochrony among dinosaurian taxa. Finally, we analyze the evolution of these features in Dinosauria.

Introduction

Dinosaurs have long been characterized as enormous, oversized, and supergigantic, so much so that the question "how did they get so big?" has come to loom quite large in studies of dinosaurian biology. It is surprising, then, that there has been little research relating size and growth rates, particularly in terms of (1) size-related ecological, behavioral, and developmental correlates of increases in an organism's fitness (life history syndromes), and (2) the alteration of somatic growth relative to reproductive maturation (heterochrony).

The business of having offspring and forging their contribution to future pools of individuals is often viewed in terms of life history strategies, tactics, and traits (here collectively referred to as life history syndromes (LHS) (Stearns, 1976, 1977; Gould 1977; Alberch et al., 1979; Calder, 1984; Harvey, Read, and Promislow, 1989). As these papers attest, LHS are important to our understanding of evolutionary dynamics in both extant and extinct organisms.

Often described as one of the developmental correlates of life history syndromes, heterochrony comprises phylogenetic changes in the expression or rate of development of particular features. Thus, descendant ontogenies are altered relative to ancestral ontogenies. One important aspect of such developmental alteration is change in the relationship between size and age.

We speculate on the phylogenetic distribution of what is known of dinosaur life history syndromes and heterochrony, not so much to provide specific answers to questions about dinosaur size and/or developmental biology (which we do not). Instead, this chapter outlines what we consider important avenues in dinosaur research that presently exist in the interstices of systematic, paleoecologic, and functional studies. Although we take a much simplified tack, we feel that by bringing the subjects of heterochrony and life history strategies to the fore we might provide a more dynamic view of the evolutionary biology of these animals.

Life history syndromes and heterochrony

How brood size, relative maturity and size of young, relative age distribution of reproductive effort, and interactions of reproductive effort with adult mortality relate to increases in an organism's fitness comprise the basis for life history syndromes. From both theoretical and analytical studies, we have a good understanding of the endpoints of the evolutionary continuum of LHS strategies "employed" by organisms: the so-called r- and K-strategies. The r-strategists are those organisms exposed principally to density-independent natural selection (r-selection), whereby a population expands with relatively little negative feedback on growth rate by dwindling resources. Ecological situations favoring r-selection include large, frequent, and unpredictable environmental fluctuations; frequent catastrophic mortality; superabundant resources; and lack of "crowding."

Life history traits related to r-selection are those favoring early and rapid production of large numbers of rapidly-developing young. These include high fecundity, early maturation, short life span, limited parental care, rapid development, and a greater proportion of available resources committed to reproduction. In contrast K-strategists are those organisms exposed predominately to density-dependent selection (K-selection), whereby one kind of organism increases at the expense of another. Low reproductive effort with late maturation, longer life, and a tendency to invest a great deal of parental care in small broods of late maturing offspring are among the life history traits thought to be evolutionary solutions to the crowded, stable, and benign environments behind K-selection.

Heterochrony is evolutionary change in the developmental timing of features (including changes in body size, age of reproductive maturity, etc.); that is, alteration of the expression of traits in descendants relative to the more primitive expression. There are six kinds of heterochrony that derive from changes in only three timing parameters: onset timing, offset timing, and growth rate (Alberch et al., 1979; McNamara, 1986).

In theory, each can be analyzed independently. Early growth onset in a descendant produces predisplacement relative to the ancestral condition, while late descendant onset produces postdisplacement. That is, a descendant ''starting earlier'' than its ancestor in overall development will have features that appear earlier (all other things being equal), thus yielding a predisplacement pattern. The opposite pole, postdisplacement, results in the delay of the expression of features because their development in descendants ''starts later'' than it did in their ancestors. The consequence of early growth offset between descendant and ancestor is called ''progenesis.'' Here development of descendant features is truncated or arrested earlier than it was in the ancestral condition.

Progenesis contrasts with hypermorphosis, the condition of late offset in which overall development is continued beyond the ancestral condition. Finally, elevated growth rates (acceleration) increase the rate at which features are realized during the ontogeny of descendants relative to the rate of realization in ancestors. In contrast, decreased growth rates produce neoteny, in which the appearance of features in descendants is protracted relative to their appearance in ancestral ontogenies. In this heterochrony scheme, the time of onset and offset, as well as the growth rates themselves, need to be scaled against absolute time as expressed as the age of individuals. Size is often a poor surrogate for age unless it can be shown that the organisms under examination grew at the same rate (Shea, 1983; Emerson, 1986). For fossils, absolute time or age is a particularly knotty problem, as pointed out by a number of authors, among them McKinney (1988) and McNamara (1988). As we will demonstrate, there are some potential remedies for this problem among Dinosauria.

Gould (1977) and others have argued that, in combination, heterochrony status and life history syndromes can reflect the impact of natural selection on survivorship and reproductive biology. Variable longevity, fecundity, and timing of reproduction – all life history traits – among closely related species may be synergistically linked to heterochronic changes during the phylogeny of these organisms. Thus, the evolution of altriciality and precociality may represent such a combination of LHS and heterochrony. Altricial organisms are those that have larger litters and shorter gestations than their ancestors and that give birth to relatively poorly developed young (Nice, 1962). It has been argued (Gould, 1977) that neoteny, with its retention of juvenile morphology in adult descendants via retardation of somatic development, is linked by K-selection to the evolution of altriciality. In contrast, precocial organisms give birth to smaller litters, exhibit slower development and more extended gestation, and give birth to relatively better developed young. Progenesis, with its accelerated sexual maturation, has likewise been argued to constitute a link with r-selection in the development of precociality. In both cases, the attribution of precociality and altriciality depend on comparisons of ancestral and descendant conditions, ultimately provided by a phylogenetic context. The same requirement will apply here in our discussion of dinosaurian heterochrony and LHS.

Data

Life span and the timing of somatic and sexual maturity, are some of the most difficult, if not impossible information to to establish for extinct species. Consequently, primary evaluations of LHS and/or heterochrony in Dinosauria have been rare. One of the few exceptions is the important study by Dunham et al., (1989), in which aspects of the life histories of four species of dinosaur are analyzed with respect to physiological and ecological constraints on feeding and growth. Like this study, we use what data are available, as well as a number of proxies (Table 14.1). Together, these include information from eggs, nests, embryos, and hatchlings; from bone histology; and from ontogenetic series and their morphometric analysis.

The skeleton provides the largest data set for LHS and heterochrony. Individual specimens provide data on bone histology and relative rate of formation. For example, lamellar–zonal primary bone indicates relatively slow growth rates, while fibrolamellar primary bone appears indicative of elevated growth rates (Ricqlès, 1976; Currey 1984). From an ontogenetic series of specimens come the raw materials for the study of morphometric changes throughout growth (viz., Rohlf & Bookstein, 1990). Just as egg size and the smallest individuals provide an approximation on hatchling size, large individuals are commonly thought to represent sexually mature and older stages (see also Dunham et al., 1989).

Joint morphology (i.e., articular surfaces of long bones, fusion of cranial and vertebral sutures) in perinatal individuals allows a qualitative assessment of skeletal maturity early in life (Horner & Weishampel, 1988).

Table 14.1. Life history syndrome and heterochrony data sources for Dinosauria

Taxon	Eggs and/or nests	Bone histology	Joint maturity in hatchlings and juveniles	Skeletal morphometrics	Ontogenetic series
Therapoda					
Ceratosauria					
Syntarsus rhodesiensis		Plexiform Fibrolamellar	Unanalyzed	Bivariate	Patchy
Coelophysis bauri		Dense fibrolamellar	Unanalyzed	Bivariate	Patchy
Carnosauria					
Allosaurus fragilis		Laminar and plexiform fibrolamellar, incomplete and dense Haversian	Unanalyzed		Patchy
Tyrannosaurus rex		Dense Haversian			Patchy
Tarbosaurus bataar					Patchy
Ornithomimosauria					
Gallimimus bullatus					Patchy
Maniraptora					
Troodon formosus	Eggs, nest	In progress			Reasonably good (study in progress)
Sauropoda					
Camarasauridae					
Camarasaurus lentus					Patchy
Diplodocidae					
Apatosaurus excelsus					Patchy
Titanosauridae					
Titanosaurus indicus	Eggs and nests questionably referred to *T. indicus*				
Hypselosaurus priscus	Eggs and nests questionably referred to *H. priscus*				
Magyarosaurus dacus			?Immature		Patchy
Thyreophora					
Stegosauridae					
Dacentrurus armatus					Patchy
Kentrosaurus aethiopicus		Dense Haversian			Patchy
Lexovisaurus durobrivensis					Patchy

Table 14.1. (*continued*)

Taxon	Eggs and/or nests	Bone histology	Joint maturity in hatchlings and juveniles	Skeletal morphometrics	Ontogenetic series
Stegosaurus stenops		Plexiform Fibrolamellar, incomplete and dense Haversian			Patchy
Ankylosauria					
Ankylosauridae					
Pinacosaurus grangeri					Patchy
Ornithopoda					
Hypsilophodontidae					
Orodromeus makelai	Eggs, nests	Fibrolamellar, dense Haversian	Mature	In progress	Good
Iguanodontia					
Tenontosaurus tilletti					Fair
Rhabdodon priscus		Plexiform fibrolammellar, incomplete and dense Haversian			Fair
Dryosaurus lettowvorbecki		Plexiform fibrolammellar	Mature	Bivariate	Good
Dryosaurus altus	Eggs, nests				Good
Camptosaurus dispar		Dense Haversian			Good
Iguanodon atherfieldensis		Plexiform fibrolamellar, incomplete and dense Haversian			Patchy
Telmatosaurus transsylvanicus	Eggs questiona- bly re- ferred to *T. trans- sylvanicus*	Dense Haversian			Fair
Maiasaura peeblesorum	Eggs, nests	Fibrolamellar, dense Haversian	Immature	Landmark	Good
Bactrosaurus johnsoni	Eggshell			Fair	
Corythosaurus casuarius				Bivariate	Good
Lambeosaurus lambei				Bivariate	Good
Hypacrosaurus altispinus				Landmark	Good
Hypacrosaurus n. sp.	Eggs, nests	Fibrolamellar, dense Haversian	Immature	Landmark	Good
Undescribed Mongolian hadrosaurid					Patchy

Table 14.1. (*continued*)

Taxon	Eggs and/or nests	Bone histology	Joint maturity in hatchlings and juveniles	Skeletal morphometrics	Ontogenetic series
Marginocephalia					
Ceratopsia					
Psittacosaurus mongoliensis		Dense Haversian			Good
Protoceratops andrewsi	Eggs, nests	Dense Haversian		Bivariate, multivariate	Good
Bagaceratops rozhdestvenskyi					Patchy
Avaceratops lammersi					Patchy
Chasmosaurus mariscalensis				Bivariate	Good
Centrosaurus apertus					Good
Living Aves	Eggs, nests	Fibrolamellar, dense Haversian	Variable	Bivariate, multivariate	Complete
Prosauropoda					
Thecodontosaurus antiquus					Patchy
Massospondylus carinatus	Eggs, nest		Unanalyzed		Fair
Yunnanosaurus huangi					Patchy
Plateosaurus engelhardti		Laminar and plexiform fibrolamellar, incomplete Haversian		Multivariate	Reasonably Good
Lufengosaurus huenei					Patchy
Ammosaurus major					Patchy
Mussaurus patagonicus	Eggshell		Patchy		Patchy
Sellosaurus gracilis					Patchy
Riojasaurus incertus					Fair

Notes: Data sources given in text.

Bone beds produced by mass mortality can provide not only a wide range of ontogenetic material, but also an absolute time scale for aging (Currie & Dodson, 1984; Horner in preparation). The discrete size classes of skeletal metrics from these bonebeds are likely indicative of the age structure of the herd at the time of death and hence should supply direct information on age calibration of size, shape, and relative maturity. While conditions providing this sort of information are rare, it should be possible to use absolute time to scale ontogenetic size and shape changes when they are available. In doing so, estimates of the maturity of the skeletal system with respect to age can be made.

Other data relevant to LHS and heterochrony come from the sedimentary context of the skeletal material and more especially the preservation of eggs and egg nests (see Horner 1984, 1987; Coombs 1989). Nests themselves obviously provide information on the minimal number of eggs that were once laid. They may also provide data on the state of hatchling development and behavior. For example, nests that preserve eggshell fragments versus those that retain intact, but hatched eggs have been used to identify the degree to which hatchlings were nestbound for some or all of their neonatal development (Horner & Makela, 1979; Horner, 1984; Horner & Weishampel, 1988). In contrast, the presence of intact, inplace, yet hatched eggs strongly suggests that young were not resident within the nest after hatching.

In summary, we look to changes in the timing of expression of features (including size) to provide information on heterochrony. We also look to bone histology and its evolutionary changes to adduce relative growth rates across dinosaurian clades. Finally, we look to specimens of eggs and nests, and analyses of their sedimentological disposition to provide information on hatchling size, fecundity, and aspects of parental investment in offspring. Each of these sources is influenced by scaling over a rather wide range of body sizes (for adult Dinosauria, over five orders of magnitude; Thulborn, 1982). Thus, the phylogenetic context of our database also requires scaling of appropriate comparisons (e.g., fecundity, hatchling size, aspects of bone histology, etc.).

Although often very patchy, the kinds of data described here exist for a wide array of dinosaurian taxa. Among basal Dinosauria (i.e., *Staurikosaurus pricei*, Herrerasauridae; Sues 1990), there are no published data relevant to LHS and heterochrony (including histology and morphometrics), even for better known taxa like *Staurikosaurus pricei* and *Herrerasaurus ischigualastensis*.

We turn to Saurischia, Theropoda, and Sauropodomorpha (but not Segnosauria) to provide restricted data relevant to LHS and heterochronic changes. Among extinct Theropoda, ontogenetic series are taxonomically very patchy (a sample of 7 out of over 100 species) and often unanalyzed. In addition to the usual adult material, some juvenile specimens are known for *Syntarsus rhodesiensis, Coelophysis bauri, Allosaurus fragilis, Tyrannosaurus rex, Tarbosaurus bataar, Gallimimus bullatus*, and *Troodon formosus* (Madsen 1976; Colbert 1989, 1990; Raath 1990; Rowe & Gauthier, 1990; Molnar, Kurzanov, & Dong 1990; Barsbold & Osmólska 1990; Osmólska & Barsbold, 1990; Varricchio in press).

Only in *T. formosus* is nest and hatchling material possibly available (Horner & Gorman, 1988; Varricchio, personal communication). Clutch size for *?T. formosus* appears to be at least six eggs, arranged in side-by-side pairs (Horner, 1987). Given the size of these eggs, hatchlings probably weighed approximately 0.5 kg or 1–2% adult mass. Skeletal development at this early stage (based on ossification of the distal femur) suggests that *T. formosus* was postnatally reasonably mature. In all known cases, histology (fibrolamellar primary bone and Haversian secondary bone) indicates that growth rates were elevated in theropods (Ricqlès, 1980; Rowe & Gauthier, 1990; Chinsamy 1990, 1992).

Basal sauropodomorphs (considered the monophyletic Prosauropoda by Galton, 1990, and Sereno, 1989, but as paraphyletic sequential outgroup taxa to Sauropoda by Gauthier, 1986, and Benton, 1990) are somewhat better represented in data relevant to LHS and heterochrony than are extinct theropods. Ontogenetic series are better represented (nine out of sixteen species), principally because of the relatively common occurrences of monospecific bone beds (viz., *Plateosaurus engelhardti, Lufengosaurus huenei, Riojasaurus incertus*; Galton, 1990); these are beginning to be analyzed morphometrically (Weishampel & Chapman, 1990; in preparation). All basal sauropodomorphs for which there are data have fibrolamellar primary bone and Haversian secondary bone (Ricqlès, 1980; Chinsamy, 1991).

Only for *Massospondylus carinatus* have there been analyses of ontogenetic variation in bone histology (Chinsamy, 1991, 1992); these indicate proportionately higher growth rates among juvenile individuals. *Massospondylus carinatus* is also known from juvenile material (Gow, 1990; Gow, Kitching, & Raath, 1990) and possibly embryonic material as well (Kitching, 1979; weight approximately 1 kg). The type material of *Mussaurus patagonicus* consists of hatchling material (Bonaparte & Vince, 1979; weight approximately 1 kg), with other material, earlier referred to *Plateosaurus* sp. (Casamiquela, 1980), probably comprising adult individuals (Galton, 1990). Nest material is known for *Mussaurus patagonicus* and possibly *Massospondylus carinatus*; if properly referred, the nest of *Mussaurus patagonicus* consists of at least two eggs, while that of *Massospondylus carinatus* contains at least six eggs. These specimens indicate relatively small hatchlings (approximately 0.2–0.4 percent adult length).

Sauropoda – the public's image of the archetypal dinosaur – is exceedingly poorly known when it comes to ontogenetic series (5 out of more than 100 species). Histology most commonly indicates plexiform fibrolamellar primary bone and incomplete, dense Haversian secondary bone (Ricqlès, 1980), testifying to rapid growth in these animals. Very little is known about preadult ontogenetic stages, although subadult *Camarasaurus lentus* and *Apatosaurus excelsus* have been described (Peterson & Gilmore, 1902; Gilmore, 1925) and possible (although controversially referred) hatchling material is available for *Titanosaurus indicus* and *Magyarosaurus dacus* (Mohabey, 1987; Jain, 1989; Weishampel, Grigorescu, & Norman 1991; Norman, et al., in preparation). Although probably not early hatchlings, these small individuals are likely no more than 5 percent adult body size. A consequence of such patchy data is that ontogenetically based morphometric studies of sauropod species are nonexistent. Finally, egg clutches from southern France that consist of up to fourteen eggs arranged in semicircles have been tentatively attributed to *Hypselosaurus priscus* (Breton, Fournier, & Watté, 1985). Egg volume suggests that early-stage hatchlings weighed no more than 5 kg, at least 0.05 percent adult mass.

Segnosauria comprises the remaining saurischian clade (Barsbold & Maryańska, 1990). These animals are incompletely known at present and consequently relatively poorly understood phylogenetically. They have also failed so far to yield material that addresses either heterochrony or LHS.

Among Ornithischia, LHS and heterochrony data are known for members of Thyreophora, Ornithopoda, and Ceratopsia. No eggs, nests, or hatchlings are known for Thyreophora, but juvenile specimens have been described for taxa otherwise known from adult material in five out of forty-two species (Galton, 1981a, 1983a, 1990; see also Winkler, Murry, & Jacobs, 1992). Absolute age distributions are unknown and skeletal maturity and/or morphometric analyses have not been attempted, in large part because stegosaurs are not known from bone-bed occurrences. Ontogenetic studies of bone histology have also not been carried out, but where known in adults (all are reported as having fibrolamellar primary bone and dense Haversian secondary bone; Ricqlès, 1980), these animals had elevated growth rates.

Within Ornithopoda (fifteen out of seventy-seven species), both Hypsilophodontidae and Iguanodontia (i.e., Euornithopoda) are represented by material relevant to dinosaurian LHS and heterochrony. For *Orodromeus makelai* (Horner 1984, 1987; Horner & Weishampel, 1988), clutch size is believed to be twelve eggs (Horner in preparation). These eggs were laid in a logarithmic spiral. Hatchlings began life at about 1–2 kg, approximately 1% adult mass. Hatchling skeletons appear to be reasonably mature, given the degree of ossification of the distal femoral articular surface (Horner & Weishampel, 1988). On the basis of bone histology (fibrolamellar primary bone, dense Haversian secondary bone; Horner in preparation), *O. makelai* grew rapidly, presumably at the same order of magnitude found throughout Dinosauria. Similar information is also beginning to become available for many other euornithopods. An as-yet-unnamed hypsilophodontid from Texas (Winker & Murry, 1989; Winkler et al., 1992) is known from a single locality from which a dense ontogenetic series is being studied. Ontogenetic samples are also available for *Tenontosaurus tilletti* (Dodson, 1980; Forster, 1990), although no nesting sites are yet documented. Likewise, the suite of isolated material referred to *Rhabdodon priscus* comprises a moderate ontogenetic series (Norman & Weishampel 1990; Weishampel et al., 1991; Norman et al. MS). In addition, several egg clutches from southern France, consisting of from three to eight eggs, have been tentatively attributed to *Rhabdodon priscus* (Breton et al., 1985). No embryonic or hatchling material has yet been identified, although egg size indicates early-stage hatchlings of from 3 to 5 kg, amounting to 1–2 percent adult mass.

Dryosaurus altus and *D. lettowvorbecki* are known from bonebed occurrences and young stages (Gilmore 1925; Janensch 1955; Galton & Jensen, 1973; Galton, 1981b, 1983b; Scheetz, 1991; Heinrich, Ruff, & Weishampel in press; Carpenter, Chapter 19). These individuals weighed in excess of 1 kg (approximately 0.70 percent adult mass). Distal femoral articular surfaces are well ossified, suggesting that hatchlings were skeletally mature. No eggs for either *D. altus* or *D. lettowvorbecki* are yet known. There is only a meager record of the young stages of *Camptosaurus dispar* and *Iguanodon bernissartensis*, (Dodson, 1980; Chure, Turner, & Peterson, 1992, Chapter 20; Norman, 1980), in contrast to that of hadrosaurids (Gilmore 1933; Dodson 1975; Horner & Makela 1979; Barsbold & Perle, 1983; Horner, 1984; Weishampel & Horner, 1986, 1990; Weishampel, Norman, & Grigorescu in press). A good deal of this

ontogenetic information comes from monospecific bonebed occurrences (Hooker, 1987; Horner & Gorman, 1988), but others are known from isolated, relatively complete, and presumably sympatric young and old individuals (Dodson, 1975, 1980). Other data are provided by eggs and nests.

In *Maiasaura peeblesorum* and n.s.p. *Hypacrosaurus* (Horner, Varricchio, & Goodwin, 1992; Horner and Currie, chapter 21), clutches consist of as many as twenty nestlings. Hatchlings of these species weighed approximately 1 kg (approximately 0.15 percent adult mass). Basally in the group, *Telmatosaurus transsylvanicus* apparently produced parallel rows of spherical eggs, with generally four eggs per clutch (Grigorescu et al., 1989; Chapter 6; Weishampel et al., 1991). Hatchlings of this species also weighed in excess of 1 kg (approximately 2 percent adult mass). In the case of *M. peeblesorum*, the new *Hypacrosaurus* material, and perhaps *T. transsylvanicus*, hatchlings and nestlings were skeletally immature, based on the degree of ossification of the distal femur in postnatal individuals. Bone histology in all of these euornithopods indicates rapid growth rates, as high or higher than those found among ratite birds (Horner in preparation).

Ceratopsia is well represented in material amenable to LHS and heterochrony issues (data for six out of twenty-nine species). *Protoceratops andrewsi*, *Chasmosaurus mariscalensis*, and *Centrosaurus apertus* are known from reasonably dense ontogenetic series (Brown & Schlaikjer, 1940; Kurzanov, 1972; Dodson, 1976; Currie & Dodson, 1984; Lehman, 1990). Among these, *Psittacosaurus mongoliensis* and *Protoceratops andrewsi* are also known from early stage hatchlings (Dodson, 1976; Coombs, 1980, 1982; Sereno, 1990). Clutch size in *Protoceratops andrewsi* is approximately eighteen eggs, apparently laid in concentric circles. Claims that several female *P. andrewsi* contributed to single nests (Hotton, 1963) appear to be based on the mistaken notion that there were 30 to 35 eggs within a nest (Brown & Schlaikjer, 1940).

Hatchling *Protoceratops andrewsi* and *Psittacosaurus mongoliensis* were relatively small (approximately 1 kg), approximately 0.5 percent adult mass. Hatchling skeletons appear to be mature in view of the well-ossified nature of the distal femoral articular surface. Ontogenetic variation (as well as variation associated with sexual dimorphism) has long been part of the study of *Protoceratops andrewsi* (Brown & Schlaikjer, 1940; Kurzanov, 1972; Dodson 1976; Chapman, 1990) and growth in other species is beginning to be analyzed. Much of these data come from monospecific ceratopsid bonebeds, but isolated subadult material is also known for *Avaceratops lammersi* and *Bagaceratops rozhdestvenskyi* (Dodson, 1986; Maryańska & Osmólska, 1975). Histology (dense Haversian secondary bone; Ricqlès, 1980) indicates that these, like other Dinosauria, had rapid growth rates.

Pachycephalosauria constitutes the remaining major ornithischian clade (Maryańska, 1990). To date, these animals have yielded no material that addresses the issues discussed in this paper.

Analyses

Given the distribution of these features that are thought to relate to LHS and heterochrony, it should be possible to analyze their disposition (albeit in a patchy fashion) within the phylogeny of the clade. In particular, the mapping and optimization of LHS and heterochrony data onto the phylogeny of Dinosauria amounts to documenting the pattern in which clades these features were acquired (viz., Coddington 1988; Sillen-Tullberg, 1988; Brooks & McLennan, 1991; Harvey & Pagel, 1991).

In order to understand the phylogenetic context of these characters, we undertook a numerical cladistic analysis of a subset of the taxa under discussion using Phylogenetic Analysis Using Parsimony (PAUP 4.2). We used the TOPOL option of PAUP to set the topology of dinosaurian phylogeny, which is here taken as a combination of Gauthier's (1986) saurischian phylogeny and Sereno's (1986) ornithischian phylogeny as summarized in Benton (1990), but with two important modifications. First, *Mussaurus patagonicus* and *Massospondylus carinatus* were considered a monophyletic taxon independent of *Hypselosaurus priscus*, reflecting the recent work of Sereno (1989) that suggests that Prosauropoda is monophyletic within the context of Sauropodomorpha (see also Galton, 1990). Second, the position of *Rhabdodon priscus* is unsettled at present. It may be a basal iguanodontian as noted in passing by Sereno (1986). However, there are conflicting characters which suggest hypsilophodontid affinity (Weishampel et al. 1991). As a consequence, these authors placed the species in an unresolved tritomy at the base of Euornithopoda. For these analyses, the TOPOL option requires fully resolved trees, which forced us to run two separate PAUP runs, treating *R. priscus* first as the sister-taxon to *Orodromeus makelai* within Hypsilophodontidae (Topology 1) and second as a basal iguanodontian (Topology 2).

We also consider Hadrosauridae (as used by Weishampel & Horner, 1990) as a monophyletic clade (contra Horner, 1990). The basis for identifying Hadrosauridae as monophyletic is given in Weishampel et al. (in press).

The character–taxon matrix for the twofold PAUP analyses is provided in Appendix I; each character is described in Appendix II. Only four features, all relating to nesting and hatchling status, comprise this matrix. Other features (e.g., information from morphometric studies), that bear on aspects of heterochrony are not yet well enough understood and/or analyzed to be incorporated in this cladistic investigation. It is hoped that these features will be included in later research.

As coded, all characters exist in a multistate transformation series, with a hypothetical outgroup coded as 0. We were unable to polarize remaining characters relative to this outgroup, in large part because of the lack of relevant observations in the actual outgroup taxa. Hence, all remaining taxa were coded 1, 2, and 3. Lacking information on character polarity, the data matrix was run unordered (UNORD option). Separate runs were made with no character weighting and with the WEIGHT SCALES option to reduce the impact of features with greater number of transformation series than others. Finally, the Accelerated Transformation (ACCT-RAN) and Delayed Transformation (DELTRAN) options were used separately to evaluate the level at which reversals and convergences accrued within the tree.

Analyses of the complete data matrix result in a single 15-step tree with a consistency index of 0.600 for Topology 1 and Topology 2 (Fig. 14.1). The phylogenetic distribution of LHS and heterochrony characters resulting from these analyses, as well as other features as yet too poorly circumscribed for analyses, are discussed below.

Discussion

Since appropriate information on *Staurikosaurus pricei* and Herrerasauridae is lacking, there is no basis for assessing the distribution of LHS or heterochrony characters basally within Dinosauria. Similarly, the apomorphic states of Eudinosauria (sensu Novas, 1992; Saurischia + Ornithischia) are difficult to appraise. Because all characters in our analyses were given a nonzero coding (except for the hypothetical outgroup), each character has a transformation identified as derived for the eudinosaurian node. Thus, in our analyses this dinosaur clade is apomorphical by the acquisition of large clutch size (character 1), mature skeletal morphology (character 2), relatively small hatchling size (character 3), and relatively small adult size (character 4). Keep in mind, however, that these character transformations at the eudinosaurian node may be due in part to our lack of knowledge of the primitive condition outside the clade and our consequent running of these characters as unordered.

For those features discussed above that are not yet amenable to cladistic analyses, it appears that all dinosaurs for which there is some ontogenetic sample appear to be characterized by fibrolamellar primary bone and often dense Haversian secondary bone throughout their lives (Ricqlés, 1980; Chinsamy, 1990, 1991; Horner in preparation). Thus, dinosaurs appear to be relatively rapidly growing animals. What the extent of variation might be of these histologically-based growth rates within Dinosauria and especially how histologic architecture varies with size is unfortunately still unknown. Also unknown is the extent to which these data differ from the immediate outgroups of Dinosauria (among them, *Lagosuchus talampayensis*, *Lagerpeton chanarensis*, Pterosauria; Benton & Clark, 1985; Sereno, 1991).

In addition, all known basal dinosaurs are absolutely larger in body size (~400% linear dimensions) than successive proximal outgroups and appear to have relatively larger body proportions (for example, jaw length and pubic proportions) as well. Although never all studied, if properly analyzed such size and shape changes may be found to be heterochronic shifts in basal dinosaur skeletal anatomy at the level of Dinosauria.

The analyzed portion of the saurischian clade (i.e., *Troodon formosus*, *Hypselosaurus priscus*, *Mussaurus patagonicus*, and *Massospondylus carinatus*) appears to be characterized by an apomorphic reduction in clutch size (character 1) using accelerated transformation. This feature is then reversed to constitute an apomorphy in *H. priscus* (see below). However, while this pattern may well be true, it is equally likely that reduced

Figure 14.1 **A.** Comparison of dinosaur relationships. Phylogenetic relationships of Dinosauria (after Benton, 1990). **B, C.** Cladograms of taxa analyzed using PAUP (15 steps, C.I. = 6.00). **B.** Cladogram with *Rhabdodon priscus* as a hypsilophodontid. **C.** Cladogram with *R. priscus* as an iguanodontian. Arrow A indicates acquisition of large hatchling size (character 3) and large adult size relative to hatchling size (character 4) in *Hypselosaurus priscus*. Arrow B indicates acquisition of large hatchling size (character 3) in *R. priscus*. Arrow C indicates reduction of skeletal maturity (character 2) and large adult size relative to hatchling size (character 4) in Hadrosauridae.

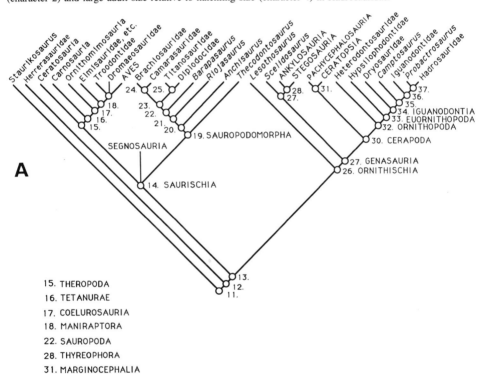

15. THEROPODA
16. TETANURAE
17. COELUROSAURIA
18. MANIRAPTORA
22. SAUROPODA
28. THYREOPHORA
31. MARGINOCEPHALIA

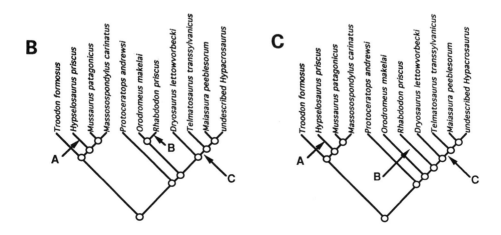

clutch size may have convergently evolved in *T. formosus* and the small clade of *Mussaurus patagonicus* and *Massospondylus carinatus* (via delayed transformation). Thus, variation in clutch size is so far phylogenetically ambiguous with respect to Saurischia.

Troodon formosus is the sole theropod for which appropriate data for our analyses exist. That said, none is identified as apomorphic for this species. Other yet-to-be analyzed aspects of LHS and heterochrony are available, albeit with limited distribution, for other theropods, including of course modern birds. Growth in *T. formosus*, *Coelophysis bauri*, and *Syntarsus rhodesiensis* appears to be relatively rapid, but no assessment as of yet has been made to identify variation in growth rates within these taxa. It might be argued that *Troodon formosus* was precocial (based on egg, nest, and hatchling data; Horner personal observation), perhaps the result of progenesis. Whether these patterns (and their explanations) can be extended to Troodontidae, Deinonychosauria, or even Maniraptora remains to be seen. Finally, extant avian species – and perhaps their fossil relatives, although data are sparsest for these organisms – display wide-ranging LHS and heterochronic shifts (Portmann, 1950; Nice 1962; James & Olson, 1983). In the context of Aves, Thulborn (1985) has even argued that *Archaeopteryx* and other basal birds were paedomorphic (?deinonychosaurian) theropods. This view, while possibly true, clearly requires a more thorough understanding of the ontogeny and phylogeny of taxa comprising the basal avian nexus.

Sauropodomorphs may yet yield some of the most interesting data on LHS and heterochrony among Dinosauria, simply because many of these animals are "pushing the size envelope" among terrestrial vertebrates. However, for the taxa under consideration, no features were identified as apomorphic at the level of Sauropodomorpha. Among its members, there are also no immediate changes at the level of Prosauropoda. There is, however, an apomorphic increase in adult size relative to hatchlings in *Massospondylus carinatus* (character 4).

As noted above, *Hypselosaurus priscus*, the sole sauropod taxon that we analyzed, potentially shows an apomorphically reversed increase in clutch size (character 1) using the ACTRAN option (see above). However, delayed transformation identifies this feature as convergently apomorphic for *Troodon formosus, Massospondylus carinatus,* and *Mussaurus patagonicus*. Unambiguous apomorphic acquisitions for *H. priscus* include an increase in hatchling mass (character 3) and a large increase in adult mass relative that of hatchlings (character 4). On the basis of other features, Bonaparte and Vince (1979) argued for the retention of juvenile features of prosauropod ancestors (as in *Mussaurus patagonicus*) into adulthood for sauropod descendants. This view of evolution by neoteny among basal sauropods may hold, but it clearly needs to be probed by

morphometric analyses and then so with an explicit phylogeny of sauropodomorph species and higher taxa in hand. Only then can the evolutionary context of these changes can be assessed.

How heterochrony and LHS status may have changed basally in Ornithischia is particularly difficult to ascertain, in large part because of the lack of information on a number of basal taxa. Similarly, thyreophorans have not yet yielded the kinds of data used in our analyses. Among those ornithischians with relevant data, Cerapoda (Ornithopoda + Marginocephalia; Sereno, 1986) bears no apomorphic changes. Nor does *Protoceratops andrewsi*, the sole representative of Marginocephalia (and consequently Ceratopsia) in our analyses. Within Ornithopoda, Heterodontosauridae lacks relevant data. Its sister-clade, Euornithopoda, is by contrast relatively well represented. Basally, PAUP analyses did not identify any apomorphic acquisitions.

The sister-taxa Hypsilophodontidae and Iguanodontia constitute the contents of Euornithopoda (Sereno, 1986; Weishampel and Heinrich in press). As earlier indicated, *Rhabdodon priscus* currently has an unresolved (probably basal) relationship within the context of Euornithopoda (Weishampel et al., 1991). With Topology 1 (*R. priscus* as the sister-taxon of *Orodromeus makelai*), there are no apomorphies that accumulate at the base of Hypsilophodontidae nor with *Orodromeus makelai* itself within the clade. In contrast, *R. priscus* can be characterized by the apomorphic decrease in clutch size (character 1) and increase in hatchling size (character 3). These two features suggest that *R. priscus* may have been a K-strategist relative to the more primitive condition. On the iguanodontian side of the euornithopod clade, Dryomorpha lacks apomorphic changes as does its basal member, *Dryosaurus lettowvorbecki*.

Hadrosauridae exhibits an apomorphic shift in reduced skeletal maturity (character 2) and increased relative adult size (character 4). These features intimate that hadrosaurid species may have evolved in the context of r-based LHS as compared with more their relatives. The evolution of skeletally immature hatchlings that grew into relatively large adults roughly corresponds to what Horner & Weishampel (1988) identified as altriciality among these species, in which extended parental care promoted the survival of nest-bound hatchlings. Within Hadrosauridae, there is a reduction of clutch size (character 1) in *Telmatosaurus transsylvanicus*, and an increase in hatchling size in Euhadrosauria (character 3). Remaining hadrosaurids show no further character shifts.

Analyses of Topology 2 (*R. priscus* as a basal iguanodontian positioned beneath *Dryosaurus lettowvorbecki*) treats *O. makelai* as the sole hypsilophodontid, for which there are no apomorphic shifts. PAUP does identify a decrease in clutch size (character 1) as apomorphic at basal levels within Iguanodontia (i.e., *R.*

priscus + Dryomorpha). *R. priscus* itself apomorphically acquires an increase in hatchling size (character 3). Both Dryomorpha and its basal species *Dryosaurus lettowvorbecki* present no apomorphies. Hadrosauridae is again characterized by a decrease in hatchling skeletal maturity (character 2) and an increase in adult size (character 4). Within Hadrosauridae, *Telmatosaurus transsylvanicus* itself belies no LHS or heterochrony apomorphies. However, for euhadrosaurs, there is an increase in clutch size (character 1) and an increase in hatchling mass (character 3). Finally, neither *Maiasaura peeblesorum* nor the new *Hypacrosaurus* species are characterized by apomorphic conditions within Euhadrosauria.

Other features of these animals are obviously relevant to both LHS and heterochrony. Among them is the miniaturization of the dentary dentition at the level of Hadrosauridae and miniaturization of the maxillary dentition at the level of Euhadrosauria. Together, these two features comprise the well-known dental batteries that are found in hadrosaurids, and may well represent the retention of juvenile tooth size into adulthood (Weishampel & Horner, 1990; Weishampel et al., 1991, in press).

Conclusions

Any numerical cladistic study of the disposition of LHS and heterochrony in dinosaurs based on 3.5% of presently recognized species is obviously fraught with a rocky future. The probability that any of these apomorphies will be superceded by further discovery and/or analysis assuredly must be high. Indeed, we would be very pleased if future discoveries or analyses were to overthrow the phylogenetic patterns we have identified here. Nonetheless, these are the patterns of acquisition as they presently stand among the examined taxa.

In its broadest terms, the goals of this paper were (1) to advance the point of view that LHS and heterochrony are important to our understanding of dinosaur evolutionary biology and (2) to identify ways that enhance this understanding. With these goals in mind, we conclude by suggesting five research approaches that ought to improve our knowledge of dinosaur evolutionary dynamics from the perspective of LHS and heterochrony:

1. More complete ontogenetic series of dinosaur species. Diverse aspects of dinosaur evolutionary biology would obviously benefit from any increase in our knowledge about their ontogenies. As pertains to this chapter, better sampling of dinosaur ontogenies should improve the quality of morphometric studies of growth and histologic changes with age.

2. Histologic distribution within these ontogenetic series (scaling of data within histologic categories). Should it be possible, the scaling of density and configuration of osteocyte lacunae and/or Haversian canals with growth rates would be immensely valuable in calibrating ontogenetic series and thus addressing absolute growth rates among individual dinosaurs.

3. Age calibration of ontogenetic series via bonebed occurrences. Monospecific bone-beds that result from catastrophic mortality have the potential for producing data on absolute rates of growth. Obviously necessary in these instances are restricted times of breeding or hatching within species in order for ages classes (and hence age-related morphology) to be clumped. Ages and growth rates can then be read directly from these clustered data. As bone-bed occurrences become better documented, we hope that these kinds of data will be used to calibrate absolute growth rates among dinosaur species.

4. Distribution of skeletal maturity over the perinatal period of development. Work has begun on the qualitative assessment of skeletal maturity among late-stage embryos and hatchlings (Horner & Weishampel, 1988). Just how development proceeds during this important stage in more taxa than those few for which some observations have been made will require similar ontogenetic research.

5. Within-species morphometric analyses. Bivariate, multivariate, and landmark approaches are the most common form of within-species dinosaur morphometrics. We would like to see these kinds of analyses extended across as many closely related species as can be analyzed and whose phylogenetic relationships are well known. If possible, these samples should be calibrated against real time (see item 3 above). With this kind of information, species-level heterochronic changes become the real stuff of dinosaurian evolutionary dynamics.

Acknowledgments

We thank the editors for providing the opportunity to speculate freely into the realm of life history syndromes and heterochrony among dinosaurs. It is our hope that such an approach, embryonic though it may be, might emerge as a significant and satisfying way to look at dinosaur evolutionary biology. If not, it was fun to do the research anyway.

Several people have been instrumental in providing information for, or comments on, this manuscript: P. Dodson, G. Erikson, R. Heinrich, D. Varricchio, and L. Witmer. We thank all of them for their help. The research described in this chapter has been supported in part by funds from the National Science Foundation (INT-8619987, EAR-8719878, and EAR-9004458, to DBW; EAR-8507031, and EAR-8705986 to JRH).

References

Alberch, P., Gould, S. J., Oster, G. F., & Wake, D. B. 1979. Size and shape in ontogeny and phylogeny. *Paleobiology* 5: 296–317.

Alexander, R. M. 1985. Mechanics of posture and gait of

some large dinosaurs. *Zoological Journal of the Linnean Society* 83: 1–25.

Anderson, J. F., Hall-Martin, A., & Russell, D. A. 1985. Long bone circumference and weight in mammals, birds and dinosaurs. *Journal of Zoology* 207: 53–61.

Barsbold, R. and Maryańska, T. 1990. Segnosauria. *In* D. B. Weishampel, P. Dodson,& Osmólska, H. (eds.) *The Dinosauria*. (Berkeley: University of California Press), pp. 408–15.

Barsbold, R., & Osmólska, H. 1990. Ornithomimosauria. *In* D. B. Weishampel, P. Dodson, and Osmólska, H. (eds.) *The Dinosauria*. (Berkeley: University of California Press), pp. 225–44.

Barsbold, R. & Perle, A. 1983. [On taphonomy of joint burying of juvenile dinosaurs and some aspects of their ecology]. *Sovmestnaya Sovetsko-Mongolskaya Paleontologicheskaya Expeditsiya*, Trudy 24: 121–5.

Benton, M. J. 1990. Origin and interrelationships of dinosaurs, *In* D. B. Weishampel, P. Dodson, & Osmólska, H. (eds.), *The Dinosauria*. (Berkeley: University of California Press), pp. 11–30.

Benton, M. J. & Clark, J. 1985. Archosaur phylogeny and the relationships of the Crocodylia. *In* M. J. Benton, (ed.), *The Phylogeny and Classification of the Tetrapods*, Vol. 1. Systematics Association Special Volume 35A (Oxford: Oxford University Press), pp. 295–338.

Bonaparte, J. F. & Vince, M. 1979. El hallazgo del primer nido de Dinosaurios Triásicos (Saurischia, Prosauropoda), Triásico superior de Patagonia, Argentina. *Ameghiniana* 16: 173–82.

Breton, G., Fournier, R., & Watté, J.-P. 1985. Le lieu de ponte de dinosaures de Rennes-le-Chateau (Aude): premiers résultats de la campagne de fouilles 1984. *Actes du Colloque Les Dinosaures de la Chine à la France 1985*: 127–40 (Toulouse: Muséum de Toulouse).

Brooks, D. R. & McLennan, D. A. 1991. *Phylogeny, Ecology, and Behavior*. (Chicago: University of Chicago Press).

Brown, B. & Schlaikjer, E. M. 1940. The structure and relationships of *Protoceratops*. *Annals of the New York Academy of Science* 40: 133–266.

Calder, W. 1984. *Size, Function, and Life History*. (Cambridge, MA: Harvard University Press).

Casamiquela, R. M. 1980. La presencia del genero *Plateosaurus* (Prosauropoda) en el Triasico superior de la Formacion El Tranquilo, Patagonia. *Actas II Congreso Argentino de Paleontología y Bioestratigrafía y I Congreso Latinoamericano de Paleontología* 1: 143–58

Chapman, R. E. 1990. Shape analysis in the study of dinosaur morphology. *In* K. Carpenter & P. J. Currie (eds.), *Dinosaur Systematics*. (New York: Cambridge University Press), pp. 21–42.

Chinsamy, A. 1990. Physiological implications of the bone histology of *Syntarsus rhodesiensis* (Saurischia: Theropoda). *Palaeontologia africana* 27: 77–82.

——— 1991. The bone histology and possible growth strategy of the prosauropod dinosaur *Massospondylus carinatus*. *In* Z. Kielan-Jaworowska, N. Heintz, & H. A. Nakrem (eds.), *Fifth Symposium on Mesozoic Terrestrial Ecosystems and Biota, Extended Abstracts University of*

Oslo, Contributions from the Paleontological Museum 364: 13.

——— 1992. Ontogenetic growth of the dinosaurs *Massospondylus carinatus* and *Syntarsus rhodesiensis*. *Journal of Vertebrate Paleontology* 12: 23A.

Chure, D. J., Turner, C., & Peterson, F. 1992. An embryo of the ornithopod dinosaur *Camptosaurus* from the Morrison Formation (Upper Jurassic) of Dinosaur National Monument, Utah. *Journal of Vertebrate Paleontology*, 12. Supplement to No. 3: 23A–24A.

Coddington, J. A. 1988. Cladistic tests of adaptational hypotheses. *Cladistics* 4: 3–22.

Colbert, E. H. 1989. The Triassic dinosaur *Coelophysis*. *Museum of Northern Arizona Bulletin* 57: 1–160.

——— 1990. Variation in *Coelophysis bauri*. *In* K. Carpenter & P. J. Currie (eds.), *Dinosaur Systematics*. (New York: Cambridge University Press), pp. 81–90.

Coombs, W. P. 1980. Juvenile ceratopsians from Mongolia the smallest known dinosaur specimens. *Nature* 283: 380–1.

——— 1982. Juvenile specimens of the ornithischian dinosaur *Psittacosaurus*. *Palaeontology* 25: 89–107.

——— 1989. Modern analogs for dinosaur nesting and parental behavior. *Geological Society of America Special Paper* 238: 21–53.

Currey, J. 1984. *The Mechanical Adaptations of Bone*. (Princeton: Princeton University Press).

Currie, P. J., & Dodson, P. 1984. Mass death of a herd of ceratopsian dinosaurs. *In* W.-E. Reif & F. Westphal, (eds.). *3rd Symposium on Mesozoic Terrestrial Ecosystems*. (Tübingen:. Attempto Verlag), pp. 61–6.

Dodson, P. 1975. Taxonomic implications of relative growth in lambeosaurine hadrosaurs. *Systematic Zoology* 24: 37–54.

——— 1976. Quantitative aspects of relative growth and sexual dimorphism in *Protoceratops*. *Journal of Paleontology* 50: 929–40.

——— 1980. Comparative osteology of the American ornithopods *Camptosaurus* and *Tenontosaurus*. *Société Géologique de France Mémoire* 139: 81–5.

——— 1986. *Avaceratops lammersi*: a new ceratopsid from the Judith River Formation of Montana. *Proceedings of the Academy of Natural Sciences of Philadelphia* 138: 305–17.

Dunham, A. E., Overall, K. L., Porter, W. P., & Forster, C. A. 1989. Implications of ecological energetics and biophysical and developmental constraints for life-history variation in dinosaurs. *Geological Society of America Special Paper* 238: 1–19.

Emerson, S. B. 1986. Heterochrony and frogs: the relationship of a life history trait to morphological form. *American Naturalist* 127: 744–71.

Forster, C. A. 1990. Evidence for juvenile groups in the ornithopod dinosaur *Tenontosaurus tilletti* Ostrom. *Journal of Paleontology* 64: 164–5.

Galton, P. M. 1981a. A juvenile stegosaurian dinosaur, "*Astrodon pusillus*" from the Upper Jurassic of Portugal, with comments on Upper Jurassic and Lower Cretaceous biogeography. *Journal of Vertebrate Paleontology* 1: 245–56.

——— 1981b. *Dryosaurus*, a hypsilophodontid dinosaur from the Upper Jurassic of North America and Africa. Postcranial skeleton. *Palaeontologisches Zeitschrift* 55: 271–312.

1983a. A juvenile stegosaurian dinosaur, *Omosaurus phillipsi* Seeley from the Oxfordian (Upper Jurassic) of England. *Géobios* 16: 95–101.

1983b. The cranial anatomy of *Dryosaurus*, a hypsilophodontid dinosaur from the Upper Jurassic of North America and East Africa, with a review of hypsilophodontids from the Upper Jurassic of North America. *Geologica et Palaeontologica* 17: 207–43.

1990. Basal Sauropodomorpha–prosauropods. *In* D. B. Weishampel, P. Dodson, & H. Osmólska (eds.), *The Dinosauria.* (Berkeley, University of California), pp. 320–44.

Galton, P. M. & Jensen, J. A. 1973. Small bones of the hypsilophodontid dinosaur *Dryosaurus altus* from the Upper Jurassic of Colorado. *Great Basin Naturalist* 33: 129–132.

Gauthier, J. A. 1986. Saurischian monophyly and the origin of birds. *In* K. Padian (ed.), *The Origin of Birds and the Evolution of Flight. Memoirs of the California Academy of Science* 8: 1–55.

Gilmore, C. W. 1925. A nearly complete articulated skeleton of *Camarasaurus*, a saurischian dinosaur from the Dinosaur National Monument. *Memoirs of Carnegie Museum* 10: 347–84.

1933. On the dinosaurian fauna of the Iren Dabasu Formation. *Bulletin of the American Museum of Natural History* 67: 23–78.

Gould, S. J. 1977. *Ontogeny and Phylogeny* (Cambridge, MA: Belknap Press).

Gow, C. E. 1990. Morphology and growth of the *Massospondylus* braincase (Dinosauria, Prosauropoda). *Palaeontologia africana* 27: 59–75.

Gow, C. E., Kitching, J. W., & Raath, M. A. 1990. Skulls of the prosauropod dinosaur *Massospondylus carinatus* Owen in the collections of the Bernard Price Institute for Palaeontological Research. *Palaeontologia africana* 27: 45–58.

Grigorescu, D., Seclamen, M., Norman, D. B., & Weishampel, D. B. 1990. Dinosaur eggs from Romania. *Nature* 346: 417.

Harvey, P. H., & Pagel, M. D. 1991. *The Comparative Method in Evolutionary Biology.* (Oxford: Oxford University Press).

Harvey, P. H., Read, A. F., & Promislow, D. E. L. 1989. Life history variation in placental mammals: unifying the data with the theory. *Oxford Surveys in Evolutionary Biology* 6: 13–31.

Heinrich, R. E., Ruff, C. B., & Weishampel, D. B. In press. Femoral ontogeny and locomotor biomechanics of *Dryosaurus lettowvorbecki* (Dinosauria, Iguanodontia).

Hooker, J. S. 1987. Late Cretaceous ashfall and the demise of a hadrosaurian "herd." *Geological Society of America, Rocky Mountain Section Abstracts with Program* 19: 284.

Horner, J. R. 1984. The nesting behavior of dinosaurs. *Scientific American* 250: 130–7.

1987. Ecological and behavioral implications derived from a dinosaur nesting site. *In* S. J. Czerkas & E. C. Olson, (eds.), *Dinosaurs Past and Present*, Volume 2. (Seattle, University of Washington Press), pp. 51–63.

1990. Evidence of diphyletic origination of the hadrosaurian (Reptilia: Ornithischia) dinosaurs. *In* K. Carpenter & P. Currie (eds.), *Dinosaur Systematics: Approaches and Perspectives.* (New York: Cambridge University Press), pp. 179–87.

Horner, J. R. & Gorman, J. 1988. *Digging Dinosaurs.* (New York: Workman Publishing Co.).

Horner, J. R. & Makela, R. 1979. Nest of juveniles provides evidence of family structure among dinosaurs. *Nature* 282: 296–8.

Horner, J. R., Varricchio, D. J., & Goodwin, M. B. 1992. Marine transgressions and the evolution of Cretaceous dinosaurs. *Nature* 358: 59–61.

Horner, J. R. & Weishampel, D. B. 1988. A comparative embryological study of two ornithischian dinosaurs. *Nature* 332: 256–7.

Hotton, N. 1963. *Dinosaurs.* (New York: Pyramid Publications).

Jain, S. L. 1989. Recent dinosaur discoveries in India, including eggshells, nests and coprolites. *In* D. D. Gillette & M. G. Lockley (eds.), *Dinosaur Tracks and Traces.* (New York: Cambridge University Press), pp. 91–108.

James, H. F., & Olson, S. L. 1983. Flightless birds. *Natural History* 92: 30–40.

Janensch, W. 1955. Der Ornithopode *Dysalotosaurus* der Tendaguruschichten. *Palaeontolographica (Supplement 7)* 3: 105–76.

Kitching, J. W. 1979. Preliminary report on a clutch of six dinosaurian eggs from the Upper Triassic Elliot Formation, northern Orange Free State. *Palaeontologia africana* 22: 41–5.

Kurzanov, S. M. 1972. [Sexual dimorphism in protoceratopsids]. *Paleontologicheskii Zhurnal* 1972: 91–7.

Lehman, T. M. 1990. The ceratopsian subfamily Chasmosaurinae: sexual dimorphism and systematics. *In* K. Carpenter & P. J. Currie (eds.), *Dinosaur Systematics* (New York: Cambridge University Press), pp. 211–30.

Madsen, J. H. 1976. *Allosaurus fragilis*: a revised osteology. *Utah Geological and Mineralogical Survey Bulletin* 1091: 1–163.

Maryańska, T. Pachycephalosauria. *In* D. B. Weishampel, P. Dodson, & H. Osmólska (eds.), *The Dinosauria.* (Berkeley :University of California Press), pp. 564–77.

Maryańska, T. & Osmólska, H. 1975. Protoceratopsidae (Dinosauria) of Asia. *Palaeontologia Polonica* 33:45–102.

McKinney, M. L. 1988. Classifying heterochrony: allometry, size, and time. *In* M. L. McKinney (ed.), *Heterochrony in Evolution.* (New York: Plenum Press), pp. 17–34.

McNamara, K. J. 1986. A guide to the nomenclature of heterochrony. *Journal of Paleontology* 60: 4–13.

1988. The abundance of heterochrony in the fossil record, *In* M. L. McKinney (ed.), *Heterochrony in Evolution.* (New York: Plenum Press), pp. 287–325.

Mohabey, D. M. 1987. Juvenile sauropod dinosaur from Upper Cretaceous Lameta Formation of Panchmahals District, Gujarat, India. *Journal of the Geological Society of India* 30: 210–16.

Molnar, R. E., Kurzanov, S. M., & Dong Z. 1990. Carnosauria: *In* D. B. Weishampel, P. Dodson, & H. Osmólska (eds.), *The Dinosauria* (Berkeley: University of California Press) pp. 510–33.

Nice, M. M. 1962. Development of behavior in precocial birds. *Transactions of the Linnaean Society of New York* 8: 1–211.

Norman, D. B. 1980. On the ornithischian dinosaur *Iguanodon bernissartensis* from the Lower Cretaceous of Bernissart (Belgium). *Institut Royal des Sciences Naturelles de Belgique Mémoires* 178: 1–103.

Norman, D. B. & Weishampel, D. B. 1990. Iguanodontidae and related Ornithopoda. *In* D. B. Weishampel, P. Dodson, & H. Osmólska (eds.), *The Dinosauria* (Berkeley: University of California Press), pp. 510–33.

Novas, F. E. 1992. Phylogenetic relationships of the basal dinosaurs, the Herrerasauridae. *Palaeontology* 35: 51–62.

Osmólska, H. & Barsbold, R. 1990. Troodontidae. *In* D. B. Weishampel, P. Dodson, & H. Osmólska (eds.), *The Dinosauria* (Berkeley: University of California Press), pp. 259–68.

Peterson, O. A., & Gilmore, C. W. 1902. *Elosaurus parvus*, a new genus and species of Sauropoda. *Annals of Carnegie Museum* 1: 490–9.

Portman, A. 1950. Le developpement postembryonnaire, *In* P. Grassé (ed.), *Traite de Zoologie*, Volume XV. Oiseaux. (Paris: Masson et Cie).

Raath, M. A. 1990. Morphological variation in small theropods and its meaning in systematics: evidence from *Syntarsus rhodesiensis*. *In* K. Carpenter & P. J. Currie (eds.), *Dinosaur Systematics*. (New York: Cambridge University Press), pp. 91-106.

Ricqlès, A. de. 1976. On bone histology of fossil and living reptiles, with comments on its functional and evolutionary significance. *In* A. d'A. Bellairs & C. B. Cox (eds.), *Morphology and Biology of Reptiles*. (London: Academic Press), pp. 123–50.

1980. Tissue structures of dinosaur bone: functional significance and possible relation to dinosaur physiology. *In* R. D. K. Thomas & E. C. Olson (eds.), *A Cold Look at Warm-Blooded Dinosaurs* (Boulder: Westview Press), pp. 103–40.

Rohlf, F. J., & Bookstein, F. L. 1990. Proceedings of the Michigan Morphometrics Workshop. *University of Michigan Museum of Zoology, Special Publication* 2.

Rowe, T. & Gauthier, J. A. 1990. Ceratosauria. *In* D. B. Weishampel, P. Dodson, & H. Osmólska (eds.), *The Dinosauria* (Berkeley: University of California Press), pp. 151–68.

Scheetz, R. D. 1991. Progress report of juvenile and embryonic *Dryosaurus* remains from the Upper Jurassic Morrison Formation of Colorado. *In* W. R. Averett (ed.) *Guidebook for Dinosaur Quarries and Tracksites Tour, Western Colorado and Eastern Utah* (Grand Junction: Grand Junction Geological Society), pp. 27–30.

Sereno, P. C. 1986. Phylogeny of the bird-hipped dinosaurs (Order Ornithischia). *National Geographic Research* 2: 234–56.

1989. Prosauropod monophyly and basal sauropodomorph phylogeny. *Journal of Vertebrate Paleontology* 9, supplement to No. 3: 38A.

1990. Psittacosauridae. *In* D. B. Weishampel, P. Dodson, & H. Osmólska (eds.), *The Dinosauria* (Berkeley: University of California Press), pp. 579–92.

1991. Basal archosaurs: phylogenetic relationships and functional implications. *Journal of Vertebrate Paleontology* 11 (supplement): 1–53.

Shea, B. T. 1983. Allometry and heterochrony in the African apes. *American Journal of Physical Anthropology* 62: 275–89.

Sillen-Tullberg, B. 1988. Evolution of gregariousness in aposematic butterfly larvae: a phylogenetic analysis. *Evolution* 42: 293–305.

Stearns, S. C. 1976. Life-history tactics: a review of the ideas. *Quarterly Review of Biology* 51: 3–47.

1977. The evolution of life history traits: a critique of the theory and a review of the data. *Annual Review of Ecology and Systematics* 8: 145–71.

Sues, H.-D. 1990. *Staurikosaurus* and Herrerasauridae. *In* D. B. Weishampel, P. Dodson, & H. Osmólska (eds.), *The Dinosauria* (Berkeley: University of California Press), pp. 143–7.

Thulborn, R. A. 1982. Speeds and gaits of dinosaurs. *Palaeogeography, Palaeoclimatology, Palaeoecology* 38: 227–56.

1985. Birds as neotenous dinosaurs. *Records of the new Zealand Geological Survey* 9: 90–2.

Weishampel, D. B., & Chapman, R. E. 1990. Morphometric study of *Plateosaurus* from Trossingen (Baden-Württemberg, Federal Republic of Germany). *In* K. Carpenter & P. J. Currie (eds.), *Dinosaur Systematics* (New York: Cambridge University Press), pp. 43–52.

Weishampel, D. B., Grigorescu, D., & Norman, D. B. 1991. The dinosaurs of Transylvania: island biogeography in the Late Cretaceous. *National Geographic Research and Exploration* 7: 68–87.

Weishampel, D. B. & Horner, J. R. 1986. The hadrosaurid dinosaurs from the Iren Dabasu fauna (People's Republic of China, Late Cretaceous). *Journal of Vertebrate Paleontology* 6: 38–45.

1990. Hadrosauridae. *In* D.B. Weishampel, P. Dodson, & H. Osmólska (eds.), *The Dinosauria* (Berkeley: University of California Press), pp. 534–61.

Weishampel, D. B., Norman, D. B., & Grigorescu, D. in press. *Telmatosaurus transsylvanicus* from the latest Cretaceous of Romania: the most basal hadrosaurid?

Winkler, D. A., & Murry, P. A. 1989. Paleoecology and hypsilophodontid behavior at the Proctor Lake dinosaur locality (Early Cretaceous), Texas. *Geological Society of America Special Paper* 238: 55–61.

Winkler, D. A., Murry, P. A., and Jacobs, L. L. 1992. Differentiating adult and juvenile ornithischian dinosaurs in the Early Cretaceous. *Journal of Vertebrate Paleontology* 12, supplement to No. 3: 60A.

Appendix I: Character–taxon matrix

Outgroup	0 0 0 0
Troodon formosus	1 1 1 1
Hypselosaurus priscus	2 9 2 3
Mussaurus patagonicus	9 9 1 1
Massospondylus carinatus	1 9 1 2

Protoceratops andrewsi	2 1 1 1
Orodromeus makelai	2 1 1 1
Rhabdodon priscus	1 9 2 1
Dryosaurus lettowvorbecki	9 1 1 1
Telmatosaurus transsylvanicus	1 9 1 2
Maiasaura peeblesorum	2 2 2 2
Hypacrosaurus stebingeri	2 2 2 2

Appendix II: Characters used in cladistic analysis

1. Number of eggs per nest. The number of eggs (or more rarely, the number of hatchlings) that have been found in a single nest constitutes the minimum number of eggs that a female may have laid. At the one extreme, female ?*Troodon formosus* and ?*Massospondylus carinatus* laid at least six eggs per nest (1–10 eggs coded 1), while at the other, female *Maiasaura peeblesorum* laid at least 20 (10–20 eggs coded 2). The claim that *Protoceratops andrewsi* nests consisted of 30–35 eggs is based on a single eroded nest and suppositions about eggs that are not preserved.

2. Hatchling skeletal maturity. Development of the distal condyles of the femur vary across Dinosauria. The most common condition is for the distal condyle of hatchlings to be nearly as well developed as in adults. Thus, lateral and medial condyles are expressed to the same degree throughout the life of an individual. This character is coded 1. In contrast, in the hadrosaurs *Maiasaura*, *Hypacrosaurus*, and probably *Telmatosaurus*, the distal condyles are only partially ossified. The region is dominated by an extensive pit that would have been occupied by a large cartilaginous epiphysis. This feature is coded 2.

3. Hatchling size. The mass of hatchlings is estimated in two ways. For skeletal material, limb dimensions (diameter, circumference, cross-sectional area) are used to estimate mass in conjunction with scaling analyses developed for other organisms (Anderson, Hall-Martin, & Russell, 1985; Alexander, 1985; Heinrich et al. in press). For species with tentatively referred eggs and nests (but no hatchlings), egg volume provides the upper limit on hatchling size. Calculation of hatchling mass requires conversion of egg volume by an estimated tissue density of 1.0 g/cm^3. Hatchling mass under 3 kg is coded 1; over 3 kg it is coded 2.

4. Adult size relative to hatchling size. Adult mass is estimated in the same way as the mass of hatchlings based on skeletal material. Species with relatively very large adults (1,000 to 2,000 times hatchling size) are coded 3; species with adults 500 to 1,000 times larger than hatchlings are coded 2; and species with adults 40 to 500 times larger than hatchlings are coded 1.

15 Dinosaur reproduction in the fast lane: Implications for size, success, and extinction

GREGORY S. PAUL

Abstract

Large dinosaurs were apparently r-strategy egg layers with reproductive outputs much higher than those of K-strategy large mammals. Dinosaur populations may have consisted mostly of posthatchling juveniles that were not dependent upon their parents for survival. That dinosaurs grew as rapidly as mammals suggests similarly elevated metabolic rates. Dinosaur population recovery and dispersal potentials were probably much higher than those observed in giant mammals. Low population levels of adult dinosaurs enabled individual adults to consume relatively large portions of the available resource base, thereby becoming larger than mammals. The r-strategy-based survival potential of the dinosaurs helps explain their long period of success, and exacerbates the problem of explaining their final extinction. Some of the problems inherent in modeling the population dynamics of r-strategy dinosaurs are examined.

Introduction

Most investigations of the evolutionary forces behind the gigantism of some dinosaurs have focused upon the great size of the adults (Figs. 15.1 and 15.2; see also Weishampel & Horner, Chapter 14). Little attention has been paid to the fact that even the most colossal dinosaurs began life as nidifugous hatchlings no bigger than the chicken. Can hatchling dinosaurs provide clues toward understanding why they so often grew so large?

Large terrestrial mammals are K-strategists that emphasize low juvenile mortality, starvation resistance, and high dispersal to maintain population levels. As shown below, dinosaurs, on the other hand, were fast breeding r-strategists. The dramatic difference between these two reproductive strategies was briefly mentioned by Kurten (1953), but was only recently examined in detail by Carrano and Janis (1990), Janis and Carrano (1992), and Farlow (1993).

Many studies of dinosaur extinction tacitly assume that dinosaurs were broadly similar to large mammals in population dynamics, genetic information processing, and vulnerability to disruptive events. For example, Jablonski (1991) lists large size as a possible factor in the extinction of the dinosaurs, but also acknowledges that juveniles and small adult forms also vanished. The apparent fecundity of dinosaurs suggests that in some ways their reproductive strategies were more like those of small mammals rather than large mammals. The production of many nidifugous juveniles may help explain how and why dinosaurs often grew to gigantic dimensions and why they were such a stable and successful group for such a long time, and also makes it more difficult to explain their ultimate extinction.

Much of the data, discussion, and conclusions presented must be considered tentative, and are offered in the hope of encouraging further consideration on dinosaur size and reproduction. The prefix ''mega-'' refers to animals with a body mass of one or more tons. Dinosaurs are not considered to be reptiles in the typical sense, and birds are treated as a separate group to better examine reproductive differences. I also assume that dinosaurs reproduced on an annual basis, although multi-annual and semi-annual breeding cannot be discounted. Because dinosaurs probably grew at rates similar to those of birds and mammals (see below), it is assumed that their life and reproductive spans were also broadly similar to those of birds and mammals of similar mass.

Reproduction, growth, and recovery potential of dinosaurs
Reproductive method and potential

Recently, Bakker (1986) has suggested that sauropods gave birth to live young. This hypothesis runs contrary to the fossil evidence of sauropod nests, eggs,

Figure 15.1. Size comparisons of giant continental dinosaurs, mammals, and reptiles drawn to the same scale. Extinct taxa based on largest known specimens, and masses from volumetric models. Extant and recent taxa based on large adult males. A. 60- to 80-ton titanosaur (Bonaparte, 1989). B. 55-ton *Supersaurus*. C. 45-ton *Brachiosaurus* (= *Ultrasaurus*). D. 13-ton *Shantungosaurus*. E. 6-ton *Triceratops*. F. 7-ton *Tyrannosaurus*. G. 16-ton *Indricotherium*. H. 2-ton *Rhinoceros*. I. 5-ton *Megacerops*. J. 10-ton *Mammuthus*. K. 6-ton *Loxodonta*. L. 0.3-ton *Panthera*. M. 1-ton *Scutosaurus*. N. 1-ton *Megalania*. Human figure 1.62 m tall. Scale bar = 4 m.

Figure 15.2. *Brachiosaurus brancai* as a representative of large dinosaurs. Size distribution based on associated parallel sauropod trackways, and includes a few large adults, some subadults, and juveniles. Size of juveniles is believed to be the minimum size able to keep up with the herd.

and hatchlings (Fig. 15.3; Erben, Hoefs, & Wedepohl, 1979; Mohabey, 1987; Cousin et al., 1989; Hirsch, 1989; Paul, 1991; Moratalla and Powell, Chapter 3; however Jain, 1989, disputes Mohabey's report of a sauropod hatchling). All known dinosaur eggs had hard calcified shells, and it is possible that this adaptation barred a shift to live birth (Packard, 1977). If this is true, then dinosaurs must have been oviparous.

Prosauropods, small to large theropods, small ornithopods, some hadrosaurs (such as *Maiasaura*), and small ceratopsians had egg volumes of 0.1 to 0.5 L (Case, 1978b; Horner & Makela, 1979; Horner, 1987; Horner & Gorman, 1988; Horner & Weishampel, 1988; Winkler & Murry, 1989). Eggs of the hadrosaur *Hypacrosaurus* were more volumous at 4 liters (Horner and Currie, Chapter 21), while those assigned variously to iguanodonts and to sauropods were 2 to 3 liters in volume (Case, 1978b; Cousin et al., 1989; Dodson, 1990).

Reproductive rates for five types of Late Cretaceous dinosaurs can be determined. These are hypsilophodontid (*Orodromeus* and an unnamed species from Proctor Lake, Texas), small theropod (?*Troodon*), ceratopsian (*Protoceratops*), hadrosaur (*Maiasaura* and *Hypacrosaurus*), and sauropod (*Hypselosaurus*).

Hypsilophodontid

Nests are known for the small ~40-kg *Orodromeus* and Proctor Lake hypsilophodontid (Horner, 1982, 1984, 1987; Horner & Gorman 1988; Horner & Weishampel 1988; Coombs, 1989, 1990; Winkler & Murry, 1989; Hirsch & Quinn, 1990). The nests contain either one or two dozen eggs or nestlings. These numbers suggests that each female laid a dozen eggs, and that two dozen eggs represent the output of two females (Horner, 1987).

Small theropod

Egg clutches have been questionably assigned to *Troodon*, a small 50-kg theropod (Horner, 1987; Hirsch & Quinn, 1990). The clutches contain two to three pairs of eggs a linear row. Four eggs is considered to represent the minimal reproductive potential. It is possible that multiple clutches were deposited.

Protoceratops

I follow Thulborn (1991) in assigning clutches of about 18 eggs to this ~80-kg ceratopsian.

Hadrosaur

Nests of *Maiasaura* and *Hypacrosaurus* are known for these 2,000 to 2,500-kg hadrosaurs. There are between eighteen and twenty-four eggs in a clutch (Horner & Makela, 1979; Horner, 1982, 1984, 1987; Horner & Gorman, 1988; Horner & Weishampel, 1988; Coombs, 1989, 1990; Hirsch & Quinn, 1990; Horner and Currie, Chapter 21). The modest volume of these eggs (5 to 100 liters) relative to the size of the adults suggests that one female laid all the eggs in a nest.

Sauropod

Large arcs of eggs have been identified as *Hypselosaurus* (Case, 1978b; Hirsch, 1989; Cousin et al.,

Figure 15.3. Small size of dinosaur hatchlings (3 kg titanosaur sauropod hatchling shown here) made possible large clutch sizes and high reproductive rates that rendered even the largest dinosaurs r-strategists with exceptional population recovery and dispersal potential.

1989, Chapter 5; Dodson 1990; Vianey et al., Chapter 11). This sauropod is poorly known, but the adult mass is believed to be about 5000 kg. The arcs contain up to fourteen eggs, and this is taken as the minimum potential reproductive output of the sauropod. Case (1978b) suggested a maximum of 100 eggs per individual, a number that is reasonable for an animal of this size.

In Figure 15.4 the estimated breeding rates for the dinosaurs just discussed is compared to the known rates of extant female vertebrates. In extant taxa with an adult mass of 1 g to 10 kg, the annual number of young produced is broadly similar in both egg layers and live bearers. Above 10 kg the number of young of the two types diverge significantly, with many oviparous taxa being much more prolific than mammals. This divergence is mainly due to a decline in fecundity in mammals with a mass of over 5 tons.

Dinosaur reproduction is essentially an extension of the reptilian–avian pattern, with many eggs produced as part of an r-strategy mode of reproduction. Small dinosaurs had reproductive potentials overlapping those of both viviparous and oviparous animals. The annual reproductive output of herbivorous dinosaurs equals or exceeds that of rodents, lagomorphs, and other small mammals, and was much higher than that of large herbivorous mammals.

Reproductive potential over a lifetime can also be compared. For example, female elephants between age ten to fifty produce about a dozen young (Owen-Smith, 1988). During the same 40-year interval, a sauropod could have produced 500–4,000 eggs. Reproductive rates for dinosaurs and mammals also scale differently. Large mammals breed less rapidly than small mammals (Western, 1979; Eisenberg, 1981; Owen-Smith, 1988). In dinosaurs, however, reproductive rates remain constant or increase with size. This conclusion is similar to that of Janis and Carrano (1993) that the reproductive output of large mammals diverges from that of large dinosaurs.

Dinosaurs could breed every year or two, replacing clutches that were destroyed. Female mammals with a mass over 10–100 kg are constrained by the long gestation period or time spent by the progeny in a pouch or nursing. Weaning takes 3–12 months in most ungulates, 2–3 years in megaungulates, and 3–18 months in large marsupials (Langman, 1982; Owen-Smith, 1988; Nowak, 1991). Therefore, most cannot reproduce faster than once every year; megamammals are limited to reproducing every 2 or more years.

Parental care and feeding of juveniles

The amount of parental care given by dinosaurs appears to have varied widely, from wholly dependent (precocial) nestlings to fully independent (altricial) hatchlings (see Paul, Chapter 18).

Horner (1988; Horner & Gorman, 1988) has suggested that the short, "cute" snouts of altricial dinosaur

hatchlings may have been display characters designed to invoke parental care. Certainly hadrosaur hatchlings appeared to have been nest bound and dependent upon their parents for 1 or 2 months (Horner & Makela, 1979; Horner, 1982, 1984; Horner & Gorman, 1988; Horner & Weishampel, 1988; Coombs, 1989; Lambert, 1991).

Trackway and bone-bed evidence indicate that young dinosaurs did not join herds with adults until they were sufficiently large enough to keep up. The minimal sizes of juveniles in mixed size herds are the following: sauropods, over one third adult size (Fig. 15.2, Bird, 1985; Farlow, Pittman, & Hawthorne, 1989); hadrosaurs, nearly one half adult size (Horner & Gorman, 1988); ceratopsids, one fifth adult size (Currie & Dodson 1984); hypsilophodonts, one quarter adult size (Thulborn & Wade, 1984). Before joining the herds, the young may have formed into pods that were either independent or under the care of one or more adults (Horner & Makela, 1979; Horner & Gorman, 1988; Forster, 1990; Lockley, 1991). Pod behavior in some theropods may be seen in juvenile trackways that are not in association with the tracks of larger adults, suggesting that the youngsters were independent (Bird, 1985; Lockley 1991).

As these young became less dependent upon the parents, they were able to meet their own food, thermoregulatory, and defensive requirements which may have differed significantly from those of the adults. In these respects dinosaurs were like typical reptiles in that by pursuing a distinctly different niche, the juveniles avoid competing with and being a burden upon their parents.

Rates of growth

Juvenile dinosaurs grew rapidly at rates comparable to those of birds and mammals (Ricqles, 1980; Currie, 1981; Bakker, 1980, 1986; Currie & Dodson, 1984; Paul, 1988b; Reid, 1990; Dunham et al., 1989; Russell, 1989; Farlow, 1990; Lambert, 1991; Leahy & Paul, 1991; Chinsamy, 1992; Varricchio, 1992). Rates of growth are plotted in Figure 15.5. The very rapid growth of hadrosaur nestlings matches the highest rates in altricial bird nestlings. These rates, as well as those of extant birds, show that lactation is not a prerequisite for high rates of growth (contra Pond, 1983). Instead, high minimal and maximal metabolic rates are required (Case, 1978a; Leahy & Paul, 1991). The gap between the growth rates of terrestrial reptiles and mammals *increases* with increasing size! This fact contradicts the possibility that the growth rates of gigantic ectotherms converge with those giant endotherms.

Adult/juvenile population ratios and total populations

Richmond (1965) suggested that the scarcity of juvenile dinosaur fossils reflected a reptilian pattern of high juvenile mortality and slow growth. In point of fact, juvenile dinosaurs are abundantly represented in the fossil record (Dodson, 1975 Carpenter, 1982). Up to 80 percent of the dinosaur skeletal remains in the Two Medicine Formation are of juveniles less than half adult size (Horner & Makela, 1979). The high percentage of young individuals in the Two Medicine Formation is not abnormal, based on the many eggs laid by female dinosaurs. In fact, adult/juvenile population ratios should be skewed towards younger age classes. Support for such a prediction is available from the abundant trackway data of Texas and Korea (Lockley, 1991, Chapter 23).

The adult/juvenile ratios of dinosaurs stand in marked contrast with those of K-strategy mammals. Mature adults make up 60–70 percent of stable populations of modern megaherbivores, and 45–55 percent of growing populations (Owen-Smith, 1988). Among carnivores, adults form 40–85% of the populations (Kruuk, 1972; Schaller 1972).

Population expansion, dispersal and recovery potential

In stable populations of r- and K-strategy animals, reproductive rates tend to be balanced by juvenile mortality (Colinaux, 1978). The differences between the two reproductive strategies become more apparent when the populations are less stable due to decreased competition (e.g., availability of a new habitat) or a population decrease of predators. K-strategists produce few offspring; therefore, maximum rates of population

Figure 15.4. **A**. Annual reproductive potential as a function of body mass in living and recent vertebrates and dinosaurs. **B**. Same data reorganized to compare oviparous and oviviviparous taxa with viviparous taxa and dinosaurs. Reproductive potential is measured in terms of total egg or newborn production (marsupials newborn based on available number of teats). Ranges of reproductive potential are plotted except for unusually extreme values in mammals, and domestic and captive animals. Most living groups enclosed in least area polygons, except nonpredaceous 3- to 40-kg marsupials which all fall upon the single line indicated.

Symbols and abbreviations: small open circles, continental ratites (rhea, emu, cassowary, and ostrich); circles with dots, island ratites (kiwi and moas); small solid circles, megapodes; large solid circles, sauropod (*Hypselosaurus*); large heavy circles with dots, hadrosaurs (*Maiasaura* and *Hypacrosaurus*); large half solid circles, *Protoceratops*; large light circle with heavy dot, hypsilophodonts and the theropod *Troodon*. ae, African elephants; br, black rhinoceros; ce, cetaceans; cv, carnivores; g, giraffes; H, humans; h, hippopotamus; hm, herbivorous marsupials; ie, Asian elephants; in, insectivores; ir, Indian rhinoceros; la, lagamorphs; mo, monotremes; pm, predaceous marsupials; reptiles, sphenodonts, turtles, lizards, snakes, and crocodilians; ro, rodents; st, marine turtles; su, suids; ug, ungulates; wr, white rhinoceros.

Data from Austin and Singer (1971), Grzimek (1972), Porter (1972), Ellis (1980), Nowak (1991), Campell and Lack (1985), Perrins and Middleton (1985), Seymour (1991), data for moas from Anderson (1989).

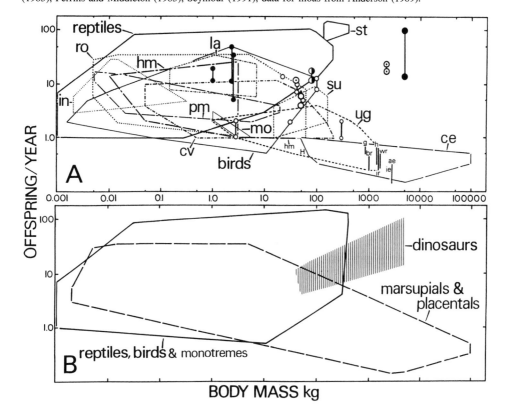

growth remain low even when juvenile mortality is reduced. In contrast, the population of r-strategists can grow much more rapidly when many juveniles survive. High rates of growth to sexual maturity increases the intrinsic rate of natural increase (McNab, 1980)

Megamammals (giraffes, rhinoceroses, hippopotomuses, and elephants), which are K-strategists, are limited to annual population expansions of only 6–12% by their maximum reproductive rates and minimal juvenile mortality rates (Owen-Smith, 1988). For this reason, at the beginning their population growth curves are rather shallow (Colinaux, 1978). Megadinosaurs, on the other hand, had steeper population growth curves because they produced many more young. How high the annual population growth of large dinosaurs might have been is unknown. Considering how many more potential, fast-growing young a megadinosaur could produce annually than a megamammal, dinosaurian population growth rates approaching or exceeding 100% per year

are hypothetically possible. In reality, high egg and juvenile mortality prevented this from happening.

The hypothetically high replacement rates of dinosaurs suggests that juvenile (post-nestling) dinosaurs could replenish decimated population levels to their former levels relatively fast, even if all the adults had been killed off! In this, megadinosaurs may have been better suited than megamammals to rebounding after devastation of their populations.

Reproduction: key to the large size and success of dinosaurs

Evolving into a large size is a common feature among vertebrates (Stanley, 1973). The great size of dinosaurs was remarkable not only because many of them grew to body sizes unmatched by terrestrial mammals (Figs. 15.1 and 15.6), but also because they were a very successful group for ~150 Myr. About three-fifths of the dinosaur species were in the mega-size range (Hot-

Figure 15.5. Growth in grams per day as a function of adult body mass in living vertebrates and dinosaurs. Only terrestrial and freshwater, noncaptive reptiles are plotted. Placental data excludes primates and edentates. Symbols and abbreviations: small solid circle, megapode; small open circle, ratite; large solid circle, subadult sauropod; large solid left half circle, ceratopsian (*Monoclonius*); large circle with black dot, hadrosaur overall growth (*Maiasaura*); large circle with white dot, hadrosaur nestling growth (*Maiasaura*); large open circle, prosauropod (*Massospondylus*); large solid lower half circle, theropod (*Syntarsus*); large solid upper half circle, theropod (*Troodon*); e, African elephant cow and bull; g, giraffe; h, hippopotamus; r, white rhinoceros. Data from Ricklefs (1968, 1973), Case (1978a), Laws (1968), Dagg and Foster (1976), Webb et al. (1978), Chabreck and Joanen (1979); Hillman-Smith et al. (1986), Hurxthal (personal communication), Reid (1981, 1990), Currie (1981), Currie and Dodson (1984), Horner and Gorman (1988), Russell (1989), Chinsamy (1992), Varricchio (1992), and Paul (unpublished notes).

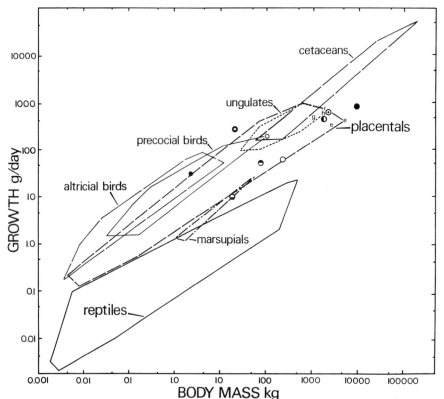

ton, 1980). Among ectotherms, the inability to grow rapidly to maturity may be one reason why they are limited to adult sizes of about 1 ton. As Dunham et al. (1989) have noted, giant animals must grow rapidly to reach sexual maturity in a reasonably short time (about two decades). Slow growth, slow generational turnover rates, and limited locomotion dispersal may inhibit even fast-breeding reptiles from exceeding 100 kg, and any reptiles from breaking the 1-ton barrier.

Dinosaurs were able to evolve large sizes because of their high metabolism and r-strategy reproductive system. Only a few adults are needed to produce the many juveniles needed to keep the population viable. With fewer adults in the population, each can claim an unusually large share of the resource base. This would allow the evolution of larger adult body sizes than could be reached by high metabolic, K-strategy giants living on the same resource base. Any serious depletion of the adult population is not critical to the survivability of the species as long as enough of the juveniles survive and mature to reproduction.

The predatory dinosaurs benefited from the increased size of their prey because their resource base was expanded. This expansion in combination with their r-strategy reproduction allowed them to achieve larger biomass/area ratios and body sizes than they otherwise could. Nevertheless, the megapredators cannot become as large as the megaherbivores because of the latter's broader energy base (various plants). This restricts the predators to sizes not exceeding elephantine masses (Farlow, 1993).

The reproductive strategy of the dinosaurs also has implications for predator/prey ratios. Much of the herbivore mortality due to predation by megapredators probably occurred among one-third to half-grown juveniles. With a large portion of the herbivore biomass made up of juveniles, predators can cull a larger portion of the herbivore biomass. If true, it is possible (but not proven) that predator/prey ratios might be higher (~1.5–3.0 percent) than are observed in modern communities of large endotherms (~0.2–1.5 percent, Farlow, 1990).

An "arms race" may have occurred between prey and predator dinosaurs. Adult herbivores are under selective pressure to improve their defensive performance to enhance the ability of each adult to protect their own high reproductive value. This might entail developing better defensive weaponry or armor, to increasing size, or a combination of both. Predators would also increase

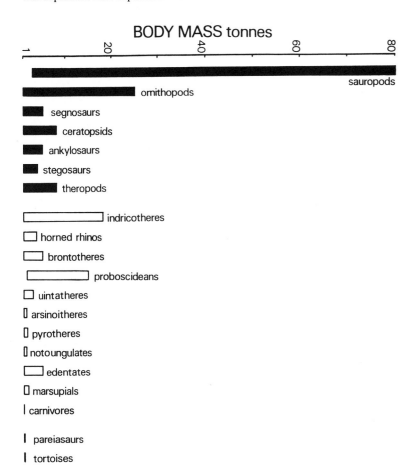

Figure 15.6. Mass ranges over 1 ton in continental tetrapod groups. Maximum values represent typical large adults for extant groups and for extinct groups of the largest known specimens (either skeletons or footprints). Data in part from Paul (1988a, 1992).

their size to better overcome the prey's defenses. The resulting feedback system could result in very large body sizes.

Another possible cause for large adult dinosaur sizes was the sheer numbers of juveniles. Small adult dinosaurs could have found themselves facing competitive or exclusionary pressures from large populations of juveniles their own size. Many small dinosaur niches may, therefore, have been filled by juveniles of larger species.

Janis and Carrano (1992) conclude that high rates of reproduction, population growth, and population recovery were important reasons that so many dinosaurs became so large. In addition, juvenile independence, genetic flexibility, high speciation rates, and the positive attributes that derive from fast growth and high metabolism also contributed to their great size and success.

Dinosaur reproduction and extinction

During the Mesozoic several events occurred that were potentially lethal to the dinosaurs. Periods of low sea level and low topography may have suppressed speciation rates among the dinosaurs and heightened extinction rates via competitive interchange between faunas and increased disease vectors (Bakker, 1977, 1986). Dinosaurs survived some of these events because of the potentially high speciation rate of their r-strategy reproduction. High rates of genetic processing should have increased the ability of dinosaurs to evolve disease-resistent strains. Wildebeest, cape buffalo, and cattle had little trouble recovering after the devastating rinderpest epidemic of the 1890s (Sinclair, 1979). The faster-breeding dinosaurs should have been even better able to recover from similar events.

The high rate of genetic processing by the dinosaurs should also have enhanced the species' ability to adapt to changing climates and floras. The high reproduction and dispersal potential of those dinosaurs that could not adapt enabled them to shift their populations to more favorable climes and habitats during the many climatic shifts of the Mesozoic.

It is doubtful that changing climate affected the ability of dinosaur eggs to hatch or skewed the sex ratios of the embryos in a dysfunctional manner. Today, birds and reptiles living in a wide variety of temperature regimes are able to maintain the nest temperatures for proper egg viability and embryonic sex selection. There is no reason to suppose that dinosaurs could not also have kept the nest temperatures within tolerable limits, contrary to the arguments of Erben et al. (1979) and Paladino et al. (1989). After all, dinosaurs successfully nested in a variety of different and changing thermal conditions for over 150 Myr.

Competition by early ungulates and other small herbivorous mammals has been suggested by Van Valen and Sloan (1977) as a possible cause for the extinction of the dinosaurs. However, large K-strategy mammals have endured well with the competition from small r-strategy mammals. In fact, fast reproduction should have given dinosaurian megaherbivores an even better ability to meet the challenge posed by the tiny herbivores.

Another hypothesis attributes the extinction of the dinosaurs at the end of the Cretaceous to the formation of extensive volcanic traps (Duncan & Pyle, 1988; Officer et al., 1987; McLean, 1988; Kerr, 1991b; Chatterjee, 1992). The intense volcanism produced extremely adverse climatic effects and pollution that caused periodic collapses of the global flora communities. However, small dinosaur species and the juveniles of larger dinosaur species might have survived to recover along with the flora. Again, the genetic flexibility of the dinosaurs improved the chances of their survival.

Finally, evidence has been growing for the impact of an asteroid coincident with the global termination of the dinosaurs globally (Melosh et al., 1990; Kerr 1991a, Izett, Dalrymple, & Snee, 1991; Pope, Ocampo, & Duller 1991; Sigurdsson et al., 1991; Smit 1991; Hildebrand et al. 1991; Chatterjee 1992). However, the effects of the impact on the dinosaurs needs to be examined in light of the reproductive system of dinosaurs.

In the worst case scenario, the impact resulted in a total collapse of the terrestrial ecosystem, resulting in the death of the dinosaurs (see discussions by Crutzen, 1987; Waldrop, 1988; Paul, 1989; Melosh et al., 1990; Kerr, 1991a; Sigurdsson et al., 1991; Smit, 1991; Wolfe, 1991; Hildebrand et al., 1991). Burrowing and freshwater vertebrates were the least affected, as were many birds (Olson, 1985).

The survival of the birds after the Cretaceous is important because they may hold a clue to the ways dinosaurs might have been able to resist extinction. Most birds lay fewer eggs than dinosaurs and are apparently more sensitive to environmental change than were dinosaurs. It is therefore difficult to understand how birds could have survived the impact but not the dinosaurs. If just a few dinosaurs survived the catastrophe, their rapid reproduction and genetic flexibility should have allowed them to reestablish themselves and adapt to a dramatically altered world. These surviving dinosaurs would have formed the basis of a new dinosaur radiation. That this did not happen is an unsolved puzzle.

The history of the dinosaurs is marked by remarkable success and stability during the Mesozoic. Far from being inherently vulnerable, the dinosaurs survived in spite of repeated changes in sea level and climate, enormous volcanic eruptions, and great impacts. Indeed, the dinosaurs' fecundity makes it hard to see how such resilent animals could ever have been killed off. The extinction of the dinosaurs was probably not part of the normal course of evolutionary fluctuations, nor was it just another result of random extraterrestrial disruptions.

Instead, it remains one of the most extraordinary and inexplicable events in Earth history.

Conclusion

The last quarter of a century has seen the emergence of a revisionist view that sees dinosaurs as much more like birds and mammals than the typical reptile. Dinosaurs are now seen as having been highly active on a sustained basis, fast growing, and often possessing highly developed social and parental skills. A few researchers have taken this concept to the extreme, suggesting that the sauropods dropped live young like the giant ungulates. However, a combination of very high breeding rates, high genetic turnover, nonlactation, and fast growth shows that large dinosaurs were not K-strategist large mammals. But nor did they have the slow generational turnover rate and short travel range of ectothermic reptiles.

Instead, they enjoyed a remarkable, dualistic system combining features of both r- and K-strategist endotherms. They had an r-strategy type of reproduction, high population growth, rapid recovery rates, dispersal performance, and genetic processing usually found in much smaller mammals. But they also utilized the K-strategy of survival, high dispersal performance, rapid growth, and generational turnover rates found in large endothermic mammals.

Juvenile (post-nestling) dinosaur independence and the potential ability to survive the loss of all adults is a reptilian trait that is shared by a few birds and even insects. The juveniles may have congregated into large numbers, either fending for themselves or cared for by a few adults. Mass migrations may have brought partly grown juveniles together with many adults.

The reproductive biology of dinosaurs was probably most similar to that of large ground birds. This is not surprising since the latter have retained or replicated the system of their ancestors. The rapid growth of dinosaurs is strongly indicative of their having had high metabolic rates. The r-strategy reproduction of dinosaurs was an important factor in their ability to exploit new conditions and made their capacity exceptionally high to resist and recover rapidly from severe environmental disruptions – so much so that no hypothesis of dinosaur extinction fully explains their demise.

Acknowledgments

I thank J. Farlow, L. Hurxthal, and K. Carpenter for discussions on some of the issues raised above, and J. Farlow and C. Janis for sharing their prepublished manuscripts on the subject.

References

Anderson, A. 1989. *Prodigeous Birds*. (Cambridge: Cambridge University Press).

Austin, O. L., & Singer, A. 1971. *Families of Birds*. (New York: Golden Press).

Bakker, R. T. 1977. Tetrapod mass extinctions – a model of the regulation of speciation rates and immigration by cycles of topographic diversity. *In* A. Hallam (ed.) *Patterns of Evolution*. (New York: Elsevier), pp. 439–68.

1980. Dinosaur heresy – dinosaur renaissance: why we need endothermic archosaurs for a comprehensive theory of bioenergetic evolution. *In* D. K. Thomas & E. C. Olson (eds.), *A Cold Look at the Warm-Blooded Dinosaurs. American Association for the Advancement of Science Selected Symposium* 28: 351–462.

1986. *The Dinosaur Heresies*. (New York: William Morrow and Company, Ltd.).

Bird, R. T. 1985. *Bones for Barnum Brown*. (Fort Worth: Texas Christian University Press).

Bonaparte, J. F. 1989. Un gigantesco titanosauridae de la Formacion Rio Limay (Provincia del Neuquen). *Jornadas Argentinas de Paleontologia de Vertebrados* 22: 27–9.

Campell, B., & Lack, E. 1985. *A Dictionary of Birds*. (Vermillion: Buteo Books).

Carpenter, K. 1982. Baby dinosaurs from the Late Cretaceous Lance Formation USA and a description of a new species of theropod dinosaur. *University of Wyoming, Contributions to Geology* 20: 123–4.

Carrano, M., & Janis, C. 1990. Scaling of reproductive turnover in archosaurs and mammals, why are large terrestrial mammals so rare? *Journal of Vertebrate Paleontology* 10 (supplement to No. 3): 17A.

Case, T. J. 1978a. On the evolution and adaptative significance of postnatal growth rates in the terrestrial vertebrates. *The Quarterly Review of Biology* 53: 243–82.

1978b. Speculations on the growth rate and reproduction of some dinosaurs. *Paleobiology* 4: 320–8.

Chabreck, R. H., & Joanen, T. 1979. Growth rates of American alligators in Louisiana. *Herpetologica* 35: 51–7.

Chatterjee, S. 1992. A kinematic model for the evolution of the Indian plate since the Late Jurassic. *In* S. Chatterjee & N. Hotton III (eds.), *New Concepts in Global Tectonics* (Lubbock: Texas Tech University Press), pp. 33–62.

Chinsamy, A. 1992. Ontogenetic growth of the dinosaurs *Massospondylus* and *Syntarsus*. *Journal of Vertebrate Paleontology* 12 (supplment to No. 3): 23A.

Coombs, W. P. 1989. Modern analogs for dinosaur nesting and parental behavior. *In* J. Farlow (ed.), *Paleobiology of the Dinosaurs. Geological Society of America Special Paper* 238: 21–53.

1990. Behavior patterns of dinosaurs. *In* D. B. Weishampel, P. Dodson, & H. Osmolska (eds.), *The Dinosauria*. (Berkeley: University of California Press), pp. 32–42.

Colinvaux, P., 1978. *Why Big Fierce Animals Are Rare: An Ecologist's Perspective*. (Princeton: Princeton University Press).

Cousin, R., Breton, G., Fournier, R., & Watte, J.-P. 1989. Dinosaur egg-laying and nesting of an Upper Maastrichtian site at Rennes-Le-Chateau (Aude, France). *Historical Biology* 2: 157–67.

Crutzen, P. J. 1987. Acid rain at the K/T boundary. *Nature* 330: 108–9.

Currie, P. J. 1981. Hunting dinosaurs in Alberta's great bone bed. *Canadian Geography* 101(4): 34–9.

Currie, P. J., & Dodson, P. 1984. Mass death of a herd of ceratopsian dinosaurs. In W. E. Reif & F. Westphal (eds.), *Third Symposium of Mesozoic Terrestrial Ecosystems.* (Tubingen: ATTEMPTO Verlag), pp. 61–5.

Dagg, A. I., & Foster, J. B. 1976. *The Giraffe, Its Biology, Behavior and Ecology.* (New York: Van Nostrand Reinhold).

Dodson, P. 1975. Taxonomic implications of relative growth in lambeosaurine hadrosaurs. *Systematic Zoology* 24: 37–54.

1990. Sauropod paleoecology. In D. B. Weishampel, P. Dodson, & H. Osmólska (eds.), *The Dinosauria.* (Berkeley: University of California Press), pp. 402-7.

Duncan, R. A., & Pyle, D. G. 1988. Rapid eruption of the Deccan basalts at the Cretaceous/Tertiary boundary. *Nature* 333: 841–3.

Dunham, A. E., Overall, K. L., Porter, E. P., & Forster, C. A. 1989. Implications of ecological energetics and biophysical and development constraints for life-history variation in dinosaurs. In J. Farlow (ed.), *Paleobiology of the Dinosaurs. Geological Society of America Special Paper* 238: 1–20.

Eisenberg, J. F. 1981. *The Mammalian Radiations.* (Chicago: University of Chicago Press).

Ellis, R. 1980. *The Book of Whales.* (New York: Alfred A. Knopf).

Erben, H. K., Hoefs, J., & Wedepohl, K. H. 1979. Paleobiological and isotopic studies of eggshells from a declining dinosaur species. *Paleobiology* 5: 380–414.

Farlow, J. O. 1990. Dinosaur energetics and thermal biology. In D. B. Weishampel, P. Dodson, & H. Osmolska (eds.), *The Dinosauria* (Berkeley: University of California Press), pp. 43–62.

1993. On the rareness of big, fierce animals: speculations about the body sizes, population densities, and geographic ranges of predatory mammals and large carnivorous dinosaurs. *American Journal of Science,* 239: 167–99.

Farlow, J. O., Pittman, J. G., & Hawthorne, J. M. 1989. *Brontopodus birdi,* Lower Cretaceous sauropod footprints from the U.S. gulf coastal plain. In D. D. Gillette & M. G. Lockley (eds.), *Dinosaur Tracks and Traces.* (Cambridge: Cambridge University Press), pp. 371–94.

Forster, C. A. 1990. Evidence for juvenile groups in the ornithopod dinosaur *Tenontosaurus tilleti* (Ostrom). *Journal of Paleontology* 64: 164–5.

Grzimek, B. 1972. *The Animal Life Encyclopedia.* (New York: Van Nostrand Reinhold).

Hildebrand, A. R., Penfield, G. T., Kring, D. A., Pilkington, M., Camargo, Z. A., Jacobsen, S. B., & Boynton, W. V. 1991. Chicxulub Crater: A possible Cretaceous/Tertiary boundary impact crater on the Yucatan Peninsula, Mexico. *Geology* 19: 867–71.

Hillman-Smith, A. K. K., Owen-Smith, N., Anderson, J. L., Hall-Martin, A. J., & Seladai, J. P. 1986. Age estimation of the White rhinoceros (*Ceratotherium simum*). *Journal of Zoology, London* 210: 355–79.

Hirsch, K. F. 1989. Interpretations of Cretaceous and pre-

Cretaceous eggs and eggshell fragments. In D. D. Gillette & M. G. Lockley (eds.), *Dinosaur Tracks and Traces.* (Cambridge: Cambridge University Press), pp. 89–97.

Hirsch, K. F. & Quinn, B. 1990. Eggs and eggshell fragments from the Upper Cretaceous Two Medicine Formation of Montana. *Journal of Vertebrate Paleontology* 10: 491–511.

Horner, J. R. 1982. Evidence of colony nesting and "site fidelity" among ornithischian dinosaurs. *Nature* 297: 675–6.

1984. The nesting behavior of dinosaurs. *Scientific American* 250(4): 130–7.

1987. Ecologic and behavioral implications derived from a dinosaur nesting site. In S. J. Czerkas & E. C. Olson (eds.), *Dinosaurs Past and Present,* Volume II. (Seattle: University of Washington Press), pp. 50–63.

1988. Cranial allometry of *Maiasaura peeblesorum* (Ornithischia; Hadrosauridae) and its behavioral significance. *Journal of Vertebrate Paleontology* 8 (supplement to No. 3): 18A.

Horner, J. R. & Gorman, J. 1988. *Digging Dinosaurs.* (New York: Workman Publishing).

Horner, J. R. & Makela, R. 1979. Nest of juveniles provides evidence of family structure among dinosaurs. *Nature* 282: 296–8.

Horner, J. R., & Weishampel, D. B. 1988. A comparative embryological study of two ornithischian dinosaurs. *Nature* 332: 256–7.

Hotton, N. 1980. An alternative to dinosaur endothermy: the happy wanderers. In D. K. Thomas & E. C. Olson (eds.), *A Cold Look at the Warm-Blooded Dinosaurs. American Association for the Advancement of Science Selected Symposium* 28: 311–50.

Izett, G. A., Dalrymple, G. B., & Snee, L. W. 1991. ^{40}Ar/^{39}Ar age of Cretaceous-Tertiary boundary tektites from Haiti. *Science* 252: 1539–42.

Jablonski, D. 1991. Extinctions: A paleontological perspective. *Science* 253: 754–7.

Jain, S. L. 1989. Recent dinosaur discoveries in India, including eggshells, nests and coprolites. In D. D. Gillette & M. G. Lockley (eds.), *Dinosaur Tracks and Traces.* (Cambridge: Cambridge University Press), pp. 99–108.

Janis, C. M. and Carrano, M. 1992. Scaling of reproductive turnover in archosaurs and mammals: why are large terrestrial mammals so rare? *Acta Zoologica Fennica,* 28: 201–6.

Kerr, R. A. 1991a. Yucatan killer impact gaining support. *Science* 251: 160–2.

1991b. Did a volcano help kill off the dinosaurs? *Science* 252: 1496–7.

Kruuk, H. 1972. *The Spotted Hyena* (Chicago: University of Chicago Press).

Kurtén, B. 1953. On the variation and population dynamics of fossil and recent mammalian populations. *Acta Zoological Fennica* 76: 1–122.

Lambert, W. D. 1991. Altriciality and its implications for dinosaur thermoenergetic physiology. *Neues Jahrbuch für Geologie und Palaontologie* 182: 73–84.

Langman, V. A. 1982. Giraffe youngsters need a little bit 0f maternal love. *Smithsonian* 12: 95–103.

Laws, R. M. 1968. Dentition and ageing of the hippopotamus. *East African Wildlife Journal* 6: 19–52.

Leahy, G. D., & Paul, G. S. 1991. Long erect legs and rapid growth require high maximal and minimal metabolisms in dinosaurs and *Archaeopteryx*. *Journal of Vertebrate Paleontology* 11(supplement to No. 3): 42A.

Lockley, M. G. 1991. *Tracking Dinosaurs*. (Cambridge, Cambridge University Press).

McLean, D. M. 1988. K-T transition into chaos. *Journal of Geological Education* 36: 237–43.

McNab, B. K. 1980. Food habits, energetics, and the population biology of mammals. *American Naturalist* 116: 106–24.

Melosh, H. J., Schneider, N. M., Zahnle, K. J., & Latham, D. 1990. Ignition of global wildfires at the Cretaceous/ Tertiary boundary. *Nature* 343: 251–3.

Mohabey, D. M. 1987. Juvenile sauropod dinosaur from Upper Cretaceous Lameta Formation of Panchmahals District, Gujarat, India. *Journal of the Geological Society of India* 30: 210–16.

Nowak, R. M. 1991. *Walker's Mammals of the World*. (Baltimore: Johns Hopkins University Press).

Officer, C. B., Hallam, A., Drake, C. L., & Devine, J. D. 1987. Late Cretaceous extinctions and paroxysmal Cretaceous/Tertiary eruptions. *Nature* 326: 143–9.

Olson, S. L. 1985. The fossil record of birds. *In* D. S. Farner, J. R. King, & K. C. Parkes (eds.), *Avian Biology*. (New York: Academic Press), pp. 80–238.

Owen-Smith, R. N. 1988. *Megaherbivores: The Influence of Very Large Body Size on Ecology*. (Cambridge: Cambridge University Press).

Packard, G. C. 1977. The physiological ecology of reptilian eggs and embryos, and the evolution of viviparity within the Class Reptilia. *Biological Review* 52: 71–105.

Paladino, F. V., Dodson, P., Hammond, J. K., & Spotila, J. R. 1989. Temperature-dependent sex determination in dinosaurs? Implications for population dynamics and extinction. *In* J. Farlow (ed.), *Paleobiology of the Dinosaurs. Geological Society of America Special Paper* 238: 63–70.

Paul, G. S. 1988a. The brachiosaur giants of the Morrison and Tendaguru, and a comparison of the world's largest dinosaurs. *Hunteria* 2:1–14.

1988b. *Predatory Dinosaurs of the World*. (New York: Simon and Schuster).

1989. Giant meteor impacts and great eruptions: dinosaur killers? *Bioscience* 39: 162–72.

1991. The many myths, some old, some new, of dinosaurology. *Modern Geology* 16: 69–99.

1992. The size and bulk of extinct giant land herbivores. *Journal of Vertebrate Paleontology* 12 (supplement to No. 3): 47A.

Perrins, C. M., & Middleton, L. A. 1985. *The Encyclopedia of Birds* (New York: Facts on File Publications).

Pond, C. M. 1983. Parental feeding as a determinant of ecological relationships. *Acta Palaeontologica Polonica* 28: 215–24.

Pope, K. E., Ocampo, A. C., & Duller, C. E. 1991. Mexican site for K/T impact crater. *Nature* 351: 105.

Porter, K. R. 1972. *Herpetology*. (Philadelphia: W. B. Saunders Company).

Reid, R. E. H. 1981. Lamellar-zonal bone with zones and annuli in the pelvis of a sauropod dinosaur. *Nature* 292: 49–51.

1990. Zonal "growth rings" in dinosaurs. *Modern Geology* 15: 19–48.

Richmond, N. D. 1965. Perhaps juvenile dinosaurs were always scarce. *Journal of Paleontology* 39: 503–5.

Ricklefs, R. E. 1968. Patterns of growth in birds. *Ibis* 110: 419–51.

1973. Patterns of growth in birds. II. Growth rate and mode of development. *Ibis* 115: 177–210.

Ricqles, A. J. de. 1980. Tissue structures of dinosaur bone: functional significance and possible relation to dinosaur physiology. *In* D. K. Thomas & E. C. Olson (eds.), *A Cold Look at the Warm-Blooded Dinosaurs. American Association for the Advancement of Science Selected Symposium* 28: 103–39.

Russell, D. A. 1989. *An Odyssey in Time*. (Toronto: University of Toronto Press).

Schaller, G. B. 1972. *The Serengeti Lion*. (Chicago: University of Chicago Press).

Sigurdsson, H., D'Hondt, S., Arthur, M. A., Bralower, T. J., Zachoes, J. C., Fossen, M. van, & Channell, J. E. T. 1991. Glass from the Cretaceous/Tertiary boundary in Haiti. *Nature* 349: 482–7.

Sinclair, A. R. E. 1979. Dynamics of the Serengeti ecosystem. *In* A. R. E. Sinclair & M. Norton-Griffiths (eds.), *Serengeti; Dynamics of an Ecosystem*. (Chicago: University of Chicago Press), pp. 1–30.

Smit, J. 1991. Dinosaurs and friends snuffed out? *Nature* 349: 461–2.

Stanley, S. M. 1973. An explanation for Cope's Rule. *Evolution* 27: 1–26.

1987. *Extinction* (New York: Scientific American).

Thulborn, T. 1991. The discovery of dinosaur eggs. *Modern Geology* 16: 113–26.

Thulborn, R. A., & Wade, M. 1984. Dinosaur trackways in the Winton Formation (mid-Cretaceous) of Queensland. *Memoirs of the Queensland Museum* 21: 413–517.

Van Valen, L., & Sloan, R. E. 1977. Ecology and extinction of the dinosaurs. *Evolutionary Theory* 2: 37–64.

Varricchio, D. J. 1992. Taphonomy and histology of the Upper Cretaceous theropod dinosaur *Troodon formosus*: life implications. *Journal of Vertebrate Paleontology* 12 (supplement to No. 3): 57A.

Waldrop, M. M. 1988. After the fall. *Science* 239: 977.

Western, D. 1979. Size, life history and ecology in mammals. *African Journal of Ecology* 17: 185–204.

Winkler, D. A. and Murry, P. A. 1989. Paleoecology and hypsilophodontid behavior at the Proctor Lake dinosaur locality (Early Cretaceous), Texas. *In* J. Farlow (ed.). *Paleobiology of the Dinosaurs. Geological Society of America Special Paper* 238: 55–61.

Wolfe, J. A. 1991. Palaeobotanical evidence for a June "impact winter" at the Cretaceous/Tertiary boundary. *Nature* 352: 420–3.

16 An embryonic *Camarasaurus* (Dinosauria, Sauropoda) from the Upper Jurassic Morrison Formation (Dry Mesa Quarry, Colorado)

BROOKS B. BRITT AND
BRUCE G. NAYLOR

Abstract

A small premaxilla from the Morrison Formation is referred to *Camarasaurus* sp. and is hypothesized to have belonged to an embryonic individual. Assignment to the Sauropoda is based on the presence of high-crowned teeth with rugose enamel and the absence of root resorption. Referral to *Camarasaurus* is based on the presence of spatulate teeth and a rectangular premaxillary body with a laterally broad, nearly vertical nasal process. The embryonic interpretation is based on size, the presence of embryonic-type fibers of unremodeled bone with no cortex, and the absence of erupted teeth.

This is the first positively identified record of an embryonic sauropod. Ratios derived from a nearly complete juvenile *Camarasaurus* specimen suggest that the embryonic skull was approximately 70 mm long. The total body length of the embryo was almost 1,090 mm. A full-scale model based on these measurements has a volume of approximately 7.5 L, indicating an egg size well below the maximum known for birds. This specimen demonstrates the existence of sauropods small enough to fit within known dinosaur eggs and casts serious doubt on hypotheses of sauropodian viviparity. Occurrence of the embryonic premaxilla in the conglomeratic sandstone of Dry Mesa Quarry suggests minimal transportation from a nearby nesting site, pointing to the possibility of future discoveries of sauropod nesting sites in the Morrison Formation.

Introduction

A small premaxilla (Figs. 16.1 and 16.2) was discovered during preparation of a jacket collected from Dry Mesa Quarry (Fig. 16.3). Located in west central Colorado, this quarry, in the Morrison Formation, contains the most diverse North American Upper Jurassic dinosaur fauna known from a single quarry (Britt, 1991). The quarry is most famous for its giant diplodocid and brachiosaurid sauropods, and the robust megalosaurid (*sensu* Britt, 1991) theropod, *Torvosaurus tanneri*. In addition to the enormous quantity of large dinosaurs, the quarry also contains abundant and varied microvertebrates (Jensen & Padian, 1989).

The dinosaurian fauna of the Morrison Formation has been reviewed by Dodson et al. (1980) and Britt (1991). Britt (1991) and Miller et al. (1991) provided faunal lists for the Dry Mesa Quarry.

Marsh (1883) suggested that a small (~2 m) specimen of *Apatosaurus* was a fetus because of its small size and imperfect ossification, and because it was found associated with an adult specimen of the same genus. The bones belonged to a very small juvenile, but there is no justification for the fetal interpretation.

Bakker (1980) has suggested, on the basis of their large pelvic canals, that sauropods and other unnamed dinosaurian groups gave live birth to relatively large babies. Bakker's argument was dismissed by Dunham et al. (1989) as based on either negative or equivocal evidence. Furthermore, several major dinosaurian groups are demonstrably oviparous because identifiable embryos have been found within, or intimately associated with, eggs. Dodson (1990) assumed that sauropods and all other dinosaurs were oviparous, a view we believe most paleontologists would accept.

Dinosaurs known to be oviparous include theropods (Currie, personal communication, 1989) and ornithopods (Horner & Weishampel, 1988; Horner & Currie, Chapter 21). The tiny (approximately 250–300 mm long) prosauropod *Mussaurus*, interpreted as a hatchling by Bonaparte and Vince (1979), was associated with eggshell fragments, suggesting that prosauropods were oviparous. Based on size and structure, we suspect the type of *Mussaurus* is an embryo. Currie (personal communication, 1991) identified small hadrosaurid dinosaurs from Mongolia (described by Barsbold & Perle, 1983) as embryonic. The specimens were found in a nest, a fact not recognized until after they were identified as embryos. Currie's identification is based on size, absence of fusion between vertebral components, and other osteological characters. The more soluble eggshell appears to have dissolved, but the bones of the embryos were preserved.

The abundant eggs of Aix-en-Provence, France, have long been considered to be of sauropod origin, specifically *Hypselosaurus priscus* (Lapparent, 1957; Buffetaut & Le Loeuf, Chapter 2). Other than the presence of sauropod remains in the same region, no direct evidence supports this claim. None of the eggs contained embryos, nor have any associated hatchlings been reported.

Large subspherical eggs from nest sites in the Upper Cretaceous Lameta Formation of Gujarat, India, were assigned to the Sauropoda (Jain, 1989), based on the common occurrence of sauropods in the formation (Mohabey 1987; Sahni et al., Chapter 13). Tiwari (unpublished M.S. thesis, Punjab University, as cited in Jain, 1989) supported this referral on the basis of similarities of size, shape, surface ornamentation, and microstructural details between the eggs from another Lameta Formation locality and the eggs from Aix-en-Provence, France. As with the Aix-en-Provence eggs, the Gujarat eggs were identified as sauropod based on the occurrence of sauropod bones in the same formation.

Recent discoveries of lambeosaurine embryos in large, round eggs (Horner & Currie, Chapter 21), with surface ornamentation similar to those of the French eggs, suggest that the French and Indian eggs could be of hadrosaurid origin. All large, subspherical eggs from the Upper Cretaceous, therefore, are not necessarily of

sauropod origin, and the true affinity of the Indian and French eggs must await discovery of embryos or hatchlings.

Mohabey (1987) reported an extremely young (hatchling or embryonic) sauropod from the Lameta Formation. Associated bones and segments of articulated vertebrae found in a clutch of eggs were identified as those of a juvenile sauropod. Jain (1989), however, questioned Mohabey's identification because the vertebrae bore zygosphene–zygantra articulations (Mohabey, 1987), a snake synapomorphy. Furthermore, the cervical vertebrae have neural spines taller than those on the dorsal vertebrae. Jain suggested the specimen may be that of a booid snake, noting that booids occur in the Lameta Formation (Sahni et al., 1982, cited in Jain, 1989).

In addition to these points, problems with identification of the vertebrae as sauropod include the extraordinarily small diameter of the neural canal and the short length of the neural arch. In known juveniles, and especially embryonic dinosaurs we have observed, the neural canal is much larger relative to the vertebra than is the case in adults (see also Chure et al., Chapter 20). Furthermore, the neural arch on the cervical and dorsal vertebrae (as identified by Mohabey) is restricted to the posterior end of the centrum, whereas in sauropods the arch extends the length of the centrum with zygapophyses extending beyond the centrum faces.

Mohabey also recognized a humerus, scapula,

Figure 16.1. *Camarasaurus* sp., BYUVP 8967, left premaxilla of embryonic individual. **A.** Posteromedial view, tooth 1 has been rotated counterclockwise slightly out of natural position, tip of tooth 2 barely extends below the interdental plates. **B.** Anteromedial view. **C.** Posterolateral view showing two unerupted teeth in the fourth alveolus (exposed by temporary removal of posterolateral edge (compare with Figure 16.2B). **D.** Anterolateral view; much of the lateral surface was removed (area between ⊢ –⊣ bracket sets) by discovery tool exposing unerupted teeth – most of the roots of teeth 1 and 2 were lost, and those of teeth 3 and 4 longitudinally sectioned. Abbreviations: en, external nares; f, foramen of nasal process; idp, interdental plate; idpf, interdental plate foramen; lp, lateral parapet; ms, medial symphysis; np, nasal process; pc, pulp cavity of tooth roots. Numbers = alveolar position; open circles = matrix; diagonal hatching = transversely broken bone. Scale = 1 cm.

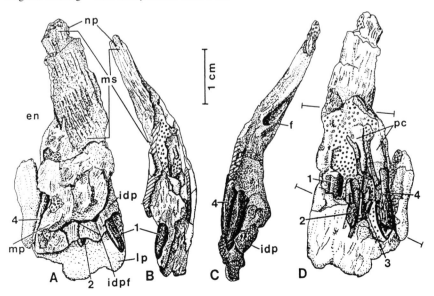

and other elements, which Jain (1989) considered to be "questionably identified." Mohabey stated that the bones are "invariably" smooth, yet most undisputed embryonic and hatchling dinosaur bone exhibits striated, unremodeled surfaces (personal observation). The report of apparently unquestioned juvenile sauropod material in the area, however, does support the identification of the eggs as sauropodian. The small articulated vertebae identified as a juvenile sauropod, however, are most likely those of a snake.

The small premaxilla from Dry Mesa Quarry, then, is the first record of an identifiable sauropod embryo. It is also the first evidence of sauropods small enough to fit in an egg. Several apomorphic characters decisively demonstrate that the specimen is assignable to *Camarasaurus* sp.

Location and geology

Dry Mesa Quarry (BYUVP Locality 725) is located on the southwest end of Dry Mesa, a part of the Uncompahgre Plateau, at an elevation of 2,242 m (7,356 ft). It overlooks the East Fork of Escalante Creek in west central Colorado (Fig. 16.3). The quarry has been worked nearly every summer since James A. Jensen first opened it in 1972 (Miller et al., 1991). The position of the quarry in the Morrison Formation is uncertain because the Morrison is difficult to distinguish from the overlying, lithologically similar Burro Canyon Formation (equivalent to the Cedar Mountain Formation west of the Colorado River, Dodson et al., 1980). The bone-bearing lithosome of the quarry consists mainly of trough cross-bedded, conglomeratic, sandstone lenses. This represents a channel-floor deposit laid down fol-

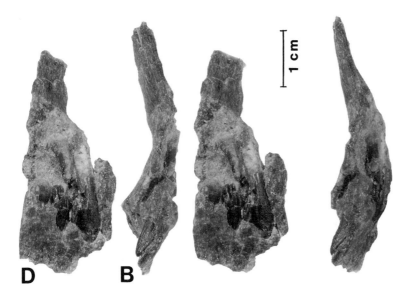

Figure 16.2. Stereophotographs of embryonic *Camarasaurus sp.*, BYUVP 8967, left premaxilla.
A. Posteromedial view; note unremodeled straps of bone near base of nasal process and interdental plates.
B. Anteromedial view.
C. Posterolateral view.
D. Anterolateral view. See Figure 16.1 for additional details. Scale = 1 cm.

lowing an erosional period. For sedimentalogical details see Britt (1991).

Quarry operations to recover macrovertebrates led to inadvertent recovery of microvertebrates. Even without the large vertebrates, the quarry would be valued for its microvertebrates. The abundance of microvertebrates at the quarry contrasts sharply with their general paucity in the Morrison Formation (Dodson et al., 1980). In addition to their recovery in the field, microvertebrates are commonly discovered in the laboratory during macrovertebrate preparation. This was how the embryonic *Camarasaurus* premaxilla was discovered.

Many microvertebrates have been recovered, including several delicate, articulated elements, such as a synsacrum and a scapulocoracoid of the pterosaur *Mesadactylus ornithosphyos* (Jensen & Padian, 1989); and three articulated, elongate pterosaur vertebrae (?pterodactyloid cervical vertebrae or ?rhamphorhycoid caudal vertebrae) and the foot of a small bipedal ?theropod. Many small, but well-preserved vertebrae have been recovered uncrushed, with fragile processes intact (Jensen & Padian, 1989). The majority of the microvertebrate sample is pterosaurian (Jensen & Padian, 1989). This material demonstrates the great potential of Dry Mesa Quarry to provide new information on the poorly known microvertebrates of the Morrison Formation, hitherto known mainly from another Colorado quarry (Callison & Rasmussen, 1980).

Description

The specimen (BYUVP 8967) is a tiny, left premaxilla measuring 37 mm (32 mm as currently preserved) from the preserved tip of the nasal process to the anteroventral margin of the premaxillary body. The vertically elongate, rectangular body measures 14 mm along its ventral border and is approximately 22 mm high. The incomplete, bladelike nasal process is approximately 15 mm long, 1.2 mm thick, and slopes slightly posteriorly. A large, well-developed foramen is located at the intersection of the posterior margin of the nasal process and premaxillary body (Figs. 16.1 & 16.2C). The anterior edges of the premaxillary body and nasal process are essentially confluent. Postdepositional deformation caused medial deflection of the process and probably a minor amount of anterior or posterior deflection.

In anterolateral view (Figs. 16.1 & 16.2D) the point of discovery is represented by an oblique Airscribe gouge, which removed the lateral face of the premaxilla and exposed unerupted teeth. The thin (about 0.2 mm) labial margin is also incomplete, especially posteroventrally. In posteromedial view (Figs. 16.1 & 16.2A) relatively large interdental plates are clearly visible as a shelf immediately above the tooth crowns and labial to the medial (lingual) parapet. The medial parapet of the tooth row is short, barely ex-

tending over the bases of the interdental plates. Adjacent interdental plates are in close contact, except between their constricted bases. The constricted proximal interdental plate margins form the margins of the interdental plate foramina (Figs. 16.1 & 16.2A), through which passed branches of the dental artery. The posterior interdental plate has been dislocated anteromedially, apparently intact, indicating that the interdental plates are not fused.

The four dental alveoli each contain one or more high-crowned, unerupted teeth. No erupted teeth are present. In anterolateral view four teeth are visible (Figs. 16.1 and 16.2D). The first tooth is the most mature, but would have been imbedded in gum tissue a couple of millimeters short of the labial alveolar margin at the time of death. The tooth crowns are robust, spoonshaped, and curve strongly lingually. Enamel surfaces are rugose. Labially (Figs. 16.1 and 16.2D), rugae and sulci are roughly aligned with the long axis of the tooth, while lingually (Fig. 16.4) rugae are shorter and somewhat randomly oriented.

A broad axial ridge, bordered on anterior and posterior margins by a trough, is present proximally on the lingual surface of the crown. Near the tooth edges, outside of the trough, rugae are longer and angle out to meet the tooth margins like the secondary veins of a leaf. The enameled crowns are approximately 9 mm long, 3 mm wide (maximum edge-to-edge dimension), and 2 mm thick (maximum labiolingual dimension). Roots are approximately the same length as the crown.

Figure 16.3. Locality map of Dry Mesa Quarry (BYUVP locality 725). Quarry is at an elevation of 2,242 m on the southwest flank of Dry Mesa, a part of the Uncompahgre Uplift, 33 km west southwest of Delta, Colorado, in section 23, T. 50 N, R. 14 W, Mesa County.

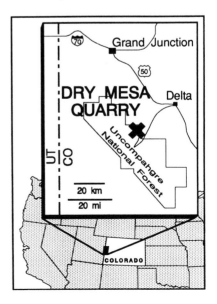

In most sauropod teeth the roots are somewhat longer than the crowns. The relative shortness of the roots in BYUVP 8967 is a function of the immaturity of the teeth. The roots would have continued to grow in length after the crowns reached maximum size [demonstrated in a sectioned *Diplodocus* dental alveolus (Marsh, 1896)]. All crowns are undistorted except the slightly crushed medial tooth in alveolus 4. The longitudinal section of the root of the fourth tooth reveals that a conical pulp cavity occupied most of the root (Figs. 16.1 and 16.2D). Two teeth are closely appressed in alveolus 4, as seen in posterolateral view (Figs. 16.1 and 16.2C). Both have descended about the same distance within the alveolus, and the younger, medial tooth has not induced resorption of the older tooth root.

In other saurischians and many other reptiles, replacement teeth induce root resorption in the old teeth, then migrate directly under the old tooth through its pulp cavity to eject the old tooth (Edmund, 1960). In sauropods, replacement teeth remained deep in the jaw to erupt rapidly following the loss of the old tooth (Edmund, 1960). This hypothesis is supported by the sectioned *Diplodocus* maxilla referred to above. There is no evidence that fully erupted teeth were ever present in the premaxilla BYUVP 8967. The juvenile premaxilla indicates a moderately broad rostrum with high, anteriorly located, laterally facing external nares.

Figure 16.4. Microphotograph of lingual surface of premaxillary tooth 1 of *Camarasaurus* embryo, BYUVP 8967. Rugae cover the entire crown. Scale = 1 mm.

Discussion

The small premaxilla has four alveoli occupied by high-crowned teeth with rugose enamel. Such enamel and the absence of root resorption (Edmund, 1960) demonstrate that the premaxilla is assignable to the Sauropoda. The number of teeth in sauropod premaxillae is very conservative, with four teeth present in all taxa (McIntosh, 1990), as in most theropods (Madsen, 1976; Chatterjee, 1978). This suggests that four is the primitive premaxillary tooth count for saurischians.

The number of sauropod genera in the Morrison Formation is a matter of debate (Dodson et al., 1980; Jensen, 1985; Paul, 1988; McIntosh, 1990), but only six genera are commonly recognized: *Diplodocus, Apatosaurus, Barosaurus, Brachiosaurus, Haplocanthosaurus*, and *Camarasaurus*. Jensen (1985) reported three new genera: *Ultrasaurus* (sensu Jensen, 1985), *Dystylosaurus*, and *Supersaurus*, all from Dry Mesa Quarry. McIntosh (1990) referred *Ultrasaurus* and *Dystylosaurus* to the Brachiosauridae and *Supersaurus* to the Diplodocidae. Except for *Haplocanthosaurus* and *Brachiosaurus*, all of these genera have been found in the Dry Mesa Quarry (Britt, 1991; Miller et al., 1991). Paul (1988) considered *Ultrasaurus* to be a junior synonym of *Brachiosaurus*, and McIntosh (1990) indicated that proper evaluation of the two genera required preparation of more *Ultrasaurus* material.

Diplodocus, Apatosaurus, Supersaurus, and *Barosaurus* are all diplodocids (McIntosh, 1990), a family characterized by elongate skulls, external nares just anterior to the orbit, and slender, pencil-shaped teeth (Fig. 16.5B). The body of the premaxilla is derived in facing mainly anteriorly rather than laterally. Furthermore, the nasal process of the diplodocid premaxilla is extraordinarily elongate, extending posteriorly at a low angle to terminate above the anterior orbital margin. These characters demonstrate that the juvenile premaxilla is not diplodocid.

No skull or tooth material has been assigned with any certainty to *Haplocanthosaurus*, a cetiosaurid (McIntosh, 1990), so no comparisons are possible. Even at the family level, comparison is difficult because only fragmentary cetiosaurid skull material has been described. Fragments of the skull of the cetiosaurid *Patagosaurus* described by Bonaparte (1979) indicate that the external nares were "relatively low" (McIntosh, 1990). The teeth of *Amygdalodon* and *Patagosaurus* are similar, with broad, heart-shaped, spatulate crowns (McIntosh, 1990). If *Amygdalodon* and *Patagosaurus* are cetiosaurids and if *Haplocanthosaurus* had similar skull and dental morphology, the juvenile premaxilla cannot be *Haplocanthosaurus* because the teeth differ.

Although no brachiosaurid skulls have been found in the Western Hemisphere, excellent skulls of *Brachiosaurus* from Africa have been described by Janensch (1935–36). In this genus, external nares are positioned largely behind the midpoint of the skull, anterodorsal to the orbit (Fig. 16.5C). The premaxilla and maxilla form a shelf anterior to the external nares. The premaxillary nasal process is elongated posteriorly and does not rise dorsally until well back on the maxilla. The juvenile premaxilla does not match these features, so it is obviously not brachiosaurid.

The embryonic premaxilla's blocklike body, its near vertical, anteroposteriorly wide and laterally thin nasal process, and its spoon-shaped tooth crowns are known only in *Camarasaurus*, the most common sauropod in the Morrison Formation (Dodson et al. 1980). Consequently, the premaxilla is confidently referred to this genus. Comparison of the premaxillary nasal process of the juvenile *Camarasaurus* (Fig. 16.5A), CMNH 11338, with that of adult specimens demonstrates that a significant reduction in the anteroposterior width of the process relative to the premaxillary body occurs with increased ontogenetic age. This same allometric trend is readily apparent when comparing CMNH 11338 with BYUVP 8965; the nasal process of the latter is considerably broader than that of the former.

The small size of the premaxilla alone suggests that it belonged to an embryo or hatchling. Other characters, however, indicate that it was an embryo at the time of death. For example, despite excellent preservation, there is no evidence of erupted teeth. Assuming sauropods were not hatched toothless – a fairly safe assumption because reptiles have functional, erupted teeth at birth – the individual must have died before hatching. Straps or needles of bone parallel to the major growth axis give the surface a striated, unfinished appearance. The bone of embryonic hadrosaurs found in eggs have similarly striated surfaces on both dermal and replacement bones (personal observation). We are not aware of the developmental factors resulting in this texture. It may be simply that the bone had not been remodeled nor had bone built up laterally (normal to the major growth axis), similar to the perichondrial layers of replacement bone. These three features of the premaxilla suggest the specimen was an embryo: (1) striated, unfinished bone surfaces, (2) small size, and (3) absence of erupted teeth.

Bones of small juvenile sauropods have been discovered in the Kheda District, Gujarat, India (Mathur & Pant 1986, as cited in Jain, 1989); Maryland (Marsh, 1888; Hatcher, 1903); Dry Mesa, Colorado (J. A. Jensen, personal communication, 1992); Cleveland-Lloyd, Utah (J. H. Madsen, personal communication, 1991) and

the San Rafael Swell, Utah (personal observation by Britt); Sheep Creek, Wyoming (Peterson & Gilmore, 1902; Carpenter & McIntosh, Chapter 17) and Como Bluff, Wyoming (Marsh, 1883 Carpenter & McIntosh, Chapter 17); and the Oklahoma panhandle (Carpenter & McIntosh, Chapter 17). Except for the Oklahoma and Wyoming specimens, these discoveries consist of isolated elements. The best-known juvenile sauropod is the nearly complete, 5 m long *Camarasaurus* (CM 11338) collected from what is now Dinosaur National Monument (Gilmore, 1925).

Measurements of the embryo BYUVP 8967, the juvenile CMNH 11338, and adult specimens of *Camarasaurus* are compared in Table 16.1. Estimates for the skull length and total (skull + body) length of an embryonic *Camarasaurus* are based on the known length of the embryonic premaxilla (14 mm) and ratios of the same measurement to skull and body lengths as determined from the juvenile *Camarasaurus*, CMNH 11338. The measurements for CMNH 11338 are these: premaxilla length 64 mm, skull length 330 mm, and total length 5,000 mm. These measurements yield a premaxilla:skull ratio of 1:5 and a premaxilla:total body length ratio of 1:78. Based on these ratios, the embryonic *Camarasaurus* is estimated to have had a 70-mm skull and a total length of 1,092 mm. CMNH 11338 exhibits juvenile characters, such as a relatively large head with a short neck and tail, and these allometric characters were probably even more pronounced in an embryo. The embryonic skull would have been slightly larger, and the overall length less, than the ratios based on CMNH 11338 indicate, perhaps less than 1,000 mm.

A clay model of a *Camarasaurus* embryo was constructed based on the skeleton of CMNH 11338. The model is 1,092 mm long, with most of the length being in the neck and tail. The model displaces less than 7,500 ml. Alexander (1989) noted that the density of most animals is about the same as that of water. These data combined with the volume of the model yields an estimated mass of 7.5 kg for an embryo. Assuming, for the sake of simplicity, that the embryo occupied the entire egg and that the egg was spherical, the inside diameter of the egg for an embryo of this size is approximately 240 mm. This is well within the size of the eggs of the Madagascar elephant bird, *Aepyornis*, which are 340 × 240 mm with a volume over 9 L (Gilbert, 1979). An Aix-en-Provence egg, TMP 84.66.12, is estimated to have measured 110 × 150 mm after accounting for

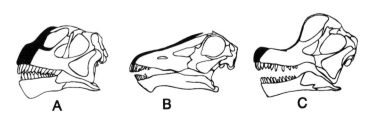

Figure 16.5 Sauropod skulls in left lateral view with premaxilla shown in black. **A.** *Camarasaurus lentus*, juvenile. **B.** *Diplodocus longus*. **C.** *Brachiosaurus branchai*. (A, after Gilmore, 1925; B, Holland, 1906; C, Janensch, 1935–36).

A   B   C

crushing. This is roughly one-half the size of the hypothesized *Camarasaurus* egg, suggesting the "*Hypselosaurus*" eggs may be too small to be of sauropod origin.

Even without embryos, other evidence suggests sauropods were oviparous. Packard, Tracy, and Roth (1977) noted that crocodilians (which along with birds are the nearest living relatives of sauropods) and turtles are restricted to oviparity because they obtain most of the calcium for ossification from the inner surface of the eggshell. Thus, they could not decrease eggshell thickness sufficiently to allow gaseous exchange to occur within the uterus. The same probably held true for nonavian saurischians. All living birds are oviparous, and this is the only extant vertebrate class for which this is true.

Duncker (1989) noted that birds are restricted to oviparity because of the developmental requirements of their specialized lungs. The lungs, specifically the paleopulmo, of birds are composed mainly of parabronchi. Parabronchi are so small, and the effects of surface tension at that scale are so great, that muscular action cannot overcome surface tension to inflate the parabronchi. Furthermore, bird lungs are rigid, while those of other amniotes are quite flexible. Therefore, bird lungs must gradually be filled with air by the slow absorbtion of pulmonary fluid into the air–capillary walls before hatching. Nonavian amniotes, on the other hand, fill their lungs with air by muscular actions that expand the lungs at the moment of birth, an action impossible for birds.

Postcranial pneumatic bones occur only in extant birds, nonavian theropods, segnosaurs, and sauropods (Britt, 1990). These bones are pneumatized by diverticula and air sacs of the lungs, suggesting these saurischians may have had lungs similar to those of birds. If this is correct, sauropods and Saurischia may also have been developmentally restricted to oviparity. Postcranial

pneumaticity among archosaurs is currently under study by one of us (Britt).

Conclusions

The *Camarasaurus* premaxilla is significant as the first evidence for the existence of sauropods small enough to fit within an egg. The premaxilla is excellently preserved, even though some of the bone is less than 0.2 mm thick. This indicates it had undergone little transportation and leaves open the possibility that sauropod eggs and nesting sites will be discovered in the Morrison Formation.

Considering the range of reproductive strategies in extant reptiles, there may be no *a priori* reason to exclude ovoviparity or vivipary as a reasonable mode for some dinosaurs. However, there is now no evidence to suggest that any dinosaur was other than oviparous. Indeed, work by Duncker (1989) demonstrates that birds are restricted to oviparity due to lung-development constraints. The presence of pneumatic vertebrae in saurischians implies they may have possessed lungs similar to birds, suggesting that saurischians too were obligatory egg-layers.

The embryonic *Camarasaurus* from which the premaxilla came was probably close to or at its final hatching size. The tooth crowns would certainly not have increased in size, and the bones were already ossified. The small size of the hatchling casts further doubt on the viviparous hypothesis for sauropods. Presumably one of the major functions of live birth is to allow the production of large, well-developed young. If sauropods, or at least Camarasaurus, produced such small young, doubts might even be raised about the feasibility of parental care. It seems difficult for a 17-m-long mother to care for a 1-m-long baby, or even be aware of its existence.

Embryos do not have to be found within eggs or associated with eggshell fragments to be correctly iden-

Table 16.1. *Comparison of adult (composite), juvenile, and embryonic* Camarasaurus *specimens.*

Specimen	Maturity	Tooth width (mm)	Premaxilla length (mm)	Skull length (mm)	Total length (mm)
AMNH 467 (skull) and other AMNH specimens	Adult	25	100	600	17,500
CMNH 11338	Juvenile	15	64	330	5,000
BYUVP 8967	Embryo	3	14	~ 70	~ 1,092

Sources: Adult [Osborn & Mook, 1921, based on AMNH 467 (skull) from Wyoming and postcranial skeletons of several individuals from Cope's Cañon City, Colorado locality (postcranium and overall length); juvenile – measurements of cast of CMNH 11338 (skull) and Gilmore (1925) (overall length); embryonic – BYUVP 8967 (premaxilla). Non-premaxillary sizes are estimates based on ratios derived from CMNH specimen.

tified as such. Dinosaur embryos may be identified by (1) the absence of bone remodeling (also in hatchlings), (2) the degree of fusion of axial elements, particulary those of the neural arch, (3) their size, especially when mature specimens are available for comparison, and (4) the absence of erupted teeth (only in very young embryos).

Acknowledgments

The *Camarasaurus* premaxilla was discovered by Dee A. Hall and exquisitely prepared by Kenneth L. Stadtman, both of Brigham Young University's Earth Sciences Museum. Special thanks are due James A. Jensen for his many years of dedication to vertebrate paleontology and to the Dry Mesa Quarry in particular. Wade E. Miller kindly granted access to the *Camarasaurus* specimen. We are particularly indebted to John S. McIntosh for his enthusiastic advice and help. Philip J. Currie unselfishly shared information on the juvenile dinosaurs he is currently studying. Special thanks are due Donna L. Sloan for preparing the restoration upon which the full scale-model was based. Funding and facilities were provided by the Royal Tyrrell Museum of Palaeontology and the Department of Geology and Geophysics at the University of Calgary. Illustration funding was provided in part by a generous donation from Dan W. Cooper.

References

Alexander, R. M. 1989. *Dynamics of Dinosaurs and Other Extinct Giants.* (New York: Columbia University Press).

Bakker, R. T. 1980. Dinosaur heresy – dinosaur renaissance. *In* R. D. K. Thomas, & E. C. Olson, (eds.), *A Cold Look at the Warm-Blooded Dinosaurs.* (Boulder, CO: Westview Press).

Barsbold, R., & Perle, A. 1983. On the taphonomy of the joint burial of juvenile dinosaurs and some aspects of their ecology. *Transcripts of the Joint Soviet-Mongolian Paleontological Expedition* 24: 121–5.

Bonaparte, J. F. 1979. Dinosaurs: A Jurassic assemblage from Patagonia. *Science* 205: 1377–9.

Bonaparte, J. F., & Vince, M. 1979. El Hallazgo del primer nido de dinosaurios Triàsicos (Saurischia, Prosauropoda), Trisico superieur de Patagonia, Argentina. *Ameghiniana* 16: 173–82.

Britt, B. B. 1990. The structure and significance of pneumatic vertebrae in theropods. *Journal of Vertebrate Paleontology* 10, Supplement to No. 3: 15A.

——— 1991. Theropods of Dry Mesa Quarry (Morrison Formation, Late Jurassic), Colorado with emphasis on the osteology of *Torvosaurus tanneri. Brigham Young University Geology Studies,* 37: 1–72.

Callison, G., & Rasmussen, T. E. 1980. Faunal list, Fruita Paleontological area. *In* H. J. Armstrong & A. Kihm, (eds.), *Fossil Vertebrates of the Grand Junction Area.* (Grand Junction CO: Grand River Institute).

Chatterjee, S. 1978. *Indosuchus* and *Indosaurus,* Cretaceous carnosaurs from India. *Journal of Paleontology* 52: 570–80.

Dodson, P. 1990. Sauropod paleoecology. *In* D. B. Weisham-pel, P. Dodson, & H. Osmólska, (eds.), *The Dinosauria.* (Berkeley: University of California Press).

Dodson, P., Behrensmeyer, A. K., Bakker, R. T., & McIntosh, J. S. 1980. Taphonomy and paleoecology of the dinosaur beds of the Jurassic Morrison Formation. *Paleobiology* 6: 208–32.

Duncker, H. R. 1989. Structural and functional integration across the reptile–bird transition: locomotor and respiratory systems. *In* D. B. Wake & G. Roth, (eds.), *Complex Organismal Functions: Integration and Evolution in Vertebrates.* (New York: Wiley and Sons).

Dunham, A. E., Overall, K. L., Porter, W. P., & Forster, C. A. 1989. Implications of ecological energetics and biophysical and developmental constraints for life history variation in dinosaurs. *In* J. O. Farlow, (ed.), Paleobiology of the Dinosaurs. *Geological Society of America Special Paper* 238: 1–19.

Edmund, A. G. 1960. Tooth replacement phenomena in the lower vertebrates. *Royal Ontario Museum Life Sciences Division Contribution* 52: 1–190.

Gilbert, A. B. 1979. Female genital organs. *In* A. S. King & J. McLelland, (eds.), *Form and Function in Birds,* Vol. 1. (London: Academic Press).

Gilmore, C. W. 1925. A nearly complete articulated skeleton of *Camarasaurus,* a saurischian dinosaur from the Dinosaur National Monument. *Memoirs of the Carnegie Museum* 10: 347–84.

Hatcher, J. B. 1903. Discovery of the remains of *Astrodon* (*Pleurocoelus*) in the *Atlantosaurus* beds of Wyoming. *Annals of the Carnegie Museum* 2: 1–72.

Holland, W. J. 1906. The osteology of *Diplodocus* Marsh. *Memoirs of the Carnegie Museum* 2: 225–78.

Horner, J. R., & Weishampel, D. B. 1988. A comparative embryological study of two ornithischian dinosaurs. *Nature* 332: 256–7.

Jain, S. L. 1989. Recent dinosaur discoveries in India, including eggshells, nests and coprolites. *In* D. D. Gillette & M. G. Lockley, (eds.), *Dinosaur Tracks and Traces.* (Cambridge: Cambridge University Press).

Janensch, W. 1935–1936. Die Schädel der Sauropoden *Brachiosaurus, Barosaurus* und *Dicraeosaurus* aus den Tendaguruschichten Deutsch-Ostafrikas. *Palaeontographica Supplement* 7, 2: 147–298.

Jensen, J. A. 1985. Three new sauropod dinosaurs from the Upper Jurassic of Colorado. *Great Basin Naturalist* 45: 697–709.

Jensen, J. A., & Padian, K. 1989. Small pterosaurs and dinosaurs from the Uncompahgre fauna (Brushy Basin Member, Morrison Formation: ?Tithonian), Late Jurassic, western Colorado. *Journal of Paleontology* 63: 364–74.

Lapparent, A. F. de 1957. Les oeufs de dinosauriens fossiles de rousset (Bouches-du-Rhône). *Comptes rendus des séances de l'Académie des Sciences* 245: 546–8.

Madsen, J. H. Jr. 1976. A second new theropod dinosaur from the Late Jurassic of east central Utah. *Utah Geology* 3: 51–60.

Marsh, O. C. 1883. Principle characters of American Jurassic dinosaurs. Part VI: Restoration of *Brontosaurus. American Journal of Science* (Series 3) 26(152): 80 .

——— 1888. Notice of a new genus of Sauropoda and other new dinosaurs from the Potomac Formation. *American*

Journal of Science (Series 3) 35: 89–94.

1896. Dinosaurs of North America. *Annual Report of the United States Geological Survey* 16(1): 133–244.

Mathur, U. B., & Pant, S. C. 1986. Sauropod dinosaur humeri from Lameta Group (Upper Cretaceous–?Palaeocene) of Kheda District, Gujarat. *Journal of the Palaeontological Society of India* 31: 22–5.

McIntosh, J. S. 1990. Sauropoda. *In* D. B. Weishampel, P. Dodson, & H. Osmólska, (eds.), *The Dinosauria.* (Berkeley: University of California Press).

Miller, W. E., Baer, J. L., Stadtman, K. L., & Britt, B. B. 1991. The Dry Mesa dinosaur quarry, Mesa County, Colorado. *In* W. R. Averett, (ed.), *Guidebook for Dinosaur Quarry and Tracksite Tour: Western Colorado and Eastern Utah* (Grand Junction, CO: Grand Junction Geological Society).

Mohabey, D. M. 1987. Juvenile sauropod dinosaur from Upper Cretaceous Lameta Formation of Panchmahals District, Gujarat, India. *Journal of the Geological Society of India* 30: 210–16.

Osborn, H. F., & Mook, C. C. 1921. *Camarasaurus, Amphicoelias,* and other sauropods of Cope. *Memoirs of the American Museum of Natural History* (New Series) 3: 247–87.

Packard, G. C., Tracy, C. R., & Roth, J. J. 1977. The physiological ecology of reptilian eggs and embryos, and the evolution of viviparity within the class Reptila. *Biological Reviews* 52: 71–105.

Paul, G. S. 1988. The brachiosaur giants of the Morrison and Tendaguru with a description of a new subgenus, *Giraffatitan,* and a comparison of the world's largest dinosaurs. *Hunteria* 2: 1–14.

Peterson, O. A., & Gilmore, C. W. 1902. *Elosaurus parvus*; a new genus and species of Sauropoda. *Annals of the Carnegie Museum* 1: 490–9.

Sahni, A., Kumar, K., Hartenberger, J. L., Jaeger, J. J., Rage, J. C., Sudre, J., & Vianey-Liaud, M. 1982. Discovery of terrestrial microvertebrates in the Palaeocene of the Deccan Traps, India: Geodynamic implications. *Bulletin de la Société Géologique de France* 24: 1093–9.

17 Upper Jurassic sauropod babies from the Morrison Formation

KENNETH CARPENTER AND
JOHN MCINTOSH

Abstract

Very small, young sauropod remains from the Upper Jurassic Morrison Formation are described from Oklahoma and Wyoming. The specimens from Oklahoma, identified as *Camarasaurus* sp. and *Apatosaurus* sp., are so small and the joint surfaces so poorly developed that they are believed to belong to babies. The specimens from Wyoming have previously been described by Marsh and by Peterson and Gilmore. A revision of their taxonomy shows that *Pleurocoelus montanus* Marsh should be referred to *Camarasaurus grandis*, and *Elosaurus parvus* Peterson and Gilmore to *Apatosaurus excelsus*.

Introduction

Sauropod remains are the most common fossil found in the Morrison Formation (Dodson et al., 1980), however, baby specimens (arbitrarily set as less than one-sixteenth adult size) are very rare. The reasons for this discrepancy is probably taphonomic rather than habitat selection, because of the geographically widespread occurrence of babies in different lithofacies (see below, and Britt & Naylor, Chapter 16). As Behrensmeyer, Western, and Dechant Boaz (1979) have noted, small bones are less likely to be preserved because of the ease by which they are destroyed through weathering, scavenging, and trampling. Besides their small size, juvenile bones are less well ossified and are more readily destroyed.

Another reason for the rarity of baby sauropods, as well as baby dinosaurs in general, is a failure to recognize the bones as being those of a baby, or if recognized, to report them. One of the earliest descriptions, albeit brief, of a baby sauropod is by Marsh (1883). The specimen was identified as a "foetal" dinosaur. It was reported to have been found in association with the holotype of "*Morosaurus*" (=*Camarasaurus*) *grandis* (YPM 1901) at Reed's Quarry 1 in the Morrison Formation at Como Bluff, Wyoming (Ostrom & McIntosh 1966). The specimen (YPM 1908) was later made the holotype of *Pleurocoelus montanus* (Marsh 1896) and two vertebrae were figured. Marsh had earlier named *Pleurocoelus nanus* for some small, including baby-sized, bones from the Arundel Formation of Maryland (Marsh, 1888). Most of this material was described and figured by Lull (1911).

Peterson and Gilmore (1902) named and described *Elosaurus parvus* for some small sauropod bones from the Morrison Formation at Sheep Creek, Wyoming. The specimen was found associated with a large specimen of *Apatosaurus excelsus*, the skeleton of which is now on display at the University of Wyoming. Additional material from Sheep Creek was described by Hatcher (1903) as *Astrodon johnstoni*.

Dinosaur bones were excavated between 1935 and 1942 from several quarries in the Morrison Formation near Kenton in the Oklahoma panhandle. The excavations were conducted by the Work Project Administration (WPA) under the direction of J. Willis Stovall (see Hunt & Lucas 1987). At that time, baby sauropod bones were collected from mudstones at Pit 1, which was worked from May 1935 until September 1938 (Hunt & Lucas, 1987). These baby bones were never described or figured in any of the brief descriptions of the Morrison dinosaurs by Stovall (e.g., Stovall, 1938). He did hint, however, that they represent a new species of small sauropod (Stovall, 1943).

More recently, a fragmentary skeleton purportedly that of a hatchling sauropod was described from the Lameta Formation near Dholi Dungri, India (Mohabey, 1987). The material is fragmentary and poorly preserved, and Jain (1989) suggested that it might actually be a booid snake. We accept this suggestion, considering the presence of the zygosphene–zygantrum articulation of the vertebrae.

Britt (1988; Britt & Naylor, Chapter 16) briefly described a small *Camarasaurus* premaxilla from the

Morrison Formation at Dry Mesa Quarry, Colorado. He suggested that based on the small size, it might belong to a hatchling.

The Oklahoma baby sauropod material was damaged during preparation, much of which was done by whittling the matrix away with knives and files (Langston, personal communication). Because it was often difficult to distinguish between the surrounding concretionary material and bone, the outer surface of the bone was often shaved away, resulting in the loss of some taxonomically important structures. For example, the medial corner of the humeral head is smoothly carved away on humerus OMNH 1278 (see Fig. 17.5I). How much this alteration has affected our identification of the baby sauropod material is unknown. However, it is puzzling that most of the appendicular material seems to belong to *Apatosaurus*, while most of the vertebrae seem to belong to *Camarasaurus*. We cannot rule out a taphonomic bias, but it is also possible that the vertebrae have been so modified by preparation as to resemble those of *Camarasaurus*. Measurements are given in Tables 17.1–17.3.

Systematics
Order Sauropoda
Family indeterminate
Material. OMNH 1247, cervical centrum (Fig. 17.1B). OMNH 1250, cervical centrum (Fig. 17.1A). OMNH 1206, caudal centrum (Fig. 17.2E). OMNH 1230, caudal centrum (Fig. 17.2F). OMNH 1233, caudal centrum (Fig. 17.2G). OMNH 1412, sacral rib (not figured).

Description. These centra (Figs. 17.1A, B & 17.2E–G) are severely damaged; they are anterior caudals belonging to either *Apatosaurus* or *Camarasaurus*. The sacral rib lacks any feature that would permit positive identification as either *Apatosaurus* or *Camarasaurus*.

Family Diplodocidae Marsh 1884
Apatosaurus sp.
Material. OMNH 1245, cervical centrum (Fig. 17.1E). OMNH 1246, cervical centrum (Fig. 17.1D). OMNH 1251, cervical centrum (Fig. 17.1C). OMNH 1210, dorsal centrum (Fig. 17.2B). OMNH 1217, dorsal centrum (Fig. 17.2C). OMNH 1219, dorsal centrum (Fig. 17.2A). OMNH 1226, incomplete dorsal centrum (Fig. 17.2D). OMNH 1255, incomplete dorsal centrum (not figured). OMNH 1300, right scapula (Fig. 17.4A). OMNH 1274, right humerus lacking both proximal and distal ends (not figured). OMNH 1275, right humerus lacking the medial corner of the distal end (Fig. 17.5H). OMNH 1276, left humerus lacking part of distal end (Fig. 17.5J). OMNH 1277, left humerus (Fig. 17.5E–G). OMNH 1278, left humerus lacking the medial corner of the proximal end (Fig. 17.5I). OMNH 2115, right humerus (Fig. 17.1A–D). OMNH 1289, right ulna (Fig.

17.6A, B, E). OMNH 1287, left radius (Fig. 17.6I). OMNH 1288, left radius (Fig. 17.6H). OMNH 1290, right radius (Fig. 17.6E–G). OMNH 4019, left ilium (Fig. 17.7A). OMNH 1294, left pubis (Fig. 17.7B, C). OMNH 1293, proximal end of right pubis (Fig. 17.7D). OMNH 1297, right ischium (Fig. 17.7F, G). OMNH 1298, right ischium (Fig. 17.7H). OMNH 1279, distal half of right femur (Fig. 17.8B). OMNH 1280, proximal half of right femur (Fig. 17.8A). OMNH 1281, distal half of left femur (Fig. 17.8C). OMNH 1285, right tibia (Fig. 17.8G). OMNH 1286, right tibia (Fig. 17.8H). OMNH 1907, right tibia (not figured). OMNH 1291, left tibia (Fig. 17.8I). OMNH 1292, proximal half of left tibia (not figured). OMNH 1299, proximal two-thirds of the left tibia (not figured). OMNH 1295, right fibula

Table 17.1. *Measurements of baby sauropod centra (in millimeters).*

Catalog No.	L	AW	AH	PW	PH	PAW
			Cervicals			
OMNH 1239	53	33	23	40	23	38
OMNH 1242	51	41	31	47	36	52
OMNH 1244	53	26	22	29	23	–
OMNH 1245	59	28	21	34	30	38
OMNH 1246	55	25	15	29	27	35
OMNH 1247	55	–	16	30	15	–
OMNH 1248	50	37	27	44e	40e	–
OMNH 1249	57	40	27	45	31	48
OMNH 1250	45	28	21	27	29	–
OMNH 1251	45	22	18	21+	23	29
OMNH 1252	44	31	20	30	26	32
OMNH 1253	35	27	19	26	23	–
OMNH 1254	62	36	25	45	35	–
			Dorsals			
OMNH 1210	56	55	47	60	57	
OMNH 1217	55	48	44	–	–	
OMNH 1219	52	49	44	52+	42	
OMNH 1226	63	–	–	–	–	
OMNH 1241	48	49	32	56	40	
OMNH 1255	53	–	–	–	–	
YPM 1908	45	35	32	38	31	
YPM 1908	62	43+	40	49	40	
			Caudals			
OMNH 1206	24	35	30+	33	33	
OMNH 1230	41	33	33	34	32	
OMNH 1233	19	21	18	18	17	
YPM 1908	25	49	43	44	39	

Notes: L, length; AW, anterior width of centrum; AH, anterior height of centrum; PW, posterior width of centrum; PH, posterior height of centrum; PAW, width across parapophyses; e, estimated.

(Fig. 17.8J). OMNH 1296, right fibula (Fig. 17.8K).

Description. The three cervical centra (Fig. 17.1C–E) are referred to *Apatosaurus* because the parapophysis is directed ventrolaterally. The pleurocoels are smaller than those in *Camarasaurus* (see below). The centra are too small to belong to the skeletons represented by the appendicular elements, indicating the existence of individuals of different ontogenetic stages. The dorsal centra (Fig. 17.2A–D) are referred to *Apatosaurus* because of the shallowness of the pleurocoels relative to the size of the centra. However, this feature must be used with caution because it appears to exist in YPM 1908, here referred to *Camarasaurus*, and it may constitute a juvenile character.

The right scapula (Fig. 17.4A) is almost complete, lacking only the distal portion of the scapular plate above the glenoid. The blade is long and narrow, identifying it as *Apatosaurus*. The blade is moderately expanded distally, but less so than in any other sauropod. The proximal third is bowed medially and is thickest at the glenoid. The ridge on the proximal plate is almost perpendicular to the long axis of the scapular blade, in sharp contrast to *Diplodocus* where it is about 65°.

Three pairs of humeri (Fig. 17.7A–J) are present, but given their lengths, it is probable that more than three individuals are represented. They are robust, with the distal end expanded transversely almost as much as the proximal end. This feature identifies them as belonging to *Apatosaurus*; in *Camarasaurus*, the distal end is narrower than the proximal end. The distal condyles are about the same size, although the articular surfaces are not well developed. The deltopectoral crests are prominent and limited to the upper half of the outer margin of the anterior face.

The right ulna (Fig. 17.6A, B, E) is bowed forwards and has a poorly developed ligamental ridge on the lower half of the shaft for connection with the radius. The proximal end of the ulna is triangular, with a well-developed radial notch on the anterior margin. As a result, the proximal end of the ulna is more expanded than the distal end.

The radii (Fig. 17.6E–I) are robust and thus may be identified as *Apatosaurus* rather than *Camarasaurus*. The ridge by which the ulna and radius are bound together by ligaments in adult animals is not developed in these juvenile animals. There is considerable variation in the shapes of the articular ends, which are poorly preserved, possibly due to preparation.

The incomplete left ilium (Fig. 17.7A) lacks most of the iliac crest and ischial peduncle. Only the iliac crest above the pubic peduncle is preserved, and this flares laterally. The articular surface of the pubic peduncle is "D" shaped in ventral view. Portions of the

Table 17.2. Measurements of baby sauropod pectoral and forelimb bones (in millimeters).

Catalog No.	L	SPH	W	P	S	D	C	C/L
			Scapula					
OMNH 1300	357	145	–	82	44	–	–	–
			Coracoid					
OMNH 1302	89	–	122e	–	–	–	–	–
			Humeri					
OMNH 1274	–	–	–	–	41	–	115	–
OMNH 1275	257	–	–	112	41	89e	128	50
OMNH 1276	233	–	–	102e	45	–	141	–
OMNH 1277	233	–	–	102	37	94	114	49
OMNH 1278	254	–	–	–	47	102	142	56
OMNH 2115	227	–	–	106	39	84e	116	51
YPM 1908	–	–	80	–	–	–	–	–
			Ulnae					
OMNH 1289	142	–	–	54	21	39	73	–
			Radii					
OMNH 1287	142	–	–	42	18	45	65	46
OMNH 1288	128	–	–	48	18	46	65	51
OMNH 1290	142	–	–	45	19	42	64	45

Notes: L, length; W, width; P, proximal width; S, shaft height; D, distal height; C, circumference; C/L, circumference/length.

first and third sacral rib facets are seen on the medial surface.

A possible matching pair of pubes is present (Fig. 17.7B–D). These bones are more slender than the pubes in *Camarasaurus*, and thus appear more like those in *Apatosaurus*. Furthermore, there is no evidence for an anterior hooklike ambiens process below the iliac peduncle instead, this area is rounded as in *Apatosaurus*. Although damaged, there are traces of the pubic apron developed medially connecting the two pubes. The distal end of the left pubis is complete and shows that it is thick. The pubic foramen is damaged, but appears to have been open, a typical juvenile feature.

One of the two right ischia (OMNH 1297) is almost complete (Fig. 17.7F–H). The distal end is expanded in *Apatosaurus* fashion so it had a long contact with the distal end of the other ischium. However, this contact is not as long as it is between the ischium and pubis. The shaft has a slight posterior bow but is not twisted.

Two, possibly three, femora are present (Fig. 17.8A–C). One femur may be represented by proximal

and distal halves, but there is the possibility that each half belongs to different femora because there is no good contact between the halves. The other femur lacks the proximal third of the bone and is lacking the tibial condyle. The tibial condyle of the other femur is larger than the first tibial condyle and is also directed obliquely when viewed ventrally. The distal condyles are not well developed, indicating the immaturity of the individuals. The femoral head is damaged and offers little information. The fourth trochanter is located on the medial side of the posterior surface as in all sauropods, but is ontogenetically underdeveloped.

Three pairs of tibiae (Fig. 17.8G–I) are present in the collection, but it is uncertain whether any two belong to one individual. The very prominent cnemial crest identifies these bones as belonging to *Apatosaurus*. The proximal end is oval in cross section, being wider than long, and the distal end is considerably expanded, especially posteriorly. Several characters of the tibia differ from those in the adult *Apatosaurus* and are interpreted as juvenile states. These include the larger circumference-to-length ratio (juvenile 0.50-0.56 versus

Table 17.3. Measurements of baby sauropod pelvic and hindlimb bones (in millimeters)

OMNH 1300	L	P	S	D	C	C/L	LP	HP
			Ilia					
OMNH 4019	–	–	–	–	–	–	163	87
			Ischia					
OMNH 1297	199	–	29	51	–	–		
OMNH 1298	180+	–	–	–	–	–		
			Pubes					
OMNH 1293	–	–	–	–	–	–		
OMNH 1294	246	82	76	70	–	–		
			Femora					
OMNH 1279	329+	–	53	103	–	–		
OMNH 1280	–	96	–	–	–	–		
OMNH 1281	–	–	51	93	150	–		
YPM 1908 (left)	330	80	48	132	132	0.40		
YPM 1908 (right)		–	106	49	138	138		
			Tibia					
OMNH 1285	206	85	44	69	116	56		
OMNH 1286	224	82	40	–	114	51		
OMNH 1291	226	82	40	71	11	50		
OMNH 1292	–	78	43	–	113	–		
OMNH 1299	–	74	34	–	99	–		
YPM 1907	205	64	37	70	103	50		
			Fibula					
OMNH 1295	235	60	32	45	84	36		
OMNH 1296	236	67	31	46	81	34		

Figure 17.1. Baby sauropod cervical and anterior dorsal centra in left lateral view. Indeterminate cervicals; **A**. OMNH 1250. **B**. OMNH 1246. *Apatosaurus* sp. cervicals. **C**. OMNH 1251. **E**. OMNH 1246. *Camarasaurus* sp. cervicals. **F**. OMNH 1245. **G**. OMNH 1253. **H**. OMNH 1252. **I**. OMNH 1239. **J**. OMNH 1248. **K**. OMNH 1242. **L**. OMNH 1249. **M**. OMNH 1254. **N**. OMNH 1243. **O**. OMNH 1241. **P**. CM 578. **Q**. *Apatosaurus excelsus* anterior dorsal, YPM 1908. Scale = 50 mm.

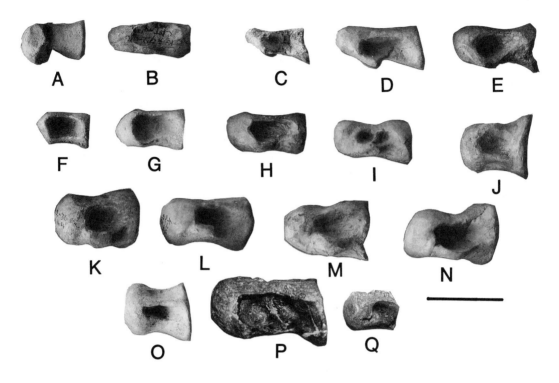

Figure 17.2. Baby sauropod dorsal, sacral and caudal centra in left lateral view. *Apatosaurus* sp. dorsals. **A**. OMNH 1219. **B**. OMNH 1210. **C**. OMNH 1217. **D**. OMNH 1226. **E**. OMNH 1206. **F**. OMNH 1230. **G**. OMNH 1233. **H**. CM 578, sacral. **I**. CM 566, caudal. **J**. CM 585. *Apatosaurus excelsus* dorsal. **K**. YPM 1908, caudal. **L**. YPM 1908. Scale = 50 mm.

Figure 17.3. Baby sauropod neural arches. **A–K.** *Camarasaurus* cervical neural arches. OMNH 1268 **A.** anterior view, **B.** left lateral view, **C.** posterior view; OMNH 1273: **D.** anterior view, **E.** left lateral view; OMNH 1266: **F.** anterior view, **G.** left lateral view; posterior dorsal neural arch of OMNH 1264: **H.** anterior view, **I.** left lateral view; sacral neural arch of OMNH 1269: **J.** anterior view, **K.** left lateral view. **L–P.** *Apatosaurus excelsus* neural arches. YPM 1908: **L.** anterior view, **M.** right lateral view, **N.** posterior view, dorsal neural arch in anterior **O.** and lateral views **P.** Scale = 50 mm.

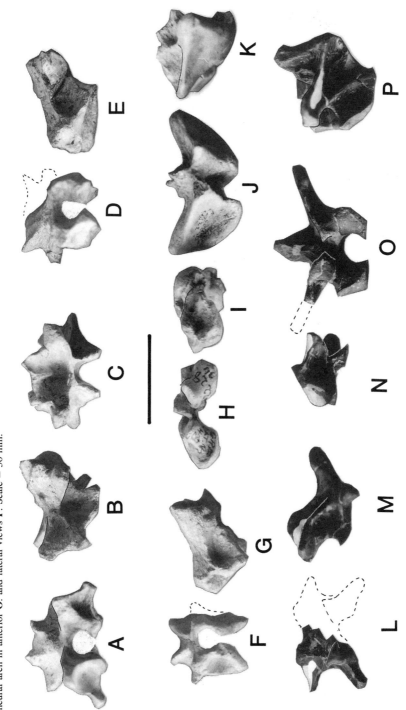

adult 0.49), poorly developed astragalar facet, and the sharper lateral border of the cnemial crest.

The two right fibulae (Fig. 17.8J, K) are complete. The shaft is straight, laterally compressed, and expanded anteroposteriorly at both the proximal and distal ends. This is especially true of the posterior corner of the proximal end. In addition, the distal end is expanded transversely. The tibial articular scar on the medial side is not well developed, nor are the muscle scars on the medial and lateral surfaces.

Apatosaurus excelsus

Material. Elosaurus parvus, CM 566, incomplete neural arch of cervical 5 or 6 (Fig. 17.3L–N); dorso-sacral neural arch (Fig. 17.3O, P); three incomplete right dorsal ribs (not figured); one sacral centrum (Fig. 17.2I); proximal end of left pubis (Fig. 17.7E); right scapula (Fig. 17.4B); right and left humerus (Fig. 17.5M–S); right ulna (Fig. 17.6C, D); right femur (Fig. 17.8D); and left fibula (Fig. 17.8L). Collected by C. W. Gilmore in 1901, Quarry E, Sheep Creek, Albany County, Wyoming.

Description. See Peterson and Gilmore (1902). The humeri (Fig. 17.5M–S) are relatively robust, with the distal ends, almost as wide as the proximal ends, thus identifying them as *Apatosaurus*. Identification of the neural arch (Fig. 17.3O, P) as that of the dorso-sacral

is based upon the parapophysis (Peterson and Gilmore's capitulum facet) being at the very base of the arch where it also extended onto the centrum. Three rib fragments from the right side of the body are also present. The anteriormost and heaviest of the three may be the third rib. It is represented by most of the head, complete tuberculum and two-fifths of the shaft. Another rib, represented by the head and tuberculum and only a small portion of the shaft, is a posterior rib, possibly the ninth. The other rib is represented by about half of the shaft. It lacks diagnostic features and cannot be identified more specifically than as a posterior rib.

The element Peterson and Gilmore (1902) identified (incorrectly) as the distal end of a right pubis is the proximal end of the left pubis (Fig. 17.7E). The pubic foramen is apparently not closed as it is in other young sauropods (e.g., *Camarasaurus lentus* YPM 1910, Ostrom & McIntosh, 1966, Plate 70), but this cannot be stated with certainty because of damage. The origin for the ambiens muscle on the proximal anterior border of the pubis extends forward slightly as in the adult *Apatosaurus*. It does not form a hooklike process typical of *Diplodocus* and *Barosaurus*. The other North American sauropods, *Camarasaurus*, *Brachiosaurus* and *Haplocanthosaurus*, do not possess the forwardly projecting ambiens process. Finally, the tibial and fibular condyles of the femur appear better developed

Figure 17.4. Baby sauropod pectoral girdle, **A**. *Apatosaurus* sp., OMNH 1300 right scapula. **B**. *Apatosaurus excelsus* CM 566 right scapula. **C**. *Camarasaurus* sp. OMNH 1302 coracoid. Scale = 50 mm

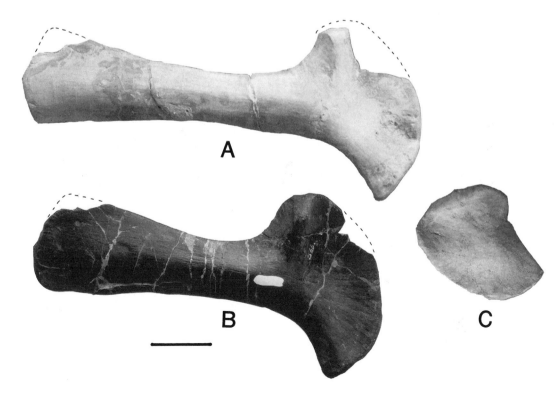

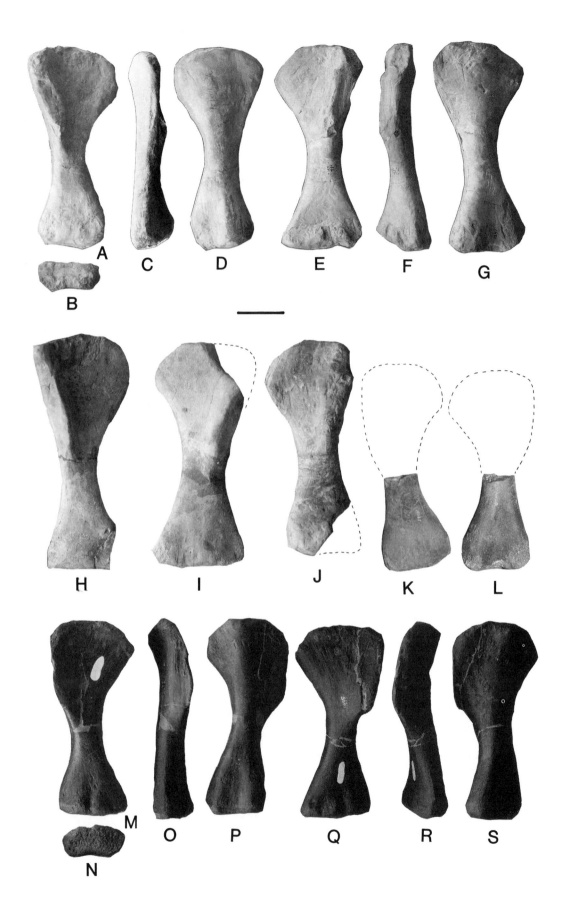

than in the femur of YPM 1908 (*Camarasaurus grandis*) discussed below, despite being the same lengths.

Because the specimen was found commingled with *Apatosaurus excelsus* (Gilmore, 1936), McIntosh (1981) referred it to *A. excelsus*.

Family Camarasauridae Cope 1877
Camarasaurus sp.

Material. Astrodon johnstoni, CM 578, cervical centrum (Fig. 17.1P) and dorsal centrum (Fig. 17.2H)

from Quarry C. CM 585, caudal centrum (Fig. 17.2J) from Quarry E. Collected 1901, Sheep Creek, Albany County, Wyoming. OMNH 1239, cervical centrum (Fig. 17.1I). OMNH 1241, cervical centrum (Fig. 17.1O). OMNH 1242, cervical centrum (Fig. 17.1K). OMNH 1243, cervical centrum (Fig. 17.1N). OMNH 1244, cervical centrum (Fig. 17.1H). OMNH 1248, cervical centrum (Fig. 17.1J). OMNH 1249, cervical centrum (Fig. 17.1L). OMNH 1252, cervical centrum (Fig. 17.1G). OMNH 1253, cervical centrum (Fig. 17.1F). OMNH

Figure 17.5. (*facing*) Baby sauropod humeri. *Apatosaurus* sp.. **A–C.** OMNH 1215 in anterior (A), distal (B), right lateral (C), and posterior views. **E, F.** OMNH 1277 in anterior (E), left lateral (F), and posterior (G) views. **H.** OMNH 1275 in anterior view. **I.** OMNH 1278 in anterior view. **J.** OMNH 1276 in anterior view. *Camarasaurus grandis.* **K, L.** YPM 1908 in anterior (K) and posterior views (L). *Apatosaurus excelsus.* **M–S.** Right CM 566 in anterior (M), distal (N), right lateral (O), and posterior views (P); left CM 566 in anterior (Q), left lateral (R), and posterior views (S). Scale = 50 mm.

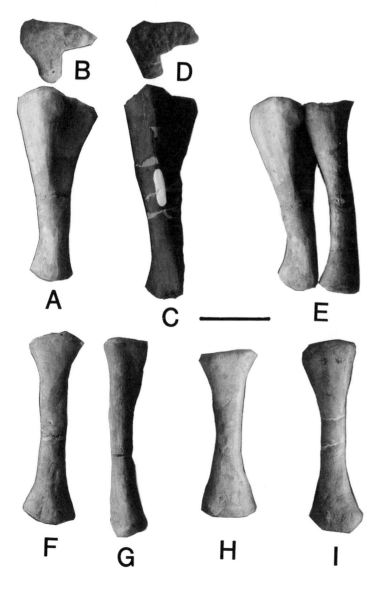

Figure 17.6. Baby sauropod ulnae and radii. *Apatosaurus* sp. **A, B.** OMNH 1289 ulna in right lateral (A) and proximal views (B). *Apatosaurus excelsus.* **C, D.** CM 566 in right lateral (C) and proximal views (D). *Apatosaurus* sp. **E.** OMNH 1289 ulna and OMNH 1290 radius in articulation. **F, G.** OMNH 1290 radius in anterior (F) and medial views (G). **H.** OMNH 1288 radius in anterior view. **I.** OMNH 1287 radius in anterior view. Scale = 50 mm.

1254, cervical centrum (Fig. 17.1M). OMNH 1266, cervical neural arch (Fig. 17.3F, G). OMNH 1268, cervical neural arch (Fig. 17.3A–C). OMNH 1273, cervical neural arch (Fig. 17.3D, E). OMNH 1264, posterior dorsal neural arch (Fig. 17.3H, I). OMNH 1263, sacral neural arch (not figured). OMNH 1269, sacral neural arch (Fig. 17.3J, K). OMNH 1302, left coracoid (Fig. 17.4C). OMNH 1256, metatarsal I (Fig. 17.8M, N). OMNH 1413, right metatarsal IV.

Description. Hatcher's (1903) reference of the cervical and dorsal centra (Figs. 17.1F & 17.2H) to the Early Cretaceous *Astrodon johnsoni* (a nomen dubium, Langston, personal communications) appears unlikely. The very large pleurocoel in each suggests that they belong to *Camarasaurus*. The caudal centrum (Fig.

17.2J) is also referred to *Camarasaurus* because it is not elongated as in *Diplodocus* and *Apatosaurus*, but is rather short.

The cervical centra (Fig. 17.1F–O) from Oklahoma are referred to *Camarasaurus* because the parapophyses project laterally. Most have deep pleurocoels, but it is uncertain how much this depth has been exaggerated by preparation. Preparation has certainly destroyed most of the laminae within the pleurocoels that may have once existed. The ball-and-socket articular faces of the centra are well developed as in adult *Camarasaurus*. An axis centrum (OMNH 1253, Fig. 17.1G) displays a dorsally placed odontoid and an extremely large, deep pleurocoel.

The cervical neural arches (Fig. 17.3A–G) are iden-

Figure 17.7. Baby sauropod pelvic girdle elements. *Apatosaurus* sp. **A.** OMNH 4019 lateral view of left ilium. **B, C.** OMNH 1294 right pubis in lateral (B) and medial views (C). **D.** OMNH 1293 left pubis in lateral view. *Apatosaurus excelsus.* **E.** CM 566 left pubis in lateral view. *Apatosaurus* sp. **F, G.** OMNH 1297 right ischium in lateral (dashed proximal area was discovered after this figure was completed) (F) and distal views (G). **H.** OMNH 1298 left ischium in medial view. Scale = 50 mm.

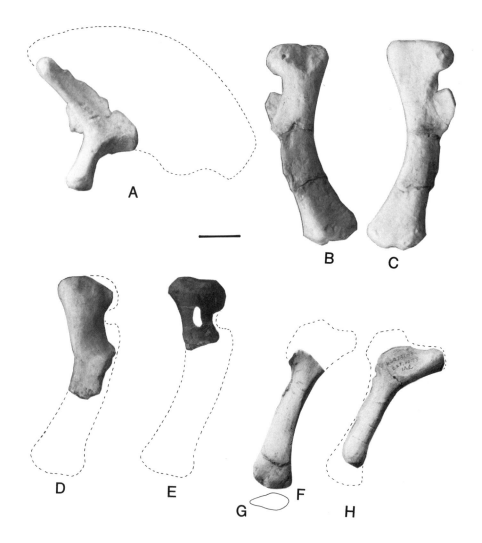

tified as *Camarasaurus* because of their apparently low bifurcated neural spine. They apparently come from the anterior middle portion of the neck. The dorsal (Fig. 17.3H, I) and sacral neural arches (Fig. 17.3J, K) are low and have robust neural pedicles typical of *Camarasaurus*.

The coracoid (Fig. 17.4C) has the typical oval shape seen in *Camarasaurus*. It has a small, shallow, subglenoid notch on the posteroventral border.

The metatarsal I (Fig. 17.8M, N) lacks the hook-like process on the posterolateral corner of the distal articular surface, thus identifying it as belonging to *Camarasaurus*. It also has a distal articular surface that is wider than long, while the proximal articular surface is longer anteroposteriorly than wide. The right metatarsal IV has a slender shaft, suggesting that it belongs to *Camarasaurus* rather than *Apatosaurus*. Its distal articular surface is almost flat, suggesting that a cartilaginous cap was present in life or that the specimen was damaged during preparation.

Camarasaurus grandis

Material. Pleurocoelus montanus, YPM 1908, complete left femur (Fig. 17.8E), fragmentary right femur (Fig. 17.8F), distal end of right humerus (Fig. 17.5K, L), anterior dorsal (Fig. 17.1Q), dorsal centrum (Fig. 17.2K), another anterior dorsal centrum (not figured), and an anterior caudal centrum (Fig. 17.2L). Collected from Quarry 1, Como Bluff, Albany County, Wyoming.

Description. See Hatcher, 1903. Four centra are present, two of which were figured by Marsh (1896) as *Pleurocoelus montanus*. The specimen was found associated with the holotype of "*Morosaurus*" *grandis* and was referred to by Marsh (1883) as a foetal sauropod. Its later assignment to *Pleurocoelus* virtually without description or further comment was no doubt due to its small size and the large pleurocoels in its centra.

In YPM 1908, the pleurocoel of the anteriormost dorsal centrum (Fig. 17.1Q) is well developed, although small and shallow. In the other two, more posterior dorsals (Fig. 17.2K), the pleurocoels are depressions rather than true pleurocoels, because the sharply defined margins seen in the adult have not yet developed. The placement of the parapophysis on the centrum of the anteriormost dorsal at the side of the pleurocoel identifies it as dorsal 1 or 2. The anterior condyle of the centrum is well developed in this centrum, whereas the anterior faces of the other two centra are only slightly convex. This condition stands in sharp contrast with adult *Camarasaurus* in which a ball is well developed. The anterior caudal centrum resembles that of *C. grandis* on both its ventral and lateral faces, and lacks the rounded V-shape cross section seen in the caudals of *Apatosaurus*.

The humerus (Fig. 17.5K, L) is incomplete, but the distal end appears narrow relative to its size and shaft diameter, suggesting that it belongs to *Camarasaurus*. The better preserved left femur (Fig. 17.8E) has been somewhat crushed anteroposteriorly. The posterior face of the shaft is damaged so the location and development of the fourth trochanter cannot be determined. The femoral head is incomplete, but enough is present to indicate that the head was higher than the greater trochanter, a distinctive character of *Camarasaurus*. The tibial and fibular condyles are so badly eroded that their development cannot be accurately determined, although they appear to have been less developed than in the adult. The shaft of the femur is straight, and its transverse diameter greatly exceeds the anteroposterior diameter, a condition heightened by crushing.

The other femur (Fig. 17.8F) was originally complete, but now consists of four segments with clean fractures. All four bear the same Yale accession number *1069* but reassembling the bone still leaves gaps (Fig. 17.8F). This femur appears larger than the other, suggesting that it belongs to a different and larger individual. The bone is not as crushed as the other, and retains a nearly circular cross section about two-thirds from the top. The proximal segment consists of the greater trochanter and head. The head, which is abraded, does not significantly rise above the top of the greater trochanter. The fourth trochanter is missing and is believed to be on one of the missing fragments of the shaft. A piece containing the fibular condyle is missing, but the tibial condyle is complete and is not as well developed as in adult *Camarasaurus*.

The referral of this specimen to *Camarasaurus grandis* is based mostly upon its association with four young adult *Camarasaurus grandis* specimens. The only other sauropod bones from the quarry are two teeth and a chevron of *Diplodocus*, an animal to which YPM 1908 cannot belong. Had this specimen of baby sauropod been found isolated, the three dorsals might have been assigned to *Apatosaurus*, although the caudal centrum is clearly camarasaurid. The femoral circumference: length ratio (0.40) is less than that of an adult *Camarasaurus* (0.44), but this ratio would appear not to be unreasonable in an immature animal.

Discussion and conclusion

The collection of baby sauropods from the Morrison Formation of Oklahoma is the largest and most diverse yet known. The abundance of individuals, compared to other sites where only single individuals are known, suggests a local nesting ground. As yet, however, no evidence of nesting, including eggshells fragments, has been found. Recently, Bakker (1980, and in Morell, 1987 and Canio, 1988) suggested that the wide pelvic canal in sauropods and the apparent absence of sauropod eggs was evidence that sauropods gave "live birth." That sauropods might have been ovoviviparous, was first suggested by Marsh (1883) for the baby *Camarasaurus grandis* found associated with an adult (see

Figure 17.8. Baby sauropod hindlimb elements. *Apatosaurus* sp. **A.** OMNH 1279 right(?) proximal femur in posterior view. **B.** OMNH 1280 right femur in posterior view. **C.** OMNH 1281 right femur in posterior view. *Apatosaurus excelsus.* **D.** CM 566 right femur in posterior view. *Camarasaurus grandis.* **E, F.** YPM 1908 left femur in posterior view (E) and right femur in anterior view (F). *Apatosaurus* sp. **G.** OMNH 1285 right tibia in anterior view. **H.** OMNH 1286 right tibia in anterior view. **I.** OMNH 1291 left tibia in anterior view. **J.** OMNH 1295 right fibula in lateral view. **K.** OMNH 1296 right fibula in lateral view. *Apatosaurus excelsus.* **L.** CM 566 left fibula in lateral view. *Apatosaurus* sp. **M, N.** OMNH 1256 metatarsal I in anterior (M) and posterior views (N). Scale = 50 mm.

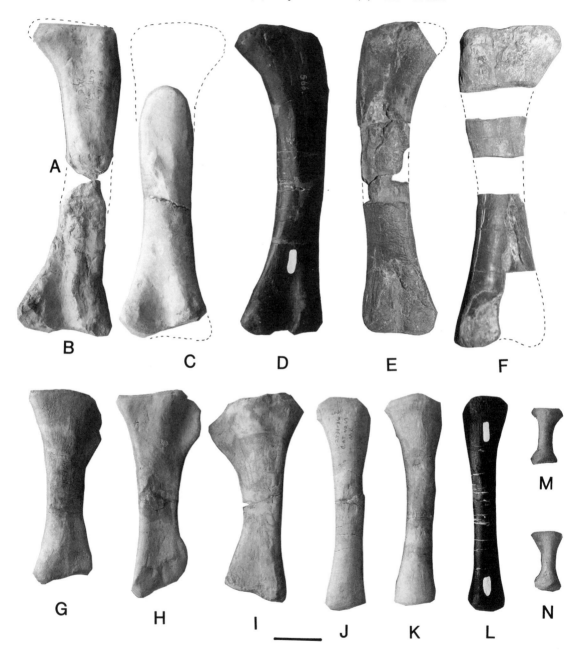

Figure 17.9. Comparison of *Apatosaurus* humeri. **A**. OMNH 2113 adult. **B**. OMNH 1215 baby. Scale = 100 mm.

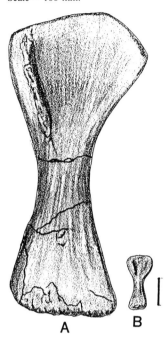

ming. The adult, UW 15556, has a humeral length of 1,100 mm, a minimal shaft circumference of 602 mm, and a circumference: length ratio of 0.55. The baby, CM 566, has humeri 220 mm long, shaft circumferences of 114 mm (right) and 116 mm (left), and circumference: length ratios of 0.52 (right) and 0.53 (left). The proportion of the humeri lengths to the adult (20 percent), circumference (19 percent%) and circumference: length ratios are similar to those for the Oklahoma specimens.

Acknowledgments

Access to the specimens used in this study was made possible by David Berman (Carnegie Museum), Richard Cifelli (Oklahoma Museum of Natural History), and John Ostrom (Peabody Museum of Natural History). Constructive comments by Wann Langston greatly improved an earlier draft.

above). If Mohabey's (1987) report of a sauropod hatchling associated with a nest of eggs is correct, it is more likely that sauropods were oviparous. However, as discussed above, this identification has been called into question by Jain (1989). Thus, we do not as yet have any proof that sauropods laid eggs, despite the discovery of numerous eggs attributed to them (see Carpenter, Horner, & Hirsch, ''Introduction''; Buffetaut & Le Loeuff, Chapter 2; Vianey-llaud et al., Chapter 11).

The Oklahoma baby sauropods afford the best evidence for the proportional changes that occur during ontogeny. This may be illustrated by comparing the largest known *Apatosaurus* humerus from Quarry 1 with the largest and smallest baby sauropod humeri from the same quarry (Fig. 17.9A, B). The adult humerus, OMNH 2113, is 1,195 mm long, has a minimum shaft circumference of 565 mm and a circumference: length ratio of 0.47. This can be compared to the largest baby *Apatosaurus* humerus, OMNH 1275, with a length of 257 mm, a circumference of 125 mm, and a circumference: length ratio of 0.49. The smallest humerus, OMNH 2115, is 227 mm long, and 116 mm in circumference, and has a circumference: length ratio of 0.51. Thus, the baby sauropod humeri are 19–21.5 percent as long as the adult humerus, and 20.5–22 percent of the circumference. During growth, the length and circumference ratios of the humeri remain the same.

These results may be compared to the baby and adult *Apatosaurus excelsus* from Sheep Creek, Wyo-

References

Bakker, R. T. 1980. Dinosaur heresy — dinosaur renaissance. *In* R. D. K. Thomas, & E. C. Olson, (eds.). *A Cold Look at the Warm-Blooded Dinosaurs.* (Boulder, CO: Westview Press).

Behrensmeyer, A., Western, D., and Dechant Boaz. 1979. New perspectives in vertebrate paleontology from a recent bone assemblage. *Paleobiology* 5: 12–21.

Britt, B. 1988. A possible ''hatchling'' *Camarasaurus* from the Upper Jurassic Morrison Formation (Dry Mesa Quarry, Colorado). *Journal of Vertebrate Paleontology, Abstracts,* 8: 9a.

Canio, M. 1988. Dinosaurs. New answers for old questions. *The Facts on File Scientific Yearbook.* (New York: Facts on File Publications).

Dodson, P., Behrensmeyer, A., Bakker, R., & McIntosh, J. 1980. Taphonomy and paleoecology of the dinosaur beds of the Jurassic Morrison Formation. *Paleobiology* 6: 208–32.

Gilmore, C. 1936. Osteology of *Apatosaurus* with special reference to specimens in the Carnegie Museum. *Memoirs of the Carnegie Museum,* 11: 175–300.

Hatcher, J. 1903. Discovery of remains of *Astrodon (Pleurocoelus)* in the Atlantosaurus Beds of Wyoming. *Annual Report Carnegie Museum* 2: 9–14.

Hunt, A., & Lucas, S. 1987. J. W. Stovall and the Mesozoic of the Cimmarron Valley, Oklahoma and New Mexico. *New Mexico Geological Society Guidebook, 38th Field Conference, Northeastern New Mexico,* pp. 139–51.

Jain, S. 1989. Recent dinosaur discoveries in India, including eggshells, nests and coprolites. *In* D. Gillette & M. Lockley (eds.), *Dinosaur Tracks and Traces.* (New York: Cambridge University Press).

Lull, R. 1911. Lower Cretaceous systematic paleontology, Vertebrata. *Maryland Geological Survey,* 1911: 181–211.

Marsh, O. 1883. Principle characters of American Jurassic dinosaurs. Part VI. Restoration of *Brontosaurus. American Journal of Science* 26: 81–5.

1888. Notice of a new genus of Sauropoda and other new dinosaurs from the Potomac Formation. *American Journal of Science* 135: 89–94.

1896. The dinosaurs of North America. *U.S. Geological Survey, Sixteenth Annual Report*, 133–415.

McIntosh, J. 1981. Annotated catalogue of the dinosaurs (Reptilia. Archosauria) in the collection of the Carnegie Museum of Natural History. *Carnegie Museum of Natural History Bulletin* 18: 1–67.

Mohabey, D. 1987. Juvenile sauropod dinosaur from Upper Cretaceous Lameta Formation of Panchmahals District, India. *Journal Geological Society of India* 30: 210–16.

Morell, V. 1987. Announcing the birth of a heresy. *Discover* 8: 26–51.

Ostrom, J., & McIntosh, J. 1966. *Marsh's Dinosaurs.* (New Haven, Yale University Press).

Peterson, O., & Gilmore, C. 1902. *Elosaurus parvus*; a new genus and species of the Sauropoda. *Annual Report Carnegie Museum* 1: 490–9.

Stovall, J. 1938. The Morrison Formation and its dinosaurs. *Journal of Geology* 46: 583–600.

1943. Stratigraphy of the Cimmarron Valley (Mesozoic rocks). *Oklahoma Geological Survey* 64: 43–132.

18 Thermal environments of dinosaur nestlings: Implications for endothermy and insulation

GREGORY S. PAUL

Abstract

Very small, altricial hadrosaur nestlings probably lived in open nests exposed to the weather. To survive and grow rapidly they needed to be insulated endothermic homeotherms. Altricial young of small ornithopods and small theropods needed similar adaptations unless they were brooded by their parents. If the parents did brood the hatchlings, they probably needed an elevated metabolism and soft insulation to insulate and warm their young.

Introduction

The dominant and most abundant dinosaurs in many Upper Cretaceous faunas are the large bodied duckbilled hadrosaurs. They grew from diminutive hatchlings with a body mass 5,000 to 10,000 times less than the adults (Fig. 18.1). These little creatures needed to thermoregulate, and how they managed this may have as much to tell us about dinosaur physiology as do studies of the adults. Until recently a lack of juvenile specimens, and the preferential interest shown for the adults have combined to keep the physiology of dinosaur babies from receiving much attention.

Thulborn (1973) and Reid (1978) have expressed the opinion that small size, naked skin, and endothermy would be a lethal combination for nonbrooded hatchling dinosaurs. Russell (1980) suggested that dinosaur hatchlings preferred densely vegetated, wet areas with minimal temperature fluctuations. In contrast, Hotton (1980) suggested that dinosaur hatchlings were large enough to be incipient inertial homeotherms. While McGowan (1979, 1984) believed that juvenile dinosaurs should have been well protected in their nests and good thermoregulators. Horner and Makela (1979) and Horner and Gorman (1988) suggested that juvenile ornithopods were endothermic, and Lambert (1991) argued that the altricial nature of some ornithopod nestlings is suggestive of endothermy.

Studies on dinosaur nesting habits and growth rates provide the basis for examining the thermal environment in hadrosaur, hypsilophodont and small theropod nests (Horner & Makela, 1979; Horner, 1982, 1984, 1987, 1988; Horner & Gorman, 1988; Horner & Weishampel, 1988; Coombs, 1989; Kurzanov & Mikhailov, 1989; Russell, 1989; Winkler & Murry, 1989; Currie, 1990; Lambert, 1991; Horner & Currie, Chapter 21). Conditions within the open nests of these dinosaurs were probably often harsh. If this is correct, then the hatchlings that inhabited these nests may have needed "sophisticated" nonreptilian physiologies in order to thrive. Alternately, if small ornithopod and theropod adults were endothermic, then they could have used their warm bodies to brood their young.

Thermal strategies of modern juvenile vertebrates

Reptiles

Reptilian hatchlings are precocial ectothermic heterotherms. They are not brooded by their parents, and immediately disperse from the nest (Bellairs, 1970). Thermoregulation occurs by behavioral modification and the use of refuges. Reptiles have a narrow preferred body temperature range (30°–39°C in most species, Bellairs, 1970) and a broader tolerance range (up to 45°–47°C in desert forms; Bellairs, 1970). When the upper or lower limits of the tolerance range is reached, reptiles seek refuge in shade, burrow, or water, for death may result from exposure. In addition, at extreme body temperatures growth stops.

Birds

Bird chicks practice a wide variety of metabolic and thermoregulation strategies to achieve ideal conditions for survival and growth. There are, however, some generalized patterns. Many altricial nestlings begin life as poikilotherms, and metabolism slows with falling ambient temperatures (Dawson & Hudson, 1970; O'Conner, 1975; Whittow, 1976a; Steen et al., 1989). They are naked upon hatching and have lower resting metabolic rates than adult birds of the same mass. Consequently, the very survival of the altricial chicks depends upon the shelter provided by the nest and/or brooding by their parents. Brooding adults provide feathery insulation and warmth at night, and shade and insulation against excessive heat during the day.

Most altricial chicks experience difficulty thermoregulating outside a narrow body temperature range of 30°–40°C. Growth in this temperature range is rapid, but it slows or stops when this range is exceeded (Whittow, 1976b). Some species become torpid and stop growing when their parents leave the nest on long foraging trips (Steen et al., 1989). Air temperatures that drop much below 10°C can be lethal to an unprotected chick (Steen et al., 1989). At high temperatures, a few taxa can use evaporative cooling (Whittow, 1976a), but they remain vulnerable to direct sunlight as long as they are naked. As the chick grows, metabolic rates rise to adult levels, and downy insulation appears. Thermoregulation improves (Whittow, 1976b) and growth slows down, but it is still fast.

Precocial chicks are tachymetabolic while still in the egg, and hatchlings are good thermoregulators (Whittow 1976a, 1976b; Dawson & Hudson 1970; Dawson, Hudson, & Hill, 1972; Freeman, 1971). The body has an insulation of down feathers to maintain a high, constant temperature. Furthermore, the metabolism rises in response to declining external temperatures, and temperatures below freezing can be tolerated for a short time. Heat production in the cold includes both shivering and nonshivering thermogenesis. In addition, elevated temperatures up to 47°C can be tolerated for brief periods (Whittow, 1976a, 1976b; Dawson & Hudson, 1970: Dawson et. al., 1972). Growth, however, slows or stops when body temperatures exceed 40°C or drop below 30°–35°C (Whittow, 1976b). Nidifugous chicks

may leave the nest immediately or soon after hatching. They seek optimal thermal microenvironments in the landscape for refuge.

Some birds form nesting colonies in open areas having little or no vegetation to screen the nests from the sun or rain (Dawson & Hudson 1970, 1972; Whittow, 1976a; Perrins & Middleton, 1985). Thermal conditions can become severe due to sunlight, darkness, or storms. Most chicks in such habitats are brooded. One exception is the nidicolous chicks of some gulls. They remain in or near the nests and, because they are not brooded, seek shelter in the surrounding terrain when the nests become too hot (Dawson & Hudson 1970; Dawson et al., 1972).

Among all bird chicks, there is a general, positive correlation between metabolic rate and growth rate (Whittow, 1976b). On the other hand, growth of poikilothermic altricial nestlings is very energy efficient because most of the energy is devoted to growth; metabolic heat is obtained from the brooding parents. As a result, altricial chicks often grow more rapidly than precocial chicks (Case, 1978). Both altricial and precocial nestlings benefit from a steady and abundant supply of food from their parents. This allows them to convert energy into growth rather than use it for foraging.

Mammals

Mammalian young also exhibit diverse thermal strategies. In addition, when they are about one-quarter grown, their metabolic rates are about one third higher than those of adults of equal size (Brody, 1974). The young of most ungulates are precocial, have fairly large bodies, are endothermic, and are insulated with fur. These ungulates are thus able to cope with a wide range of thermal extremes. *Sus* piglets, with a mass of 0.4 kg, have an insulation of light

Figure 18.1. Comparison of a 2.5 metric ton adult *Maiasaura peeblesorum*, 0.3-kg hatchling, and 20-kg juvenile. Adult stands next to a schematic cross section of an open-mound nest containing two hatchlings, while a juvenile only a few weeks old leaves the nest. Nest restored after data in Horner and Makela (1979) and Coombs (1989). Scale bar = 1 m.

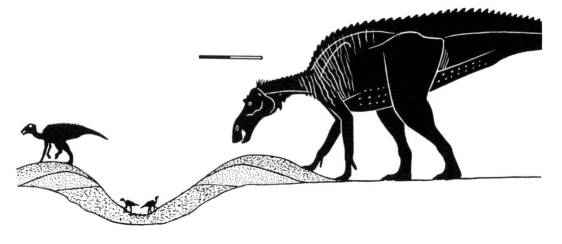

body fat and fur. They are able to cope with still, dry air temperatures as low as 5°C (Hull, 1973). Other mammals have altricial young. These are often naked and with underdeveloped endothermy. Consequently, the young receive close attention and warmth from their parents.

Growth
Modern vertebrates
Juvenile reptiles in the wild grow only about one tenth to one thirtieth as rapidly as mammals and birds (Case, 1978). Even in captivity where food is more abundant, they grow much more slowly than poikilothermic altricial bird chicks (Case, 1978; Coulson, Coulson, & Hernandez, 1973). These facts suggest that fast-growing animals must either have high metabolic rates and high, stable body temperatures or else be brooded by warm-bodied parents.

Speculations by Dunham et al. (1989), Houck, Gauthier, and Strauss (1989), and Reid (1990) that unbrooded juvenile vertebrates with reptilian or "intermediate" physiologies could grow as rapidly as birds and mammals remain unproven. For unbrooded juveniles to grow rapidly in a nest they need adaptations to cope with thermal extremes. To cope with high ambient temperatures, the ability to tolerate an elevated body temperature would be advantageous. An insulation of pelage would also protect the skin from absorbing heat. Fat is not a suitable insulation because it does not protect the skin.

Mechanisms for active cooling, such as sweating or panting, are also needed to keep the body temperatures within tolerable limits. On the other extreme, if the ambient temperature is too low, the body temperature can be kept elevated by having a high resting metabolic rate supplemented by nonshivering thermogenesis and shivering. Again, insulation is important because it allows as much energy as possible to be dedicated to growth rather than to keeping warm. Fat insulates well in cold, dry conditions. Fur or feathers insulate in cold, wet conditions because they shed rain and protect the skin from heat loss by evaporation.

Dinosaur nestlings
Maiasaura nestlings grew rapidly and were able to increase their mass forty- to sixtyfold in 8 weeks or less (Horner & Makela, 1979; Horner & Gorman, 1988; Russell, 1989). Such rapid growth is similar to that observed in altricial bird chicks (see Paul, Chapter 15, Fig. 15.5). Rapid growth is also supported by histological studies of hadrosaurs, hypsilophodonts and small theropod bones (see references in Paul, Chapter 15). The capacity for rapid growth was probably made possible by parental care that allowed the nestling to devote its energy to growth.

Structure of dinosaur breeding colonies and nests
Hadrosaurs and hypsilophodonts apparently nested in colonies consisting of a half dozen to perhaps hundreds of nests (Horner & Makela, 1979; Horner, 1982, 1984, 1987; Horner & Gorman 1988; Horner & Weishampel, 1988; Coombs, 1989; Russell, 1989; Winkler & Murry, 1989; Currie, 1990; Horner & Currie, Chapter 21). These nesting sites appear to have been used repeatedly for many years. The colonies may have resembled those of ground nesting birds being devoid of shade trees (Horner & Makela, 1979; Horner, 1982, 1984, 1987, 1988; Horner & Gorman, 1988; Winkler & Murry, 1989; Currie, 1990; Horner, Chapter 8). Such an interpretation is supported by the apparent absence of large plant roots in the colony sites. The presence of plant material within nest structures suggests that the lack of fossilized plant material between the nests is not an artifact of preservation. The close spacing of the nests, about one adult body length apart, suggests that the nests were placed in open areas.

Without the protection of plants, the nests were exposed to the weather. Some protection, however, might have been provided by the nests themselves. The hadrosaur nest appears to have been a conical depression with a raised rim about 3.0 m across and 1.0 m deep; the sides were about ~30° (Figs. 18.1, 18.2; Horner & Makela, 1979; Horner, 1984; Horner & Gorman, 1988; Coombs, 1989).

Hypsilophodont nests appear to have been smaller and shallower, and therefore more exposed to the weather than those of hadrosaurs (Horner, 1984, 1987; Horner & Gorman, 1988; Coombs, 1989; Winkler & Murry, 1989). A possible theropod nest described by Kurzanov and Mikhailov (1989) appears to have been shallow.

Environmental conditions in open dinosaur nests
The raised rims of the open dinosaur nests partially screened the interior from breezes and low-angle solar radiation. Otherwise, the nests probably provided little protection for their inhabitants. The bottom of the nest was further subjected to infrared radiation reflected and radiating from the walls, creating an ovenlike effect. At night heat loss would be high. Precipitation would have dropped directly into the nest at any time. The exposed dinosaur nests parallels the situation observed in many bird colonies, contradicting McGowan's (1979) assertion that the dinosaur nest provided important thermal protection.

In the nesting grounds, the daytime near surface air temperatures may have exceeded 45°C during the dry season of the temperate and tropical summers (Dunham et al., 1989). Unshaded ground can be up to 30°C hotter than the air (Finch, 1972), so the temperature within the nest could have been as high as or higher than 70°C.

Figure 18.2. Restoration of a breeding season in a *Hypacrosaurus* nesting colony (based on Devil's Coulee egg site, Alberta, in Currie, 1990). The central depressions of the nests are exposed to the elements. Nestlings shown restored with a downlike body insulation to protect against thermal extremes (see text).

Direct sunlight could have also heated the surface of an animal to similarly high levels as observed by Schmidt-Nielsen et al. (1957). At night, when exposed to the open sky, nest temperatures could have dropped dramatically. Temperatures may have dipped as low as 10–15°C when the night was clear and dry and the overhead weather system was cool. Any breezes swirling into the pit would have added a wind-chill factor.

The worst cooling conditions would have occurred during rainstorms, when wind chill and evaporative cooling would have caused a severe heat drain in the nestlings and driven effective skin temperatures to 0°C or less (data from National Weather Service wind chill tables). Nocturnal temperatures during the Cretaceous would have been lowest in high-latitude nests during the winter night, when occasional frosts and storms were possible (Paul, 1988; Spicer & Parrish, 1990). The actual daily fluctuations would have been dependent on cloud cover and humidity. In general, the lesser the cloud cover and the lower the humidity, the higher the daily fluctuations would have been. Dunham et al. (1989) calculated that daily temperature fluctuations for the Campanian in northern Montana exceeded 20°C.

The geographic locale of the nesting sites also had an effect on the thermal extremes inflicted on the nestlings. The *Maiasaura* and *Orodromeus* nesting colonies were located in a dry upland habitat (Horner, 1984, 1987; Wolfe et al., 1987; Horner & Gorman, 1988) that probably experienced wide fluctuations in daily temperatures (Dunham et al., 1989). The *Hypacrosaurus* nesting sites (Horner & Currie, Chapter 21) in Montana and Alberta were closer to the coastline and thus probably under a more moderate maritime influence (Wolfe & Upchurch, 1987). The Proctor Lake nests in central Texas reported by Winkler and Murry (1989) were located in a semiarid environment. Finally, the small theropod nest described by Kurzanov and Mikhailov (1989) was located in the interior of Asia in an area subject to temperature extremes (Barron & Washington, 1982).

The Alberta and Montana nesting sites were probably at higher latitudes during the Upper Cretaceous (Paul, 1988; Russell, 1989). At noon the winter sun rose as little as 6°–12° above the horizon. Winter temperatures were cool, with a mean of 15°C and lows occasionally below 10°C (Paul, 1988 and references therein; Spicer & Parrish, 1990). The coolness was accentuated by the long nights, and chilling storms may have been frequent. The low angle of the sun would probably have suppressed floral growth, so it would have been difficult to feed the nestlings. The low light would also probably have been insufficient to heat the interior of the nests directly. Given these conditions, it is unlikely that ornithopod eggs were hatched at this time of year.

At the opposite extreme, the summer day was long and the sun high in the sky (up to 52°–58° above the horizon). With the nests exposed to the sun for long periods and under such intensity, the heat loads in the nests would have been high. This effect would have been abetted by the subtropical conditions prevalent at the time and with mean summer temperatures of approximating 30°C (Russell, 1989; Dunham et al., 1989). Floral growth was reduced due to the dry conditions. The combination of heat and reduced plant production suggests that the ornithopods did not nest at this time of year.

Russell (1989) suggested that hadrosaurs tried to time incubation of eggs to coincide with the end of the dry season and hatching to coincide with the floral flush at the beginning of the wet season. Horner (1987) suggested that the rainy season in Montana occurred during the spring. He thought that *Orodromeus* nested at this time, but if true, then their nests were often exposed to rain and storms. The noonday sun at this time would have been 29°–35° high at the solstice and higher afterward. On sunny days the interior of the nests could have received high levels of solar energy. Thermal fluctuations could have been very high at this time as weather conditions changed in 24 hours from a chilly and rainy spring night to a hot sunny day. The Proctor Lake and Mongolian nesting sites were farther south than the Alberta and Montana colonies. The nests would have received more warmth in the winter but might have became unbearable between late spring and early autumn.

Based on the above discussions of the various nesting locales and their thermal extremes, certain climatic conditions could be ameliorated by the choice of nesting season. Spring probably offered the best weather conditions for raising nestlings. However, there was no way for nesting dinosaurs to ensure that their young would not experience one or more episodes of severe weather at this time.

Thermal adaptations and strategies of dinosaur nestlings

Hadrosaurs

Hadrosaurs, such as *Maiasaura* and *Hypacrosaurus*, laid between 18–24 eggs in a nests (Horner & Makela, 1979; Horner, 1982, 1984; Horner & Gorman, 1988; Coombs 1989; Russell, 1989; Currie, 1990; Horner & Currie, Chapter 21). Upon hatching, *Maiasaura* chicks weighed only 0.3–0.5 kg and *Hypacrosaurus* 3–4 kg judging from the volumes of the eggs and hatchling specimens (Fig. 18.1). Hadrosaur hatchlings, like those of most birds, were nidicolous, in that they were bound to the nest for an extended period of time (Horner & Makela, 1979; Horner, 1982, 1984, 1987, 1988; Horner & Gorman, 1988; Horner & Weishampel, 1988; Coombs, 1989; Russell, 1989; Currie, 1990). Evidence for this includes the relatively poor ossification of the limb joints; the apparent trampling of eggshells into small bits and shards, indicating that the nest was inhabited after the eggs hatched; the discovery of juvenile remains, which would have had a mass of about 20 kg, in a few of the nests; and the extremely rapid rates of growth discussed above.

The underdeveloped joints of nestlings indicate that locomotor performance may have been too poor for the altricial chicks to leave the nest until they weighed over 20 kg. Remains of chicks that size have been found near to but outside the nests, suggesting that a degree of freedom from the nest began at that growth stage. This weight is believed to have been reached at about 8 weeks or less (Horner & Makela, 1979; Horner, 1984; Horner & Gorman, 1988; Horner & Weishampel, 1988; Coombs, 1989; Russell, 1989; Lambert, 1991). The depletion of the flora surrounding nesting colonies by foraging adults, and the need for the adults to cover increasingly larger areas in search of food, makes it doubtful that a longer nesting season could have been possible (Horner & Gorman, 1988; Russell, 1989).

Hadrosaur parents probably carried food to their young within their anterior digestive tracts, and regurgitated it into the nest (Horner, 1984; Horner & Gorman, 1988; Coombs, 1989; Russell, 1989). This would have been more efficient than carrying leafy browse in the mouth. In addition, at present there is no fossil evidence of twigs that may have supported leafy greens within the sediments filling the central pits (that this absence is not due to adverse soil chemistry is implied by the presence of woody material within a nest, Horner & Makela, 1979).

Being restricted to the nest for up to 8 weeks would have exposed the nestlings to potential thermal stress. The options for avoiding this stress appear to have been very limited. If altricial, then they could not leave the nests when conditions became too difficult. The locomotor performance of the nestlings would have been inadequate for them travel to and from vegetation shelter. Restricted to the nest, the nestlings could have burrowed into the soil of the nest; however, the poor ossification of their joints indicates this was probably not possible. Another alternative is for the hadrosaur parents to brood the nestlings. But at 2–3 tons, the adults were far too big and heavy to brood their young safely. In addition, the adults lacked the soft insulation needed for such a task.

Russell (1989) suggested that adult hadrosaurs screened nestlings from the elements, but how this was accomplished is not stated. I am not aware of any modern examples in which adults deliberately screen their young; instead the young seek the shade of their parents or other large objects (Langman, 1982). Nest-bound hadrosaur chicks were probably unable to seek shade in this manner. Instead, they were adapted to tolerate harsh conditions, and this reveals something of their physiology.

Lacking insulation and a cooling system, nestling hadrosaurs operating with a reptilian or intermediate physiology, would have been in danger of succumbing to the direct rays of the sun. Furthermore, evaporative cooling by wind and rain would have been threatening to bradymetabolic chicks by suddenly driving body temperatures well below ambient air levels. Chilled, torpid hadrosaur chicks would also have been in danger of drowning if the nest flooded. Inertial homeothermy was probably not possible in newly hatched *Maiasaura* or *Hypacrosaurus* because their individual mass would have been considerably less than the 10 kg threshold value suggested by Hotton (1980). Not until body masses of 500 to 1,000 kg were reached after a few weeks was the inertial homeothermy adequate to protect against very high air temperatures and direct sunlight (Spotila et al., 1973; Dunham et al., 1989).

On the other hand, an endothermic physiology and outer surface insulation would have given the hadrosaur nestlings a much greater chance of surviving temperature extremes. Heat flow analyses (Dunham et al., 1989) of an uninsulated hadrosaur chick with a low metabolic rate showed that the body temperature could vary from 20°–45°C on a summer day. Under winter conditions, body temperature fluctuations could be 2°–20°C. The body temperature of an insulated endothermic nestling would have been much more stable, between 38°–45°C under the same conditions.

At present, there is no direct evidence for the integument of small juveniles. Nevertheless, as argued here some form of superficial insulation is needed for protection against solar radiation (including ultraviolet) and high external heat loads. This insulation may have been in the form of down feathers that were later shed (see comments in Paul, 1988). Other possible juvenile integuments are elongated scales with soft bristled edges or a hairlike pelage. A light colored downy plumage is typical of nonbrooded bird chicks in hot nesting colonies, and a similar pelage is advocated for hadrosaur chicks. Preening with glandular oils may have enhanced water shedding by the insulation.

Hadrosaur chicks may have cooled themselves by resonant frequency panting and/or gular fluttering. Water might have been obtained from the regurgitation by the parents of moist, predigested plant material. As long as this water was available, the thermoregulatory adaptations detailed above should have allowed the unshaded hadrosaur nestlings to survive daily peak temperatures as high as 70°C on the surface of their insulatory coats. Periods in which wind chill and rain drove the temperatures close to or below freezing could have also been tolerated as long as the chicks were well fed.

Hypsilophodonts

At Proctor Lake, Texas, 40-kg adult hypsilophodonts appear to have laid one or two dozen eggs per nest. The mass of the hatchlings was about 0.07–0.15 kg (Winkler & Murry, 1989). The presence of ~1.5 kg chicks in some of the nests suggest that they were either altricial, or semiprecocial in that they periodically returned to the nest. At what point they permanently left the nest is not known. Some form of parental care is suggested by the altricial nature of the juveniles, their presence in the nests, and the presence of adult skeletons near the nests (Winkler & Murry, 1989).

In contrast, the young of the Montana hypsilophodont *Orodromeus* appear to have had led a more mobile life-style (Horner, 1982, 1984, 1987; Horner & Gorman, 1988; Horner & Weishampel, 1988; Coombs, 1989). Intact eggshell bases, well-ossified limb joints in the hatchlings, and an absence of posthatchling juveniles in the nests are indicative of precocial chicks that left the nest soon after hatching – a pattern similar to most reptiles and various birds (Perrins & Middleton, 1985). The presence of half grown juveniles within the nest horizon sediments suggests that *Orodromeus* chicks remained in the area for a few weeks. A lack of adult skeletons near the nests implies, but does not demonstrate, that the precocial young were independent immediately upon hatching (Horner & Gorman, 1988). Lambert (1991), however, argues that the parents fed their young in the colony area.

Because *Orodromeus* chicks were apparently able to leave the nest to seek shade and other thermal microhabitat refuges, then their thermodynamic performance need not have been as well developed as that of hadrosaur chicks. However, the chicks could have grown rapidly only if they were tachymetabolic and insulated, or if they were brooded. The shallowness of the Proctor Lake hypsilophodont nests and the evidence for parental care are compatible with parental brooding, but do not prove that this occurred.

Brooding would have reduced or eliminated the need for well-developed thermoregulation in the nestlings. However, brooding makes sense only if the adults were themselves insulated endothermic homeotherms. A pelage would be necessary to keep the adults warm on cool nights and dry during storms, and to protect them from the sun. They may have also had a naked brood patch to facilitate warming the eggs and chicks. Insulation surrounding the patch would help screen the brood from the elements and trap heat when conditions were cool. The compact bodies of hypsilophodonts were probably not suited for brooding with their bodies in the manner of some pythons (Bellairs, 1970).

Small theropods

A nestlike structure, tentatively assigned to a small theropod by Kurzanov and Mikhailov (1989), contained the remains of many eggs. The size of the nestlings is unknown because they had evidently vacated the nest. However, they were probably similar in mass to hypsilophodont nestlings. Well-crushed eggshells and numerous small bones of possible prey suggest that the young were fed at the nest for a period of time. This care within the nest suggests that the chicks were either altricial or semiprecocial, and had little or no life outside the nest.

It is not known whether or not the theropod nest was shaded during the day or protected during storms. Regardless, if the chicks were either altricial or semiprecocial, they would have had the same need for well-developed endothermy and insulation as in hypsilophodonts. The close relationship of the theropods to birds reinforces the possibility that they had elevated metabolisms. It also suggests that, if the juveniles had insulation, it was true feathers.

Other dinosaurs

Nests of prosauropods, protoceratopsids, and iguanodonts have been reported, but not enough information is yet available to assess whether the babies in these cases were altricial or precocial. It is unlikely that sauropods gave live birth to large inertial homeothermic young (Paul, Chapter 15). The eggs of some dinosaurs assigned to sauropods and troodont theropods are laid in unstructured nests that suggest the young were fully precocial upon hatching and could retreat to thermal refuges (Horner 1987; Cousin et al., 1989).

Conclusion

All dinosaur nestlings, regardless of whether they were altricial, precocial, or semiprecocial, probably lived in exposed nests. The physiological model that best explains their rapid growth is tachymetabolic, endothermic homeothermy with an insulating pelage. Those hatchlings, such as hypsilophodonts and small theropods in which this physiology may not have been fully developed, may have been brooded by insulated, endothermic homeotherm adults.

This insulated endothermic model for the hatchlings is conservative in that it applies a standard physiological type to the subjects. There is no evidence that contradicts this conclusion, but it cannot be proven with the current data. The alternatives are that hadrosaur nestlings did not live in exposed nests, or that they managed to tolerate exposure with naked skin and a reptilian or some other physiology.

The first alternative would require a revision in our current understanding of the nest environment or the behavior of the nestlings. There might have been shade plants over the nests; the nests might have been filled with vegetation screens; or the juveniles might have been able to leave the nests when conditions were poor. Supportive evidence, however, is lacking for any of these alternatives. The second alternative would require an uninsulated nestling with a low metabolic rate to survive and grow rapidly without a refuge from weather extremes. This would require invoking a new type of physiology that would be difficult to test. That this alternative may be improbable is suggested by the fact that no reptile or altricial bird chicks live in open nests. We are left to conclude that the hypothesis of hadrosaur nestlings being insulated and endothermic best fits Ockham's razor and is therefore most probable (Jefferys & Berger, 1992).

As the young dinosaurs grew, they may have gradually shifted from an endothermic to a more reptilian physiology in response to having a more thermally stable body mass (Lambert, 1991). It may have taken 1 or 2 years for the larger species of dinosaurs to reach the mass needed for thermal stability, and about 4 to 6 years to reach the higher thermal stabilities of masses over 1 ton (Paul, Chapter 15). Rapid growth may have continued for several years or decades after this stage in some of the largest dinosaurs.

It is doubtful, however, that once adult size was reached the metabolic rates dropped significantly. A moderate decline is seen in large mammals and birds (about 30 percent relative to mass $^{0.75}$; Brody, 1974). More dramatic declines in metabolic rates do not occur because heart rates would drop, as would blood pressure. This would seriously degrade the aerobic capacity of the animal. Tachymetabolic endothermy, combined with insulation provided by either a large size or a pelage, would have also improved the ability of adult dinosaurs to stand by and guard their nesting young in shadeless colonies by day and in the cold of the night.

Acknowledgments

I thank J. Horner, P. Currie, and K. Carpenter for discussions and information on the subjects of dinosaur nesting and skin, and L. Hurxthal for the same on ostriches.

References

Barron, E. J., & Washington, W. M. 1982. Cretaceous climate: A comparison of atmospheric simulations with the geological record. *Palaeogeography, Palaeoclimatology, Palaeoecology* 40: 103–33.

Bellairs, A. 1970. *The Life of Reptiles.* (New York: Universe Books).

Brody, S. 1974. *Bioenergetics and Growth.* (New York: MacMillan Publishing Co.).

Case, T. J. 1978. On the evolution and adaptive significance of postnatal growth rates in the terrestrial vertebrates. *Quarterly Review of Biology* 53: 243–82.

Coombs, W. P. 1989. Modern analogs for dinosaur nesting and parental behavior. *In* J. Farlow, (ed.), *Paleobiology of the Dinosaurs. Geological Society of America Special Paper* 238: 21–53.

Coulson, T. D., Coulson, R. A., & Hernandez, to 1973. Some observations on the growth of captive alligators. *Zoologica* 58: 47–52.

Cousin, R., Breton, G., Fournier, R., & Watte, J.-P. 1989. Dinosaur egg-laying and nesting of an Upper Maastrichtian site at Rennes-Le-Chateau (Aude, France). *Historical Biology* 2: 157–67.

Currie, P. J. 1990. Dinosaurs. *Collier's Encyclopedia Year Book for 1990*: 62–71.

Dawson, W. R., & Hudson, J. W. 1970. Birds. *In* G. C. Whittow (ed), *Comparative Physiology of Thermoregulation*, Vol. I. (London: Academic Press), pp. 223–310.

Dawson, W. R., Hudson, J. W., & Hill, R. W. 1972. Temperature regulation in newly hatched laughing gulls (*Larus atricilla*). *Condor* 74: 177–84.

Dunham, A. E., Overall, K. L., Porter, W. P., & Forster, C. A. 1989. Implications of ecological energetics and biophysical and developmental constraints for life-history variation in dinosaurs. *In* J. Farlow (ed.), *Paleobiology of the Dinosaurs. Geological Society of America Special Paper* 238: 1–19.

Finch, V. A. 1972. Energy exchanges with the environment of two East African antelopes, the eland and the hartebeest. *Symposium of the Zoological Society of London* 31: 315–26.

Freeman, B. M. 1971. Body temperature and thermoregulation. *In* D. J. Bell, & B. M. Freeman, (eds.), *Physiology and Biochemistry of the Domestic Fowl*, Vol. II. (London: Academic Press), pp. 1115–51.

Horner, J. R. 1982. Evidence of colony nesting and "site fidelity" among ornithischian dinosaurs. *Nature* 297: 675–6.

    1984. The nesting behavior of dinosaurs. *Scientific American* 250(4): 130–7.

    1987. Ecologic and behavioral implications derived from a dinosaur nesting site. *In* S. J. Czerkas, & E. C. Olson, (eds.), *Dinosaurs Past and Present* Vol. II. (Seattle: University of Washington Press), pp. 50–63.

    1988. Cranial allometry of *Maiasaura peeblesorum* (Ornithischia; Hadrosauridae) and its behavioral significance. *Journal of Vertebrate Paleontology* 8 (Supplement to No. 3): 18A.

Horner, J. R., & Gorman, J. 1988. *Digging Dinosaurs.* (New York: Workman Publishing).

Horner, J. R., & Makela, R. 1979. Nest of juveniles provides evidence of family structure among dinosaurs. *Nature* 82: 296–8.

Horner, J. R., & Weishampel, D. B. 1988. A comparative embryological study of two ornithischian dinosaurs. *Nature* 332: 256–7.

Hotton, N. 1980. An alternative to dinosaur endothermy: the happy wanderers. *In* D. K. Thomas and E. C. Olson (eds.), *A Cold Look at the Warm-Blooded Dinosaurs. American Association for the Advancement of Science Selected Symposium* 28: 311–50.

Houck, M. A., Gauthier, J. A., & Strauss, R. A. 1989. Allometric scaling in the earliest fossil bird, *Archaeopteryx lithographica. Science* 247: 195–8.

Hull, D. 1973. Thermoregulation in young mammals. *In* G. C. Whittow, (ed.), *Comparative Physiology of Thermoregulation*, Vol. II. (London: Academic Press), pp. 167 -200.

Jefferys, W. H., & and Berger, J. O. 1992. Ockham's razor and Bayesian analysis. *American Scientist* 80: 64–72.

Kurzanov, S. M., & Mikhailov, K. E. 1989. Dinosaur eggshells from the Lower Cretaceous of Mongolia. *In* D. D. Gillette, & M. G. Lockley, (eds.). *Dinosaur Tracks and Traces.* (Cambridge: Cambridge University Press), pp. 109–13.

Lambert, W. D. 1991. Altriciality and its implications for dinosaur thermoenergetic physiology. *Neues Jahrbuch fur Geologie und Palaeontologie* 182: 73–84.

Langman, V. A. 1982. Giraffe youngsters need a little bit Of maternal love. *Smithsonian* 12: 95–103.

McGowan, C. 1979. Selective pressure for high body temperature: implications for dinosaurs. *Paleobiology* 5: 285–95.

1984. *The Successful Dragons*. (Toronto: Samuel Stevens and Co.).

O'Conner, R. J. 1975. Growth and metabolism in nestling passerines. *Symposia Zoological Society of London* 35: 277–306.

Paul, G. S. 1988. Physiological, migratorial, climatological, geophysical, survival, and evolutionary implications of Cretaceous polar dinosaurs. *Journal of Palaeontology* 62: 640–52.

Perrins, C. M., & Middleton, L. A. (1985). *The Encyclopedia of Birds*. (New York: Facts on File Publications).

Reid, R. E. H. 1978. Discrepancies in claims for endothermy in therapsids and dinosaurs. *Nature* 276: 757–8.

1990. Zonal growth rings in dinosaurs. *Modern Geology* 15: 19–48.

Russell, D. A. 1980. Reflections on the dinosaurian world. *In* L. L. Jacobs, (ed.), *Aspects of Vertebrate History*. (Flagstaff: Museum of Northern Arizona Press), pp. 257–68.

1989. *An Odyssey in Time: The Dinosaurs of North America*. (Toronto: University of Toronto Press).

Schmidt-Nielsen, K., Schmidt-Nielsen, B., Jarnum, S. A, & Houpt, T. R. 1957. Body temperature of the camel and its relation to water economy. *American Journal of Physiology* 188: 103–12.

Spicer, R. A., & Parrish, J. T. 1990. Latest Cretaceous woods of the central North Slope, Alaska. *Palaeontology* 3: 225–42.

Spotila, J. R., Lommen, G. S., Bakken, G. S., & Gates, D. M. 1973. A mathematical model for body temperatures of large reptiles: implications for dinosaur ecology. *American Naturalist* 107: 391–404.

Steen, J. B., Krog, J. O., Toyen, O., & Bretten, S. 1989. Poikilothermy and cold tolerance in young house martins (*Delichon urbica*). *Journal of Comparative Physiology* 159: 379–82.

Thulborn, R. A. 1973. Thermoregulation in dinosaurs. *Nature* 245: 51–2.

Whittow, G. C. 1976a. Regulation of body temperature. *In* P. D. Sturkie (ed.), *Avian Physiology*. (New York: Springer-Verlag), pp. 146–76.

1976b. Energy metabolism. *In* P. D. Sturkie (ed.), *Avian Physiology*. (New York: Springer-Verlag), pp. 175–84.

1978. Short ungulates. *Natural History* 87: 44–8.

Winkler, D. A., & Murry, P. A. 1989. Paleoecology and hypsilophodont behavior at the Proctor Lake dinosaur locality (Early Cretaceous), Texas. *In* J. Farlow (ed.), *Paleobiology of the Dinosaurs. Geological Society of America Special Paper* 238: 55–61.

Wolfe, J., & Upchurch, G. 1987. North American nonmarine climates and vegetation during the Late Cretaceous-Tertiary boundary. *Palaeogeography, Palaeoclimatology, Palaeoeciology* 61: 33–77.

19 Baby *Dryosaurus* from the Upper Jurassic Morrison Formation of Dinosaur National Monument

KENNETH CARPENTER

Abstract

A complete skull and partial skeleton identified as the dryosaurid *Dryosaurus altus* is redescribed. The specimen was collected from the Upper Jurassic Morrison Formation at Dinosaur National Monument, Utah. The very small size coupled with large orbits, a short, deep preorbital region, open cranial sutures, and poorly ossified articular surfaces of limbs identify the specimen as that of a baby. The specimen is referred to *Dryosaurus* because of the characteristic anterior intercondylar groove and squared shape of the distal end of the femur. Ontogenetically, the greatest amount of change occurs in the posterior portion of the skull, which deepens, and the orbital region, which becomes reduced. The ontogenetic changes documented between the baby and adult *Dryosaurus* provide a framework in which to assess the taxonomy of other closely related ornithopods. This framework indicates that the criteria used to separate *Yandusaurus multidens* from *Y. hongheensis* are ontogenetic, and that *Y. hongheensis* is the senior synonym.

Introduction

In 1922, a skeleton of a juvenile ornithischian was discovered by a crew working under the direction of Earl Douglas from the Carnegie Museum of Natural History in the Morrison Formation at what is now Dinosaur National Monument. The specimen was apparently articulated and included the majority of the skeleton. When found, the specimen lay on its back with the ribs and forelimbs distended. Unfortunately, the rear portion of the skeleton was severely damaged or destroyed, and only portions of the limbs, the anterior part of the vertebral column, and skull remain (Fig. 19.1). The specimen was briefly described by Charles Gilmore in 1925 as *Laosaurus gracilis*, and more recently as *Dryosaurus altus* by Galton (1981, 1983).

Fragmentary bones from the Morrison Formation in Colorado referable to baby *Dryosaurus altus* were described by Galton and Jensen (1973). Additional material from the same locality is currently under study by Rodney Scheetz, who has also given a preliminary description of some of this material (Scheetz, 1991).

Owing to the importance of the specimen from Dinosaur National Monument – it has the only complete baby dryosaurid skull known from the Upper Jurassic of North America – the specimen is illustrated here in detail for the first time. In comparing the baby *Dryosaurus* skull with that of the adult from the same quarry (Dinosaur National Monument), it became clear that a new reconstruction of the adult skull was needed. The result is that the skull looks less paedomorphic than that shown by Galton (1983, Fig. 2), especially because the muzzle is longer and lower (Fig. 19.2).

Rather than repeat the observations of Galton (1981, 1983) and Sues and Norman (1990) for *Dryosaurus*, only those points in which the baby differs from the adult will be mentioned. Measurements are given in Table 19.1.

Systematic paleontology

Order: Ornithischia

Family Dryosauridae

Dryosaurus altus

Material

CM 11340 (Field Number 360, Accession Number 7195, Box Number 409) consisting of a complete skull with lower jaws in occlusion; nine(?) cervicals in articulation; 13(?) partial and complete dorsals plus a gap for three more in the pectoral region; ribs; both scapulae, coracoids, sternal plates, and humeri; distal ends of both femora; shaft of an ischium; and proximal end of the right tibia and fibula (Figs. 19.1, 19.3-19.5).

Locality

Far east side of the quarry face, Brushy Basin Member of the Morrison Formation, Dinosaur National Monument, Uintah County, Utah.

Skull

The skull exhibits several characteristics of a baby dinosaur: enlarged orbits, unfused frontals, a short snout, posteriorly sloped parietal region, and fibrous cortical bone (see Carpenter, Horner & Hirsch, Chapter 24). The skull is complete, although damaged by a fracture across the suspensorial region (Figs. 19.1A, and 19.4). The tip of the snout was apparently damaged when the specimen was collected. A portion of the bone surface was damaged during preparation, but where it remains, it shows the coarse fibrous texture of cortical bone typical of baby dinosaurs. This bone must have been very soft in life because sand grains of the encasing matrix have been pressed into the bone surface. As a result, separation of matrix from bone is not very clean and may explain why so much of the bone surface was lost during initial preparation. This loss of bone surface makes it difficult to discern many of the sutural patterns except where the bones have separated (e.g., frontals, Fig. 19.4D).

The posterior portion of the premaxilla in the baby skull extends up and back, separating the maxilla from the nasal (Fig. 19.4C). A small gap exists between the premaxilla and maxilla, just in front of the dorsal process of the maxilla (Fig. 19.4A). A similar but larger gap is present in the same position of the adult skull CM 3392, although Galton (1983) does not mention this (Fig. 19.1 and 19.5). There is reason to suspect that this gap is real and not due to postmortem crushing, because there are short buttresses on the anterior portions of the premaxilla and maxilla that meet, leaving the gap posteriorly (Figs. 19.2, 19.4A, 19.5). A similar gap occurs

in *Hypsilophodon*, although Galton (1974) thought this might be variable among individuals. However, the *Hypsilophodon* skull in which this feature is absent is obliquely crushed, the bones slightly disarticulated, and the premaxillae partially overlapping the maxillae. Thus, the possibility that this gap is closed due to postmortem causes cannot be ruled out.

Both nasals are preserved, although damaged at their anterior tip (Fig. 19.4C). The nasals have separated along their midline, with the right nasal offset so the medial margin is displaced ventrally into the narial cavity. Nevertheless, it appears that the nasals are depressed along their midline into a shallow trough (Fig. 19.5). The same condition appears to be present in the adult (although not reported by Galton, 1983) and in *Yandusaurus* (see He & Cai, 1984, Fig. 3).

The frontals have slightly separated along their medial edge (Fig. 19.4D). They are considerably longer than wide, bearing a resemblance to those of the hypsilophodontids *Hypsilophodon* (Galton, 1974), *Yandusaurus* (He & Cai, 1984) and *Zephyrosaurus* (Sues, 1980). It is possible, however, that this is a juvenile feature in *Dryosaurus*, because in the adult skull the frontals are almost as broad as long. This raises the possibility that the long, narrow frontals in *Hypsilophodon*, *Yandusaurus*, and possibly *Zephyrosaurus* (all known from immature individuals) are also a juvenile condition. Long, narrow frontals are apparently present in the adult *Thescelosaurus* (Weishampel, personal communication).

Both prefrontals are present, but the shape is best seen on the right side of the skull where it has shifted

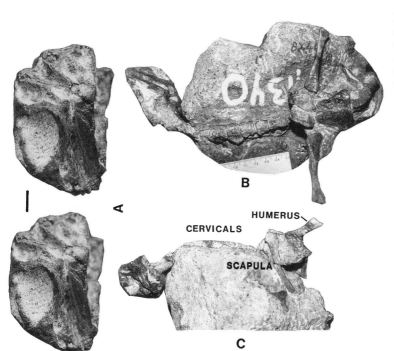

B

A

HUMERUS

CERVICALS

SCAPULA

C

Figure 19.1. Baby *Dryosaurus altus*, CM 11340. **A.** Left side of skull in stereoscopic view. **B.** Skeleton in ventral view. **C.** right side of skeleton. Scale in A = 1 cm, in B and C = 5 cm.

from contact with the nasal (Fig. 19.4C, D). This side shows that the prefrontal is triangular, tapering posteriorly over the orbits; anteriorly, it wedges between the palpebral and lachrymal.

A palpebral is present as a bar of bone in the upper part of the orbit (Fig. 19.4A, C). Unlike in the adult, the palpebral does not extend completely across the orbit (Fig. 19.5). Instead, it extends across about three-quarters the orbital length as in *D. lettowvorbecki*

Figure 19.2. Restoration of adult *Dryosaurus altus* skull. Based on CM 3392.

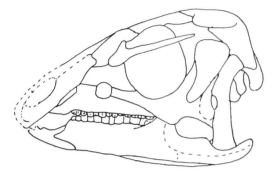

Table 19.1. Measurements of *Dryosaurus altus*, CM 11340

	Length (mm)	Width/height (mm)
Premaxilla–occipital condyle	72e	
Orbit (left)	25.25	22.3
Orbit (right)	26.2	24.6
Antorbital fenestra (left)	5.5	5.6
Antorbital fenestra (right)	8	3.8
Between orbits		22.51
Lateral temporal fenestra (left)		22
Lateral temporal fenestra (right)		18.8
Mandible (right)	69e	
Quadrate (right)		29
Scapula (right)	70	
Coracoid (left)	24.88	14.4
Coracoid (right)	21.75	–
Sternal (left)	30.8	18.4
Sternal (right)	28.4	16.5
Humerus (left)	64.2	
Humerus (right)	62.4	
Femur distal end (left)	23	26.1
Femur distal end (right)	21.7	21.6
Tibia proximal end (left)	28.4	18.1
Fibula proximal end (left)	15.9	6.9

Note: e, estimated.

from the Tendaguru (Galton 1983). In *Orodromeus* and *Yandusaurus*, it extends across about half of the orbit (Horner & Weishampel, 1988; He & Cai, 1984). The palpebral in the baby *Dryosaurus* is wedged between the posterior parts of the lachrymal and prefrontal. The bone tapers posteriorly into a slender bar, while in the adult, the bar does not taper significantly.

The lacrimal forms a triangular wedge in front of the orbit (Fig. 19.4A), while in the adult, the lacrimal is more rectangular (Fig. 19.2). The lacrimal is not overlapped by the dorsal process of the maxilla in the adult, contrary to Galton (1983). Even its overlap in *D. lettowvorbecki* is suspect because Janensch's (1955) interpretation is based on disarticulated cranial material, and alternative cranial reconstructions are possible.

Both maxillae are well preserved, although the right one has rotated medially due to crushing (Fig. 19.4A, C). The tooth rows are inset from the lateral margins, indicating that a cheek was present. The teeth are all damaged from the initial preparation of the specimen, so it is not possible to give a tooth count.

Both jugals are present, although damaged (Fig. 19.4A, C). Apparently the posteroventral margin is emarginated in both the baby (Fig. 19.5) and adult (Fig. 19.2) skulls, contrary to Janensch (1955) and Galton (1983). This condition may be seen in *Dryosaurus lettowvorbecki* (Janensch, 1955, Plate. 10, Fig. 8), although not shown in his skull reconstruction (Janensch 1955, fig. 1). A slight emargination of the jugal is present in *Orodromeus* near the posteroventral border of the orbit. A large rugosity or boss is not present on the jugal of *Dryosaurus* in marked contrast with *Zephyrosaurus* and *Orodromeus*.

A complete postorbital is present on the left side of the skull of the baby (Fig. 19.4A). It is missing on the right side and appears to have been broken either before burial or after discovery. The suture with the frontal is displaced, but indicates a long, firm union of these two bones.

The parietals are damaged, making it dificult to determine their overall shape. However, enough of the anterior portion of the parietals is present to show that they are not fused together anteriorly (Fig. 19.4D). The parietals slope posteriorly, resulting in a short cranium. The squamosals are also damaged, although the left one is in better condition than the one on the right (Fig. 19.4D). Their sutures with the surrounding bones are difficult to discern.

Both quadrates are present, although that on the left side of the skull is incomplete (Fig. 19.4A, C). The quadrate is bowed anteriorly (Fig. 19.4C) as it is in *Orodromeus* and other hypsilophodontids. The proximal end of the quadrates is expanded as in *Yandusaurus* (He & Cai, 1984), rather than tapering as in other hypsilophodontids (e.g., *Othnielia, Hypsilophodon, Zephyrosaurus, Parksosaurus,* and *Orodromeus,* see Galton, 1983). The notch in the quadrate near its suture with the quadratojugal as reported by Galton (1983) is not pronounced in the baby skull. This suggests that its development is ontogenetic. A fragment of the quadratojugal appears to be present on the right side of the skull, but its exact shape cannot be determined because the element is incomplete (Fig. 19.4A, C).

The braincase is visible on the left side (Fig. 19.4A) but the surface is damaged, making it difficult to identify most individual elements. Of the braincase elements, only the laterosphenoid is relatively complete except in the region of its suture with the presphenoid. The groove for the ramus ophthalmicus profundus is not as pronounced as in the larger, adult specimens of *Dryosaurus* described by Galton (1983), and its depth is probably an ontogenetic feature. Another feature of the braincase that is visible is a calcite-infilled posterior semicircular canal, but this also is damaged, and little can be said about the canal's shape.

At the rear of the skull, the exoccipitals are apparently not differentiated from the opisthotic as they are in *Zephyrosaurus* (Sues, 1980); the condition in *Or-odromeus* is unknown. The supraoccipital is wing-shaped, with a keel dividing the bone into right and left halves (Fig. 19.4B). The area on each side of the keel is not as deeply excavated as it is in the adult (visible only in *Dryosaurus lettowvorbecki*, Galton 1983, fig. Plate. 2, Fig. 4). This deepening is apparently due to the posterolateral expansion of the squamosals in the adult. The supraoccipital in the baby retains its anteriorly sloped position above the foramen magnum.

Both mandibles are present in the baby, although damaged (Fig. 19.4A, C). The predentary is crushed and parts are missing, making it difficult to compare with that described by Galton (1983). What is preserved suggests that it resembles the adult, but is proportionally smaller compared to the size of the skull; whether or not the ventral process was bilobed cannot be determined. The left dentary is more complete than the right one; both closely resemble those of the adult. The tooth rows are damaged from the initial preparation and nothing can be said about tooth numbers or morphology.

Only a portion of the right surangular is pre-

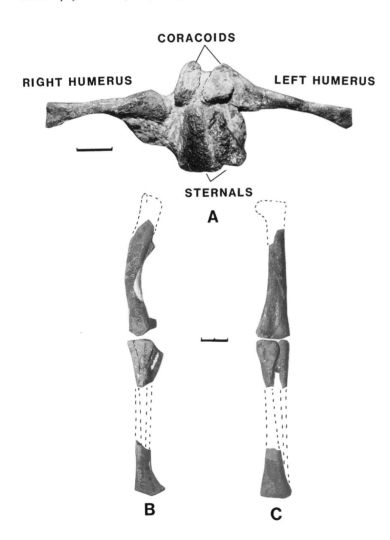

CORACOIDS

RIGHT HUMERUS   **LEFT HUMERUS**

STERNALS

A

B   **C**

Figure 19.3. Girdle and limb elements of *Dryosaurus altus*, CM 11340. **A.** Pectoral region in ventral view. **B.** Left femur, tibia and fibula in lateral view. **C.** Left femur, tibia, and fibula in anterior view. Scale in A = 1 cm, in B, C = 2 cm.

Figure 19.4. Skull of baby *Dryosaurus altus*, CM 11340, after addition preparation. **A**. Left lateral view. **B**. Posterior view. **C**. Right lateral view. **D**. Dorsal view. **E**. Ventral view. Scale = 1 cm. Abbreviations: AO, antorbital fenestra; BT, basitubera; C,

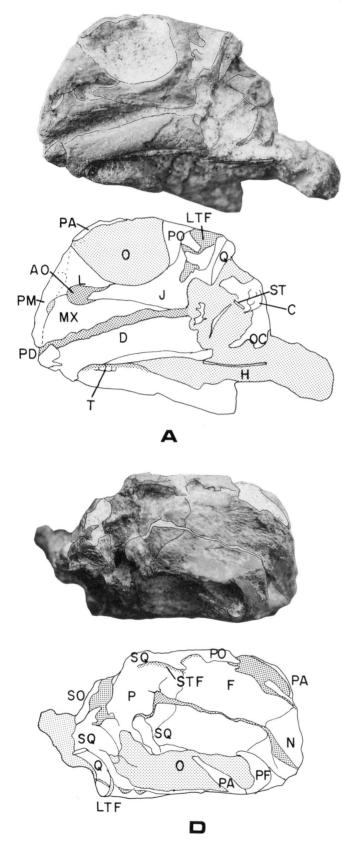

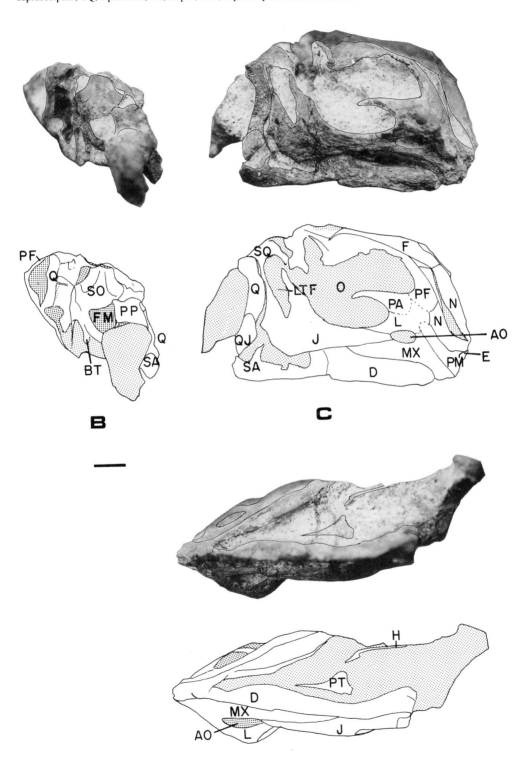

Figure 19.4. (*continued*)
semicircular canal; D, dentary; E, external nares; F, frontal; FM, foramen magnum; H, hyoid; J, jugal; LA, lacrimal; LTF, lateral temporal fenestra; MX, maxilla; N, nasal; OC, occipital condyle; P, parietal; PA, palpebral; PF, prefrontal; PM, premaxilla; PO, postorbital; PP, paroccipital process; PT,pterygoid; Q, quadrate; QJ, quadratojugal; SA, surangular; SO, supraoccipital; SQ, squamosal; ST, stapes; STF, supratemporal fenestra; T, teeth.

served, and even that is damaged; it appears to resemble that in the adult as described by Galton (1983).

A single, partial, rodlike hyoid is present (Fig. 19.4E). Its location suggests that it lay medial to the angular, which is missing.

Postcrania

The cervical series was probably complete at the time of discovery, but has since suffered some damage. Plaster of Paris tinted brown was used to attach the skull to the cervicals. Removal of this plaster reveals that the contact of the bones is not very good, and that the atlas–axis complex is severely damaged. The exact cervical count is unknown, but is probably nine or more, rather than eight as suggested by Galton (1981). This count would be in keeping with the hypsilophodontids, such

Orodromeus (personal observation), or the iguanodontids (Norman & Weishampel, 1990). The cervicals of the baby are still mostly embedded in matrix, and only the centra are visible. These are identical to those previously described by Shepard, Galton, and Jensen (1977). The centra are slightly compressed, with a rounded keel ventrally connecting the anterior and posterior articular faces. This is similar to *Hypsilophodon* (Galton, 1983) and *Zephyrosaurus* (Sues, 1980), but in marked contrast with *Orodromeus* (personal observation) and *Yandusaurus* (He & Cai, 1984) in which the centra are strongly compressed and the keel very sharp. However, unlike *Orodromeus*, the eighth cervical is not wedge shaped, where the centrum is wider at the bottom than at the top. As a result, the neck does not flex sharply upwards. This type of neck bend is also seen in

Figure 19.5. Restoration of baby *Dryosaurus altus* skull. Based on CM 11340.

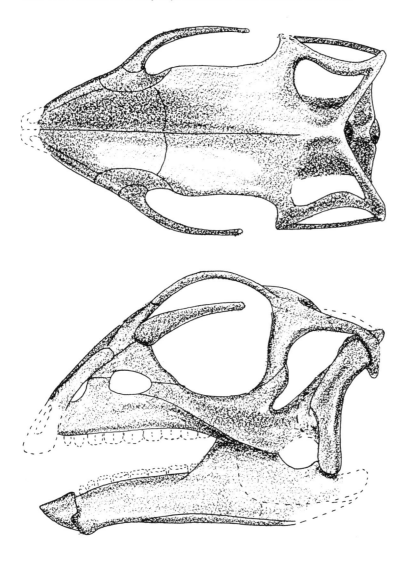

Othnielia (see Galton & Jensen, 1973, Plates 1 and 2) and *Yandusaurus* (He & Cai, 1984, Plate 1, Fig. 19.7).

The number of dorsals is unknown because the series is incomplete. As with the cervicals, only the centra are visible and these incompletely. The ventral part of the centra are gently rounded, except the anterior-most. The anterior-most (D1-2?) resembles the cervicals in having a keel, although this is not as pronounced. Nothing of the posterior dorsals, sacrals, or caudals is preserved. Several partial dorsal ribs are present, but provide little information about their shape or dimensions.

The pectoral girdle is represented by all or parts of both scapulae, coracoids, and sternal plates (Figs. 19.1B, C, 19.3A). The bone surface of these is damaged from initial preparation. Nevertheless, much of the morphology is still preserved. The right scapula is almost complete (Fig. 19.1C), although the posteroventral margin of the blade is damaged. Thus, it is not known if the distal end of the scapular blade is flared as it is in the adult and as in *Orodromeus* (personal observation), *Yandusaurus* (He & Cai, 1984), and *Othnielia* (Galton & Jensen, 1973). The biceps tubercle, prominently developed above the glenoid in the adult (Galton, 1981), is not developed in the baby. This is undoubtedly due to the immaturity of the individual. The left scapula is preserved as the proximal part and as a thin portion of the distal part of the blade embedded in matrix; the intervening portion is missing.

The left coracoid is more complete than the right (Fig. 19.3A). It is discoid in shape and is not fused to the scapula. The area near the coracoid foramen is damaged, but it appears that the foramen is not enclosed posterolaterally. A ridge present on the ventral margin of the coracoid in the adult (Galton, 1981, Fig. 19.6M) is not present in the baby. This condition suggests that the ridge on the coracoid of the large, adult *Yandusaurus hongheensis* and its absence on the coracoid of the small, juvenile *Yandusaurus multidens* may not be a valid species criteria, contrary to He and Cai (1984).

The sternal plates are preserved in situ (Fig. 19.3A) and show that they meet along the midline for most of their length, contrary to those illustrated by Galton (1983, Fig. 19.6M). This also calls into question the diverging sternal plates illustrated by Dodson and Madsen (1981; Fig. 19.3) for a generalized ornithopod (based on *Camptosaurus* and *Tenontosaurus*. It is more likely that the plates met along their midline.

Both humeri are also present and slightly damaged (Fig. 19.3A). They are anteroposteriorly compressed, with a very low deltopectoral crest. The crest is considerably less developed than in the holotype of *Orodromeus*, although both humeri are the same length. There is some restoration of both the humeral shafts, thus exact measurements are equivocal. Nevertheless, the shaft is apparently straight, rather than curved as in the adult and as in *Orodromeus* and *Yandusaurus*. The straight shaft in the baby calls into question the taxo-

Figure 19.6. Comparison of *Dryosaurus altus* skulls from the Morrison Formation, Dinosaur National Monument. **A.** Adult CM 3392. **B.** Baby CM 11340. Scale = 2 cm.

nomic significance of the greater medial curvature of the humerus in *Y. hongheensis* than in *Y. multidens* (He & Cai, 1984). Apparently this curvature of the humerus is ontogenetic.

The squared-off articular ends of the baby humeri are very poorly developed. This suggests that a large cartilaginous cap was present. The left(?) radius is represented by the distal end. It is too incomplete to provide any new information.

Both femora are represented by their distal ends, that of the left including the base of the fourth trochanter (Fig. 19.3B, C). In lateral profile, the femur is bowed anteriorly as in all small ornithopods (see Sues & Norman, 1990). The distal articular ends are not well developed. Instead they are nearly flat. This also suggests the presence of a large cartilaginous cap, similar to that on the humeri. As first noted by Galton (1981), a well-developed anterior intercondylar groove for the knee extensor (iliotibialis) muscle is present distally. This feature is characteristic of *Dryosaurus*. In most other small ornithopods, such as *Orodromeus* and *Yandusaurus*, this feature is poorly developed (Sues & Norman, 1990).

The proximal most part of the left tibia and fibula are present in articulation (Fig. 19.3B,C). In proximal view, the tibia is more triangular than in the adult illustrated by Galton (1981, Fig. 19.16E). In lateral profile, the cnemial crest is not well developed, and the medial condyle is not flat but rounded. The fibula resembles that of the adult (Galton, 1981; Fig. 19.17E), except the cartilaginous grooves are absent. Also, there is a greater overhang of the anterior edge of the proximal surface.

Summary of ontogenetic changes

A comparison of the baby *Dryosaurus* with the adult provides clues about the type and degree of ontogenetic change expected in other ornithopods, ultimately providing a test of taxonomic identifications as will be discussed below.

Figure 19.7. Overlays of restored *Dryosaurus altus* skulls showing relative direction and amount of ontogenetic changes. Adult, solid line; baby, dashed line.

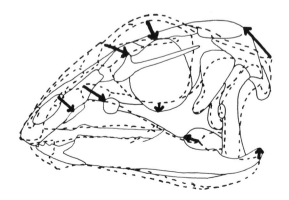

The greatest ontogenetic changes occur in the skull (compare Fig. 19.6A with Fig. 19.6B) and, as may be seen in Figure 19.7, most prominently affect the orbit and posterior portion of the skull. Less significant changes occur in the fenestra between the premaxilla and maxilla, and in the size and position of the antorbital fenestra.

The large orbit is typical of a baby vertebrate because of the large size of the eyeball. Ontogenetically, the orbit grows at a lesser rate than the rest of the skull. As a result, the orbit in *Dryosaurus* is about 16 percent smaller vertically and about 30 percent smaller anteroposteriorly relative to skull length in the adult than in the baby (see Fig. 19.7). The palpebral lengthens considerably so that it laps onto the postorbital.

The other area of major change occurs as a result of growth of the posterior portion of the skull. This includes a change in the orientation of the quadrate (less curved) and an increase in the size of the squamosal. There may also be an increase in the size of the parietals, although evidence is equivocal. The result is that the slope of the cranium changes from being posteriorly directed to being anteriorly directed. This modification is accompanied by little change in the proportion of the quadrate height relative to skull length (therefore little change in lateral temporal fenestra size), but a slight increase in the relative size of the quadrate notch. The squamosal increases in relative size and depth, possibly to offset the increased force generated by the larger M. adductor mandibulae externus.

The increase in muzzle length is accompanied by a posteroventral shift of the antorbital fenestra and fenestra between the premaxilla and dorsal process of the maxilla. The antorbital fenestra decreases in relative size, while the fenestra between the premaxilla and maxilla becomes proportionally elongated. The change in the size and position of the antorbital fenestra may reflect a decrease in the overall size of the jugal.

Postcranial changes in *Dryosaurus* are harder to document because of the fragmentary nature of the baby skeleton. Nevertheless, the straight humeral shaft and underdevelopment of the deltopectoral crest of the baby humerus are two major differences from the adult (see also Galton, 1981, Figs. 8A, C, G, J). One further difference is the lack of a ridge on the ventral side of the coracoid in the baby, a ridge that is present in the adult.

Discussion

Despite the fragmentary nature of the baby *Dryosaurus* specimen, it provides important information about ontogenetic changes in this dinosaur. This in turn gives insight into the changes that might occur in other, closely related ornithopods, specifically the hypsilophodontids, *Tenontosaurus*, and iguanodontids, and allows reassessment of the criteria used to separate closely related species.

For example, among the criteria cited by He and

Cai (1984) to separate *Yandusaurus multidens* from *Y. hongheensis* are the smaller size of *Y. multidens*, the scapula shorter than the humerus, the absence of a ridge on the ventral surface of the coracoid, and a humerus with a narrower proximal end and less curvature of the shaft. Based on the ontogenetic changes in *Dryosaurus*, all of these features are hallmarks of an immature individual. This interpretation is supported by the enlarged orbits and short muzzle of the holotype of *Y. multidens*, suggesting a very immature individual, possibly a baby.

Acknowledgments

Access to the *Dryosaurus* specimens used in this study and permission for additional preparation of the baby was made possible by David Berman, Carnegie Museum. Review comments by David Weishampel helped clarify the text, and his assistance is greatfully acknowledged.

References

Dodson, P., and Madsen, J., 1981. On the sternum of *Camptosaurus*. *Journal of Paleontology* 55: 109–12.

Galton, P. 1974. The ornithischian dinosaur *Hypsilophodon* from the Isle of Wight. *Bulletin of the British Museum (Natural History), Geology Series* 25: 1–152.

——— 1981. *Dryosaurus*, a hypsilophodontid dinosaur from the Upper Jurassic of North America and Africa: postcranial skeleton. *Palaeontologica Zeitschrift* 55: 21–312.

——— 1983. The cranial anatomy of *Dryosaurus*, a hypsilophodontid dinosaur from the Upper Jurassic of North America and East Africa, with a review of hypsilophodontids from the Upper Jurassic of North America. *Geologica et Palaeontologica* 17: 207–43.

Galton, P., and Jensen, J. 1973. Small bones of the hypsilo-

phodontid dinosaur *Dryosaurus altus* from the Upper Jurassic of Colorado. *Great Basin Naturalist* 33: 129–32.

Gilmore, C. 1925. Osteology of ornithopodous dinosaurs from the Dinosaur National Monument, Utah. *Memoirs of the Carnegie Museum* 10: 385–411.

He X. & Cai K. 1984. *The Middle Jurassic dinosaurian fauna from Dashampu, Zigong, Sichuan, Vol. 1. The Ornithopod Dinosaurs.* (Sichuan China: Sichuan Scientific and Technological Publishing House).

Horner, J., & Weishampel, D. 1988. A comparative embryological study of two ornithischian dinosaurs. *Nature* 332: 256–7.

Janensch, W. 1955. Der ornithopode *Dysalotosaurus lettowvorbecki*. *Palaeontographica, Supplement* 7: 105–76.

Norman, D., and Weishampel, D. 1990. Iguanodontidae and related ornithopods. *In* D. Weishampel, P. Dodson, & H. Osmólska (eds.), *The Dinosauria*. (Berkeley: University of California Press).

Scheetz, R. 1991. Progress report of juvenile and embryonic *Dryosaurus* remains from the Upper Jurassic Morrison Formation of Colorado. *In* W. Averett (ed), *Guidebook for Dinosaur Quarries and Tracksites Tour: Western Colorado and Eastern Utah*. (Grand Junction: Grand Junction Geological Survey).

Shepard, J., Galton, P., and Jensen, J. 1977. Additional specimens of the hypsilophodontid dinosaur *Dryosaurus altus* from the Upper Jurassic of Western North America. *Brigham Young University Geology Studies* 2: 11–15.

Sues, H.-D. 1980. Anatomy and relationships of a new hypsilophodontid dinosaur from the Lower Cretaceous of North America. *Palaeontographica, Abt. A* 169: 51–72.

Sues, H.-D., & Norman, D. 1990. Hypsilophodontidae, *Tenontosaurus*, Dryosauridae. *In* D. Weishampel, P. Dodson, & H. Osmólska, (eds.), *The Dinosauria*. (Berkeley: University of California Press).

20 An embryo of *Camptosaurus* from the Morrison Formation (Jurassic, Middle Tithonian) in Dinosaur National Monument, Utah

DANIEL CHURE, CHRISTINE TURNER AND FRED PETERSON

Abstract

A small partial skeleton of an ornithopod dinosaur was collected from bentonitic mudstone in the upper part (Tithonian) of the Brushy Basin Member of the Morrison Formation. The specimen consists of two posterior cervical and six anterior dorsal vertebrae, five anterior caudal vertebrae, scapula, coracoid, humerus, tibia, fibula, and dorsal ribs. On the basis of the distinctively shaped coracoid, the specimen is referred to *Camptosaurus*, a common Morrison ornithopod which reached lengths of 8 m. Bone texture, extreme small size, and some morphological features indicate that the specimen is the remains of an advanced embryo, even though there is no associated eggshell. The unossified epiphyses of the long bones suggest that *Camptosaurus* hatchlings were altricial. The estimated total length of the embryo is 240 mm.

Beds at or near the fossil horizon are characterized by smectitic overbank and lacustrine mudstones with scattered isolated fluvial channel sandstones. The locality is 10 m above the stratigraphic level of the famous Carnegie Dinosaur Quarry. Both the Dinosaur National Monument specimen and embryonic and hatchling *Dryosaurus* specimens from western Colorado are in alluvial plain sediments of the Morrison Formation. The sites are some 200–300 km downstream from inferred highland sediment source areas to the west.

Introduction

The Morrison Formation is an Upper Jurassic continental sequence of fluvial, lacustrine, paludal, and floodplain deposits originally covering approximately 1,000,000 sq. km in the western interior of the United States. It is famous for its rich and diverse dinosaurian fauna, which has been intensively collected over the last 125 years. In contrast to the dinosaur record, microvertebrates are poorly known from the Morrison, although that situation has improved somewhat over the last decade (Callison, 1987; Chure & Engelmann, 1989; Chure, Engelmann, & Madsen 1989; Engelmann, Chure, & Madsen, 1989; Engelmann et al., 1990).

Juvenile dinosaur remains are rare in the Morrison, but juveniles of most groups are known (e.g., sauropods: Peterson & Gilmore, 1902; Gilmore, 1925a; Britt & Stadtman, 1988; Britt & Naylor, Chapter 16; Carpenter & McIntosh, Chapter 17; theropods: Madsen, 1976; ornithopods: Gilmore, 1909, 1925b; Carpenter, Chapter 19; stegosaurs: Galton, 1982). The paucity of juveniles is somewhat surprising considering the abundant adult dinosaurs known from the formation (Dodson, et al., 1980). Poorer still is the record of dinosaur eggs and hatchlings in the Morrison. Scheetz (1991) briefly described embryonic and hatchling *Dryosaurus* from the Morrison, and Young (1991) illustrated unidentified small bones (which may be embryonic) from a dinosaur nest site. Hirsch et al. (1990) identified an early stage "embryo-like" object within a dinosaur egg from the Cleveland–Lloyd Quarry.

Recently, an embryonic dinosaur skeleton has been discovered at Dinosaur National Monument (DNM) (Fig. 20.1). We attempt to determine its systematic affinities, and explore its implications for the breeding biology of Morrison ornithopods.

Discovery of Specimen

The specimen was collected by Scott K. Madsen on 16 September 1991 from a site in the Brushy Basin Member of the Morrison Formation, approximately 1 km (0.5 mi) west of the Carnegie Quarry in the Utah portion of Dinosaur National Monument (detailed records of the site, DNM 315, are in the permanent files at the Monument). Two of us (Turner and Peterson) had earlier measured a detailed section through the Morrison Formation in this area and several carbonaceous mudstones had been discovered. During sampling of these sediments for plant remains, the embryo was discovered as a result of random sampling; there was no surface indication that vertebrate material was present.

Similar discoveries have been made in other areas of the Monument while trenching for sections, including two partial skulls of sphenodontids and the mandible of a dryolestid mammal. It may be that microvertebrate and small dinosaur material is more common in the Morrison than is thought, but lack of surface expression of the material means their discovery is serendipitous.

Only a small excavation (approximately 3 m along the strike) has been made at DNM 315, and few specimens other than the hatchling have been recovered. These include a small maxillary or dentary tooth of a theropod, a diplodocid tooth, an adult ornithopod tooth, an isolated sphenodontid palatine, a few scattered lizard postcranial bones, and a fish scale. Unfortunately, while the locality holds promise for additional material, the topography and exposures are such that additional excavation probably will not be very extensive.

Geologic setting
Stratigraphy

The Morrison Formation at Dinosaur National Monument consists of four members: from base to top these are the Windy Hill (Peterson in press), Tidwell, Salt Wash, and Brushy Basin Members (Fig. 20.2). The Windy Hill unconformably overlies the Redwater Member of the Stump Formation and consists of marginal marine sandstone. The Tidwell consists largely of gray mudstone and scarce sandstone beds deposited in overbank, fluvial, and possibly marginal marine environments. The Salt Wash Member consists of fluvial sandstone interbedded with thin (0–3 m thick) red and gray overbank mudstone. The Brushy Basin is divided into two parts distinguished by clay composition and color. The lower part contains dominantly red, non-swelling mudstones. The upper part consists predominantly of greenish-gray mudstone that contains swelling (smectitic) clays up to the level of the Carnegie Quarry sandstone bed. The upper part then becomes variable in both smectite content and color up to the base of the Lower Cretaceous Cedar Mountain Formation. The upper contact of the Brushy Basin Member is tentatively interpreted to be an unconformity.

Age

Recent isotopic dates from the Morrison Formation indicate that it is entirely Upper Jurassic in age, based on comparison with the geologic time scale of Harland et al. (1990), modified to place the Jurassic–Cretaceous boundary at 141.1 Ma (Bralower et al. 1990). A single-crystal laser-fusion $^{40}Ar/^{39}Ar$ date of 154.0 ± 1.5 Ma (latest Oxfordian–earliest Kimmeridgian) by J. D. Obradovich (written communication, 1991) was obtained from sanidine grains. These grains were separated from a bentonite bed about 2.4 m (8 ft) above the basal unconformity of the Tidwell Member near Notom, Utah.

Dates from the Brushy Basin at Montezuma Creek in southeastern Utah (equivalent to the upper part of the Brushy Basin Member at Dinosaur National Monument), range from 145.2 to 149.4 Ma (Kowallis, Christiansen, & Deino 1991). These dates indicate an early-to-middle Tithonian age and suggest that the base of the Tithonian may be close to the top of the lower part of the Brushy Basin Member. The isotopic dates indicate that the Morrison is Kimmeridgian and Tithonian in age, that the basal beds could possibly be latest Oxfordian, and that the top of the formation probably is no younger than about middle Tithonian. Because the site is stratigraphically high in the upper part of the Brushy Basin Member, the embryo skeleton is most likely middle Tithonian in age.

Climate

Lithologic features in the Morrison indicate that the overall climate was generally semi-arid to arid. Gypsum deposits in the Tidwell Member require that evaporative conditions prevailed during deposition of the lower part of the formation. Eolian sandstone deposits in the lower half of the formation also support the interpretation of a dry climate. The presence of alkaline, saline lake deposits in the Brushy Basin Member in the eastern Colorado Plateau region require that evaporation exceeded runoff and precipitation, and therefore indicate that the dry climate persisted throughout deposition of the formation (Turner & Fishman, 1991).

The local presence of ferns and horsetails in rare laminated claystone beds of the Brushy Basin Member attests to wetter conditions, either locally in an overall dry environment or during brief wet climatic intervals.

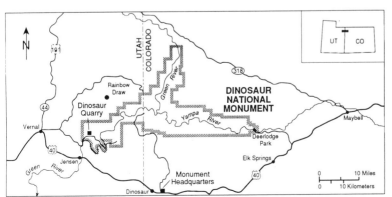

Figure 20.1. Index map showing location of Dinosaur Quarry (Carnegie Quarry) in Dinosaur National Monument. Modified from National Park Service brochure.

The depositional record is thus best interpreted as one characterized by extended dry periods and briefer wet periods. Such a scenario is analogous to the present-day semi-arid to arid climate of the Lake Eyre basin, southwestern Australia (Turner & Fishman, 1991).

Sedimentology of the Brushy Basin Member

Lenticular sandstones and red mudstones in the lower part of the Brushy Basin Member were deposited in stream channels and adjacent overbank floodplain environments. The streams derived sediments from upland source regions several hundred kilometers to the west.

During deposition of the upper part of the Brushy Basin Member, airborne volcanic ash from a magmatic arc to the west was incorporated in the sediments, either as discrete ash beds or as admixtures with clastic sediments. The greenish gray mudstones in this part of the member are largely smectitic as a result of the alteration of ash to smectitic mixed-layer clays. The presence of carbonized plant remains, charophytes, ostracodes, gastropods, and conchostracans suggests a lacustrine origin

for some of the mudstones. These lacustrine deposits formed in small lakes or ponds that developed on a broad floodplain. Fluvial channel sandstones, including the quarry sandstone interval, are also present in the upper part of the Brushy Basin Member. Coalesced fluvial sandstones in the quarry sandstone interval appear to represent deposits of perennial streams, based on the abundance of unionid bivalves. However, it is uncertain whether other fluvial channel sandstones in the upper part of the member were ephemeral, intermittent, or perennial.

Lithology of the site

The embryo site is located in the upper part of the Brushy Basin Member in the Dinosaur Quarry West measured section (Fig. 20.2), about 1 km (1/2 mi) west–northwest of the Visitor Center Quarry building. Stratigraphically, the site is about 10 m (34 ft) above the quarry sandstone interval and about 26 m (86 ft) below the top of the Brushy Basin Member within an interval that consists predominately of mudstone with numerous thin bentonite beds. The mudstones are typically mas-

Figure 20.2. Measured section of the Morrison Formation showing the age of the formation and stratigraphic position of locality DNM 315.

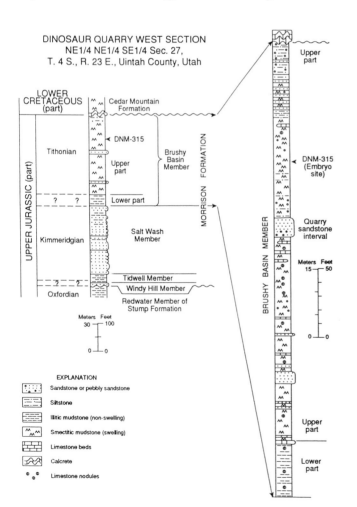

sive, although some exhibit burrow traces and root markings, the result of deposition primarily in an overbank environment. Even those beds that may have been deposited in a lacustrine environment probably were altered postdepositionally by burrowing or soil-forming processes so they resemble overbank deposits. The embryonic remains were found in a greenish-gray smectitic mudstone bed that was probably deposited in an overbank environment.

Description

The specimen (DINO 15624) is a partial skeleton scattered among five small blocks, four of which fit together and show that the specimen is preserved in a small area (Fig. 20.3). Block A contains an impression of the centrum of the last cervical, and six dorsal vertebrae, the left scapula and coracoid, a dorsal rib head, and an incomplete bone that may be part of the pubis. Block B contains two cervicals, the anterior vertebra of the caudal series, three dorsal ribs, and half a dorsal neural arch. Block C contains five caudal neural arches and parts of dorsal ribs. Block D contains four caudal centra and fragments of dorsal ribs. Block E was found a short distance from the rest of the specimen and contains the tibia–fibula. Based on size, morphology, and style of preservation, the tibia–fibula is referred to the skeleton. Overall, the bone of the specimen has an unfinished look, with the surfaces of the long bones and the scapula being fibrous. Measurements of select bones are given in Table 20.1.

Of the anterior vertebrae, there are two articulated posterior cervicals followed by the impression of the centrum of the first dorsal and dorsals 2 through 6. The neurocentral suture is open in all these vertebrae.

The centrum of the penultimate cervical is shield-shaped in outline in posterior view (Fig. 20.4A). It has no neural spine, but does have a well developed diapophysis on a pedicle and a parapophysis at the neurocentral suture. The last cervical is only partly preserved and yields little information.

The first dorsal centrum is preserved only as an impression of the centrum. Dorsals 2 through 6 (Fig. 20.5) show the following features in common: centra are square in lateral view (measuring 2.5 × 2.5 mm), there is no parapophysis on the centrum, the lateral surface of the centra are slightly concave, there is a sharp ventral keel on the centra, the centra are shield shaped in end view. Only the fourth dorsal arch is complete and visible in lateral view (Figs. 20.4B & 20.5). The contact between the arch and centrum is not a tight fit (much less so than the condition in the cervicals) and was probably cartilaginous. Although the arch is tall, the neural spine is low. The diapophysis is high on the arch.

Dorsal neural arches 2, 5, and 6 are missing their left halves. The medial surface of these arches is visible (Figs. 20.5 and 20.6). The right half of neural arch 3 is present but incomplete, and does not add any information to observations made on dorsal four. An isolated half neural arch is preserved in medial view, under a group of dorsal ribs, a short distance from this vertebral

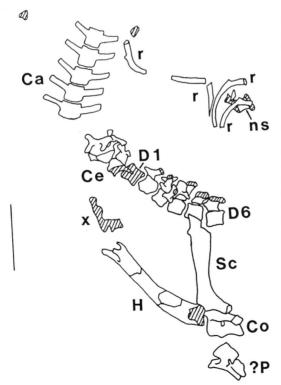

Figure 20.3. Embryo of *Camptosaurus* (DINO 15624) from Dinosaur National Monument. Composite of four blocks containing all of the specimen except the tibia–fibula. Cervical vertebrae are shown as they appear in right lateral view because the left side cannot be exposed. Caudal vertebrae are diagrammatic because not all centra are visible in ventral view. Abbreviations: Ca, caudal vertebra; Ce, cervical vertebra; Co, coracoid; D, dorsal vertebra; H, humerus; ?P, possible fragment of pubis; ns, neural spine; r, dorsal rib; x, unidentified bone. Hatching = broken bone. Scale bar = 1 cm.

series. There is no evidence of ossified epaxial tendons along the dorsal vertebrae.

A second series of five articulated vertebrae are proximal caudals. They are preserved in three blocks and are separated along the bases of the neural arches. Thus, in one block the neural arches are in ventral view (Fig. 20.7), while the other blocks show the dorsal surface of the centra.

The caudal centra are low in end view, although this may be due to slight crushing. The ventral surface of the centra bear a prominent low keel on the first two caudals in the series, but the keel is absent on the third. There are no chevron facets visible on any of the exposed centra. The neurocentral suture is open and more closely resembles the close-fitting condition seen in the cervicals than the looser contact of the anterior dorsals. The transverse processes are long and flattened dorsoventrally; they rise at a low angle from the arch and become shorter posteriorly (Fig. 20.7).

The dorsal surface of the first two neural arches

Table 20.1. Measurements of *Camptosaurus* embryo (DINO 15624)

Element	Height (mm)	Length (mm)	Width (mm)
Scapula	17.0	–	–
Coracoid	3.0	5.5	–
Humerus	–	18.0	–
Tibia–fibula	–	23.0	–
Dorsal centrum 2	2.5	2.5	–
Dorsal centrum 3	2.5	2.5	–
Dorsal centrum 4	2.5	2.5	–
Dorsal centrum 5	2.5	2.5	–
Dorsal centrum 6	2.5	2.5	–
Caudal centrum 2	–	2.5	–
Caudal centrum 5	–	2.5	–
Caudal 2 (across transverse processes)	–	–	9.5
Caudal 5 (across transverse processes)	–	–	8.0

lack neural spines (Fig. 20.8A). In fact, a gap down the middle of the dorsal surface of the neural spine indicates that the two halves were not yet fused. This is similar to the condition seen in the dorsal vertebrae in which the medial surface of the neural arches is visible. A moderately developed neural spine is present on the last vertebra of this series (Fig. 20.8B). Intervening vertebrae have not been prepared because of the fragile nature of the bone. The posterior cervicals, anterior dorsals, and caudals all show a very large neural canal (in cross section) relative to the centrum.

Chevrons occur on the first caudal vertebra of *Camptosaurus* "*medius*" (Gilmore, 1925b), whereas it occurs on the second chevron in *C. dispar* (Gilmore, 1925b), and *C.* "*nanus*," and *C.* "*browni*" (Gilmore, 1909). Flattened transverse processes occur on caudals 1–11 in *C.* "*medius*," caudals 1–12 in *C.* "*browni*," and caudals 1–13 in *C. dispar* (Gilmore, 1925b). The well-developed transverse processes in DINO 15624 indicate that this series comes from the anterior region of the tail. The apparent lack of chevron facets may be due to incomplete preparation or, more likely, to the fact that chevron facets were poorly developed in an individual this small. This underdevelopment is analogous to the poorly defined parapophyses of the anterior dorsals.

The left scapula is exposed in lateral view (Fig. 20.6). It is long and narrow, and is bowed along its length. There is a slight flaring near the incomplete dorsal edge, and a small tubercle is present along the anterodorsal margin. This margin is relatively straight, while the posterior margin is concave along its ventral edge. The shaft is constricted ventrally and flares considerably near its articular end. The humerus overlies the scapula, obscuring the acromion process and clavicular facet. The scapula is broken and slightly displaced near its contact with the coracoid. The glenoid faces posteriorly and forms a large portion of the glenoid fossa. The maximum preserved length of the scapula is 17 mm.

The left coracoid (Fig. 20.6) is quadrangular in outline, with a well-defined craniodistal angle and short caudal process. The coracoid foramen is located at midlength along its contact with the scapula.

The left humerus is exposed in medial view (Fig.

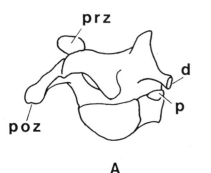

A

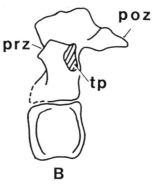

B

Figure 20.4. **A**. Posterodorsal view of penultimate cervical vertebra, DINO 15624. **B**. Fourth dorsal vertebral centrum and neural arch restored, DINO 15624. Abbreviations: d, diapophysis; p, parapophysis; poz, postzygopophysis; prz, prezygopophysis; tp, base of transverse process. Hatching = broken bone. Scale bar = 5 mm.

20.6). The proximal half is bent and at an angle to the distal portion. This shape indicates that the deltopectoral crest (which is not exposed in this view) is slightly less than half the length of the humerus. The proximal and distal ends are not ossified. The maximum preserved length is 18 mm.

Next to the coracoid is an incomplete bone (Fig. 20.6). It bears some resemblance to the proximal part of a camptosaur pubis, but it is too incomplete for positive identification.

Only a few ribs are preserved. The most complete are the proximal halves of two dorsal ribs near the cervico–dorsal series. The capitulum is well developed but the tuberculum is very reduced. The heads of these ribs

Figure 20.5. Left lateral view of dorsal vertebrae 1–6, DINO 15624. Abbreviations as in Figure 20.3. Cross hatching = articular surfaces on right neural arches for left half of neural arch. Scale bar = 5 mm.

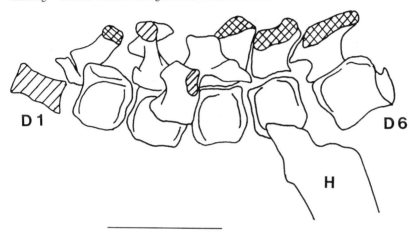

Figure 20.6. Anterior dorsals, scapula, coracoid, humerus, and ?fragment of pelvic girdle of embryo of *Camptosaurus* (DINO 15624). Scale in millimeters.

are nearly at a right angle to the shaft. Several dorsal rib fragments are present along the left side of the anterior caudal series, and one shows a well developed capitulum (Fig. 20.3).

An articulated tibia and fibula (Fig. 20.9) was found within a meter of the rest of the specimen. The bones are closely adpressed, have unfinished ends, a fibrous surface texture, and measure 23 mm in length.

Discussion

A specimen such as the embryo poses several identification problems, especially considering its incomplete nature and the lack of any cranial, mandibular, or dental material. There are two questions to be answered in identifying DINO 15624. First, what are the systematic affinities of the specimen? Second, what ontogenetic stage does it represent, that is, is this a hatchling of a large dinosaur or is it a juvenile of a dinosaur of small adult size? Although poorly known, dinosaurs with very small adult size are present in the Morrison (Callison & Quimby, 1984), so the latter question is not a theoretical one.

Systematic affinities

While clearly reptilian, DINO 15624 shows no affinities to turtles, lizards, sphenodonts, crocodilians, champsosaurs, or pterosaurs, the groups of nondinosau-

rian reptiles known from the Morrison Formation.

Benton (1990) lists nine synapomorphies for the clade Dinosauria. Because of the incomplete nature of DINO 15624, the presence or absence of most of these features could not be established. However, two of Benton's synapomorphies are present in the DNM specimen: (1) a scapulo–coracoidal glenoid facing fully backward and, (2) a low deltopectoral crest that runs one third or one half of the distance down the shaft of the humerus.

Of the dinosaurs groups known from the Morrison, DINO 15624 shows no synapomorphies with theropods, sauropods, stegosaurs, or nodosaurs. However, the vertebrae, scapula, and humerus do bear strong similarities to those elements in some ornithopod ornithschians: long narrow scapular blade with gentle distal expansion, proximal half of the humerus at an angle to the distal part, and overall vertebral morphology. Ornithopods are a moderately diverse group in the Morrison and include undescribed "fabrosaurids" (Callison & Quimby, 1984); the hypsilophodontid *Othnielia rex*, the dryosaurid *Dryosaurus altus*; and the camptosaurids *Camptosaurus dispar*, *C. amplus*, and *C. depressus* (following the *Camptosaurus* systematics of Sues & Norman, 1990).

With the pectoral girdle oriented such that the scapulo–coracoidal suture is horizontal (Fig. 20.10C),

Figure 20.7. DINO 15624, anterior caudal neural arches in ventral view, anterior to right. Scale in millimeters.

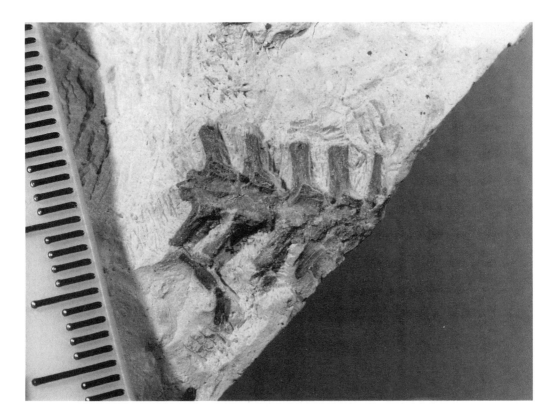

the scapula of DINO 15624 differs from *Othnielia rex* (Fig. 20.10F) in having a straight (i.e., vertical) orientation rather than a posterodorsally curving one. The scapula of *Dryosaurus altus* (Fig. 20.10F) resembles that of *Othnielia rex*, although the scapula of the closely related *Dryosaurus lettowvorbecki* (Fig. 20.10D) from the Upper Jurassic of Tanzania more closely resembles that of the Dinosaur National Monument specimen. The scapula of *Camptosaurus dispar* (Figs. 20.10A, B) resembles that of DINO 15624. The scapula is unknown for *C. amplus* and *C. depressus*.

The coracoid of *Dryosaurus altus* (Fig. 20.10E), and *Dryosaurus lettowvorbecki* (Fig. 20.10D) is semilunate, with a curved anterior border and a tapering caudal process. The coracoid of *Othnielia rex* (Fig. 20.10F) is incompletely known, although the caudal process is tapering and the outline appears to be semilunate. The coracoid of DINO 15624 is unusual in being quadrangular in outline (Fig. 20.10C). It has a well-defined craniodistal angle and a short caudal process, two features cited by Dodson (1980) as diagnostic of *Camptosaurus* (Fig. 20.10B). However, the coracoid in *Camptosaurus* is sometimes subrectangular (Fig. 20.10A). The coracoid is unknown for *C. amplus* and *C. depressus*. The similarities between DINO 15624 and *Camptosaurus* are corroborated by the very weak or absent cervical neural spines in the DNM specimen, a synapomorphy of the Ankylopollexia, shown in Figure 20.12 (Sereno, 1986), a diverse group of advanced ornithopods which includes *Camptosaurus* but not any other known Mor-

Figure 20.8. DINO 15624. **A**. Anterodorsal view of first two neural arches of mid-dorsal series midline gap between unfused right and left halves. **B**. Posteroventral view of last neural arch in mid-dorsal series showing moderately developed neural spine. Abbreviations as in Figure 20.3. Scale bar = 1 cm.

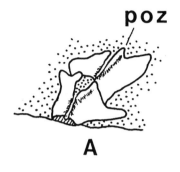

A

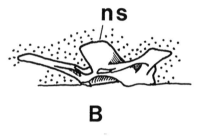

B

Figure 20.9. DINO 15624, tibia–fibula. Scale in millimeters.

rison ornithopods.

Drinker nistii is a small ornithopod closely related to *Othnielia*. Only a preliminary description has been published to date (Bakker et al., 1990) and although several partial skeletons are known, no elements comparable with the DNM specimen have been described or figured. However, the postcrania are briefly described as being very similar to those of *Othnielia*, and presumably the scapula and coracoid show the same differences from the DINO 15624 as does *Othnielia*.

There are also several taxa of ornithopods in the Morrison Formation that are poorly known. Callison (1987) reported "fabrosaurids" from the Fruita Paleontological Area in western Colorado, but the specimens have not yet been described, and it is not clear what material is present. However, the Fabrosauridae is now recognized as a paraphyletic grouping of primitive ornithopods with no taxonomic standing (Sereno, 1991). Thus, it is not clear what taxa of "fabrosaurs" should be used for comparison.

Callison (1987) stated that one of the Fruita fabrosaurs has been positively identified as *Echinodon*, a taxon known only from lower jaws found in England and Portugal. Because there is no mandibular material in the DNM material, any comparison with the Fruita *Echinodon* must await description of that material. A further complication is that one of the Fruita *Echinodon* mandibles bears an enlarged canine at its anterior end (J. I. Kirkland, personal communication, 1991), a feature unknown in "fabrosaurids" but present in heterodontosaurids. The latter group has not previously been recognized in the Morrison Formation.

Because the Morrison "fabrosaurid" (and ?heterodontosaurid) specimens have not been described, DINO 15624 must be compared with other "fabrosaurids" and heterodontosaurids from other ages and formations. The only well-known "fabrosaurs" are *Lesothosaurus diagnosticus* (Thulborn, 1972; as *Fabrosaurus australis*, Santa Luca, 1984; as "fabrosaur," Sereno, 1991) and *Scutellosaurus lawleri* (Colbert, 1981). The scapula of *Lesothosaurus* is erect (Fig. 20.10H), although its anterior margin differs from the DNM specimen in not being straight (Thulborn, 1972, Fig. 6a). The coracoid is rectangular (Fig. 20.10I; Santa Luca, 1984, Fig. 7) and is similar to DINO 15624, although not as similar in outline as is *Camptosaurus*

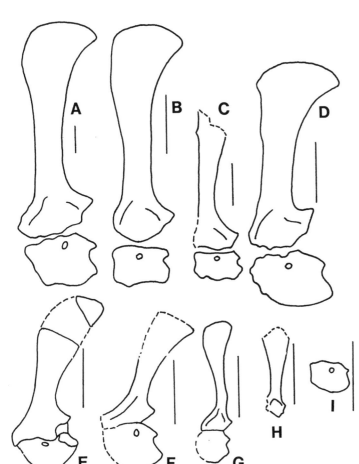

Figure 20.10. Scapulae and coracoids of select ornithopods, in left lateral view. **A.** *Camptosaurus "browni"* (after Gilmore, 1909). **B.** *Camptosaurus "nanus"* (after Gilmore, 1909, reversed). **C.** *Camptosaurus* sp. DINO 15624. **D.** *Dryosaurus lettowvorbecki* (after Janensch, 1955). **E.** *Dryosaurus altus* (after Galton, 1981, reversed). **F.** *Othnielia rex* [after Galton & Jensen, 1973, as *"Nanosaurus"*(?) *rex*; reversed]. **G.** *Heterodontosaurus tucki* (after Santa Luca, 1980). **H.** Scapula of *Lesothosaurus australis* (after Thulborn, 1972). **I.** Coracoid of unidentified "fabrosaur" (after Santa Luca, 1984). Scale bar = 5cm, except for C where bar = 5 mm.

(Fig. 20.10A). Santa Luca (1984) noted that the posterior cervical vertebrae in his "fabrosaur" material had well developed neural spines, a condition that differs from the present specimen. The scapula and coracoid of *Scutellosaurus* are incomplete (Colbert, 1981, Fig. 18), but preserved portions show a rounded coracoid and a very broad and distally expanded scapula, conditions in contrast to the DNM material.

Heterodontosaurus tucki is the only well-known heterodontosaurid. This species differs from DINO 15624 in having well-developed neural spines on the posterior cervical vertebrae, a scapula lacking a straight anterior border (Fig. 20.10G), and a semicircular coracoid (Santa Luca, 1980, Figs. 5a, 10D).

We are aware of the possibility that embryonic morphology might not resemble adult morphology. However, embryonic bones of *Dryosaurus* are nearly identical with those of adults (Scheetz, personal communication, 1991). Because the scapula and coracoid of DINO 15624 show some morphological features that are unique to *Camptosaurus*, we refer the specimen to that genus.

Ontogenetic stage

DINO 15624 is a very small ornithopod dinosaur, but what ontogenetic stage does it represent? Morrison ornithopods attained a wide range of adult sizes (*Othnielia*: nearly 1.5 m; *Drinker*: 2.0 m; *Dryosaurus*: 2.5 m; *Camptosaurus*: 8.0 m). Although exact lengths are not available, Callison (1987) reports that the Fruita "fabrosaurs" grew no larger than domestic chickens. Thus, a small skeleton such as this could be a juvenile or subadult of a Morrison ornithopod with a very small adult size. However, if DINO 15624 is referable to *Camptosaurus*, then it must be an extremely young individual. The question is this: Is it an embryo or a hatchling? Several lines of evidence (bone texture, morphology, and size) strongly suggest that DINO 15624 is an embryo.

Bone texture

There is very little if any lamellar bone on the specimen, giving the bone an "unfinished" look. The scapula, humerus, tibia, fibula, and neural arches of the caudals have a fibrous surface texture. All centra lack a smooth lamellar finish and appear spongy, although similarly sized embryonic *Dryosaurus* centra do show well-developed lamellar bone (personal observation of Chure). From the brief description of bone texture in embryos of *Orodromeus* (Horner & Weishampel, 1988), the texture seems similar to that seen in the specimen from DNM.

Morphology

A number of features in advanced embryos of hadrosaurs (Horner & Currie, Chapter 21; Currie, personal communication, 1992) can also be seen in DINO 15624. This strongly suggests that the present specimen is embryonic.

The right and left halves of the dorsal neural arches are not fused and, the contact for the left halves can be seen on the preserved right halves (Fig. 20.6). There is a marked zone of weakness between the two halves of the neural arches of the first two vertebrae of the caudal series. The one completely preserved cervical shows a midline contact between the neural arch halves, which we interpret to indicate that the halves are not fused. In cross section the neural canal in all vertebrae is proportionally much larger in DINO 15624 than in an adult camptosaur. The same relative enlargement is seen in hadrosaur embryos.

Some features, such as the loose contact between the dorsal neural arches and their centra are features that are also present in hatchling, juvenile, and some adult dinosaurs. However, the surface for contact with the neural arch in the DNM specimen differs from that seen in juveniles and adults in being smooth rather than being covered with a complex system of ridges and grooves. This may be an embryonic feature.

Size

The DNM specimen is an extremely small skeleton. The third dorsal centrum has a length of 3 mm, compared with a length of 72 mm for the same vertebrae in adult *Camptosaurus browni* (Gilmore, 1909, Fig. 15). The tibia–fibula in the DNM specimen has a length of 23 mm, compared with a tibia length of 48 mm in embryonic *Orodromeus* (Horner & Weishampel, 1988). *Orodromeus* reaches a much smaller adult body length (2.5 m) than *Camptosaurus* (8 m, Russell in Erickson, 1988). Adjusting a skeletal restoration of *Camptosaurus* (Carpenter in Case, 1982, Fig. 35-3) to a size at which scapular and humeral lengths are equal to those in the DNM specimen yields an estimated total length of 240 mm, approximately 3 percent of the adult length of 8 m (Fig. 20.11).

The skeletons of advanced embryos of dinosaurs are well enough ossified to be preserved. Unfortunately, except for hadrosaur and hypsilophodontid embryos from the Upper Cretaceous of Montana (Horner & Weishampel, 1988), most embryonic remains are very fragmentary. As far as can be determined from the literature, DINO 15624 is close to the Montana embryos in degree of ossification, although it is somewhat smaller. Scheetz (1991) reported embryonic *Dryosaurus* associated with eggshell fragments in the Morrison. Although no elements comparable to the DNM specimen were figured by Scheetz, the size is close to that of DINO 15624.

Dinosaur eggs are known from the Morrison Formation, although the record is poor. A single egg is known from the Cleveland–Lloyd Dinosaur Quarry (Hirsch et al., 1988; Madsen, 1991), and a complete egg was discovered in a nest near Grand Junction

Figure 20.11. Preserved elements of DINO 15624, showing positions of bone in living animal. Scale bar = 2.5 cm.

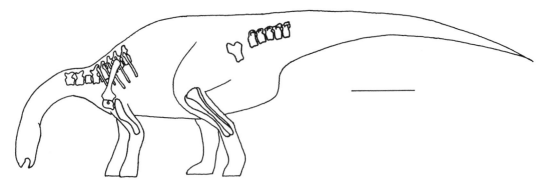

Figure 20.12. Phylogeny of ornithischian dinosaurs (after Sereno, 1986).

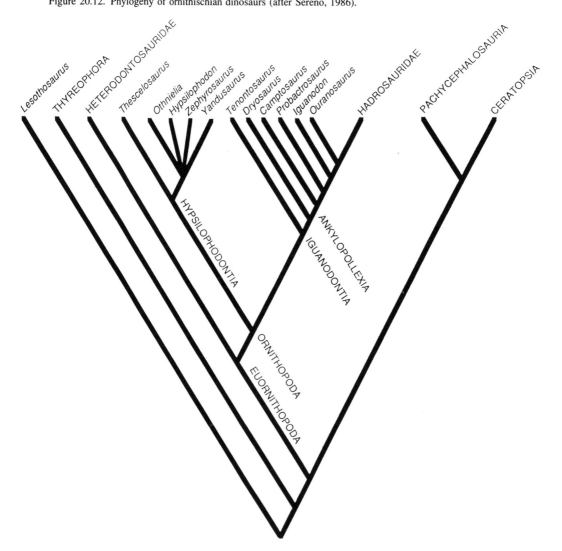

(Young, 1991). Eggshell fragments in the Morrison are more common, and some are associated with embryonic remains (Hirsch et al., 1987; Hirsch et al., 1990, Scheetz, 1991, Young, 1991; Hirsch, Chapter 10).

The Cleveland–Lloyd egg measures 110 mm × 60 mm and the Grand Junction egg 110 mm × 80 mm (however, the latter egg is badly crushed). Both these eggs are large enough to accommodate an embryo the size of DINO 15624. Unfortunately, other than being dinosaurian, the systematic affinities of these eggs is uncertain, so it is not clear what the size of a *Camptosaurus* egg would be.

What is the ontogenetic stage of the DNM specimen? At most, it is a *Camptosaurus* hatchling that could be no more than a few days old. However, nothing precludes it from being an advanced embryo; it is comparable in size to known embryonic ornithopods and could fit within eggs known from the Morrison. In fact, some features are strongly indicative of an embryonic condition.

The only serious objection to DINO 15624 being an embryo is the lack of any eggshell at the site. Limited screenwashing at the locality did not turn up even small fragments of eggshell. However, hadrosaur embryos have been found in China which also lack any associated eggshell (Currie, personal communication, 1992), possibly due to preservational bias (see also Carpenter, Hirsch, & Horner, "Introduction" of this volume). Nevertheless, the present specimen compares well in size, bone texture, and morphology with known embryonic dinosaurs. The lack of associated eggshell may be a consequence of the limited nature of the excavations at the site or an artifact of preservation. Thus, we would interpret DINO 15624 as the remains of an advanced embryo of *Camptosaurus*.

Paleobiological implications

Nearly all published records for embryonic and hatchling dinosaurs in the Morrison are restricted to ornithopods (Scheetz, 1991). The only exception is a small sauropod described by Britt and Stadtman (1988) as a hatchling but may be embryonic (Britt & Naylor, Chapter 16). Hirsch et al. (1990) described an early stage "embryo-like" object with a dinosaur egg, but the systematic affinities of that specimen are unknown. The small bones illustrated by Young (1991) may not belong to the egg found at that site, and we reject as unfounded his determination that the nest and eggs may be attributed to the theropod *Ornitholestes hermanni*.

Except for the DNM specimen, all hatchling and embryonic ornithopods in the Morrison belong to the genus *Dryosaurus*, an ornithopod very closely related to *Camptosaurus* (Fig. 20.12; Sereno, 1986). In spite of their close phylogenetic relationships, the hatchlings of these taxa seem to have quite different locomotor abilities. Hatchling and even embryonic *Dryosaurus* have well finished and ossified ends on their long bones. In contrast, the hatchlings of *Camptosaurus* have unossified ends on the long bones. Horner and Weishampel (1988) noted a similar pattern in the Late Cretaceous (?Late Campanian) hadrosaur *Maiasaura peeblesorum* and hypsilophodontid *Orodromeus makeli* from Montana. *Orodromeus* attained an adult length of approximately 2.5 m, has well formed and finished bones in embryos and hatchlings, and is interpreted as being precocial. In contrast, *Maiasaura* reached an estimated adult length of 8 m and has been interpreted as being altricial. Being bound to the nest, the young required parental care until they reached a length of 1.5 m and left the nest. This was corroborated by the more poorly ossified ends of the limb bones in *Maiasaura*. By inference from the work by Horner and Weishampel, *Dryosaurus* can be interpreted as being precocial, whereas *Camptosaurus* hatchlings were altricial. Such a scenario implies that *Camptosaurus* provided parental care to altricial hatchlings.

The Hypsilophodontia form one major clade of the Ornithopoda, and the Iguanodontia (dryosaurs + camptosaurs + iguanodonts + hadrosaurs) form another (Fig. 20.12; Sereno, 1986). Based on the limited data from Horner and Weishampel (1988), these two clades show overall differences in neonatal behavior (i.e., Hypsilophodontia being precocial and the Iguanodontia being altricial). However, data from the Morrison show that this is not the case because Iguanodontia has both precocial (*Dryosaurus*) and altricial (*Camptosaurus* and *Maiasaura*) young. This pattern supports the observations of Horner and Weishampel that such differences are related to adult size because the precocial *Dryosaurus* and *Orodromeus* are relatively small, whereas the larger *Camptosaurus* and *Maiasaura* are altricial.

Within the Euornithopoda (Fig. 20.12), altriciality occurs only in the Ankylopollexia (*Camptosaurus*, iguanodonts, and hadrosaurs), a group more derived than the Hypsilophodontia and *Dryosaurus*. This might be interpreted as indicating that altriciality is a specialization of the Ankylopollexia. However, ankylopollexians are much larger than other euornithopods (with the possible exception of *Tenontosaurus*). This pattern of precociality and altriciality may be a function of the body size of adults (as suggested by Horner & Weishampel, 1988)) and therefore of no phylogenetic significance. Data on hatchlings and neonate behavior of other euornithopods and their outgroups (pachycephalosaurs, ceratopsians, and thyreophorans) may help clarify trends within the Ornithischia and resolve questions concerning phylogenetic and adult size controls on the development of precocial and altricial young in ornithischian dinosaurs.

Acknowledgments
We thank Brooks Britt, Philip Currie, James Kirkland, John McIntosh, and James Madsen for fruitful discussions. Sue Ann Bilbey, Kenneth Carpenter, and Peter Dodson reviewed

the manuscript. Philip Currie kindly provided unpublished information on skeletal features of embryonic dinosaurs that was critical to our determination of the specimen described here as embryonic. Scott Madsen and Sidney Ash assisted in the field leading to the discovery of the embryo. Scott Madsen has shown his usual wizardry with a carbide needle and microscope on this small, difficult, and fragile specimen. The radiology department of the Ashley Valley Medical Center kindly X-rayed the specimen blocks to help locate bones for preparation. Field work by Christine Turner and Fred Peterson was conducted under NPS contract CA-1463-5-0001 for sedimentologic and stratigraphic analysis of the Morrison Formation within Dinosaur National Monument. Linda West provided the draft of Figure 20.12.

References

Bakker, R. T., Galton, P. M., Siegwarth, J., & Filla, J. 1990. A new latest Jurassic vertebrate fauna, from the highest levels of the Morrison Formation at Como Bluff, Wyoming, with comments on Morrison biochronology. Part IV. The dinosaurs: a new *Othnielia*-like hypsilophodontoid. *Hunteria* 2: 8–13, 15–19.

Benton, M. J. 1990. Origin and interrelationships of dinosaurs. *In* D. B. Weishampel, P. Dodson, & H. Osmólska (eds.), *The Dinosauria*. (Berkeley: University of California Press), pp. 11–30.

Bralower, T. J., Ludwig, K. R., Obradovich, J. D., & Jones, D. L. 1990. Berriasian (Early Cretaceous) radiometric ages from the Grindstone Creek Section, Sacramento Valley, California. *Earth and Planetary Science Letters* 98: 62–73.

Britt, B., & Stadtman, K. L. 1988. A possible ''hatchling'' *Camarasaurus* from the Upper Jurassic Morrison Formation (Dry Mesa Quarry, Colorado). *Journal of Vertebrate Paleontology* 8 (supplement to No. 3): 9A–10A.

Callison, G. 1987. Fruita: a place for wee fossils. *In* W. R. Averett (ed.) *Paleontology and Geology of the Dinosaur Triangle. Guidebook for the 1987 Field Trip.* (Grand Junction, CO: Museum of Western Colorado), pp. 91–6.

Callison, G. & Quimby, H. M. 1984. Tiny dinosaurs: are they fully grown? *Journal of Vertebrate Paleontology* 3: 200–9.

Case, G. R. 1982. *A Pictoral Guide To Fossils*. (New York: Van Nostrand Reinhold Co.).

Chure, D. J. & Engelmann, G. F. 1989. The fauna of the Morrison Formation in Dinosaur National Monument. *In* J. J. Flynn (ed.) *Mesozoic/Cenozoic Vertebrate Paleontology: Classic Localities, Contemporary Approaches. 28th International Geological Congress, Field Trip Guidebook* T322: 8–14.

Chure, D. J., Engelmann, G. F. & Madsen, S. K. 1989. Non-mammalian microvertebrates from the Morrison Formation (Upper Jurassic, Kimmeridgian) of Dinosaur National Monument, Utah–Colorado, USA. *Journal of Vertebrate Paleontology* 9 (supplement to No. 3): 16A–17A.

Colbert, E. H. 1981. A primitive ornithischian dinosaur from the Kayenta Formation of Arizona. *Museum of Northern Arizona Bulletin* 53: 1–61.

Dodson, P. 1980. Comparative osteology of the American ornithopods *Camptosaurus* and *Tenontosaurus*. *Memoirs de la Societe geologique de France* (N.S.) 59: 81–5.

Dodson, P., Behrensmeyer, A. K., Bakker, R. T., & McIntosh, J. S. 1980. Taphonomy and paleoecology of the dinosaur beds of the Jurassic Morrison Formation. *Paleobiology* 6(2): 208-32.

Engelmann, G. F., Chure, D. J., & Madsen, S. K. 1989. A mammalian fauna from the Jurassic Morrison Formation of Dinosaur National Monument. *Journal of Vertebrate Paleontology* 9 (supplement to No. 3): 19A.

Engelmann, G. F., Greenwald, N. S., Callison, G., & Chure, D. J. 1990. Cranial and dental morphology of a Late Jurassic multituberculate mammal from the Morrison Formation. *Journal of Vertebrate Paleontology* 10 (supplement to no. 3): 22A.

Erickson, B. R. 1988. Notes on the postcranium of *Camptosaurus*. *Scientific Publications of the Science Museum of Minnesota* 6(4): 1–23.

Galton, P. M. 1981. *Dryosaurus*, a hypsilophodontid dinosaur from the Upper Jurassic of North America and Africa: postcranial skeleton. *Palaeontologische Zeitschrift* 55: 271–312.

1982. Juveniles of the stegosaurian dinosaur *Stegosaurus* from the Upper Jurassic of North America. *Journal of Vertebrate Paleontology* 2: 47–62.

Galton, P. M. & Jensen, J. A. 1973. Skeleton of a hypsilophodontid dinosaur (*Nanosaurus*(?) *rex*) from the Upper Jurassic of Utah. *Brigham Young University Geology Studies* 20: 137–57.

Gilmore, C. W. 1909. Osteology of the Jurassic reptile *Camptosaurus*, with a revision of the genus, and description of two new species. *Proceedings of the United States National Museum* 36: 197–332.

1925a. A nearly complete articulated skeleton of *Camarasaurus*, a saurischian dinosaur from the Dinosaur National Monument. *Memoirs of the Carnegie Museum* 10: 347–84.

1925b. Osteology of ornithopodous dinosaurs from the Dinosaur National Monument, Utah. Part I. On a skeleton of *Camptosaurus medius* Marsh. Part II. On a skeleton of *Dryosaurus altus* Marsh. Part III. On a skeleton of *Laosaurus gracilis* Marsh. *Memoirs of the Carnegie Museum* 10: 385–409

Harland, W. B., Armstrong, R. L., Cox, A. V., Craig, L. E., Smith, A. G., & Smith, D. G. 1990. *A Geologic Time Scale, 1989*. (New York: Cambridge University Press).

Hirsch, K. F., Stadtman, K. L., Miller, M. E., & Madsen, J. E. 1988. A pathological Jurassic dinosaur egg containing an early age embryo from Central Utah. *Journal of Vertebrate Paleontology* 8 (supplement to No. 3): 17A.

1990. Upper Jurassic dinosaur egg from Utah. *Science* 243: 1711–13.

Hirsch, K. F., Young, R. G., & Armstrong, H. J. 1987. Eggshell fragments from the Morrison Formation of Colorado. *In* W. R. Averett (ed.), *Paleontology and Geology of the Dinosaur Triangle. Guidebook for the 1987 Field Trip.* (Grand Junction, CO: Museum of Western Colorado), pp. 79–84.

Horner, J. R. & Weishampel, D. B. 1988. A comparative embryological study of two ornithischian dinosaurs. *Nature* 332: 256–7.

Janensch, W. 1955 Der ornithopode *Dysalotosaurus* der Tendaguruschichten. *Palaeontographica* (Supplement 7 Part 3): 105–76.

Kowallis, B. J., Christiansen, E. H. and Deino, A. L. 1991. Age of the Brushy Basin Member of the Morrison Formation, Colorado Plateau, western USA. *Cretaceous Research* 12: 483–93.

Madsen, J. H., Jr. 1976. *Allosaurus fragilis*: a revised osteology. *Utah Geological and Mineral Survey Bulletin* 109: 1–163

Madsen, J. H., Jr. 1991. Egg-siting (or) first Jurassic dinosaur egg from Emery County, Utah. *In* W. R. Averett (ed.), *Guidebook for the Dinosaur Quarries and Tracksites Tour, Western Colorado and Eastern Utah.* (Grand Junction, CO: Grand Junction Geological Society), pp. 55–6.

Peterson, F. in press. Paleogeography of a widespread continental sequence – The Morrison Formation and related beds of the Colorado Plateau and vicinity. *In* D. D. Gillette (ed.), *Chemistry and Preservation of Dinosaur Bone.* (New York: Springer-Verlag Publishers).

Peterson, O. A. & Gilmore, C. W. 1902. *Elosaurus parvus*; a new genus and species of Sauropoda. *Annals of the Carnegie Museum* 1: 490–9.

Santa Luca, A. P. 1980. The postcranial skeleton of *Heterodontosaurus tucki* (Reptilia: Ornithischia) from the Stormberg of South Africa. *Annals of the South African Museum* 79: 159–211.

1984. Postcranial remains of Fabrosauridae (Reptilia: Or-

nithischia) from the Stormberg of southern Africa. *Palaeontologia Africana* 25: 151–80.

Scheetz, R. D. 1991 Progress report of juvenile and embryonic *Dryosaurus* remains from the Upper Jurassic Morrison Formation of Colorado: 27-29. *In* W. R. Averett (ed.), *Guidebook for the Dinosaur Quarries and Tracksites Tour, Western Colorado and Eastern Utah.* (Grand Junction, CO: Grand Junction Geological Society), pp. 27–9.

Sereno, P. 1986. Phylogeny of the bird-hipped dinosaurs (Order Ornithischia). *National Geographic Research* 2: 234–56.

1991. *Lesothosaurus*, ''Fabrosaurids'', and the early evolution of the Ornithischia. *Journal of Vertebrate Paleontology* 11: 168–97.

Sues, H.-D., & Norman, D. B. 1990. Hypsilophodontidae, *Tenontosaurus*, Dryosauridae. *In* D. B. Weishampel, P. Dodson, & H. Osmólska (eds), *The Dinosauria.* (Berkeley: University of California Press), pp. 498–509.

Thulborn, R. A. 1972. The post-cranial skeleton of the Triassic ornithischian dinosaur *Fabrosaurus australis*. *Palaeontology* 15(1): 29–60.

Turner, C. E. & Fishman, N. F. 1991. Jurassic Lake T'oo'dichi': A large alkaline, saline lake, Morrison Formation, eastern Colorado Plateau. *Geological Society of America, Bulletin* 103: 538–58.

Young, R. G. 1991. A dinosaur nest in the Jurassic Morrison Formation, western Colorado. 1–15. *In* W. R. Averett, W. R. (ed.), *Guidebook for the Dinosaur Quarries and Tracksites Tour, Western Colorado and Eastern Utah.* (Grand Junction, CO: Grand Junction Geological Society).

21

Embryonic and neonatal morphology and ontogeny of a new species of *Hypacrosaurus* (Ornithischia, Lambeosauridae) from Montana and Alberta

JOHN R. HORNER AND
PHILIP J. CURRIE

Abstract

Cranial and postcranial elements of embryonic and nestling specimens are described from Alberta and Montana, for a new lambeosaurid species named *Hypacrosaurus stebingeri*. Ontogenetic changes between embryonic and nestling individuals include incipient development of the nasal crest, increase in the rows of teeth, changes in the proportion of the orbits to skull size, deepening of articular rugosities at union junctions, and changes in osteohistological structures. For the majority of elements, the morphology of the nestlings resembles that of the adults. Circumference-to-length ratios of the femora and tibiae change during ontogeny, with the elements becoming less robust with age. Worn teeth of the embryos indicate that these animals ground their teeth in the eggs, and that the teeth were functional upon hatching. Histological studies show that this species experienced very rapid growth during its nest-bound period. Phylogenetically, *Hypacrosaurus stebingeri* appears to be an intermediate taxon between a species of *Lambeosaurus* and *Hypacrosaurus altispinus*.

Introduction

Baby dinosaurs and dinosaur eggs have long been a source of fascination to both paleontologists (Jepsen, 1964) and the public (Andrews, 1932). The first reports of embryonic dinosaurs were from alleged protoceratopsian eggs found in Mongolia (Andrews, 1932). Although these reports have not been substantiated by further preparation of the eggs, embryonic bones of *Protoceratops* were nevertheless recovered by the American Museum of Natural History expedition (Brown & Schlaikjer, 1940). Embryonic remains of an unknown dinosaur were also reported from the eastern Gobi by Sochava (1972). Additional undescribed embryonic bones from central Asia were collected by the Canada–China Dinosaur Project and the Soviet–Mongolian Expeditions (see also Carpenter & Alf, Chapter 1).

In 1955, C. M. Sternberg described a small ''hadrosaurine'' skull that he referred to as a very young individual. Size comparison with embryonic specimens from Devil's Coulee suggests that the specimen (CMN 8917) is small enough to have been either embryonic or

nepionic (postembryonic) when it died. Other isolated bones from the Judith River Formation of Alberta and Montana in the collections of the Royal Tyrrell Museum of Palaeontology, Canadian Museum of Nature, Museum of the Rockies, University of California at Berkeley, The Academy of Natural Sciences in Philadelphia (see Fiorillo, 1987), and the Yale Peabody Museum (Princeton Collections) clearly represent embryonic, hatchling, or nestling ''hadrosaurs.''

The Two Medicine Formation of northern Montana and southern Alberta has yielded an abundance of baby dinosaur remains (Horner & Makela, 1979), primarily found on nesting horizons (Horner 1982 and Chapter 8). A new species of *Hypacrosaurus* (see Appendix 1) is represented by the largest collection of baby material from the uppermost part of the Two Medicine Formation. Three nesting horizons, one in Alberta and two in Montana (Fig. 21.1) have yielded numerous eggs and the remains of baby individuals ranging from embryos to large nestlings (Currie & Horner, 1988). Juvenile and adult remains of this species are also very common.

Geological setting

All of the described and referred specimens, including the holotype of this new species of lambeosaurid, were found in the uppermost 100 of the 650-m thick Two Medicine Formation. A bentonite located within a few meters of the Devil's Coulee sites has been dated at 75.05 ±0.08 Ma (D. Eberth & A. Deino, personal communications, 1992). The nests and isolated bones from Montana and Alberta were found in greenish-grey mudstones, some of which contained abundant caliche. Terrestrial gastropods and small wood fragments are common on the Montana nesting horizons.

Specimens

Specimens from Montana were derived from two major nesting horizons (see Horner, Chapter 8). One of the horizons, located near Landslide Butte, Glacier County, extends for approximately 3 km. Hundreds of skeletal elements have been found, many associated with crushed eggs. None of the specimens represent individuals larger than 1 m in length. Sites along this horizon include Egg Baby Butte (MOR Locality Number TM-035), Egg Baby West (MOR TM-036), Baby Slide (MOR TM-037), Egg Explosion Hill (MOR TM-038), North Dome (MOR TM-039), and Egg Baby North (MOR TM-051). Specimens were found strewn randomly on the nesting horizon at each of these sites (Horner, Chaptrer 8). No associated skeletons were found, although a few associated but crushed eggs were collected.

Another nesting horizon, located on Blacktail Creek, Glacier County (MOR TM-066), yielded a nest

Figure 21.1. Map of Montana and Alberta showing geographic relationship of nest localities.

Figure 21.2. *Hypacrosaurus stebingeri.* Map of egg clutch from Devil's Coulee (RTMP 89.79.53).

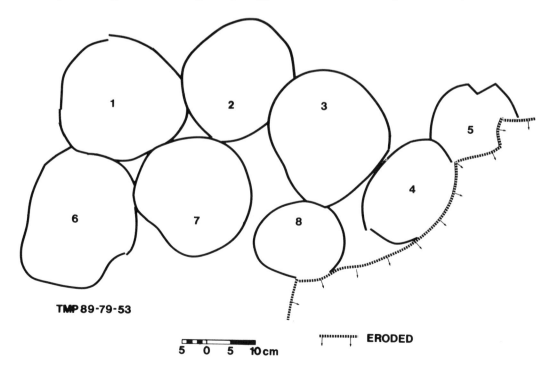

with broken eggs, a clutch of eggs with embryos (MOR 559), and a fluvial concentration of disarticulated skeletal remains (MOR 548) representing at least twelve individuals. An isolated embryonic or nepionic individual (MOR 562) was collected from Badger Creek, Glacier County (TM-065S).

Devil's Coulee in southern Alberta has yielded several nests from a variety of sites. Most of the specimens used in this study of embryonic individuals came from a single nest (No. 2) at a locality known as Little Diablo's Hill. Four eggs (RTMP 88.79.36) excavated from Nest No. 2 have not been prepared, but embryonic bones were seen in at least one of the broken eggs at the time they were collected. Disarticulated bones from broken eggs at the edge of the nest represent at least three individuals, and other bones washed out of the nest by recent erosion were cataloged under accession number RTMP 87.79. A single associated individual, RTMP 89.79.52, was recovered from another part of the nest.

Little Diablo's Hill has yielded four additional nests, two of which contain embryonic specimens, whereas the remaining two contain eggs without embryonic material. One of these nests (RTMP 89.79.53) contained eight eggs (Fig. 21.2). The eggs are large and almost round. A reconstruction (RTMP 90.130.1) of an egg from the nest measures 18.5 × 20 cm and has a volume of 3,900 ml. Erosion had destroyed one side of the nest so the original clutch size is unknown. Two other sites in Devil's Coulee (North Baby Butte and Kiddie's Corner) have also yielded embryonic material.

The embryonic and nestling material from Montana is in relatively good condition, although some cranial elements such as the premaxillae have been crushed. In contrast, nearly all of the embryonic material from Devil's Coulee is uncrushed and in very good condition.

Cranial morphology

Every bone of the skull, except the supraoccipital and the vomer, have been identified among the embryonic material from Devil's Coulee. Some elements, however, are hidden from view within articulated skulls. The estimated skull length of RTMP 89.79.52 is 7.5 cm (Figs. 21.3 & 21.4). For the large nestling, all the major cranial elements are known except the vomer, pterygoid, ectopterygoid, angular, splenial, and articular. The composite skull of nestling MOR 548 from Blacktail Creek, Montana, is 20 cm in length (Fig. 21.5). For descriptive convenience the skulls of the embryos and nestlings are divided into the segments (complexes) originally defined and used by Ostrom (1961).

Neurocranial complex

The parietals of the smallest embryos show no midline suture between left and right sides. Embryonic specimens of *Maiasaura peeblesorum* also show no sign of a midline suture. This suggests that the bone formed from a single ossification center in contrast with the ancestral state of reptiles, in which the parietals are paired (Weishampel & Horner, 1990). The embryonic parietal (RTMP 87.79.241) of the hypacrosaur is broader than long and broadly expanded (Fig. 21.6). A strong midline ridge separates the squamosals posteriorly, but this ridge does not extend anteriorly into a sagittal crest. The anterior margin of the parietals are notched on the midline, presumably a center for growth that is not seen in mature individuals (Gilmore, 1937). The posterior ends of the frontals overlapped onto a digitated ledge on the anterolateral surfaces of the parietals. In the nestlings (MOR 548) the parietal remains broad but has developed a relatively high sagittal crest (Fig. 21.7 A,B). Two posteroventral processes of the parietals attach to the dorsal surface of the supraoccipital, and the posterior

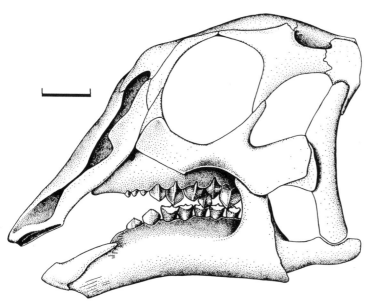

Figure 21.3. *Hypacrosaurus stebingeri.* Reconstruction of embryonic skull based on RTMP 89.79.52, RTMP 87.79.206 (frontal), RTMP 87.79.241 (parietal), RTMP 87.77.92 (prefrontal) and RTMP 87.79.333 (postorbital). Scale bar = 1 cm.

end of the posterodorsal process extends between the squamosals. The anterior ends of the parietals of the nestlings are more rugose where the frontals meet than those of embryonic specimens.

The embryonic frontal (Fig. 21.8) is thickened laterally where it meets the prefrontal and postorbital. There is a distinct posterolateral depression on the ventral side that received a portion of the postorbital condyle of the laterosphenoid. In hadrosaurids, the postorbital condyle of the laterosphenoid does not contact the frontal (Horner, 1992). The medial edge of the embryonic frontal is extremely thin, and there is no evidence of a sutural union. There may actually have been an unossified region between the paired elements. In the

Figure 21.4. *Hypacrosaurus stebingeri*. **A**. RTMP 89.79.52, enlargement of left maxillary tooth, labial view. **B**. RTMP 89.79.52 in left lateral view. **C**. RTMP 89.79.52 in right lateral view. **D**. Right squamosal in external view. **E**. Right squamosal in internal view. Abbreviations: an, angular; ar, articular; b, basioccipital; bs, basisphenoid–parasphenoid complex; d, dentary; ec, ectopterygoid; eo, exoccipital; h, hyoid; j, jugal; ls, laterosphenoid; m, maxilla; n, nasal; pal, palatine; pd, predentary; pm, premaxilla; po, postorbital; pr, prootic; prf, prefrontal; pt, pterygoid; q, quadrate; sa, surangular.

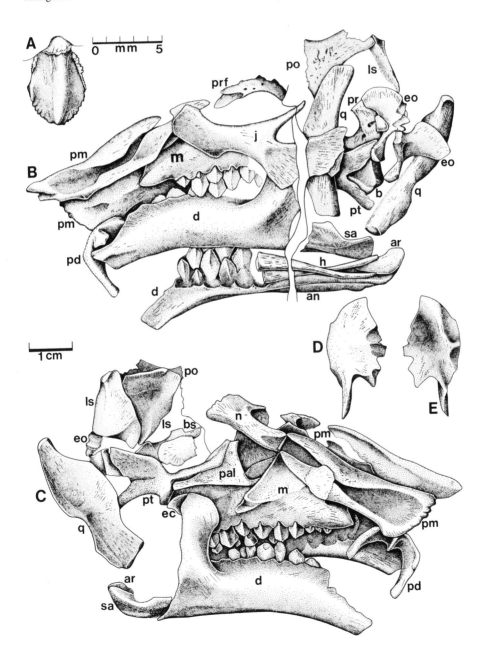

nestlings (Fig. 21.7C), the medial surface is thickened and slightly interdigitated where it clearly unites the elements. In both the embryonic and nestling individuals, the frontals are greatly excavated ventrally and domed dorsally as in other juvenile lambeosaurids. The frontal surface for union with the nasal is indistinct in the embryonic specimen, but very distinct in the nestling and all juveniles.

The supraoccipital of the nestling (Fig. 21.7D) is almost identical to the supraoccipital of juvenile or adult lambeosaurids (see Gilmore, 1937; Langston, 1960), except that the supraoccipital bosses are poorly developed, and the shelf over the foramen magnum is extremely short. The supraoccipital contacts the squamosals and parietals laterally, the exoccipitals ventrolaterally, and the prootic anterolaterally. Within the anterolateral process, which contacts the prootic, is an opening that represents the dorsoposterior end of the otic vestibule. In the nestlings, the nuchal notch, located posteriorly between the supraoccipital and parietal, is an opening that extends from the posterior end of the brain cavity to the exterior surface of the skull. This notch closes with maturity.

The exoccipitals (Figs. 21.7E and 21.9) apparently fuse with the opisthotics early in embryonic development, because there is no evidence of a suture between them as has been suggested by Langston

(1960). As in hadrosaurids, the exoccipital unites with the squamosal, supraoccipital, basioccipital, prootic, and basisphenoid. An extremely narrow wing contacts the paired exoccipitals on the midline of the posteroventral surface of the supraoccipital. On the anterodorsal surface of the basisphenoid process is located the opening to the posterior end of the otic vestibule. This opening is particularly large in the embryos and nestlings, suggesting that the babies had a very good sense of hearing.

The basioccipital (Figs. 21.7G and 21.10) is an indistinct flattened element with a broad, slightly concave depression that forms the base of the brain cavity. The basioccipital unites with the exoccipitals, prootics, and basisphenoid. The floor of the foramen magnum is proportionately very broad in both the embryonic and nestling individuals.

The prootic (Figs. 21.7H and 21.11) attaches to the anterior margin of the exoccipital and to the basioccipital, supraoccipital, laterosphenoid, and basisphenoid. The posterior end of the prootic houses the anterior end of the otic vestibule (Fig. 21.11). Anteriorly, the prootic is excavated where it forms the posterior wall of the trigeminal foramen (cranial nerve V, Fig. 21.4B). In the embryos, this opening is relatively much larger than in mature individuals. The foramen for cranial nerve VII also penetrates the prootic. Between the posterior end of the prootic and the anteroventral end of the exoccip-

Figure 21.5. *Hypacrosaurus stebingeri*, composite skull of MOR 548 in right lateral view. Scale in centimeters.

ital is located a large open area that was apparently occupied by cartilage and the fenestra ovalis (Langston,
1960).

The laterosphenoid (Figs. 21.4, 21.7I, and 21.12)
unites with the prootic, basisphenoid, frontal, postorbital, parietal, orbitosphenoid, and in larger specimens, the
opposite laterosphenoid. In the embryos (Figs. 21.4,
21.12), the laterosphenoid is a triangular, platelike bone
with a postorbital process that is short relative to those
of mature specimens. The posteroventral angle forms
the anterior margin of the trigeminal foramen and is
separated from the rest of the bone by a deep groove
for the ophthalmic branch of the trigeminal. The laterosphenoid of the nestling (Fig. 21.7I) has much the
same morphology as in the mature specimens. The dorsal process of the laterosphenoid fits into a socketlike
depression on the ventral surface of the postorbital and
frontal. In hadrosaurids this union is restricted to the
postorbital (Horner, 1992).

The basisphenoid and parasphenoid are united to
form a single element, herein referred to as the parasphenoid–basisphenoid complex (Figs. 21.7F, 21.13).
Fusion of the two bones apparently occurred very early

Figure 21.6. *Hypacrosaurus stebingeri.* RTMP
87.79.241. Embryonic parietals in ventral and dorsal
views. 1, frontal suture. Top is anterior. Scale = 5
mm.

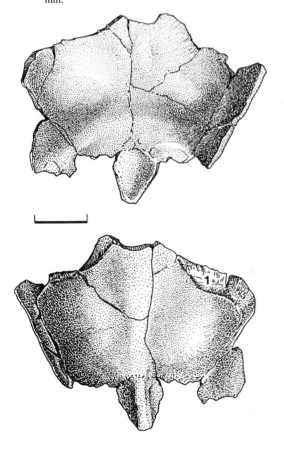

in embryonic development, because they are fused in
the smallest embryonic individuals. This complex unites
with the laterosphenoids, prootics, opisthotics (exoccipitals), basioccipital, and pterygoids. It forms the ventral
margin of the trigeminal foramen and plays a minor role
in the formation of the fenestra ovalis. The sella turcica
is deep and well demarcated.

Shallow canals on either side of the dorsum sellae
mark the passage of the sixth cranial nerves, which
passed out of the braincase anteromedial to the laterosphenoid–basisphenoid suture. The basipterygoid processes are broken in RTMP 87.79.335, but they appear
to have been relatively shorter than those of nestlings,
juveniles, and adults. The suture with the basioccipital
is transverse, but the basioccipital is also supported by
a pair of short, posteriorly directed processes in the position of the basal tubera. The cultriform process is U-
shaped in section where it supported the interorbital
septum, and it is separated from the pituitary region by
a distinct step.

The round orbitosphenoid attaches to the anterior
edge of the laterosphenoid. It also attaches to the frontal,
the presphenoid, and the opposite orbitosphenoid. The
nestling orbitosphenoid (Fig. 21.7J) is a subrounded,
flattened bone that is pierced by two foramina apparently for cranial nerves III and IV. A wall for the foramina of cranial nerve II most likely bordered the
element (see Horner, 1992).

None of the ethmoids appear to have been ossified in the embryonic specimens. They were not found
nor identified in the nestlings as well.

Maxillary complex

The premaxilla (Figs. 21.4, 21.14A,B, & 21.15)
is one of the largest bones in the skull, even though it
is relatively smaller compared to the skull than in mature individuals. The attachment surface for the keratinous bill (ramphothaeca) is a distinct transverse groove
with dorsal projections located along the anterior border
of the premaxilla (Fig. 21.15). The nasal groove is restricted just posterior to the external nares, similar to
Hypacrosaurus altispinus. Posteriorly, the groove expands, then constricts before meeting the nasal. There
is a distinct tapering process on the posteromedial end
of the premaxilla. The lateral surface of this process
meets the medial side of the nasal. The lateroventral
process of the premaxilla meets the anterolateral end of
the nasal.

One of the few differences between the embryonic and nestling premaxillae is the presence of an anteriorly directed excavation on the posteroventral
surface of the nestling premaxilla. This excavation re-
sides between the dorsal and ventral posterior processes
and apparently represents the initial development of the
S-loop (Weishampel, 1981a). The nestling premaxilla
also clearly shows a pair of shallow grooves that extends
anteriorly along the medial surface from the premaxil-

lary–nasal union. The dorsal groove extends anteriorly to the attachment groove for the ramphothaeca. The ventral groove passes down and exits on the ventromedial surface of the premaxilla. These grooves, distinct in juveniles and adults, apparently carried vessels and nerves to the keratinous bill and to the roof of the anterior end of the mouth.

The maxilla of both embryos and nestlings is basically triangular in lateral aspect (Figs. 21.4, 21.14D, and 21.16). The apex of the triangle, at the dorsal limit of the jugal suture, is approximately one third the distance from the anterior margin of the maxilla in the embryo. The apex in juveniles and adults is located in the middle of the maxilla. The premaxillary shelf of the maxilla is angled downward very steeply in the embryos (47°) and nestlings (42°). In adults, this angle is about 30°. Embryonic specimen RTMP 87.79.286 appears to have twelve rows of teeth over its 45 mm length. The nestlings have about twenty teeth in maxillae, which average 85 mm in length. As in mature individuals, the teeth of the nestlings are closely packed, although the

packing is looser because the teeth being shed are smaller than those replacing them. The largest teeth in the embryonic maxillae average about 4 mm in anteroposterior length (Fig. 21.4A). The teeth have several marginal denticles on both the anterior and posterior edges. The nestling teeth are about 5 mm in length, and show only a slight indication of marginal denticles.

The jugal of the embryo (Figs. 21.4, and 21.17) has an elongate maxillary process and a very broad orbital margin. The jugal of the nestling (Fig 21.14E) is almost identical to the jugal of an adult *Hypacrosaurus altispinus* (Gilmore, 1924), except that the orbital margin remains proportionately larger in the nestling. In addition, the anterior process is not as anteroposteriorly compressed, and the postorbital process is more elongate than in the adult.

The quadratojugal of the nestling (Fig. 21.14F) is a thin triangular bone with a characteristic anteroventrally projecting jugal process. The medial side of the quadratojugal fits like a flap over the quadratojugal notch in the quadrate. The quadrate is nearly covered

Figure 21.7. *Hypacrosaurus stebingeri*. MOR 548. **A**. Parietal in dorsal view. **B**. Parietal in left lateral view. **C**. Right frontal in dorsal view. **D**. Dorsal view of supraoccipital (D). **E**. Posterior aspect of exoccipitials. **F**. Left lateral view of basisphenoid. **G**. Dorsal view of basioccipital (anterior left). **H**. Lateral view of left prootic. **I**. Lateral view of left laterosphenoid. **J**. Lateral aspect of left orbitosphenoid. Scale = 4 cm.

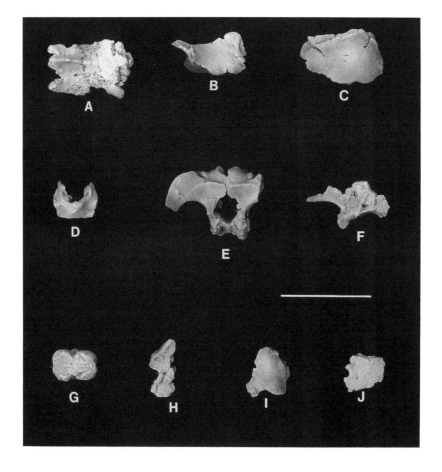

by the quadratojugal in the embryonic RTMP 89.79.52 (Fig. 21.4B).

The quadrates of the embryos and nestlings (Figs. 21.4, 21.14G, and 21.18) are nearly identical with the corresponding adult quadrates. The only differences are that the embryonic specimen has a much less distinctive quadratojugal notch and articular condyle.

The squamosals of embryos and nestlings (Figs. 21.4B,C, and 21.14H) are very similar to those of the adults but do not form as broad a roof over the posterior portion of the supratemporal fenestra. This condition is, however, variable among the nestlings. The embryonic squamosal is thin along most margins, and the contact with the supraoccipital has not developed into the pronounced process seen in juveniles and adults.

The postorbital of the embryo (Fig. 21.19) is similar to that of the nestling (Fig. 21.14I), except that the prefrontal and frontal processes are not nearly as broad or rugose. In the embryo, the prefrontal process of the postorbital is pointed and thin, whereas in the nestling it is blunt and thickened. Also in the nestling the posterior wall of the orbit is much broader than in the embryo. Both the postorbital and squamosal processes are similar in the embryos and nestlings, although in the nestling these processes possess much more distinctive articulations. One difference between the embryos and nestlings is the embryo's proportionally smaller supra-

temporal fenestra. In the nestlings, the fenestra is proportionally similar to mature individuals.

The prefrontal (Figs. 21.4B, 21.14J, and 21.20) is an elongated bone with a slightly raised medial rim. A longitudinal groove along the median side receives the lateral edge of the nasal. A shallow V-shaped groove on the anterolateral end of the bone receives the lacrimal. On the anteromedial end of the bone is a shallow longitudinal groove that receives the posterior end of the premaxilla. In mature individuals, the medial rim of the prefrontal is very steep where it laps onto the expanded premaxillae of the nasal crest (Gilmore, 1937). The prefrontal becomes foreshortened with maturity. At least one foramen penetrates the orbital rim of the prefrontal. In juvenile lambeosaurids, the entire medial face of the prefrontal rests on the posterolateral surface of the nasal. In hadrosaurids, the medial surface of the prefrontal is separated into dorsal and ventral surfaces by a longitudinal ridge. The dorsal surface receives the nasal, and the ventral surface forms the lateral wall of the internal nares (Horner, 1992).

The lacrimal is a triangular bone in both the embryos and nestlings (Fig. 21.14K). It contacts the premaxilla, maxilla, jugal, and palatine. In juveniles and adults, the lacrimal becomes more elongate and narrows anteroventrally as the crest develops.

RTMP 89.79.52 includes most of the left nasal,

Figure 21.8. *Hypacrosaurus stebingeri.* RTMP 87.79.206. Left frontal in dorsal and ventral views. Laterosphenoid suture (1). Scale = 5 mm.

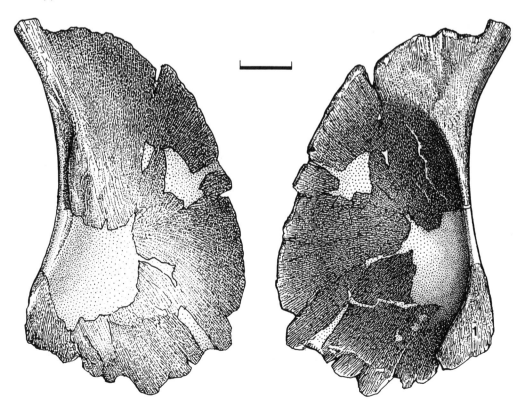

Figure 21.9. *Hypacrosaurus stebingeri.* RTMP 87.79.307. Left exoccipital–opisthotic. **A.** Posterior view. **B.** Dorsolateral view. **C.** Anterior view. **D.** Medial view. Supraoccipital contact (1); prootic suture (2). Scale = 5 mm.

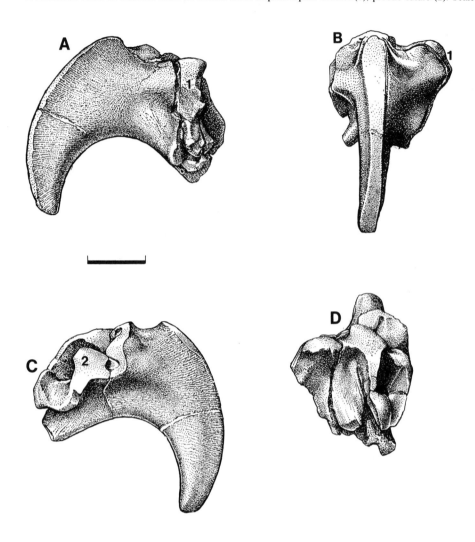

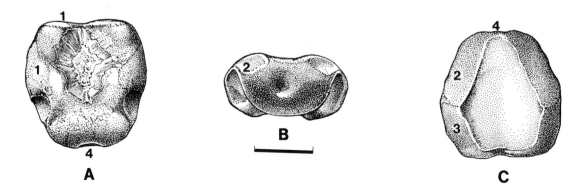

Figure 21.10. *Hypacrosaurus stebingeri.* RTMP 87.79.157. Basioccipital. **A.** Ventral view. **B.** Posterior view. **C.** Dorsal view. Basisphenoid contacts (1); exoccipital suture (2); prootic suture (3); occipital condyle (4). Scale = 5 mm.

which has been flipped over and exposed on the right side of the skull (Fig. 21.4C). Anteriorly, a pair of processes form the posterior margin of the narial opening. The shorter lateroventral process is largely covered by the premaxilla, whereas the more extensive anteromedial process has a complex, forked contact with the premaxilla. In both the embryo and nestling (Fig. 21.14C), the nasal is a flattened, elongated bone that has a concave ventral surface and shows slight convexity on the dorsal surface posterior to the external narial boundary. The paired nasals are separated anteriorly by a pair of dorsoposterior processes of the premaxillae. In larger specimens, the lateral margin of the nasal is a smooth, ventrolaterally angled surface that is sutured to the prefrontal. The ventral surface of the posterior process rests on the anterodorsal shelf of the frontal.

Figure 21.11. *Hypacrosaurus stebingeri.* RTMP 87.79.201. Left prootic in posterior aspect. Otic vestibule (ov). Scale = 5 mm.

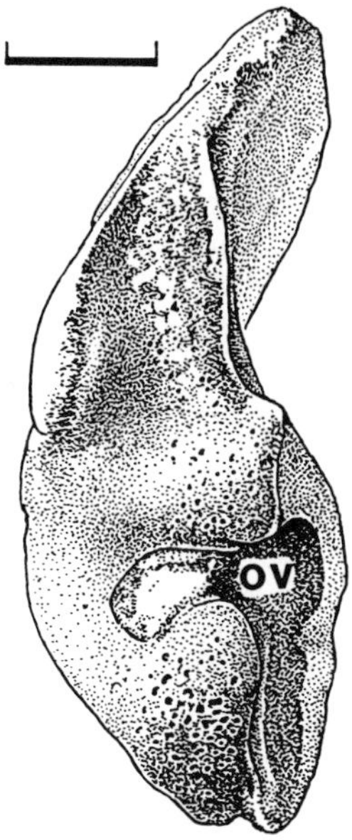

The embryonic pterygoid (Fig. 21.4B,C) is similar to that of other lambeosaurids but is relatively longer and lower than in mature individuals. The saddlelike groove (Heaton, 1972) is more pronounced, and the dorsal margin of the posterior alar projection is more arched in lateral view. The sutural contact with the palatine is not fused.

The embryonic ectopterygoid (Figs. 21.4 and 21.21) is an elongate, gently curving bone that separates the pterygoid from the maxilla. The distinctive lambeosaurid shape is already established in the ectopterygoid even at this stage of development. The medial surface is divided by a sharp but thin ridge into an elongate dorsal surface that posteriorly contacts the pterygoid and a ventral concavity for the maxilla. The posteroventral end of the ectopterygoid is wrapped around the ventral end of the ectopterygoid ramus of the pterygoid.

The embryonic palatine (RTMP 89.79.52) is exposed in labial view (Fig. 21.4C) and is relatively longer and shorter than the same element of mature lambeosaurids (Heaton, 1972). The ventral margin is almost straight, rather than convex in ventral outline, and rests on the median, posterodorsal surface of the maxilla. The anterolateral flange of the nestling palatine (Fig. 21.14L) articulates with the ascending maxillary apex ventrally and apparently a very small portion of the jugal dorsally (Heaton, 1972). This junction is very different in hadrosaurids, and there is no junction of the palatine and lacrimal as in *Prosaurolophus* (Horner, 1992). Along the dorsal margin, on the posterior half of the palatine, is located a deep excavation for contact with the anterior end of the pterygoid. In the anterior half, a process extends medially, and together with the anteriormost end of the pterygoid forms the roof of the palate. As in hadrosaurids, the posterior process of the vomer fits between the anteriormost ends of the palatine and pterygoid (Horner, 1992). In lambeosaurids, the anterior pterygoid process rests on the dorsolateral surface of the

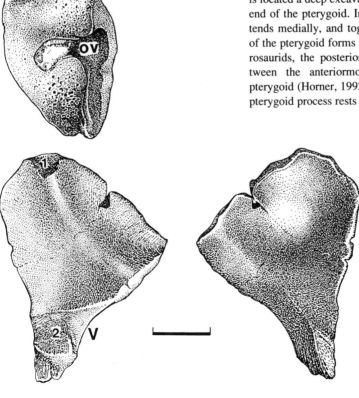

Figure 21.12. *Hypacrosaurus stebingeri.* RTMP 87.79.227. Right laterosphenoid in lateral and medial views. Postorbital articulation (1); groove for ophthalmic branch of trigeminal (2); anterior margin of opening for fifth cranial nerve (V). Scale = 5 mm.

medial process of the palatine, whereas in hadrosaurids the palatine rests on the anterior process of the pterygoid.

Mandibular complex

The predentaries of the embryo (Fig. 21.4B,C) and nestling (Fig. 21.22B) are identical in morphology and proportions to the corresponding element in juveniles and adults.

The embryonic dentary (Figs. 21.4B,C and 21.23) is similar in morphology to those of more mature specimens, although the ratio of height/length of the dental battery decreases with maturity. The paired dentaries of RTMP 87.79.266/267 are 54 mm long. There are eleven rows of teeth in the 41 mm long dental batteries, with two teeth per row. New rows are added at both the front and back of the battery and the antero-posterior crown lengths of unerupted teeth are less than 3 mm. In contrast, an unerupted tooth in the middle of the jaw is 5.2 mm long and 9.8 mm high. Most of the erupted teeth show occlusal tooth wear. The dentaries of RTMP 89.79.52 are shorter (49 mm), but the largest unerupted teeth are the same size as those of RTMP 87.79.266/ 267. The smallest embryonic dentary, RTMP 87.79.6, recovered from Nest No. 5 on Little Diablo's Hill, is less than 40 mm long and had only eight or nine rows of teeth. An embryonic jaw from nest MOR 559 has a length of 40 mm with eight tooth rows.

The nestling dentary (Fig. 21.22A) is very similar in morphology to the adult dentary, except that the ratio of height/length of the dental battery is deeper in the nestling. There are twenty rows of teeth within an 80 mm long dental battery. Two teeth per row are preserved, although the lower portion of the dental battery is crushed as is the case with most hadrosaurid and lambeosaurid jaws (Horner, 1983). The largest unerupted tooth, found in the midsection of the dentary, has a crown height of 14 mm and a crown antero-posterior

length of 5 mm. The largest nestling dentary has 21 teeth in 90 mm. The largest teeth have a crown height of 17 mm and a length of 6 mm. The nestling teeth have little or no serrations along their anterior or posterior margins.

Surangulars of the embryos and nestlings (Figs. 21.4, 21.22C, D, 21.23) have proportions and morphology that are nearly identical to those of the juveniles and adults.

The splenial, angular, and articular are found in the embryonic mandibles (Figs. 21.4 & 21.23) and look much as they do in the juveniles and adults. The articular is a small, lateromedially thin bone between the splenial and surangular. As in more mature individuals, it makes only a small contribution to the jaw articulation.

A pair of hyoids (ceratobranchials) are preserved in RTMP 89.79.52 (Fig. 21.4). They extend from the back of the jaws to the coronoid region. In overall morphology, they are indistinguishable from those described in mature specimens of *Corythosaurus* (Ostrom, 1961) in being deep and flattened anteriorly and in tapering posteriorly.

Postcranial morphology

All bones of the postcranial skeleton are represented for the embryos (RTMP 89.79.52) and nestlings (MOR 548). However, none of the nestling pubii have complete anterior processes.

Axial skeleton

Vertebrae of baby hadrosaurids, such as *Maiasaura peeblesorum* and lambeosaurids, share several common features and a nearly identical early ontogeny. The centra have a notochordal pit on the anterior and posterior faces in both embryos and nestlings. In addition, embryonic and nestling hadrosaurids and lambeosaurids have proportionally large neural canals, the sizes

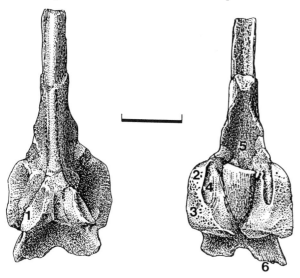

Figure 21.13. *Hypacrosaurus stebingeri.* RTMP 87.79.335. Parasphenoid–basisphenoid complex in ventral and dorsal views. Basipterygoid process (1); laterosphenoid suture (2); prootic suture (3); trough for sixth cranial nerve (4); sella turcica (5); basal tubera (6). Scale = 5 mm.

of which are age dependent. The sutural surfaces for the neural arches and transverse processes vary in degree of rugosity depending on the growth stage. Embryonic centra have shallow, smooth depressions for the neural arch and transverse processes, whereas large nestlings have shallow, slightly rugose sutural surfaces. These surfaces become progressively more rugose until finally co-ossification occurs.

The timing of fusion in the vertebrae is not the same throughout the column. In large nestlings, the cervical and dorsal neural arches are generally attached but are not yet fused; timing for fusion is not yet known. Fusion of sacral centra, on the other hand, occurs progressively beginning about the time the individuals are 2.5–3 m in length. The neural arches and transverse processes of sacrals do not begin to fuse until the juveniles are 3–4 m in length. Distal caudal arches begin to fuse when the individuals are about three meters in length.

The cervical vertebrae of the nestlings (MOR 548) have the features of the adult specimens (Figs. 21.24A, B). However, the neural canal is exceptionally

Figure 21.14. *Hypacrosaurus stebingeri*. MOR 548. **A**. Lateral view of the proximal half of the right premaxilla. **B**. Dorsal view of the distal half of the right premaxilla. **C**. Right nasal in dorsal aspect (anterior to left). **D**. Lateral view of right maxilla. **E**. Left jugal in lateral aspect. **F**. Lateral surface of left quadratojugal. **G**. Right quadrate in lateral view. **H**. Left squamosal. **I**. Postorbital. **J**. Prefrontal. **K**. Lacrimal. H–K shown in lateral aspects. **L**. Dorsal view of right palatine. Scale = 4 cm.

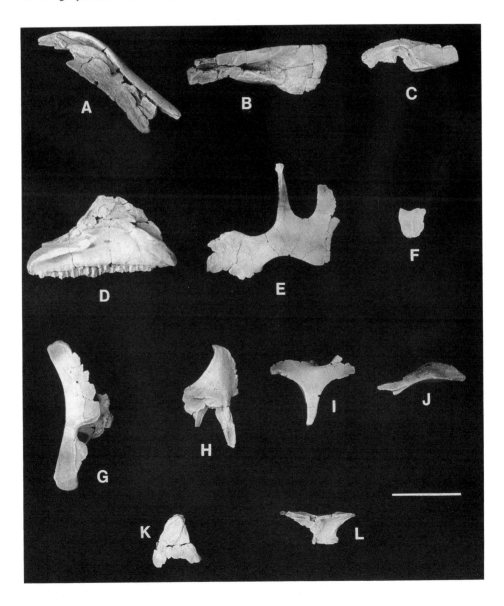

Figure 21.15. *Hypacrosaurus stebingeri*. RTMP 87.79.334. Right premaxilla in lateral, ventral, and dorsal views. Scale = 5 mm.

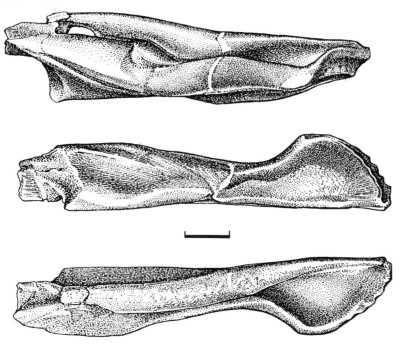

Figure 21.16. *Hypacrosaurus stebingeri*. RTMP 87.79.336. Left maxilla in medial and lateral views. Jugal suture (1); ectopterygoid contact (2). Scale = 5 mm.

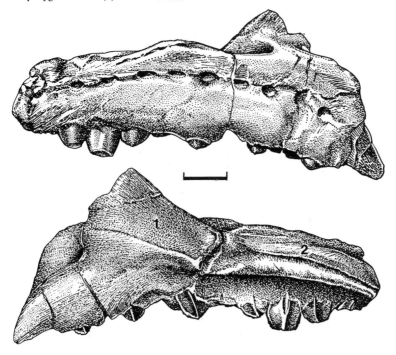

large in the anterior cervicals, and the zygapophyses are proportionally larger than in juveniles and adults. Proportions of the posterior cervicals of nestlings are nearly identical with the adults. In the embryo, the right and left halves of the axis (RTMP 87.79.242) are not co-ossified.

Like the cervicals, the dorsal vertebrae (Figs. 21.24C, D, E) have all of the morphological features found in the adults. The anterior dorsals have long diapophyses and neural spines that are proportionally much shorter those of adults. However, the neural spines of the nestlings are relatively broad anteroposteriorly. The posterior dorsals also have relatively short neural spines compared to the adult.

There is no fusion of the embryonic or nestling sacrum. Only five centra can be positively identified as sacrals (Fig. 21.24F, G, H). Two additional vertebral centra have expanded neural canal floors similar to the other sacrals and may represent the anteriormost and posteriormost sacrals of the juveniles. The adult sacrum consists of eight co-ossified vertebrae. The ventral surface of the nestling sacral centra have a median keel similar to adult lambeosaurids. All of the sacral parapophyses are disarticulated (Fig. 21.24I) and their relationships with their respective sacral vertebrae have not been determined. The sacral neural spines are very broad anteroposteriorly, and relatively high (Figs. 21.24G,H). The neural canal is exceptionally large, and

becomes progressively smaller with maturity (Weishampel & Horner 1990).

The caudal vertebrae of the nestlings (Figs. 21.24J, K, L) have proportions similar to the adults. The neural spines are relatively long and broaden anteroposteriorly near their superior ends. The chevrons (Fig. 21.24M) are very long and taken together with the tall neural spines make a very deep tail.

The ribs of the embryos and nestlings (Fig. 21.26A) have the features and proportions of the adult ribs.

Appendicular skeleton

The scapula, coracoid, humerus, ulna, and radius of the embryos and nestlings (Fig. 21.25A) have a morphology similar to the adults. The smallest embryonic scapula (MOR 559S2), with a length of 30 mm, has an extremely narrow neck and enlarged articular facets. Proportions are more like the adult in larger embryos and nestlings. The nestling coracoid (Fig. 21.25B) is nearly identical to more mature specimens, although the anteromedial blade is somewhat thinner. Also, the coracoid foramen is not enclosed but is a narrow slit that splits the scapular and humeral articulation surface. The sternal of the embryo (RTMP 87.79.205) has a short, broad anterior blade, whereas in the nestlings the blade is narrower and more elongate (Fig. 21.25C) and is similar to that of the adult.

Figure 21.17. *Hypacrosaurus stebingeri.* RTMP 87.79.332. Left jugal in lateral and medial views. Maxillary suture (1). Scale = 5 mm.

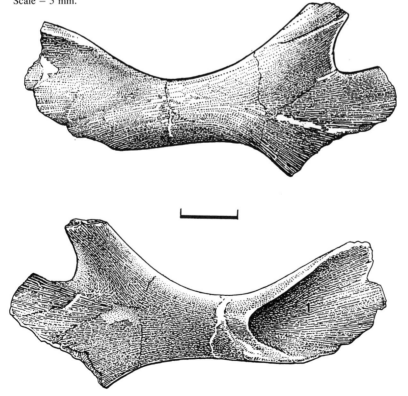

Figure 21.18. *Hypacrosaurus stebingeri*. RTMP 87.79.320. Left quadrate in posteromedial and lateral views. Scale = 5 mm.

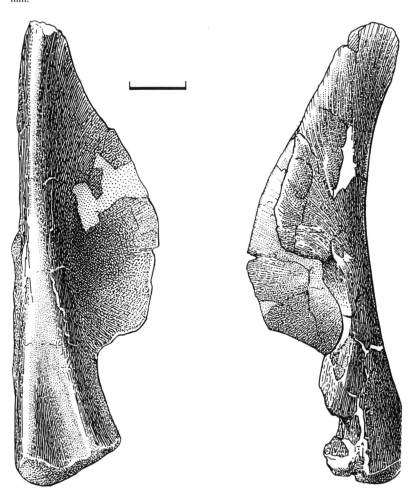

Figure 21.19. *Hypacrosaurus stebingeri*. RTMP 87.79.333. Left postorbital in internal and external aspects. Squamosal contact (1); laterosphenoid articulation (2); frontal suture (3); orbital rim (4); upper temporal fenestra (5); lateral temporal opening (6). Scale = 5 mm.

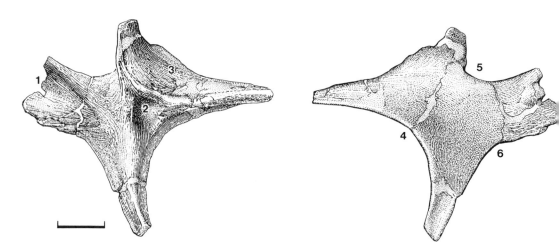

The humerus (Fig. 21.25D) is relatively stout compared to more mature specimens, but the deltopectoral crest has similar proportions to that of the adult. The ulna and radius (Fig. 21.25E) are longer than the humerus and proportionally more massive than the corresponding adult elements. Metacarpals III (Fig. 21.25F) and IV are slightly less than half the length of the radius. The elements of the manus are all very well ossified and are not

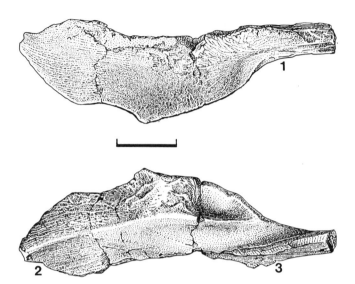

Figure 21.20. *Hypacrosaurus stebingeri*. RTMP 87.77.92. Left prefrontal in external and medial views. Orbital rim (1); frontal contact (2); nasal suture (3). Scale = 5 mm.

Figure 21.22. *Hypacrosaurus stebingeri*. MOR 548. **A**. Right dentary in medial view. **B**. Predentary in dorsal aspect. **C**. Right surangular in lateral view. **D**. Left surangular in dorsal aspect. Scale = 4 cm.

Figure 21.21. *Hypacrosaurus stebingeri*. RTMP 87.79.247. Left ectopterygoid in medial and dorsal views. Scale = 5 mm.

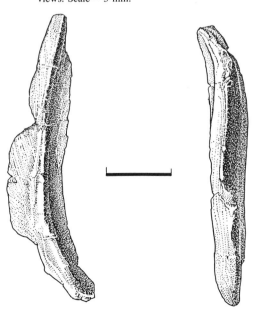

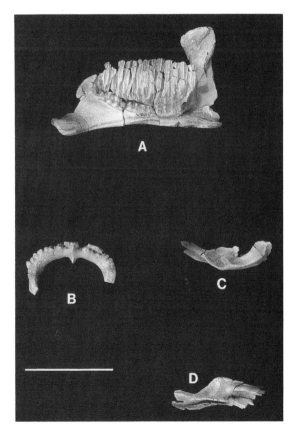

different from those of other lambeosaurids. Nothing was found that could be identified as a carpal element.

Within the embryonic pelvic girdle, the ilium is elongate and shallow dorsoventrally. Angulation between the anterior and posterior processes is minimal, but the antitrochanter is well developed. The nestling ilium (Fig. 21.26B) is similar to that of the adult. The ischium of the embryo has an expanded distal end, but it lacks the characteristic foot seen in the nestlings (Fig. 21.26C) and adults. The embryonic pubis is similar to that of the nestling (Fig. 21.26D,E), although the anterior blade is much narrower dorsoventrally. The anterior blade becomes deeper with maturity.

The femur, tibia, fibula, tarsals (including the distal tarsal), and pes elements of both the embryos and nestlings are of similar morphology to the adults. However, the nestling femora (Fig. 21.27A) do however, have a much more defined muscle scar medial to the fourth trochanter than do those of more mature individuals. The nestling tibia (Fig. 21.27B) and fibula (Fig. 21.27C) are similar in morphology to those of adults but are much more robust. Length-to-midshaft-circumference proportions of the femur and tibia change during growth (Table 21.1). Both the femur and tibia are more robust in the younger individuals than in the adults. This may have to do with the large cartilaginous caps that

Figure 21.23. *Hypacrosaurus stebingeri*. RTMP 87.79.266. Left mandible in lateral and medial aspects. Abbreviations: ar, articular; d, dentary; sa, surangular; sp, splenial. Scale = 10 mm.

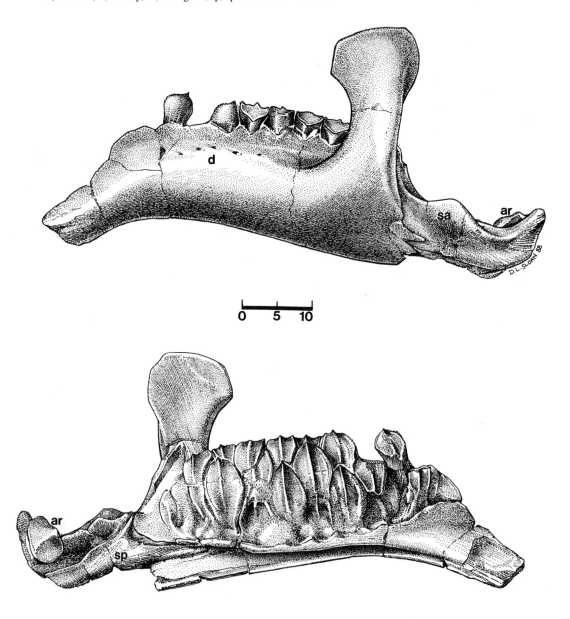

apparently existed at the ends of these elements, as explained below.

The embryonic and nestling astragalus (Fig. 21.27D) and calcaneum (Fig. 21.27E) are well ossified and morphologically similar to corresponding adult elements. The nestling metatarsals are slightly more robust than those of adults. Nestling phalanges, including the unguals (Fig. 21.27H), are nearly identical to those of adults.

Ossified tendons are present in embryonic individuals (RTMP 88.121.35). The tendons are highly vascularized, and were apparently very flexible during this period.

Osteohistology

Histologically, the cortex of embryonic bone (MOR 559) is highly vascularized, contains abundant

osteocyte lacunae (Fig. 21.28A), and has little or no fibrous structure (Fig. 21.28B). Primary osteons are in their initial stages of development. The nestling bone is also highly vascularized, with abundant osteocyte lacunae (Fig. 21.28C), but has better developed primary osteons (Fig. 21.27D). However, the primary osteons are not fully developed. The bone is fibrolamellar in structure.

All of the articular surfaces of the appendicular skeleton of the embryos are incomplete. Abundant calcified cartilage columns are found at the ends of these bones. In the nestlings, these surfaces are more complete but still have abundant calcified cartilage columns. The incomplete ends of the bones of the embryos suggests there were large cartilage cones at all joints and ends of bones (Horner & Weishampel, 1988). Calcified cartilage is also abundant at all articular joints and union surfaces in the skull. The surfaces of the bones have a very

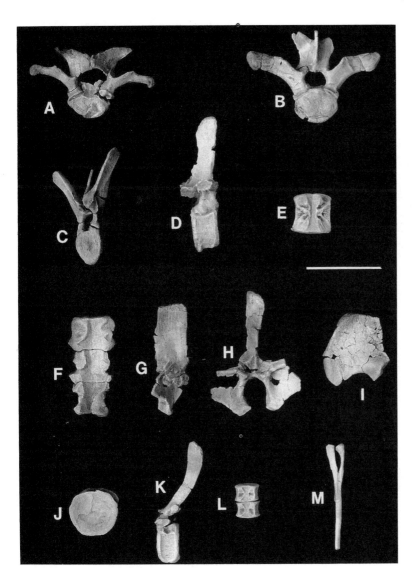

Figure 21.24. *Hypacrosaurus stebingeri*. MOR 548. **A, B**. Anterior views of cervical vertebrae. **C**. Anterior view of dorsal vertebra. **D**. Lateral view of dorsal vertebra. **E**. Dorsal centrum in dorsal aspect. **F**. Dorsal view of three sacral centra. **G, H**. Lateral and anterior views of sacral neural arch. **I**. Dorsal view of sacral rib. **J**. Posterior view of anterior caudal centra. **K**. Left lateral view of caudal vertebra. **L**. Two articulated distal caudals in dorsal view. **M**. Anterior view of a haemal arch. Scale = 4 cm.

coarse texture, with numerous vascular canals exiting on the surface, indicating very active growth at the periosteum. The combination of highly vascularized, fibrolamellar bone with an abundance of calcified cartilage indicates that these baby dinosaurs were growing very rapidly before their deaths (Ricqles, 1976; Currey, 1984).

It is unclear whether or not the femora and tibiae of the babies had different proportions than those of the adults. It may be that with the addition of the extensive cartilaginous caps at the ends of the baby bones, their proportions would have been the same as adults.

Figure 21.25. *Hypacrosaurus stebingeri.* MOR 548. **A, B**. Right scapula (A) and coracoid (B) in lateral views. **C**. Right sternal from anterior. **D**. Right humerus in lateral aspect. **E**. Right ulna and radius in medial view. **F**. Metacarpal III. **G**. Manus phalange. **H**. ungual phalange. Scale = 10 cm.

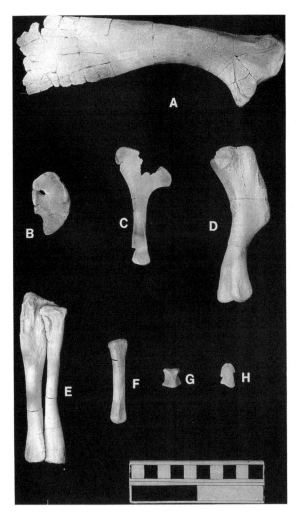

Early ontogeny

The most noticeable change that take place during the early ontogeny of *Hypacrosaurus stebingeri* (Appendix 21.1) occur in the skull due to the early development of the nasal crest. During development from the embryonic stages to the large nestling stage, the nasal crest domes slightly and the S-loop narial passage begins formation. Continued changes result in increased skull size as the crest vaults vertically. Most of the other cranial elements retain their overall morphology throughout ontogeny. The orbits remain proportionally large, but steadily decrease with size, and the snout remains short and highly angled. New tooth rows are added to both the maxilla and dentary, and there is most likely an addition of a least one tooth to each vertical row.

In the postcranial skeleton, the sutural facets of the vertebral centra and their respective neural arches become more rugose. The neural canal becomes proportionally smaller as the neural spines increase in height. The femur and tibia become less robust, a trend that continues until maturity. The ends of all bones ossify, and the processes and muscle scars become more distinct. The ossified tendons become more compact and therefore less flexible.

Phylogenetic considerations

It may be of phylogenetic importance that the premaxilla and nasal of the embryos possess no distinctive characters that would lead to the conclusion that lambeosaurids were derived from the hadrosaurids. The folds in the dorsal and ventral posterior processes of the premaxilla are apparently formed before ossification. All cranial elements of the embryos and nestlings possess each of the characters found in adult lambeosaurids. The embryos, therefore, offer no new data that would help to disprove the hypothesis that the classic "Hadrosauridae" are polyphyletic (Horner, 1990).

Discussion

The cranial elements of both the embryonic and nestling individuals of *Hypacrosaurus stebingeri* clearly show that the premaxillae/nasal crest of lambeosaurids was the major component that changed during ontogeny. Other changes include increase in the number of teeth, decrease in the relative proportion of the orbit, deepening and thickening of sutural rugosities, histological development, and possibly negative allometry in the femora and tibiae. Most other cranial and postcranial elements underwent few changes (Figs. 21.29 and 21.30). The small, slightly modified nasal capsules of the babies most likely allowed for higher frequencies of vocalization and facilitated communication with the adults, as suggested by Weishampel (1981b).

The embryonic and nestling specimens of *Hypacrosaurus stebingeri* described represent the largest collection of baby skeletal material of any single species of ''hadrosaur'' known from any area in the world. Continued work with these specimens and the more mature individuals of the species promises to yield a much clearer understanding of individual variation, development, and growth in this species.

Acknowledgments

We thank Kevin Aulenback (RTMP) for preparation of the embryonic material, and Carrie Ancell (MOR) for preparation of the nestling material. We also thank the Kashiwagi Museum of Nagano, Japan, for making specimens and casts of juvenile material available for comparative studies. We would also like to thank the Tribal Council of the Blackfeet Nation for access to tribal lands, and Seldon Frisbee, Wally Bradley, and Ricky Reagan for access to their deeded or leased lands. Figures 21.3 and 21.4 were made by one of us (Currie), and Donna Sloan (RTMP) did the remaining embryonic illustrations. Terry Panasuk (MOR) photographed the nestlings. Funding for JRH was provided by NSF Grant No. EAR 8705986 and the Museum of the Rockies. Funding for PJC was derived from the Royal Tyrrell Museum of Palaeontology.

References

Andrews, R. C. 1932. *The New Conquest of Central Asia*, Vol. 1. (New York: American Museum of Natural History).

Figure 21.26. *Hypacrosaurus stebingeri*. MOR 548. **A–C.** Dorsal rib (A), left ilium (B), and ischium (C) in lateral views. **D, E.** Left pubis in medial (D) and lateral (E) views. Scale = 4 cm.

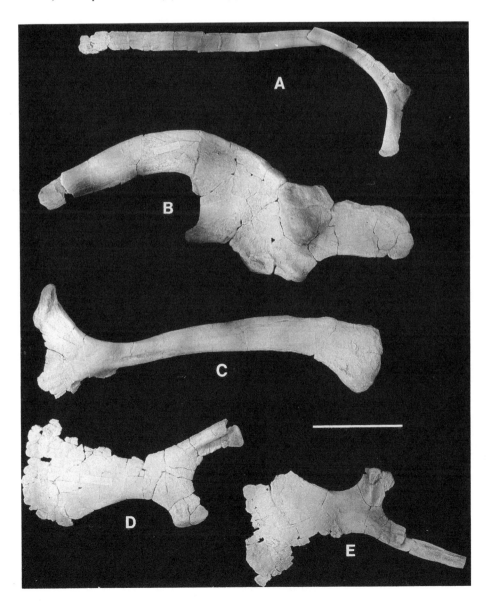

Brown, B. 1913. A new trachodont dinosaur, *Hypacrosaurus*, from the Edmonton Cretaceous of Alberta. *American Museum of Natural History Bulletin* 32: 395–406.

Brown, B., & Schlaikjer, E. M. 1940. The structure and relationship of *Protoceratops. Annals of the New York Academy of Sciences* 40: 133–266.

Currey, J. 1984. *The Mechanical Adaptations of Bone.* (Princeton: Princeton University Press).

Currie, P. J. & Horner, J. R. 1988. Lambeosaurine hadrosaur embryos (Reptilia: Ornithischia). *Journal of Vertebrate Paleontology* 8 (Supplement to No. 3): 13A.

Dodson, P. 1975. Taxonomic implications of relative growth in lambeosaurine hadrosaurs. *Systematic Zoology* 24: 37–54.

Fiorillo, A. R. 1987. Significance of juvenile dinosaurs from Careless Creek Quarry (Judith River Formation), Wheatland County, Montana. *In* P. J. Currie & E. H. Koster (eds.), *Fourth Symposium Mesozoic Terrestrial Ecosystems.* (Drumheller: Tyrrell Museum of Palaeontology), pp. 88–95.

Gilmore, C. W. 1917. *Brachyceratops*, a ceratopsian dinosaur from the Two Medicine Formation of Montana, with notes on associated fossil reptiles. *U.S. Geological Survey Professional Paper* 103: 1–45.

1924. On the skull and skeleton of *Hypacrosaurus*, a helmet-crested dinosaur from the Edmonton Cretaceous of Alberta. *Geological Survey of Canada Bulletin* 38: 49–64.

1937. On the detailed skull structure of a crested hadrosaurian dinosaur. *U.S. National Museum Proceedings* 84: 481–91.

Heaton, M. J. 1972. The palatal structure of some Canadian Hadrosauridae (Reptilia: Ornithischia). *Canadian Journal of Earth Sciences* 9: 185–205.

Horner, J. R. 1982. Evidence of colonial nesting and 'site fidelity' among ornithischian dinosaurs. *Nature* 297: 675–6.

1983. Cranial osteology and morphology of the type specimen of *Maiasaura peeblesorum* (Ornithischia: Hadrosauridae), with discussion of its phylogenetic position. *Journal of Vertebrate Paleontology* 3: 29–38.

1990. Evidence of diphyletic origination of the hadrosaurian (Reptilia: Ornithischia) dinosaurs. *In* K. Carpenter & P. J. Currie (eds.), *Dinosaur Systematics, Approaches and Perspectives.* (New York: Cambridge University Press), pp. 179–87.

1992. Cranial morphology of *Prosaurolophus* (Ornithischia; Hadrosauridae) with descriptions of two new hadrosaurid species, and an evaluation of hadrosaurid phylogenetic relationships. *Museum of the Rockies Occasional Paper* 2: 1–120.

Horner, J. R., & Makela, R. 1979. Nest of juveniles provides evidence of family structure among dinosaurs. *Nature* 282: 296–8.

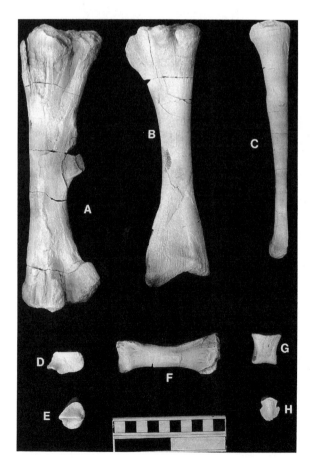

Figure 21.27. *Hypacrosaurus stebingeri.* MOR 548. **A, B**. Left femur (A) and tibia (B) in posterior views. **C**. Left fibula in medial view. **D, E**. Right astragulas (D) and left calcanium (E) in dorsal aspect. **F**. Left metatarsal III in anterior view. **G**. Left first phalange III in dorsal view. **H**. Ungual phalange from the pes in dorsal aspect. Scale = 10 cm.

Horner, J. R., Varricchio, D., & Goodwin, M. 1992. Marine transgressions and the evolution of Cretaceous dinosaurs. *Nature* 358: 59–61.

Horner, J. R. & Weishampel, D. B. 1988. A comparative embryological study of two ornithischian dinosaurs. *Nature* 332: 256–7.

Jepsen, G. L. 1964. Riddles of the terrible lizards. *American Scientist* 52: 227–46.

Langston, W., Jr. 1960. The vertebrate fauna of the Selma Formation of Alabama. Part VI. The dinosaurs. *Fieldiana: Geology Memoirs* 3: 319–63.

Lull, R. S. and Wright, N. E. 1942. Hadrosaurian dinosaurs of North America. *Geological Society of America Special Papers* 40: 1–242.

Morris, W. J. 1978. *Hypacrosaurus altispinus?* Brown from the Two Medicine Formation, Montana, a taxonomically indeterminate specimen. *Journal of Paleontology* 52: 200–5.

Ostrom, J. H. 1961. Cranial morphology of the hadrosaurian dinosaurs. *American Museum of Natural History Bulletin* 122: 33–186.

Ricqles, A. de. 1976. On bone histology of fossil and living reptiles, with comments on its functional and evolutionary significance. In A. d'A. Bellairs & C. B. Cox (eds.), *Morphology and Biology of Reptiles*. (London: Academic Press), pp. 123–50.

Sochava, A. V. 1972. The skeleton of an embryo in a dinosaur egg. *Paleontological Journal* 6: 527–31.

Sternberg, C. M. 1955. A juvenile hadrosaur from the Oldman Formation of Alberta. *National Museum of Canada Bulletin* 136: 120–2.

Weishampel, D. B. 1981a. The nasal cavity of lambeosaurine hadrosaurids (Reptilia: Ornithischia): Comparative anatomy and homologies. *Journal of Paleontology* 55: 1046–57.

Weishampel, D. B. 1981b. Acoustic analyses of potential vocalization in lambeosaurine dinosaurs (Reptilia: Ornithischia). *Paleobiology* 7: 252–61.

Weishampel, D. B., & Horner, J. R. 1990. Hadrosauridae. In D. B. Weishampel, P. Dodson, & H. Osmolska (eds.), *The Dinosauria*. (Berkeley: University of California Press), pp. 534–61.

Table 21.1. Length-to-midshaft-circumference ratios for ontogenetic series of femora and tibiae of *Hypacrosaurus stebingeri*

Specimen numbers	Length (mm)	Length/circumference ratio
Embryonic femora		
RTMP 87.79.265		2.00:1
	62.5	
RTMP 89.79.52	68	2.00:1
MOR 562	80	2.00:1
RTMP 87.79.219	84	2.10:1
Nestling femora		
MOR 548-F1	168	2.25:1
MOR 548-F2	195	2.32:1
MOR 548.F3	235	2.35:1
Juvenile femur		
MOR 35	600	2.40:1
Subadult femur		
MOR 553	870	2.61:1
Adult femur		
MOR 549 (holotype)	1,050	2.75:1
Embryonic tibiae		
RTMP 87.79.113	66	2.76:1
RTMP 87.79.115	79	2.35:1
MOR 562	80	2.28:1
RTMP 87.79.114	81	2.25:1
Nestling tibiae		
MOR 548-TI	140	2.33:1
MOR 548-T2	180	2.57:1
MOR 548-T3	220	2.77:1
Juvenile tibia		
MOR 355-8-26-5-8	600	3.19:1
Adult tibia		
MOR 355-8-25-5-4	1,060	3.31:1

Appendix 21.1. Diagnosis for *Hypacrosaurus stebingeri* n. sp. with discussion

Family Lambeosauridae (Horner, 1990)
Genus *Hypacrosaurus*; Brown 1913
Hypacrosaurus stebingeri, new species
Holotype
MOR 549, nearly complete skeleton, (Fig. 21.31).

Referred specimens
AMNH 5461, MOR 355, 455, 548, 553 (in part), 559 and 562; RTMP 88.79.36, 89.79.52, 87.79.22, 89.79.53 (for a more complete list of RTMP specimens see Appendix 21.2) and USNM 7948 and 11893.

Type locality
TM-065, Badger Creek, Glacier County, Montana.

Stratigraphic horizon
Upper Two Medicine Formation.

Age
Upper Cretaceous (Upper Campanian).

Etymology
stebingeri, to honor the late Eugene Stebinger, who first described the Two Medicine Formation and discovered the first remains of this species.

Diagnosis
This species has no autapomorphies. With *Hypacrosaurus altispinus* it shares a restricted external naris, wide narial crest, and restricted dorsal centra with very tall neural spines. With species of *Lambeosaurus* it shares a narial crest, the majority of which is composed of the premaxillae. *Hypacrosaurus stebingeri* has a long thin lacrimal intermediate in morphology between species of *Lambeosaurus* and *Hypacrosaurus altispi-*

Figure 21.28. *Hypacrosaurus stebingeri*. Histological sections of embryonic cortex (A, B) of femur (MOR 559), and nestling cortex (C, D) of femur (MOR 548). **A**. Section showing tremendous vascularization of the embryo. **B**. Section showing unorganized nature of embryonic bone. **C, D**, Sections showing woven bone and much more organized and oriented around vascular canals. Scale of A and C = 1 mm; of B and D = 0.5 mm.

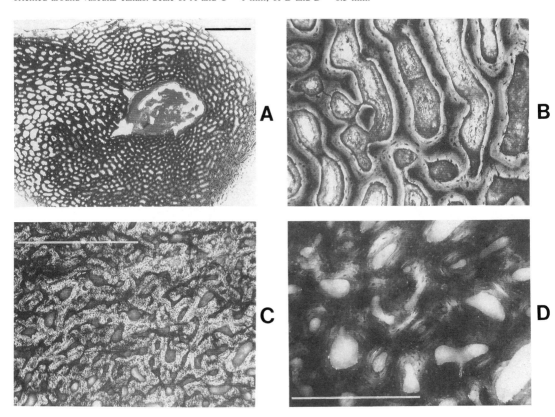

Figure 21.29. *Hypacrosaurus stebingeri*. Reconstructed composite skeleton of embryo (RTMP 88.3.2). Scale = 5 cm.

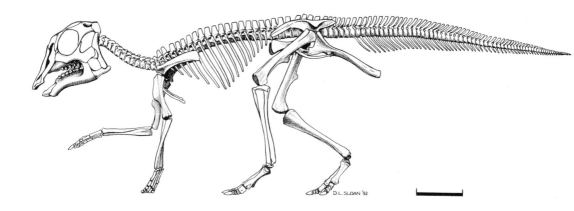

Figure 21.30. *Hypacrosaurus stebingeri*. Reconstructed skeleton of nestling (MOR 548). Scale = 21 cm.

Figure 21.31. *Hypacrosaurus stebingeri*. MOR 549. **A–D**. Holotype skull in lateral view (A), left ilium in lateral aspect (B), and middorsal vertebra in anterior (C) and lateral (D) views. Scale in centimeters.

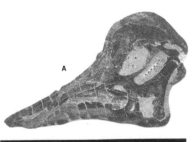

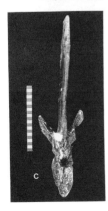

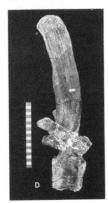

nus, and also has a narial crest with its apex directly over the anterior ends of the orbits, similar to the two latter taxa.

Discussion

Hypacrosaurus stebingeri is an unusual dinosaur taxon in that it has no autapomorphies. Phylogenetically, it appears to be an intermediate stage in the evolution of *Lambeosaurus* from the Judith River Formation to *Hypacrosaurus altispinus* from the Horseshoe Canyon Formation (Horner, Varricchio, & Goodwin, 1992). Interestingly, in juvenile specimens of *H. stebingeri* the anterior end of the nasal branch, where it meets the premaxilla, is bifurcated in a manner similar to species of *Corythosaurus*. More mature individuals of *H. stebingeri*, however, indicate a greater expansion of the premaxilla, whereas in *Corythosaurus* juveniles and subadults there is greater expansion of the nasal. Without the ontogenetic stages of these lambeosaurid taxa it would be difficult to identify some juveniles to genus and species even with an attempt as rigorous as that by Dodson (1975).

A specimen of *Hypacrosaurus altispinus*? was reported from the Two Medicine Formation of Montana by Gilmore (1917). Morris (1978) stated that there was not enough cranial material to verify whether or not Gilmore's specimen belonged to *Hypacrosaurus*, and suggested that it might instead be a species of *Lambeosaurus*. Whether coincidental or not, it is interesting that MOR 549 indicates a close relationship between *Lambeosaurus* and *Hypacrosaurus* as suggested by Morris. This conclusion stands in contrast to that of Lull and Wright (1942) that *Hypacrosaurus* was more closely related to *Corythosaurus* because the narial crests appear similar. Gilmore's specimen was found at the same stratigraphic level as the holotype (MOR 549) and other referred specimens of *H. stebingeri* and is therefore referred to as *H. stebingeri*.

With regard to the stratigraphic positions of the two species of *Hypacrosaurus*, it should be noted that *H. stebingeri* postdates the known species of *Lambeosaurus*, but predates *Hypacrosaurus altispinus* (Horner et al., 1992). *H. stebingeri* is placed in the genus *Hypacrosaurus* because it shares more characters with *Hypacrosaurus* than it does with *Lambeosaurus*.

APPENDIX 2 Specimens of *Hypacrosaurus stebingeri* from Devil's Coulee

Little Diablo's Hill, Nest No. 2

Four unprepared broken eggs with embryonic bone (RTMP 88.79.36). Disarticulated bones from at least three individuals found along the eroded edge of the nest (RTMP 89.79). A single associated individual from the nest (RTMP 89.79.52). Cranial material (* = from same individual): an almost complete skull with associated squamosal and lower jaws RTMP 89.79.52 (Fig. 21.4); articulated pair of premaxillae, *RTMP 87.79.334 (Fig. 21.15), *RTMP 87.79.370); five right maxillae, RTMP 87.79.108, .153, .155, .238, .306 and three left maxillae, RTMP 87.79.154, .286, *.336 (Fig. 21.16); left jugal, *RTMP 87.79.332 (Fig. 21.17) and right jugal, *RTMP 87.79.369; two left postorbitals, *RTMP 87.79.333 (Fig. 21.19), RTMP 87.79.364; two sets of parietals, RTMP 87.79.241 (Fig. 21.6), *RTMP 87.79.374, three left quadrates, RTMP 87.79.18, .77, *.320 (Fig. 21.18) and one right quadrate, RTMP 87.79.298; a left frontal, RTMP 87.79.206 (Fig. 21.8), a pair of frontals, *RTMP 87.79.371, .372; left exoccipital, *RTMP 87.79.307 (Fig. 21.9); left and two right exoccipitals, RTMP 87.79.158; left laterosphenoid, RTMP 87.79.227 (Fig. 21.12), right laterosphenoid, *RTMP 87.79.373, left and right prootic, RTMP 87.79.201, *.303; parasphenoid–basisphenoid, *RTMP 87.79.335 (Fig. 21.13); basioccipitals, RTMP 87.79.157 (Fig. 21.10), RTMP 87.79.193; predentary, *RTMP 87.79.263; pair of lower jaws, *RTMP 87.79.266 (Fig. 21.22, *RTMP 87.79.267; left dentary fragments, RTMP 87.79.50, 51, 52, 149, .150, .151; right dentary fragments, RTMP 87.79.53, .152, .253; left ectopterygoid, RTMP 87.79.247.

Little Diablo's Hill, Nest No. 3

Broken egg containing an articulated skeleton with skull, RTMP 87.79.22. Not yet prepared.

Little Diablo's Hill, Nest No. 5

Right prefrontal, RTMP 87.79.17; laterosphenoid, RTMP 87.79.16; right exoccipital, RTMP 87.79.18; right dentary, RTMP 87.79.6. Specimens small representing embryos arrested earlier in development.

Little Diablo's Hill, Nest RTMP 89.79.53

Specimens found several meters east and 25 cm higher than Nest No. 2. Eight eggs and probably more originally (Fig. 21.2). No bone present and CT scans of two eggs did not reveal any embryonic skeletons.

Little Diablo's Hill

Isolated eggs, RTMP 88.79.34, 35, found near Nests 2 and 5. Eggs may have washed out been expelled from these nests. No embryonic bones present in eggs, suggesting they may have been sterile.

North Baby Butte

Locality stratigraphically lower than Little Diablo's Hill. Left prefrontal, RTMP 87.77.92 (Fig. 20.20), three basioccipitals, RTMP 87.77.88, .89, .90; five dentary fragments RTMP 87.77.83, .84, .85, .86, .138.

Kiddie's Corner

Locality 100 m from and 10 m below Little Diablo's Hill. Fragments of a left jugal, RTMP 87.80.25; left dentary RTMP 87.80.26.

22 A nodosaurid scuteling from the Texas shore of the Western Interior Seaway

LOUIS L. JACOBS, DALE A. WINKLER, PHILLIP A. MURRY, AND JOHN M. MAURICE

Abstract

A small dinosaur specimen, less than 1 m in total length, has been found in the marginal marine Paw Paw Formation (Lower Cretaceous, Albian) in Fort Worth, Texas. The specimen is represented by various skeletal parts. It is a nodosaurid as determined by derived features of the skeleton, notably the displacement of the scapular spine toward the glenoid. Immature traits include the spongy texture of the bone and lack of fusion of composite bones. As in young stegosaurs, a lesser trochanter is evident on the femur. No dermal armor has been recovered with the specimen, possibly because it had not ossified at the growth stage represented.

The young nodosaurid was introduced into the marine environment where it was scavenged, probably by sharks and crustaceans. The ilium was then utilized as a substrate by oyster spat before being buried. The entire process from birth to burial of this specimen probably entailed only a few months, perhaps as brief as one.

The scuteling indicates the presence of a breeding population of nodosaurids along the southwestern shore of eastern North America after the completion of the Western Interior Seaway. The incomplete development of the long bone metaphyses suggests altricial behavior in the nodosaurids.

Introduction

The Ankylosauria, a suborder of predominantly Cretaceous, mostly northern hemisphere, armored, quadrupedal ornithischians (Coombs & Maryanska, 1990), has two families: Nodosauridae and Ankylosauridae. Only one of these, the Nodosauridae, is known from Texas. Nodosaurids are known from the Upper Cretaceous of Big Bend (Lehman, 1987; Carpenter, 1990; Weishampel, 1990) and from the Lower Cretaceous of central Texas. One specimen from the Albian Paw Paw Formation is ontogenetically the youngest nodosaurid yet to be described.

We describe this young specimen to confirm its identification as a nodosaurid. Next, we use it to assess ontogenetic change in nodosaurids. Finally, we evaluate the geologic age and taphonomic context of the specimen. Where appropriate, we make morphological comparisons with the large sample of immature individuals of an undescribed small ornithopod from the Twin Mountains Formation (Lower Cretaceous, Aptian) from Proctor Lake, Texas (Winkler et al., 1988; Winkler and Murry, 1989).

Quite young ankylosaurids have been discovered recently in China (Lessem, 1989), and a juvenile ankylosaurid has been described from Canada (Coombs, 1986). Some larger nodosaurids have been described as subadult [e.g., *Polacanthus* (? = *Hylaeosaurus*), Blows, 1987]. However, we are not aware of any other nodosaurid specimen that represents such an immature stage of ontogeny as seen in the Paw Paw specimen. Consequently, there is a dearth of comparative nodosaurid material representing the growth stage of the Paw Paw juvenile. We do not attempt a taxonomic identification and refer to the specimen simply as a nodosaurid.

Johnny Maurice discovered the nodosaurid specimen in 1989, in the Lower Cretaceous Paw Paw Formation exposed near the Motorola, Inc. plant, Fort Worth, Texas [locality information on file at Southern Methodist University (SMU)]. He and his father (JMM) surface collected the exposed, disarticulated bones and screened the matrix for additional pieces. The result was that much of a tiny skeleton was recovered, including skull parts, teeth, limb bones, and vertebrae. It was brought to SMU and catalogued (SMU 72444). Given its immature stage of development, and considering that one of the striking features of ankylosaurians in general is dermal armor, we refer to the specimen as a scuteling.

Nodosaurids were widespread in North America during the Early Cretaceous before the connection of the Gulf of Mexico to the Arctic Ocean by the Western Interior Seaway. They are present in the Cloverly Formation of Wyoming and Montana, and in the Arundel

Formation of Maryland (Ostrom, 1970; Weishampel, 1990). The best known North American Early Cretaceous nodosaurid is *Sauropelta edwardsi* described by Ostrom (1970) from the Cloverly Formation. So far as we are aware, no immature specimens have yet been found definitively attributable to that species.

In Texas, there are no occurrences of nodosaurids before the completion of the Western Interior Seaway. A previous report of a questionable scute from the Antlers Formation (SMU 72429; Thurmond 1969; mentioned by Langston, 1974) has been reinterpreted as a dermal ossicle of the turtle *Naomichelys*.

After completion of the Western Interior Seaway (approximately 100 million years ago) and the beginning of the Late Cretaceous (97.5 Ma) nodosaurids appeared in Texas. One specimen other than the scuteling has been found in the Paw Paw Formation. This specimen (Weishampel, 1990) is in the Smithsonian Institution (USNM 337987) and was examined by one of us (DAW) during this study. It is larger and thus more mature than the scuteling. The Smithsonian specimen was located at Blue Mound, while the scuteling was located approximately 5 km east of the community of Blue Mound. Thus, both Early Cretaceous records of nodosaurids in Texas are from the same formation and very close to each other geographically. These specimens are the oldest known records of nodosaurids in Texas, and they are from the eastern shore of the Western Interior Seaway. Late Cretaceous records of nodosaurids in Texas are from Big Bend, which was at that time west of the Seaway.

Stratigraphy and geologic age

The Paw Paw Formation, which contains the scuteling locality (SMU locality 241; Fig. 22.1), lies within the Washita Group. The Paw Paw is predominantly nearshore marine facies, including deltaic or estuarine (Scott et al., 1978). It has a diverse marine fauna (Table 22.1), but carbonized plant fragments as well as possible desiccation cracks suggest short periods of subaerial exposure.

The Albian–Cenomanian boundary (and therefore the Lower Cretaceous–Upper Cretaceous boundary) falls within the upper part of the Washita Group. The lowest level at which the boundary has been placed is at the base of the ammonite *Plesioturrilites brazoensis* Zone. This zone occurs within the Paw Paw Formation in north Texas along the Red River (Young, 1986). However, in the Dallas–Fort Worth area, *P. brazoensis* first occurs in the Main Street Limestone overlying the Paw Paw (Fig. 22.1). Alternatively, the base of the Cenomanian is sometimes recognized at the base of the ammonite *Graysonites adkinsi* Zone (Birkelund et al., 1984; Mancini, 1979), or using foraminifers, at the base of the *Rotalipora* s.s. Assemblage Zone (Michael, 1972). Both of these proposed boundaries fall within the Grayson Formation, which overlies the Main Street in

the Dallas–Fort Worth area. Thus, by any of these definitions, the Paw Paw Formation at the scuteling locality underlies the base of the Cenomanian and is therefore Albian. The age of the Albian–Cenomanian boundary is 97.5 Ma (Kent & Gradstein, 1985).

The completion of the Western Interior Seaway is marked by the *Inoceramus comancheanus* Zone, which represents the spread of this benthic bivalve mollusk along the seaway (Kauffman, 1984). The lowest occurrence of *I. comancheanus* in Texas is in the Kiamichi Formation, well below the Paw Paw (Perkins 1960). Jacobs, Winkler, and Murry (1991) estimate its age to be approximately 100 Ma. From these dates, we conclude that the Paw Paw scuteling is between 100- and 97.5-million-years-old, probably closer to the younger limit. It lived when the Western Interior Seaway was complete.

Description

The Paw Paw scuteling is recognized as an immature individual because of its small size (humerus length 68.4 mm, Table 22.2); unfused centra and neural arches; thin, fibrous, and plexiform periosteal bone; and spongy metaphyses. In the description and comparisons that follow, those features of the skeleton that facilitate the identification of the specimen are emphasized.

The elements of the skull are unfused. Some fragments of bone that cannot be positively identified appear most likely to be skull elements. Major bone overgrowth, as seen in adult nodosaurids and ankylosaurids, is apparently not yet fully developed, but some excrescence is present.

The basioccipital (Fig. 22.2A,C,E) is nearly as wide as it is long. The occipital condyle, formed entirely from the basioccipital, rests on a short, thick, downturned neck as in other Ankylosauria. It is hemispherical ventrally, but more acute dorsally. In the scuteling, the unique ankylosaurian occipital orientation is demonstrated by comparison with an ornithopod basioccipital from Proctor Lake (Fig. 22.2B,D,F). The dorsal surface of the scuteling basioccipital is convex and therefore must be rotated caudoventrally relative to the skull axis to allow for the passage of the spinal cord. Rostrodorsally, the basioccipital has a distinct tuberosity for articulation with the basisphenoid. The ventral surface lacks a strong median ridge. In the ornithopod from Proctor Lake, which we take to be primitive in these features; the dorsal surface of the basioccipital and condyle are concave, the ventral surface has a strong median ridge; and there is no tubercle for articulation with the basisphenoid.

The basisphenoid (Fig. 22.3A) is shorter than the basioccipital. It has a shallow pit in the dorsum sellae for the reception of the basioccipital tubercle. The basal tubera (Maryanska, 1977) at the ventral border with the basioccipital are poorly developed. Strong basipterygoid processes project ventrolaterally. A large Vidian canal pierces the dorsum sellae and enters the pituitary fossa. On the ventral surface, grooves extend rostromedially from near the entrance of the Vidian canal and around the basipterygoid processes. These are perhaps for the internal palatine arteries as suggested by Maryanska (1977).

The left maxilla (Fig. 22.3B) is nearly complete. It is broadly convex dorsoventrally. The cheek region is more strongly overhung caudally than rostrally. No part of the cheek is oriented horizontally. The dorsolateral wall of the maxilla is thin with its highest point near the middle of the maxilla. A groove is present along the caudodorsal border of the lateral process for articulation with the lacrimal. Immediately caudal and ventral to this groove is the articulation for the jugal. The medial wall is expressed in the rostral half of the maxilla with its highest point near the rostral end. The lateral and medial walls are joined near the midpoint of the maxilla but diverge rostrally.

The tooth row flares caudolaterally. There are approximately twelve tooth positions with two nearly complete and several partial replacement teeth unerupted in the maxilla. Isolated but associated teeth (Fig. 22.3C), found by sieving, match the morphology of developing teeth in the jaw. These teeth are coarsely denticulate and leaf shaped as in other thyreophorans and primitive dinosaurian herbivores generally. A cingulum is poorly defined on one side.

Both quadrates (Fig. 22.3D,E) are represented, but the pterygoid flange of each is damaged. Articulation with the lower jaw occurs through an asymmetrical condyle that is broader medially and tapers to a thin ridge laterally, extending slightly up the lateral aspect of the shaft. The pterygoid flange diverges slightly caudally from the plane of the condyle. Proximally, a smooth, rounded surface for articulation with the squamosal is much smaller than the distal condyle. Traced distally the quadrate shaft bends rostrally and laterally.

One bone is identified as an incomplete right frontal (Fig. 22.4A). It apparently formed the roof of the orbit. The dorsal surface is rugose, while the ventral surface has a fibrous texture. The lateral margin is complete, tapering to the edge. The frontal shows no evidence of articulation with a supraorbital, nor is it fused with any other bones. There is a small pit at the caudolateral margin of the frontal, probably for articulation with the postorbital. The other margins are broken. Ventrally, a keel marks the medial wall of the orbit.

The axial skeleton is represented by seventeen essentially complete centra, one complete neural arch, and

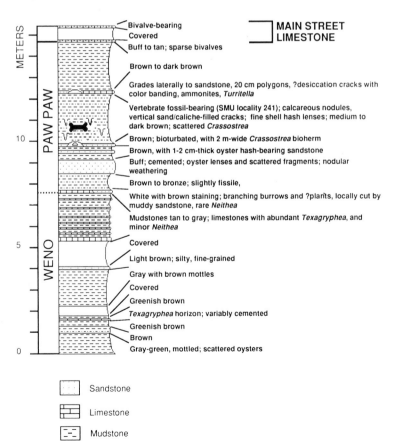

MAIN STREET LIMESTONE

Bivalve-bearing
Covered
Buff to tan; sparse bivalves

Brown to dark brown

Grades laterally to sandstone, 20 cm polygons, ?desiccation cracks with color banding, ammonites, *Turritella*

Vertebrate fossil-bearing (SMU locality 241); calcareous nodules, vertical sand/caliche-filled cracks; fine shell hash lenses; medium to dark brown; scattered *Crassostrea*

Brown; bioturbated, with 2 m-wide *Crassostrea* bioherm
Brown, with 1-2 cm-thick oyster hash-bearing sandstone
Buff; cemented; oyster lenses and scattered fragments; nodular weathering

Brown to bronze; slightly fissile,

White with brown staining; branching burrows and ?plants, locally cut by muddy sandstone, rare *Neithea*
Mudstones tan to gray; limestones with abundant *Texagryphea*, and minor *Neithea*

Covered

Light brown; silty, fine-grained

Gray with brown mottles

Covered

Greenish brown

Texagryphea horizon; variably cemented

Greenish brown

Brown

Gray-green, mottled; scattered oysters

Figure 22.1. Stratigraphic section at nodosaurid scuteling locality (SMU locality 241).

Sandstone

Limestone

Mudstone

Silty Mudstone

fragments of several more vertebrae or processes. The precise vertebral formula cannot be determined. There are five complete cervical centra, including the lunate atlas (Fig. 22.5A). The other four cervicals (Fig. 22.5B,C) bear facets for rib articulation. The axis was not recovered. Eight dorsal centra (Figs. 22.4B, & 22.5D,E) lack rib facets and increase in length caudally along the vertebral column. One sacral centrum (Fig. 22.5F,G) is complete. It is broad with a wide neural canal. Three centra have facets for hemal arches and incipient transverse processes and are therefore identified as caudals (Fig. 22.5H,I). None of these centra shows a notochordal projection (Coombs, 1978). One complete neural arch has a wide neural canal, low cau-

dally inclined spine, and short transverse processes. Compared to the size of centra, it appears to correspond to an anterior dorsal.

The pectoral girdle is represented by an unfused left proximal scapula (Fig. 22.6A) and coracoid (Fig. 22.4C). The scapular spine is displaced toward the glenoid. The coracoid foramen is not closed. A robust left humerus (Fig. 22.4D) with a broad proximal end is complete except for surface damage, particularly on the distal half. The deltopectoral crest extends distally approximately half way down the shaft. The distal end is not greatly rotated relative to the proximal end. Only the distal end of the right humerus is present.

Table 22.1. Fauna from the Paw Paw Formation at the nodosaurid scuteling locality (SMU locality 241)

Mollusca
  Bivalvia
    Stearnsia robbinsi
    Trigonia clavigera?
    Texigryphea washitaensis
    Neithea sp.
    Lopha quadriplicata
    Lima sp.
  Gastropoda
    Turritella sp.
  Cephalopoda
    Engonoceras sp.
Arthropoda
  Crustacea
    Linuparus adkinsi
    Xanthosia wintoni
    Xanthosia aspera
    Raninella sp.
Echinodermata
  Crinoidea
    Poecilocrinus dispandus
  Asteroidea indet.
  Echinoidea indet.
Vertebrata
  Chondrichthyes
    Pseudohypolophus sp.
    Paraisurus macrorhiza
    Cretolamna appendiculata
    Leptostyrax macrorhiza
    Squalicorax sp.
  Osteichthyes
    Ichthyodectiformes
      Saurodon sp. or *Saurocephalus* sp.
  Reptilia
    Testudines indet.
    Dinosauria
      Nodosauridae, not determined

Table 22.2. Skeletal measurements of Paw Paw nodosaurid scuteling (SMU 72444)

Element	Length (mm)	Width (mm)	Height (mm)
Basioccipital	12.7	12.3	–
Condyle	–	7.1	–
Basisphenoid	11.4	12.0	–
across basipterygoid processes	–	11.0	–
Quadrate (right)	22.5	19.5(d)	–
Maxilla	–	–	12.7
Tooth a	3.1	1.5	3.2
Tooth b	2.4	1.2	2.7
Tooth c	2.6	1.4	2.6
Atlas	4.4	10.8	–
Cervical centrum a	7.8	12.6	9.5
Cervical centrum b	8.8	13.1	10.0
Cervical centrum c	9.9	12.7	9.8
Cervical centrum d	10.1	12.8	9.7
Dorsal centrum e	10.5	12.5	10.7
Dorsal centrum f	11.9	11.5	–
Dorsal centrum g	14.0	–	–
Dorsal centrum h	13.8	10.9	11.6
Dorsal centrum i	14.6	11.5	11.6
Dorsal centrum j	14	11.0	11.7
Dorsal centrum k	14.8	12.0	11.1
Dorsal centrum l	14.8	13.2	10.8
Sacral centrum m	11.8	15.8	10.7
Caudal centrum n	8.3	11.0	8.6
Caudal centrum o	7.7	12.5	8.8
Caudal centrum p	8.9	8.8	8.4
Scapula cranial end	–	28.5	–
Coracoid	26.7	21.9	–
Humerus	68.4	27.2(p)	–
	–	22.8(d)	–
Ilium (incomplete)	–	–	–
Femur	73.0	18.4(d)	–
Ungual phalanx	8.0	4.5	–

Notes: d, distal; p, proximal.

The pelvic girdle is represented by the right ilium (Fig. 22.4E, 22.6B). The ilium has a broad, flat, horizontal, and bladelike anterior process. The medial margin appears to parallel the vertebral column, but the lateral margin diverges anteriorly from the midline. The ilium is not fused to any vertebrae or to the other elements of the girdle. The ilium forms the dorsal margin of the acetabulum, where it has a caudal process for articulation with the ischium and a cranial process for the pubis. Neither pubis was recovered.

The right femur (Figs. 22.4F,G, and 22.6C) is preserved. The head is damaged, but there is a distinct, columnar lesser trochanter on the craniolateral surface. A fourth trochanter is present as a low but distinct ridgelike process just proximal to the midpoint of the shaft. Distal hindlimb elements include a questionable right fibula with a corroded surface and distal end missing. There is one questionable tarsal element, possibly a calcaneum. Four metapodials are present but their position is unclear. Two phalanges have been recovered, including one ungual (Fig. 22.6D).

Ontogeny

The growth stage exhibited by the Paw Paw specimen allows testing of the effects of ontogeny on characters used in phylogeny. Characteristics primitive for ornithischians and exhibited by the scuteling are tooth shape (Fig. 22.3C) and a lesser trochanter (Fig. 22.6C) on the femur. The robust humerus (Fig. 22.4D) and large, flat, horizontally projecting blade of the ilium (Figs. 22.4E, and 22.6B), are derived features associated with quadrupedalism. The scuteling possesses unambiguous traits that mark it as an ankylosaurian, and a nodosaurid in particular. The occipital condyle (Fig. 22.2C) is oriented downward on a short thick neck, and

the basipterygoid processes (Fig. 22.3A) are prominent (Coombs & Maryanska, 1990; the latter condition possibly stands in contrast to that in *Sauropelta edwardsi*; Ostrom, 1970). The orientation of the occipital condyle reflects the downturned position of the head (e.g., Lambe, 1919).

A ridgelike fourth trochanter (Fig. 22.4F) is located on the proximal half of the femur, and the coracoid is large with a round cranioventral corner. Most significant, however, is the displacement of the scapular spine toward the glenoid (Fig. 22.6A), a synapomorphy of nodosaurids (Coombs, 1978; Coombs & Maryanska, 1990).

The total craniocaudal length of the scuteling is estimated to be about 0.7 m. An adult *Sauropelta* is estimated to have been 5.2 m in total length, with an estimated body weight of 1500 kg (Carpenter, 1984). Length measurements of *Sauropelta* humeri presented by Ostrom (1970) average about 0.57 m, while the humerus of the scuteling is less than 0.07 m. Based on these numbers, it is reasonable to suggest that the scuteling represents a growth stage between 10–15 percent of adult body size.

Many of the features observed in the scuteling are simply the result of immaturity. Its state of ossification and its size indicate that it was a neonate, although it is uncertain how old this individual was at death. The bone is quite spongy, including that of the metaphyses of the limbs (Fig. 22.4D). A major indication of immaturity is the lack of fusion among bones, such as between vertebral centra and neural arches, and among skull bones. Significant in this regard is the lack of fusion of the pelvic girdle elements into a solid synsacrum (or the fusion of the posterior dorsals into a rod extending craniad from the pelvic girdle).

In adult nodosaurids, the quadrate is fused to the

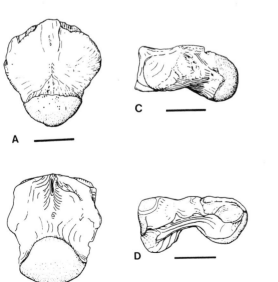

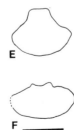

Figure 22.2. **A, C, E.** Basioccipital of the Paw Paw nodosaurid scuteling (SMU 72444) in ventral (A) and left lateral (C) views; outline of occipital condyle is shown in E. **B, D, F.** Basioccipital for a Proctor Lake juvenile ornithopod (SMU 72549) shown for comparison (ventral, B; left lateral, D; and occipital condyle, F). Entire occipital condyle is derived from the basioccipital in the scuteling, but includes portions of exoccipitals (not shown) in the Proctor Lake ornithopod. Scale bar = 5 mm.

squamosal and the pterygoid to the basisphenoid, but this is not the case in most ankylosaurids (Coombs, 1978). The lack of fusion in these elements in the scuteling is expected, given the state of development. However, the presence of a smooth articular condyle on the quadrate (Fig. 22.3D, E) for articulation with the squamosal is unexpected. It implies that cranial kinesis occurred in juvenile nodosaurids, rather than in adults. This kinesis suggests ontogenetic variation in chewing mechanics.

The scapula (Fig. 22.6A) and coracoid (Fig. 22.4C) are unfused at this stage, and the coracoid foramen is a notch not yet encircled by bone. Less than half of the twenty-five to twenty-seven adult tooth positions found in *Sauropelta* (Ostrom, 1970) are present in the scuteling. The maxilla (Fig. 22.3B) lacks the important adult character of an overhung (horizontal) cheek margin.

The ilium (Fig. 22.6B) of the scuteling demonstrates the very early ontogenetic development of a ventromedially facing anterior process, a synapomorphy of adult ankylosaurians. However, the presence of a distinct pubic peduncle is interpreted as an ontogenetic plesiomorphic holdover that was later lost in development.

The presence of a distinct lesser trochanter on the femur and its absence in most known adult femora may suggest ontogenetic variation in this feature similar to that seen in stegosaurs (Galton, 1982a, 1982b, 1983a). However, a cleft lesser trochanter, similar in morphology to the Paw Paw nodosaurid, is visible in lateral aspect on the femora of the Upper Jurassic nodosaurid *Cryptodraco* and in the Lower Cretaceous *Hoplitosaurus* (Galton, 1983b; and possibly other genera; Coombs, personal communication). In adult nodosaurids the fourth trochanter is located on the proximal half of the femur, a condition that differs from the more distal fourth trochanter of ankylosaurids, and that indicates the lack of musculature associated with a tail club (Coombs, 1979). The proximal position of the fourth trochanter in the scuteling (Fig. 22.4F) implies that little ontogenetic change occurs in this character and demonstrates its usefulness in family diagnosis. *Sauropelta* has a humerus to femur length ratio of 0.72 to 0.77 (Ostrom, 1970). That of the scuteling is 0.93, suggesting that the limb proportions may change ontogenetically with the hindlimb becoming longer relative to the forelimb.

No dermal body armor was recovered with the specimen. Because dermal armor is perhaps the most salient trait of the Ankylosauria, the apparent lack of it in the scuteling is puzzling. We cannot rule out the possibility that it simply was not preserved or collected even though it may have been present in life. However, we consider it plausible that the armor had not yet ossified in the growth stage represented by the scuteling, and that the absence may be accurate in this regard. This suggestion is supported by the preserved portions of the skull, which are rugose on the surface, but show no overgrowths of dermal bone as seen in larger ankylosaurian specimens. The lack of dermal armor in very young nodosaurids is consistent with observations on juvenile stegosaurs (Galton, 1982a), which have tail spikes but apparently no dermal plates.

No unequivocal derived traits are present in the scuteling to link it specifically with known species of nodosaurids. *Hylaeosaurus armatus* retains primitively an unfused scapula and coracoid as an adult (Coombs, 1978; Coombs & Maryanska 1990). This lack of fusion is predicted for all juvenile ankylosaurs and so can not be used to link the scuteling with *Hylaeosaurus*. A distinct lesser trochanter on the femur of the scuteling is also a presumed primitive retention in adult *Hoplitosaurus marshi* and *Cryptodraco eumerus*.

Taphonomy

The Paw Paw scuteling was found disarticulated but with all its remaining parts in close proximity (< 1 m²). Recovered bones (Table 22.2) include elements of the skull, representatives of each region of the vertebral column, portions of both the pectoral and pelvic girdles,

Figure 22.3. Cranial elements of the Paw Paw nodosaurid scuteling. **A**. Basisphenoid in ventral view. **B**. Left maxilla in lateral view. **C**. Isolated tooth. **D**. Right quadrate in lateral view. **E**. Right quadrate in rostral view. Scale for tooth (C) = 1 mm. Scale of A, B, D, E = 5 mm.

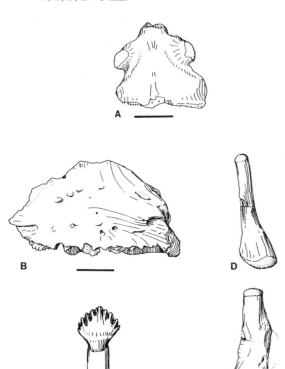

and parts of both fore- and hindlimbs, including an ungual phalanx. This distribution of recovered elements suggests that the body of the scuteling was intact when it was washed into the nearshore marine environment.

Evidence of scavenging is apparent on virtually every bone. Most common are small scratches (Fig. 22.4A,F), but some larger puncture marks (Fig. 22.4B,G) are present, for example on the femur. The scratches on each bone, and the occurrence of all the bones within a small area, suggest that the each element of the skeleton was individually picked clean of its flesh by small scavengers. Sharks and crabs (Table 22.1) are abundant at the scuteling site and would have been the most likely scavengers. Portions of the scuteling skeleton were probably removed by such animals, and this may account for the distribution in the skeleton of the

bones recovered and their lack of articulation when collected.

The defleshed bones of the scuteling provided a substrate on which oyster spat set (Fig. 22.4E). The diameters of two oysters on the ilium are 15 and 18 mm. Other than these two oysters, no other encrustations are apparent on the bones. Judging from the growth rates and mortality of modern oysters (Galtsoff, 1964), the bones remained unburied on the ocean floor for a short

Figure 22.5. Vertebral centra of Paw Paw nodosaurid scuteling in cranial and left lateral views. **A**. Atlas (cranial view only). **B, C**. Cervical centrum b. **D, E**. dorsal centrum j. **F, G**. Sacral centrum m. **H, I**. Caudal centrum n. Scale = 5 mm.

Figure 22.4. Skeletal elements of Paw Paw nodosaurid scuteling. **A**. Right frontal (?) in dorsal view, anterior to left. **B**. Dorsal vertebral centrum j (the lower case letter is a curatorial expedient) in ventral view showing puncture mark. **C**. Left coracoid in lateral view. **D**. Left humerus in caudal view. **E**. Right ilium in dorsal view showing oyster encrustation. **F**. Right femur in caudal view showing fourth trochanter and scratches. **G**. Right femur in cranial view showing the lesser trochanter, punctures, and scratches. Scale of A, B = 5 mm; scale of C, D, E, F, G = 10 mm.

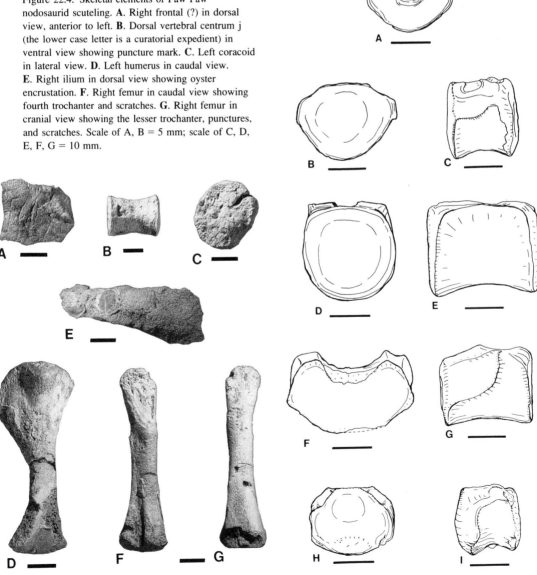

length of time, possibly between a month and a few years. During this time the oyster spat settled and grew. Burial of the bones halted the growth of the encrusting organisms.

Discussion and conclusions

Horner (1979) noted nine incidences of nodosaurids in Upper Cretaceous marine sediments of western North America and suggested a possible coastal environment for the group. The range of nodosaurids obviously did include such environments, but not to the exclusion of others, as demonstrated by the terrestrial Cloverly and Arundel occurrences. The spongy bone of the scuteling suggests altricial behavior for nodosaurid young because the ends of limb bones are not well formed as would be expected in species with more active, precocial young (Horner & Weishampel, 1988). The immature stage of the Paw Paw scuteling, the interpretation of altricial behavior, if correct, and the likelihood that it entered the marine environment intact all suggest that the scuteling was native to the coastal area near the Paw Paw sea. It was not likely to have traveled far of its own volition, and if transported by other agents, did not suffer much damage in the process. Furthermore, the immaturity of the scuteling suggest that nodosaurids reproduced along the shore. Specifically, there was a breeding population of nodosaurids along the southern shore of eastern North America.

The scuteling was transported whole into the water. It lay in a shallow area of sea where silt and clay were being deposited locally. The locality is far landward from the edge of the shelf, nearer to the shore, and at a latitude of no more than 33° north, the approximate current latitude. The climate was subtropical and water temperature was warm (Kauffman, 1984). Salinity was normal marine, not brackish, judging from the diverse invertebrate fauna, including abundant starfish, free-floating crinoids, and urchins. The scuteling was scavenged by crabs, small sharks, and perhaps other organisms until the bones were clean, after which oyster spat utilized the bones as a substrate. The oysters grew to a minimum of 15 mm before they died. Although the scuteling was not found in an oyster bank, concentrations of oysters are common in the Paw Paw. No other encrusting organisms are found on the bones indicating that the bones were buried and no longer available for that purpose. The whole process may have taken as little as a month.

The Paw Paw Formation, while marine, is generally considered a regressive phase because of its higher clastic content compared to underlying and overlying formations. It is one of four dinosaur bearing regressive sequences in north-central Texas. These are represented (in order) by the Twin Mountains Formation (Aptian), the Paluxy Formation (Albian), the Paw Paw Formation (late Albian), and the Woodbine Formation (Cenomanian). It is unclear whether the Paw Paw regression and the subsequent transgression are of local or of broader geographic significance. Nevertheless, invertebrate biostratigraphy imposes chronological constraints, and eustatic sea-level curves (e.g., Haq, Hardenbol, & Vail, 1987, 1988) can be used as a framework to facilitate broader correlations.

The age of the Paw Paw Formation is constrained by the underlying *Inoceramus comancheanus* Zone, which we have considered to be approximately 100 Ma (Jacobs et al., 1991). The younger age limit of the Paw Paw is constrained by the Albian–Cenomanian boundary, given by Kent and Gradstein (1985) as 97.5 Ma. Regardless of the absolute age of the Paw Paw, it must correlate with events and formations elsewhere that also fall between the *I. comancheanus* Zone (which is within the Kiowa–Skull Creek transgressive–regressive cycle

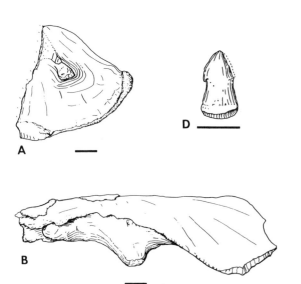

Figure 22.6. Skeletal elements of Paw Paw nodosaurid scuteling. **A.** Proximal left scapula in lateral view showing displacement of spine toward glenoid. **B.** Right ilium in lateral view. **C.** Right femur in lateral view showing cleft lesser trochanter. **D.** Ungual phalanx in dorsal view. Scale = 5 mm.

in the Interior; Kauffman, 1984) and the base of the Cenomanian. In western North America, one possible correlative terrestrial rock unit may be the J sandstone of the Denver Basin, which lies beneath an unconformity dated at 97 Ma (Weimer, 1988). Because of the uncertainties of dating and differences in time scales, this correlation is not unreasonable. In addition, the regression that controlled deposition of the Paw Paw may be part of the same large-scale process that led to the deposition of the J sandstone in the Denver Basin. Furthermore, the regression may have been at least regional in scope, if not eustatic.

Finally, at the growth stage it exhibits the scuteling probably resembled nothing in the modern world so much as an armadillo. Indeed, comparison of nodosaurids to armadillos is not without precedent (Carpenter, 1984). Armadillos are the quintessential animal of Texas (Smith & Doughty, 1984), and it is amusing that the scuteling is similar in both size and general body form to these animals. It is all the more amusing that the scuteling lived in what is now the heart of Texas so long before the construction of highways, the nemesis of many a hapless armadillo. Let us hope that armadillos can withstand their nemesis better than nodosaurids and most other dinosaurs did theirs.

Acknowledgments

We thank John C. Maurice, who found the baby nodosaurid, for bringing it to our attention and for placing it in the Shuler Museum of Paleontology. We thank S. Ray, Texas A & M University at Galveston, for information about living oysters. Acknowledgment is made to the donors of the Petroleum Research Fund, administered by the American Chemical Society, for partial support of this project. Additional support was provided by National Science Foundation grant BSR 88-16313 to (DAW and PAM) and institutional grants from Tarleton State University (to PAM).

References

Birkelund, T., Hancock, J. M., Hart, M. B., Rawson, P. F., Reman J., Robaszynski, F., Schmid, F., & Surlyk, F. 1984. Cretaceous stage boundaries – proposals. *Bulletin of the Geological Society of Denmark* 33: 3–20.

Blows, W. T. 1987. The armoured dinosaur *Polacanthus foxi* from the Lower Cretaceous of the Isle of Wight. *Palaeontology* 30: 557–80.

Carpenter, K. 1984. Skeletal reconstruction and life restoration of *Sauropelta* (Ankylosauria: Nodosauridae) from the Cretaceous of North America. *Canadian Journal of Earth Sciences* 21: 1491–8.

— 1990. Ankylosaur systematics: example using *Panoplosaurus* and *Edmontonia* (Ankylosauria: Nodosauridae). *In* K. Carpenter & P. J. Currie (eds.), *Dinosaur Systematics: Approaches and Perspectives*. (New York: Cambridge University Press).

Coombs, W. P. 1978. The families of the ornithischian dinosaur order Ankylosauria. *Palaeontology* 21: 143–70.

— 1979. Osteology and myology of the hindlimb in the An-

kylosauria (Reptilia, Ornithischia). *Journal of Paleontology* 53: 666–84.

— 1986. A juvenile ankylosaur referable to the genus-*Euoplocephalus* (Reptilia, Ornithischia). *Journal of Vertebrate Paleontology* 6: 162–73.

Coombs, W. P., & Maryanska, T. 1990. Ankylosauria. *In* D. B. Weishampel, P. Dodson & H. Osmólska (eds.), *The Dinosauria*. (Berkeley: University of California Press).

Galton, P. 1982a. Juveniles of the stegosaurian dinosaur *Stegosaurus* from the Upper Jurassic of North America. *Journal of Vertebrate Paleontology* 2: 47–62.

— 1982b. The postcranial anatomy of stegosaurian dinosaurian *Kentrosaurus* from the Upper Jurassic of Tanzania, East Africa. *Geologica et Palaeontologica* 15: 139–60.

— 1983a. A juvenile stegosaurian dinosaur, *Omosaurus phillipsi* Seeley from the Oxfordian (Upper Jurassic) of England. *Geobios* 16: 95–101.

— 1983b. Armored dinosaurs (Ornithischia: Ankylosauria) from the Middle and Upper Jurassic of Europe. *Palaeontographica* 182: 1–25.

Galtsoff, P. S. 1964. The American oyster *Crassostrea virginica* Gmelin. *United States Department of the Interior, Fish and Wildlife Service, Fishery Bulletin* 64: 1–480.

Haq, B. U., Hardenbol, J., & Vail, P. R. 1987. Chronology of fluctuating sea levels since the Triassic. *Science* 235: 1156–67.

— 1988. Mesozoic and Cenozoic chronostratigraphy and cycles of sea-level change. *Society of Economic Paleontologists and Mineralogists Special Publication* 42: 71–108.

Horner, J. R. 1979. Upper Cretaceous dinosaurs from the Bearpaw Shale (marine) of south-central Montana with a checklist of Upper Cretaceous dinosaur remains from marine sediments in North America. *Journal of Paleontology* 53: 566–77.

Horner, J. R., & Weishampel, D. B. 1988. A comparative embryological study of two ornithischian dinosaurs. *Nature* 332: 256–7.

Jacobs, L. L., Winkler, D. A., & Murry, P. A. 1991. On the age and correlation of Trinity mammals, Early Cretaceous of Texas, USA. *Newsletters on Stratigraphy* 24: 35–43.

Kauffman, E. G. 1984. Paleobiogeography and evolutionary response dynamic in the Cretaceous Western Interior Seaway of North America. *In* G. E. G. Westerman (ed.), *Jurassic-Cretaceous biochronology and paleogeography of North America. Geological Association of Canada, Special Paper* 27: 273–306.

Kent, D. V., & Gradstein, F. M. 1985. A Cretaceous and Jurassic geochronology. *Geological Society of America Bulletin* 96: 1419–27.

Lambe, L. M. 1919. Description of a new genus and species (*Panoplosaurus mirus*) of an armoured dinosaur from the Belly River beds of Alberta. *Transactions of the Royal Society of Canada* 13: 39–50.

Langston, W. 1974. Nonmammalian Comanchean tetrapods. *Geoscience and Man* 8: 77–102.

Lehman, T. M. 1987. Late Maastrichtian paleoenvironments and dinosaur biogeography in the western interior of

North America. *Palaeogeography, Palaeoclimatology, Palaeoecology* 60: 189–217.

Lessem, D. 1989. Secrets of the Gobi Desert. *Discover* June: 40–6.

Mancini, E. A. 1979. Late Albian and early Cenomanian Grayson ammonite biostratigraphy in north-central Texas. *Journal of Paleontology* 53: 1013–22.

Maryanska, T. 1977. Ankylosauridae (Dinosauria) from Mongolia. *Paleontologia Polonica* 37: 85–151.

Michael, F. Y. 1972. Planktonic foraminifera from the Comanchean Series (Cretaceous) of Texas. *Journal of Foraminiferal Research* 2: 200–20.

Ostrom, J. H. 1970. Stratigraphy and paleontology of the Cloverly Formation (Lower Cretaceous) of the Bighorn Basin Area, Wyoming and Montana. *Yale University, Peabody Museum of Natural History, Bulletin* 35: 1–234.

Perkins, B. F. 1960. Biostratigraphic studies in the Comanche (Cretaceous) series of northern Mexico and Texas. *Geological Society of America Memoir* 83: 1–138.

Scott, R. W., Fee, D., Magee, R., & Laall, H. 1978. Epeiric-depositional models for the Lower Cretaceous Washita Group. *The University of Texas at Austin, Bureau of Economic Geology, Report of Investigations* 94: 1–23.

Smith, L. L., & Doughty, R. W. 1984. *The Amazing Armadillo: Geography of a Folk Critter.* (Austin: University of Texas Press).

Thurmond, J. T. 1969. *Lower vertebrates and paleoecology of the Trinity Group (Lower Cretaceous) in northcentral Texas.* Ph.D. Thesis. (Dallas: Southern Methodist University).

Weimer, R. J. 1988. Record of relative sea-level changes,Cretaceous of Western Interior, USA. Sea-level changes – an integrated approach. *Society of Economic Paleontologists and Mineralogists Special Publication* 42: 285–8.

Weishampel, D. B. 1990. Dinosaurian distribution. *In* D. B. Weishampel, P. Dodson, & H. Osmólska (eds.), *The Dinosauria.* (Berkeley: University of California Press).

Winkler, D. A., and Murry, P. A. 1989. Paleoecology and hypsilophodontid behavior at the Proctor Lake dinosaur locality (Early Cretaceous), Texas. *In* J. O. Farlow (ed.), *Paleobiology of Dinosaurs. Geological Society of America Special Paper* 238: 55–61.

Winkler, D. A., Murry, P. A., Jacobs, L. L., Downs, W. R., Branch, J. R., & Trudel, P. 1988. The Proctor Lake dinosaur locality, Lower Cretaceous of Texas. *Hunteria* 2: 1–8.

Young, K. 1986. The Albian-Cenomanian (Lower Cretaceous-Upper Cretaceous) boundary in Texas and northern Mexico. *Journal of Paleontology* 60: 1212–19.

23 Dinosaur ontogeny and population structure: Interpretations and speculations based on fossil footprints

MARTIN G. LOCKLEY

Abstract

The recent resurgence of interest in dinosaur tracks has resulted in the accumulation of a large amount of size-frequency data derived from trackways. Such data bases are often larger and easier to obtain than most of those obtained from skeletal accumulations. Moreover, where trackways on single bedding planes represent gregarious groups, each can reasonably be assumed to represent an individual from a discrete population. Such data has obvious potential for population studies.

Trackway size–frequency data for large ornithopods and sauropods is compared with postulated growth curves for the dinosaur *Maiasaura* and known growth curves of modern vertebrates to infer the age structure of dinosaur populations. The results indicate that although footprint evidence for very small, hatchling-sized dinosaurs is not common, possibly supporting inferences of rapid early growth, many track assemblages apparently reveal significant numbers of juveniles and subadults.

A Cretaceous sauropod trackway sample from South Korea is particularly intriguing because it reveals evidence for a significant number of individuals that were no bigger than large domestic dogs. A persuasive case can be made for regarding these trackmakers as very young (post-hatchling) individuals in their first year of growth. Such animals were probably not far from their place of birth and were evidently not migrating or traveling purposefully in set directions as is often revealed by assemblages of trackways made by subadult and adult sauropods.

Introduction
Dinosaur growth rates: general models

"[T]he recognition of juvenile dinosaurs has emerged as a major theme of study," (Weishampel, Dodson, & Osmólska, 1990, p. 3 and references therein). However, dinosaur ontogeny and population structure are difficult to study (Thulborn, 1990, pp. 331–2). Despite an abundance of size–frequency data, there is currently little reliable information to help us determine the exact or absolute age of dinosaurs (see also Weis-hampel and Horner, Chapter 14). The same is true for most vertebrate fossils and for fossils in general, although as Kurtén (1953) has pointed out there is considerable potential for applying population dynamics to paleoecology, especially when dealing with fossil mammals.

To understand dinosaur ontogeny and population structure we must be able to determine the age of dinosaurs of given sizes with reasonable accuracy and confidence. Recent work by Horner and his collaborators has made some progress in this area (Horner 1987; Horner & Gorman, 1988). For example, we know the exact size of the hadrosaur and hypsilophodont hatchlings *Maiasaura* and *Orodromeus*, respectively (Horner & Weishampel, 1988) and can compare this with the maximum known size of the adults. This gives us a reliable size range and a relative age scale from zero years at hatching to estimated age when full grown. Horner (1992) estimated that *Maiasaura* reached adult size in about 5 years.

The paleontological literature is full of examples of studies that adopt the approach of equating size with age to establish relative age estimates (see Boucot, 1953; Craig & Oertel, 1966; Raup & Stanley, 1971; Richards & Bambach, 1975; Whyte, 1982; Cadee, 1988 for various invertebrate examples, and Kurtén, 1953 and references therein for vertebrate examples). The size–frequency data, although derived from dead individuals, is then examined to see how closely it compares with size–frequency distributions observed in living populations. The closer the correspondence the more the fossil assemblage is considered representative of a once–living population (that is, not reworked or taphonomically altered to a severe degree). Such studies rely to a large degree on analogy with modern organisms and their population dynamics. With invertebrates such as bivalves or corals, daily, monthly, or annual growth rings are often reliable indicators of absolute age. In vertebrates, especially mammals, teeth are usually the most reliable indicators of absolute age. Thus age can be compared with size and plotted to show a strong positive correlation.

The extent to which the expected correlation is linear or curvilinear (allometric) is a reflection of the morphological features that are measured and their growth rates throughout ontogeny. For example, some species may grow continuously throughout life and display a steady or regular growth rate. Others may grow rapidly during early ontogeny (juvenile stage) and display a marked decline in growth rate once sexual maturity (adulthood) is reached.

To date there have been few detailed studies of dinosaur growth rates. The best example is the famous ontogenetic series of *Protoceratops* specimens collected by the American Museum (Brown & Schlaikjer, 1940; Dodson, 1976), and believed to show sexual dimorphism. However, even in this sample estimates of age have proved speculative (Case, 1978). Other detailed studies of growth and variation in fossil vertebrates include the classic work of Olson (1951) on the Permian amphibian *Diplocaulus* and the equally thorough work of Kurtén (1953) on various Cenozoic mammals. The later study benefited from direct comparison between fossil and recent mammalian samples.

Growth rates have been plotted for modern vertebrates such as elephants, alligators, and ostriches (Dodson, 1975; Western, Moss, & Georgiadis 1983; Morell 1987; Obata & Tomida 1990). The results provide important models for interpretations and speculations on dinosaur ontogeny. These growth rates can be compared with growth curves for fossil mammals plotted by Kurtén (1953, Figs. 16–17). In the summaries of Horner's work, Morell (1987) and Obata and Tomida (1990) plot growth curves for *Maiasaura* and compare them with those obtained from ontogenetic studies of alligators and ostriches. They conclude that *Maiasaura* probably grew very rapidly like an ostrich rather than like an alligator, reaching sexual maturity within 5–7 years (Fig. 23.1).

Russell (1989, p. 150) evidently agreed with this model and suggested that full adult length (7 m) was attained in about 4 years. Such growth rates compare favorably with the rapid growth rates observed in mammals (Kurtén, 1953). Morell (1987, p. 35) quotes Horner as saying, "Baby hadrosaurs are thirteen inches long when they hatch, and they reach ten feet by the end of their first year." This growth rate is reminiscent of altricial birds, many of which accomplish most of their growth to full adult size within a year or less. Evidently, such growth rates can only be sustained by endotherms. Horner (1992) verified this growth rate quite explicitly, thus substantiating the reliability of the proposed growth curve (Fig. 23.1). Although these conclusions are speculative, there are several points suggest that the *Maiasaura* growth curve is valid as a general model:

1. Inferred growth rates are not the only evidence that many dinosaurs were fast growing endotherms. Latitudinal zonation, predator–prey ratios, bone histology, and mammal – and birdlike social behavior and activity levels are other lines of evidence cited for endothermy (see Ostrom 1969; Bakker, 1968, 1975, 1986 for discussion, and Farlow, 1990 for alternate arguments).

2. Large bone beds, for example of *Maiasaura* (Horner 1987; Horner & Gorman, 1988) and *Centrosaurus* (Currie & Dodson, 1984), apparently show discontinuities or clusters in size distributions. These are interpreted as annual recruitments (for example, 1, 2, and 3-year-old individuals; Russell, 1989). Such size or age group clusters are most likely to be detected in large monospecific populations (herds) of fast-growing species rather than in small assemblages or those of slow-growing species. In the former populations recruitments stand out more prominently as size clusters.

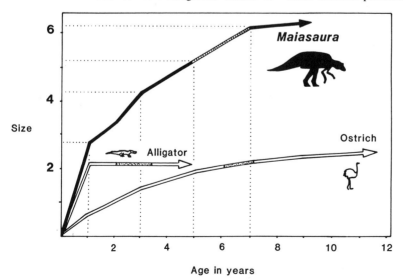

Figure 23.1. Postulated growth curve for *Maiasaura* (Hadrosauridae) based on the work of Horner (redrawn from Morell, 1987, and Obata & Tomida, 1990).

3. The scarcity of small dinosaurs (juvenile individuals) except at nest sites can be attributed, in part to taphonomic bias. However, it can also be attributed to rapid early growth rates such that most dinosaurs would be small (juvenile) only for a relatively short period in ontogeny (that is, during year one). This interpretation admittedly relies on negative evidence; however, the following nest site evidence clearly supports the rapid growth rate model proposed by Horner (Morell, 1987).

4. Juvenile hadrosaurs apparently grew from 13 inches to at least 3 or 4 ft in length before leaving their nests (Horner & Makela, 1979). This evidence clearly indicates that hatchlings and very young juveniles would rarely, if ever, have made tracks or become incorporated into fossil assemblages away from nesting colonies.

5. Because of the nest-site evidence, the lack of tracks of juvenile dinosaurs (first noted by Leonardi, 1981) of large species such as *Maiasaura* and other ornithopods is readily explained without relying purely on negative evidence. As more and more trackway data emerge, it becomes apparent that the tracks of small juveniles of most large dinosaur species are quite rare. We should note, however, that baby dinosaurs were once considered rare (Richmond, 1965), a perception we no longer uphold (Horner & Gorman, 1988; Chure, 1992). The current rarity of tracks of babies may simply reflect of lack of ichnological work in this field. It also may be a reflection of preservational and observational biases against preservation and recognition of small footprints (Lockley & Conrad, 1989).

Trackway data

To use trackway data to address questions regarding dinosaur ontogeny and population structure, we must make certain assumptions. To avoid undue speculation, these assumptions should be consistent with previous paleontological precedent and basic biological principles. First, we can assume that, within a particular ichnotaxon, dinosaur track size is a reflection of dinosaur body size and relative age. Studies of modern elephant tracks (Western et al., 1983) show that such a positive correlation holds true for large mammals (Fig. 23.2). Second, we assume that the best estimates of dinosaur growth rates are those discussed above for *Maiasaura* and that the absolute age estimates derived from Figure 23.1 are the best currently available.

We also must be aware that it is difficult to correlate tracks with particular species of dinosaurs (Lockley & Gillette, 1989; Thulborn, 1990; Lockley, 1991) or to prove that given track assemblages represent monospecific dinosaur populations. Given these uncertainties we should be selective about the track data used.

The track data that most probably represent monospecific assemblages are those obtained for a single ichnotaxon from a single bedding plane. This is especially true where there is evidence for herding, such as parallel trackways (Lockley, 1991). In such cases, the trackways give us useful size–frequency data for species that can be identified or labeled ichnotaxonomically, and attributed to particular families or orders (for example, hadrosaurs, ornithopods, ceratopsians or sauropods).

If the same ichnospecies (or ichnotaxon) occurs in a thin stratigraphic unit or at several horizons within a particular sedimentary facies or unit (cf. Lockley et al., 1992), we may infer that the same trackmaker is represented at several levels. In fact, without ichotaxonomic studies showing the presence of more than one ichnospecies belonging to a given group (e.g., hadrosaurs), it is most parsimonious to assume that a single trackmaking species is represented. Put another way, when a particular dinosaur group (e.g., sauropod or ornithopod) is represented by a single ichnotaxon, there is no paleontologic evidence to suggest that more than one species of that group is represented. This assumption is again consistent with the principles of paleoichnology and the now common observation that, within widespread ichnofacies, individual track assemblages (ichnocoenoses) repeatedly reveal similar compositions (Lockley et al., 1992).

Although the aforementioned uncertainties limit our discussion of dinosaur ontogeny and population structure to selected, unknown species (or taxa) of dinosaur, track data have some distinct advantages over size–frequency data derived from bones. First, we find that it is possible to obtain much more size–frequency data from large track assemblages than from most bone assemblages. Second, these data represent living individuals and populations (biocoenoses), not death assemblages (thanatocoenoses) as in bone beds or invertebrate shell accumulations. Third, in general track assemblages (ichnocoenoses) are less likely to be time averaged than bone accumulations. Recent studies have shown that

Figure 23.2. Relationship between elephant footprint size and age (redrawn from Western et al., 1983). Open circles represent males; black circles represent females.

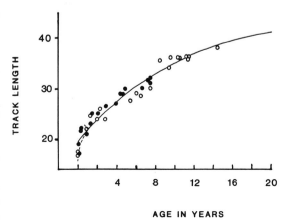

time averaging in modern track assemblages takes place very quickly, in weeks or only days (Cohen et al., 1991; Schult & Farlow 1992), and is caused, at least in part, by the activities of animals from the same population.

With track assemblages representing herds, we know that the tracks represent living populations active on a given day or during relatively short periods of geological time. Evidence of herding also strengthens the assumption that the sample represents a monospecific assemblage. Moreover, as I have stressed previously (Lockley, 1989a, 1991), it is important to integrate track and bone (trace and body fossil) data. As a source of information tracks should not be viewed merely as an alternative or a last resort when bone data are unavailable or are inadequate for the task at hand.

Given these considerations, I present trackway data to establish the size range of individuals of particular species (or taxa) at several well-studied sites. This size–frequency data can then be used to make inferences about the population structure and the age of dinosaurs represented at various localities. In larger species, the size range of footprints between hatchling-sized and fully adult individuals is much greater, and therefore easier to recognize.

Study areas

Thulborn (1990) and Lockley (1991) have indicated that much new dinosaur trackway data have emerged in recent years. We can now recognize the trackways of most major dinosaur groups, with the exception of stegosaurs. Often, the likely identity of the trackmakers is further constrained by the dating of the track assemblage at the epoch or age level.

In this study, I have selected a small number of sites from which abundant ornithopod and sauropod trackway data have been derived. Often, the trackways were recorded on a single bedding plane, with the trackmakers representing a gregarious group or herd. In other cases, trackways of a particular type occur in relatively thin stratigraphic units within the same sedimentary facies.

I have avoided theropod trackway data sets for several reasons. First, it is hard to distinguish the affinity of many tridactyl tracks. Although large broad-footed iguanodontid and hadrosaur tracks are easily recognizable, there is much debate about the affinity of smaller early Mesozoic tridactyl tracks. For example, Thulborn (1990, Fig. 11.12) referred to tridactyl trackways from the Jurassic Purgatoire site as ornithopod (Lockley, Houck, & Prince, 1986). However, a reevaluation of the tracks indicate that all are of theropod affinity (Lockley & Prince, 1988; Prince & Lockley, 1989).

Second, research into this thorny problem of tridactyl track identification is currently underway (Farlow, written communication, 1992), and it would be premature to anticipate future results. Suffice it to say that there are huge data sets available for tridactyl track-

ways from Mesozoic deposits; for example: *Grallator* and *Eubrontes* from the Lower Jurassic of the Connecticut Valley and elsewhere around the world, Upper Jurassic theropod tracks from the Purgatoire site, Lower Cretaceous theropod tracks from Texas (Farlow, 1987; Pittman, 1989), and Upper Cretaceous theropod tracks from Toro Toro in Bolivia (Leonardi, 1984).

Because the growth curve presented by Morell (1987) and Obata and Tomida (1990) was based on the ornithopod *Maiasaura*, we can begin with a consideration of ornithopod trackway data. I assume that the *Maiasaura* growth curve (Fig. 23.1) is the best model currently available.

Ornithopod trackway data
Caririchnium leonardii

Caririchnium leonardii is a distinctive ornithopod track from the Late Albian–Early Cenomanian Dakota Group of the Colorado Front Range area. Recent work has shown that the trackmaker was a quadrupedal ornithopod and that the tracks are generally well preserved in a thin stratigraphic sequence of laterally extensive deposits (Lockley, 1985, 1987a, 1988, 1990; Currie, Nadon, & Lockley 1991; Lockley et al., 1992). The sample consists of several hundred well-preserved tracks and trackways. The range of observed track size (pes length) is between 20 and about 48 cm for well-preserved bedding plane impressions. At the Alameda Avenue locality known as Dinosaur Ridge, almost the entire known range of track size can be observed from a single bedding plane (Lockley et al., 1992). The same size range has been observed at other localities, including Mosquero Creek, New Mexico.

Using the well-known model proposed by Alexander (1976, 1989) for estimating animal size from track length, the smallest *Caririchium* trackmaker had a hip height of about 80 cm (4 × 20 cm), and the largest would have had a leg length (= hip height) of about 192 cm (4 × 48 cm). Hip height using Thulborn's (1989) formulas are 96–230.4 cm (4.8 × foot length) using the small ornithopod model, or 283.2 cm (5.9 × 4.8 cm) using the large ornithopod model. One difficulty with Thulborn's method is that it requires one formula for small individuals with a foot length of less than 25 cm and a different formula for larger individuals.

If we roughly average the Thulborn methods and use a hip height estimate of 5 × foot length, we arrive at hip height estimates of between 100 and 240 cm. Using the ratio of leg length/overall size (about 1:3.5) in large iguanodontids (e.g., *Iguanodon bernissartensis*), we arrive at overall size estimates between about 3.5 m (11.5 ft) and 8.4 m (27.5 ft). The largest animals were therefore a little larger than adult *Maiasaura*. Independent estimates of body size, derived by inferring glenoacetabular dimensions from quadrupedal trackways, also give similar range and size estimates for the *Caririchnium* trackmakers (Lockley, 1985).

I conclude that a small *Caririchnium* trackmaker, with a foot length of only 20 cm, was about 42 percent of its full adult size (see Fig. 23.3). According to the Horner model, this represents an individual about 1 year old. Using the same model, 2-, 3- and 4-year-old individuals might have foot lengths of about 25–26, 32–33, and 36–37 cm, respectively. Older animals (5–7 years of age) would be 83–100 percent of full size (equivalent to foot lengths of 40–48 cm in *Caririchnium*) and would be reaching sexual maturity.

The Dinosaur Ridge and Mosquero Creek *Caririchnium* samples contain parallel trackway evidence for gregarious behavior among adults (individuals with foot lengths of about 45 cm) and subadults (individuals with much smaller foot lengths, 1–4 years old). As discussed below, the extent to which dinosaurs of different size classes (ages) engaged in unidirectional progression (herding or migration) was probably age dependent to some degree. Farlow (written communication, 1992) has inquired whether these two size groups might not represent different species. This is possible but cannot be supported on morphological (ichnotaxonomic) grounds; a monospecific interpretation is the most parsimonious.

In a recent study, Lockley et al. (1992) have shown that the *Caririchnium*-dominated ichnofacies is recognizable in a thin stratigraphic package that can be traced over an 80,000-km² area of coastal plain popu-

larly known as the "Dinosaur Freeway." The Alameda sample is from the northern part of this area and consists only of impressions. A sample of casts from the southern region is also included to show the considerable size range (Fig. 23.4). Study of a site at Mosequero Creek (Lockley et al., 1992) reveals new evidence of large gregarious groups and size ranges consistent with the two samples cited herein.

Cretaceous ornithopods from the Jindong Formation, South Korea

A large amount of size–frequency data for both ornithopod and sauropod trackways has recently been compiled (Lockley et al., 1991a; Lim, 1991). As shown in Figure 23.5, the ornithopod trackway sample is large (N = 217) and dominated by medium-sized individuals (foot-length size range, 24 to 38 cm). Although the sample was not recorded from a single horizon, many horizons yield evidence of many individuals progressing together as a group or herd. The size frequency of individuals from single horizons is similar to the pooled sample. This fact argues for a single species of trackmaker. The data suggests a situation similar to that inferred for the Dakota Group sample, that is, a scarcity of very young individuals less than 1-year old, but an abundance of older subadult–adult individuals. Out of a sample of 217 measured trackways (foot-length range

Figure 23.3. Estimated age of *Caririchnium leonardi* trackmakers. (Ornithopoda) from the Cretaceous of Colorado, based on comparison with the *Maiasaura* growth curve (Fig. 23.1). Size of tracks ranges from 20 to 48 cm (foot length); see text for details.

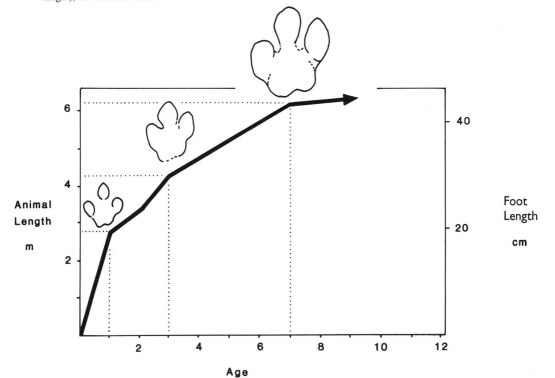

15–51 cm) only five (2 percent) are less than 24 cm long, and only four exceed 45 cm.

Experience has shown us that the larger tracks (foot length, about 45–51 cm) often have diffuse margins and may be interpreted as undertracks. Among the small tracks we have only observed one well preserved example (Fig. 23.6). This particular track (foot length: 14.5 cm, foot width: 13.0 cm) compares quite closely with an isolated track specimen from the Tetori Group of Japan (Manabe, Hasegawa, & Azuma, 1989; Azuma & Takeyama, 1991) which has a length of 16 cm and a width of 13.5 cm. Such small Lower Cretaceous ornithopod tracks are rare in most samples.

Using the same criteria as applied to the *Caririchnium* tracks, and discarding the few large, imprecisely measured tracks from the sample, the maximum Jindong ornithopod foot length is about 45 cm. The pronounced modal peak (30–32 cm) represents individuals about two thirds (67–71 percent) of maximum size, or about 3–4 years old. A few individuals have a foot length of less than 20 cm, indicating an age of less than 1 year. However, quite a few individuals fall in the size range estimated to represent 1- to 3-year olds (foot length, 24–29 cm). Similarly, many individuals fall in the size range of 4- to 7-year olds (foot length, 33–45 cm).

It is reasonable to conclude, therefore, that tracks from the Jindong Lake Basin predominantly represent those of groups or herds of subadults and adults passing through the region on purposeful local or long-distance migrations (that is, not milling around or browsing locally). Because more than one subadult age group (or recruitment) may be represented is tentatively interpreted as evidence that the area was frequented or used as a thoroughfare by different groups at different times. Certainly the high frequency of ornithopod track-bearing levels proves this to be the case. Trackway orientation data show multiple levels exhibiting groups traveling toward the southwest.

Amblydactylus, Gething Formation, western Canada

The Aptian Gething Formation of western Canada has produced several interesting ornithopod tracks and trackway assemblages indicative of the presence of both large (adult) and smaller (juvenile–subadult) individuals (Currie & Sarjeant, 1979; Currie, 1983). As shown in Figure 23.7, Currie (1983, Table 1) recorded measurements for thirty-six *Amblydactylus* trackways and a range of foot lengths between 14 and 73 cm (see Thulborn, 1990, Fig. 11-12 for histogram based on this data). The largest track is exceptionally long, and its size may be exaggerated by erosion or other preservational phenomena.

The tracks recorded were not all from a single site, but in some cases as many as eleven trackways were present in a single horizon (i.e., individual *Amblydactylus* dominated ichnocoenoses, within the Gething *Amblydactylus* ichnofacies, show that the trackmakers were abundant and gregarious). In one instance, these trackways are comprised of parallel trails indicative of the progression of a herd. This example (Currie, 1983, Fig. 2) has been frequently cited as an example of ornithopod herding (Paul, 1987; Lockley, 1989, 1991; Farlow, 1991a).

Based on the size–frequency data, the smallest *Amblydactylus* track (foot length, 14 cm) is only 19 percent of the length of the largest (foot length, 73 cm). Only two other small tracks (foot length, 26.5–27.5 cm) could could have been made by animals less than 1 year old based on the Horner growth curve. All the remaining tracks (35.5 cm and larger) are inferred to represent individuals that are at least a year old (minor modal peaks at 43 and 54 cm represent 2 year olds and 3–4 year olds according to the Horner model).

Currie and Sarjeant (1979) note that they were unable to attribute tracks to the ichnospecies *Amblydactylus gethingi* Sternberg, 1932. As a result, they named a new ichnospecies *Amblydactylus kortmeyeri* based on a large holotype (foot length, 42 cm; width, 43 cm) and five much smaller paratypes (foot length, 10.8–19 cm). In addition, the Royal Tyrrell Museum contains three more small *Amblydactulus* tracks (TMP 77.17.2, TMP 78.11.13, TMP 79.23.4) with foot lengths between 12 and 15 cm. If these specimens are added to the sample (Fig. 23.7), it increases the sample size by 25 percent and adds significantly to the proportion of small (juvenile) individuals. The smallest specimen is then only 15 percent of full adult size (10.8/73). Eleven specimens,

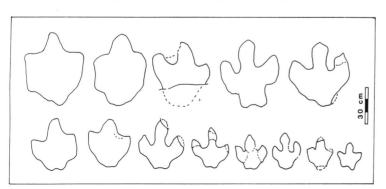

Figure 23.4. Sample of ornithopod track casts from the Cretaceous Dakota Group of Southeastern Colorado (after Lockley et al., 1992). Footprint cast length ranges from about 20 to 55 cm.

30 cm

about one quarter of the sample, represent juveniles estimated at less than a year old. Thus, we can tentatively conclude that the Gething Formation ornithopod trackway sample includes a greater representation of juve-

Figure 23.5. Size–frequency data for ornithopod trackways from the Cretaceous Jindong Formation of South Korea.

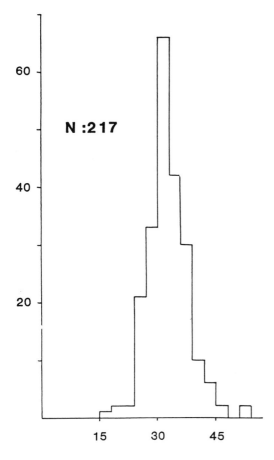

BIPEDAL (ornithopod) TRACKS

Jindong Fm. Korea

N :217

Foot length cm

niles than any other known sample of large Cretaceous ornithopods.

Some Late Cretaceous ornithopods bred and apparently migrated locally in the mid- to high latitudes of the northern hemisphere (Horner 1987; Horner & Gorman 1988). Furthermore, some ornithopods also preferred vegetated coastal plain environments (Ostrom, 1964). For these reasons, it is perhaps not surprising to find a large size range of ornithopod tracks represented in the Gething Formation coastal plain deposits. Sarjeant (1981) reported examples of large tracks overlapped by smaller ones, suggesting that juveniles followed the adults. Lockley, Young, and Carpenter (1983) noted a similar phenomenon in an assemblage of hadrosaur tracks from an Upper Cretaceous coal mine in Colorado.

Hadrosaur tracks, Mesa Verde Group, Colorado and Utah

A variety of hadrosaur track casts have been recovered from Upper Cretaceous coal mines in the Price area in eastern Utah (Wilson, 1969) and in western Colorado (Lockley et al., 1983; Parker & Rowley, 1989). These tracks range in size from 3 to about 100 cm in length (Lockley, 1986). The Wilson Collection (Lockley, 1986, Fig. 11) is particularly interesting because it contains at least thirty-two small tracks between 3.3 and 11.5 cm in length. This stands in contrast to the majority of tracks in the Prehistoric Museum (College of Eastern Utah, Price) and those illustrated by Parker and Rowley (1989). Those range from 27 to 87.5 cm in length (sample of twenty-one measured specimens), while those from Gunnison, Colorado (Lockley et al., 1983), range in length from 38 to 91 cm.

It is hard to draw substantive conclusions on the isolated tracks from the many different localities and stratigraphic levels (ichnocoenoses) in the Mesa Verde coal-bearing facies. Wilson probably collected small tracks preferentially. Large tracks were only periodically donated to the museum in Price, Utah. As indicated by Lockley (1986), some of the small tracks may be manus casts and others may have had their shapes altered or enhanced by carving. Nevertheless, the abundance of small track casts suggests that juvenile hadrosaurs are represented in the Mesa Verde footprint sample. Some are apparently

Figure 23.6. A–C. Small ornithopod tracks from Japan (A, B) and South Korea (C). A, redrawn from Manabe et al. (1989) and Azuma and Takeyama (1991).

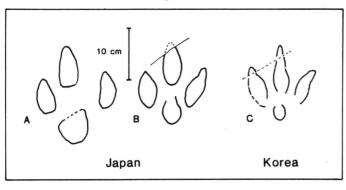

Japan   Korea

small enough to be considered hatchling-sized (Horner, 1987). See Carpenter (1992) for further discussion.

Sauropod trackway assemblages
General comments

Footprint assemblages containing sauropod trackways are known throughout the Jurassic and Cretaceous. However, only a few have yielded a large amount of data from single bedding planes or specific facies. I confine my discussion to three such sites: the Jurassic Morrison Formation of Colorado, the Cretaceous Glen Rose Formation of Texas, and the Cretaceous Jindong Formation of Korea (Fig. 23.8). The former two represent samples from single bedding planes. The latter represents the largest sauropod trackway sample currently known, but it was obtained from multiple levels in a

thin lacustrine sequence regarded as part of a single ichnofacies.

Although Case (1978) estimated that it took the sauropod *Hypselosaurus* 62 years to reach sexual maturity, this estimate may be too large an order of magnitude. These animals may have matured in only about 6 years (Dodson, 1990). Because there are no empirically derived growth curves yet proposed for sauropods, it is necessary either to utilize Horner's growth curve for large ornithopods or to employ the growth curve for an elephant or other large mammal as an alternative (see Dunham et al., 1989 for a theoretical treatment). Either method gives results more consistent with the 6-year maturation model rather than the 62-year estimate.

Applying the Horner growth curve for ornithopods to sauropods has some points in its favor as well

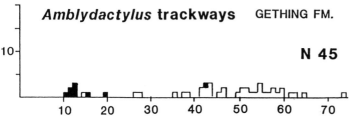

Figure 23.7. Size–frequency data for various *Amblydactylus* tracks from the Cretaceous Gething Formation of western Canada. Sample includes *A. kortmeyeri* (Currie & Sarjeant, 1979) shown in black and *Amblydactylus* sp. (Currie, 1983) shown in white.

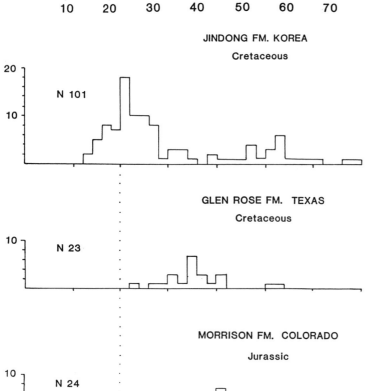

Figure 23.8. Size–frequency histograms for sauropod trackways from the Jurassic and Cretaceous (see text for details).

SAUROPOD PES LENGTH   cms

as some shortcomings. The main advantages are that Horner's model is applicable to large dinosaurs and is the only model available. Also, despite contrary, heretical opinions (Bakker 1986; Bakker in Morell, 1987), there is little or no evidence to suggest that sauropods gave birth to live young (Dodson, 1990). In fact most recent evidence (Britt, 1988; Britt & Naylor chapter 16; Mohabey, 1987; Sahni, 1991; Tandon & Sood, 1991) suggests that sauropods were egg-layers. Consequently, like *Maiasaura* and other large dinosaurs, sauropods must have hatched as very small individuals before growing to their full adult size.

The purported hatchling remains discovered by Mohabey (1987) suggest an animal about of 50 cm in

Figure 23.9. Reconstruction of a hatchling sauropod (probably titanosaur), Late Cretaceous, India (after Mohabey, 1987, and Lockley, 1989b). Although the identity of this specimen is controversial and has not been described in detail, the illustration presents the estimated size of a sauropod hatchling.

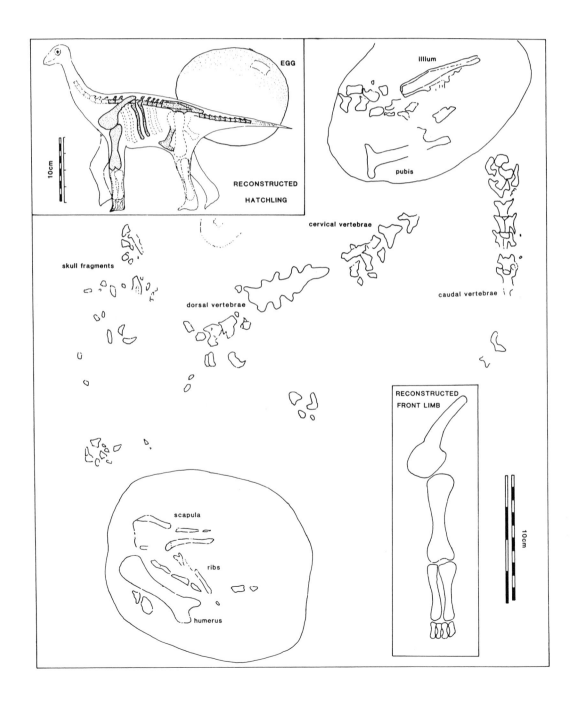

length (see reconstruction by Lockley, 1989b, Fig. 50.7 & Fig. 23.9 herein). This reconstruction may prove premature, since Mohabey has provided no further information on a very important specimen (Jain, 1989). Moreover, McIntosh (personal communication; Carpenter & McIntosh, Chapter 17) expresses some doubt about the affinity of this specimen. Dodson (1990, p. 406) postulates that hatchlings were up to 10 kg in weight and 1.5 m in length.

Despite these justifications, we need to bear in mind that there is no evidence that sauropod growth rates were similar to those of large ornithopods. If a sauropod growth curve is constructed in the future, it may be necessary to revise the following age estimates to conform to a different model. It could also be argued that trackways of small sauropods represent small species and not juvenile individuals. However, there is no evidence to support this position either. In fact, Dodson (1990) asserts that "all sauropods were large and ranged from 12 to 30 m or more in total length."

cf. *Brontopodus* sp., Morrison Formation, Colorado

Foot size measurements are available for a sample of 24 sauropod trackways (ichnogenus *Brontopodus*) from a single horizon at the Upper Jurassic, Purgatoire site in Colorado (Lockley et al., 1986, Table 23.1). Foot length ranges from 48 to 87 cm (Fig. 23.8) with up to half the sample (twelve trackways) comprising a gregarious group or herd heading westward (Fig. 23.10).

If we assume that a foot length of 87 cm represents a fully grown adult, then a foot length of 48 cms represents an individual that has already reached 55 percent of full size. According to the Horner model such an animal would be about 2 years old. Within the twelve parallel trackways, five trackways (nos. 29, 30, 33, 34, and 35 of Lockley et al., 1986, Fig. 2; Fig. 23.10) show a size range of 52–60 cm (60–69 percent of maximum size). These tracks suggest a group of subadults of about the same size and age (2–3 years). Estimates of hip height and glenoacetabular length suggest that these subadults were almost the size of the Carnegie Museum *Camarasaurus* (CM 11338) discussed below and illustrated in Figure 23.11. The regular intertrackway spacing of this group (Fig. 23.10) has been frequently cited as an example of herding (Lockley & Prince 1988; Lockley 1989, 1991; Lockley, Matsukawa and Obata 1991b). Although the Purgatoire site has only yielded a relatively small sample, the available evidence suggests a modal foot length of about 60 cm. I infer that this lake-basin paleoenvironment was a site through which both subadult and adult individuals passed on local or long-distance migrations without the company of small individuals or babies.

Brontopodus birdi, Glen Rose Formation, Texas

Tracks from the well-known Davenport Ranch site in the Albian upper Glen Rose Formation, Texas, were made famous by Roland T. Bird (1944). He published a detailed map showing the trackways of twenty-three individuals all heading in the same direction and inferred that they represented a passing herd. Later, Bakker (1968) proposed that the larger individuals were traveling on "the periphery of the herd" with the smaller ones only in the center. Although this purported occurrence of parental protection is appealing, the trackway evidence does not show such a simple scenario or support the hypothesis of a "structured herd" (Ostrom 1972, 1985; Lockley, 1987b, 1991).

Leaving these considerations aside, age estimates may help constrain speculation somewhat. The range of foot length in the Davenport Ranch sample falls between 35 and 78 cm (see Lockley, 1987b; Table 23.1). If the largest tracks are considered representative of adults, then the smallest individual (45 percent of adult size) would be a year old, according to the Horner curve. The other small individuals (foot length, 43-46 cm or 55–59 percent of full size) would have an estimated age of about 2 years. This leads to the interpretation that young subadults or juveniles of 1 to 2 years of age were traveling in a group with older (mature) individuals. The structured-herd hypothesis may be further compromised by the apparent lack of juveniles less than a year old, or juveniles so small as to need the parental care that might have been provided to newborns, babies, or juveniles near a nest site.

Additional measurements of large sauropod tracks have been recorded by Farlow, Pittman, and Hawthorne (1989) from the upper Glen Rose Formation at Dinosaur Valley State Park (DVSP). Measurements from four trackways (S1–S4) indicate animals with foot lengths between 84 and 98 cm. Pooling these data with those from the Davenport Ranch site gives a size range of 35 to 98 cm (foot lengths). The resultant data suggest a larger proportion of small individuals or juveniles. The smallest individual (foot length, 35 cm) would have been a little less than a year old with the other small individuals (43–46 cm) being just over a year in age.

Unquestionably these two alternative interpretations indicate the problems associated with small samples that do not reflect the full size range of tracks known from a particular formation. Moreover, they indicate the pattern of age-estimate revision that will probably be necessary as new data are made available or as alternate growth curves are proposed. In general, it is desirable to apply growth curve analysis only to samples where adult-sized individuals are also represented. This would make the latter of the two Glen Rose interpretations more preferable.

cf. *Brontopodus* sp. Jindong Formation, South Korea

The Cretaceous (?Aptian–Albian) Jindong Formation has yielded an enormous sample of dinosaur trackways (Lockley et al. 1991, Lockley, Matsukawu, & Obata, 1991; Lim, 1991). Measurements have been obtained for over 100 sauropod trackways from numerous different horizons in a thin lacustrine sequence (100–200 m in thickness). The tracks range from 16 to 100 cm in foot length. Thus, assuming the presence of only one species, a remarkably wide range of individual sizes is represented. The smallest tracks are by far the smallest sauropod tracks recorded anywhere in the world, while the largest ones are about the same size as the largest known, reliably measured tracks from Texas.

Application of Horner's model to the footprint

Table 23.1. *Life tables constructed from sauropod trackway data.*

Footprint size (length) interval (cm)	X Age interval*	Davenport Ranch		Purgatoire River		Jindong Formation	
		# Trackways	LX	# Trackways	LX	# Trackways	LX
0–6	0–1	–	1.00	–	1.00	–	1.00
7–12	–2	–	1.00	–	1.00	–	1.00
13–18	–3	–	1.00	–	1.00	–	1.00
19–24	–4	–	1.00	–	1.00	7	1.00
25–30	–5	–	1.00	–	1.00	15	0.93
31–36	–6	1	1.00	–	1.00	28	0.78
37–42	–7	1	0.96	–	1.00	18	0.50
43–48	–8	4	0.91	1	1.00	4	0.33
49–54	–9	8	0.74	1	0.96	4	0.29
55–60	–10	5	0.39	6	0.92	2	0.25
61–66	–11	2	0.17	5	0.67	2	0.23
67–72	12	–	0.09	4	0.46	5	0.21
73–78	–13	2	0.09	4	0.29	4	0.16
79–84	–14	–	0.00	2	0.13	7	0.12
85–90	–15	–	0.00	1	0.04	2	0.05
91–96	–16	–	0.00	–	0.00	1	0.03
97–102	–17	–	0.00	–	0.00	1	0.02
103–108	–18	–	0.00	–	0.00	1	0.01
109–114	–19	–	0.00	–	0.00	–	0.00
		N = 23		N = 24		N = 101	

Note: *Age units are *not* considered equivalent to years, nor is each age unit considered of equal duration.

Figure 23.10. Five parallel trackways attributed to subadult sauropods, Purgatoire site, Upper Jurassic of Colorado.

Figure 23.11. Top to bottom: restoration of a 5.5 m long subadult *Camarasaurus* based on CM 11338 from Dinosaur National Monument; 4.7 m long skeleton of *Bellusaurus* (after Dong, 1988, 1990); juvenile *Camarasaurus* scaled (2.5 m long) to the smallest Jindong sauropod trackway; Indian titanosaur hatchling and egg, and 1.5-m long hatchling (see Dodson, 1990) for comparison.

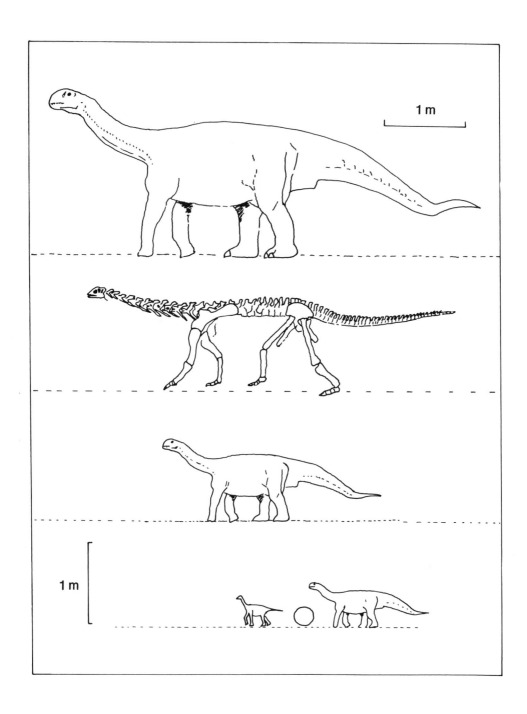

data leads to some interesting inferences. The majority of trackways (about 70 percent) represent animals that were less than a year old. The modal size, 30–32 cm, represents only 30–32 percent of maximum foot length. The smallest individuals are only 1/6 to 1/7th (16 percent or 1/6.25) of adult size (linear dimension). In crude terms this means the large individuals were between 6^3 and 7^3 times the mass of the smallest individuals. These small animals are too large to be hatchlings (sensu Mohabey, 1987; Lockley, 1989b; Fig. 23.11), but much smaller than the famous juvenile *Camarasaurus* (CM 11338) from Dinosaur National Monument. This individual stood about 1.5 m at the hip and measured about 5.5 m from head to tail (see Fig. 23.11 for restoration).

Footprints in the Jindong modal size class may represent animals about the size of those in the assemblage of seventeen juvenile *Bellusaurus sui* excavated and described by Dong (1988, 1990) from the Middle Jurassic of China. Dong Zhiming (personal communication, 1991) independently estimated that these sauropods (about 4.7 m long) were less than a year old.

Using standard size-estimation techniques on the Jindong trackways, a foot length of 16 cm suggests a hip height of only 64 (4 × 16 cm) to 80 (5 × 16 cm) cm (Fig. 23.12). Estimates of glenoacetabular lengths from well-preserved quadrupedal trackways suggest that these small individuals had a short trunk length of about 65–80 cm. From this we can infer an animal about the size of a young calf or foal, perhaps no larger than certain large domestic dogs. Even allowing for a relatively long neck and tail, the smallest individuals would not have been very large (perhaps only 2–3 m in length).

Such small sauropods were not very heavy, about $1/6^3$ to $1/7^3$ (1/216 and 1/343) of adult mass. The estimated weight is dependent on the mass of the adult. For a 10-metric ton adult (*Camarasaurus*-size), the smallest individual would weigh between 29 and 46 kg; for a 30-metric ton adult (*Apatosaurus*-size), 87–139 kg; and for a 50-metric ton adult (*Brachiosaurus*-size, as in Fig. 23.12), 146–231 kg (adult weights from Alexander, 1989; Paul, 1989). Such weight estimates are the right order of magnitude to be consistent with the size estimates and are not much more than the maximum hatchling weight of 10 kg postulated by Dodson (1990). Recently, a camarasaurid tooth was discovered in the pre-Jindong, Lower Cretaceous Hasangdong Formation of Korea, thus suggesting a possible trackmaker in the region during this epoch.

None of the small trackways from Jindong, and virtually none of the larger sauropod trackways, occur in parallel configurations indicative of gregarious progression. This stands in marked contrast to the abundance of parallel sauropod trackways at other locations. Interestingly, at least a dozen sauropod trackways curve. This suggests a pattern of milling around and changing direction that is rarely observed at trackway sites (Lockley, 1991). It is therefore tempting to conclude that the Jindong track site represents a record of mainly young sauropods (less than 1 year old) accompanied by a small number of adults and subadults.

Because the estimated age of the smallest sauropods is on the order of only a few months, using Horner's model, it is possible that the tracks were made quite near a nesting site. If this supposition is correct, then the abundance of small, randomly oriented trackways probably represents the activity of young juveniles that were old enough to wander around locally but too young to migrate. Evidently these animals were not inclined to form large gregarious groups or travel in set directions. Given that immature individuals tend to romp around (consider, for example, mammals), whereas adults tend to conserve energy, it is not unreasonable to speculate that the high proportion of curved or circuitous trackways of immature sauropods might be a reflection of juvenile behavioral characteristics. It is also possible that the population structure in the nesting area would include many small individuals in association with several mature individuals of breeding age, and few, if any, subadults. The size-frequency data for the Jindong Formation (Fig. 23.8) are consistent with such a supposition.

Life tables and survivorship

The size–frequency data presented above can be synthesized as age pyramids showing age structure or as survivorship curves derived from life tables (Table 23.1 and Fig. 23.13). The sauropod trackway size–frequency data (Fig. 23.8) from Texas and Colorado represent samples from single bedding planes indicate herds. The Korean datum is the largest known sample from any ichnofacies and contains numerous trackways of small individuals.

Using footprint length as an arbitrary measure of age, nineteen relative age intervals have been constructed for the size range observed in all three sauropod trackway samples (Table 23.1). For comparison, the Amboseli elephant track data were grouped into a dozen size classes corresponding to actual ages from 0 to 715 years (Fig. 23.14). From this exercise we can obtain survivorship curves (Fig. 23.13) that show the relative proportions of young individuals in the population or sample (Kurtén 1953; Boughey, 1971, Fig. 2.2).

It is clear from the results that the age structure of the Colorado and Texas track assemblages is very similar, comprising predominantly subadult- and adult-sized individuals. By contrast, the Korean more closely resembles the survivorship patterns observed in the elephant tracks from Amboseli National Park (Figs. 23.2 & 23.14). These tracks consists of a sample in which about 50 percent of the population represents individuals that are less than 4 years old.

Figure 23.12. Juvenile sauropod (brachiosaur) reconstruction based on trackway evidence from the Cretaceous Jindong Formation of South Korea. Note the size discrepancy between this individual and a fully grown adult. (Brachiosaur modified from Paul, 1987).

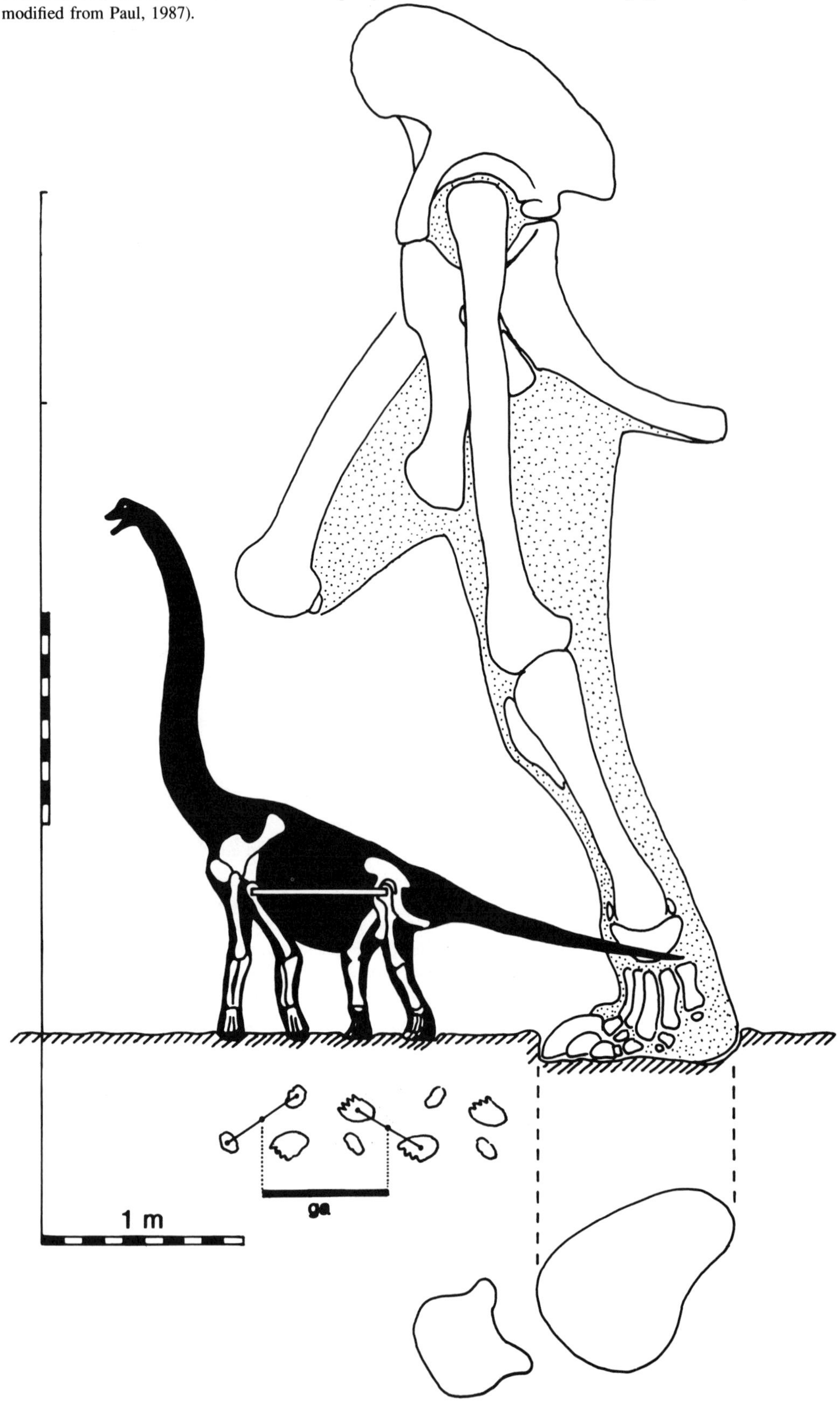

Other considerations

Most of the interpretations presented assume that Horner's *Maiasaura* growth curve is a suitable model for both ornithopod and sauropod ontogeny. Although numerous objections can be raised, no better model currently exists. I also assume there is a linear relationship between increasing foot length (size) and overall body length (size). This model is presented for the sake of simplicity and as a first attempt to interpret dinosaur population structure at particular sites based on track size frequency data. The model is not meant to imply that dinosaur growth was linear rather than allometric. Indeed, it has already been suggested that sauropod foot shape may have changed allometrically during ontogeny (Lockley et al., 1986) and new data from the Jindong Formation supports this inference (Lim, 1991). However, recognizing allometric growth using bivariate or multivariate statistics does little to improve our estimates of age. In any event many studies of population structure (survivorship) of fossil organisms are based on the use of univariate parameters.

In the discussion up to this point we have not considered the possiblity that sexual dimorphism influenced the size–frequency data obtained from tracksites (Western et al., 1983). It is quite possible that adult male dinosaurs were larger than adult females, although an opposite conclusion has been suggested by Carpenter (1990) and Raath (1990). However, because there is no way to test this inference on track data at present, we must assume that individuals of a particular size– whether male, female, or both – were on average older than individuals in smaller size classes.

It is also most parsimonious to assume that tracks of a given type from single horizons or even from multiple, closely-spaced horizons were made by single species populations. This conclusion is supported by studies on a regional scale of ichnofacies (e.g., Lockley et al., 1992). Ultimately, such suppositions may be tested by the use of bivariate and multivariate analyses of foot shape wherever suitable samplings of well-preserved tracks occur.

Finally, despite ambiguities associated with the interpretation of tracks, several points should be considered. Although juvenile dinosaur skeletal remains "are very uncommon" and the "overwhelming majority of sauropod specimens are at least 80 percent of adult size" (Dodson, 1990), rarity is definitely not the case with trackway data. Based on the three sauropod samples discussed above, the opposite may be inferred, that is, the majority of tracks are less than 80 percent of adult size. For example, based on an adult foot length of about 100 cm, only 10% of the Korean sample attains a foot length of 80 cm, while less than half attains a foot length of 50 cm (see Table 23.1). This evidence highlights the utility of tracks in providing evidence for the existence of subadults and juveniles in a population. It also supports the notion that body fossil and trace fossil records are often biased in different directions (Lockley, 1989a,

Figure 23.13. Survivorship curves for sauropod track data from Colorado, Texas, and Korea. Horizontal axis shows arbitrary age units (see Fig. 23.14).

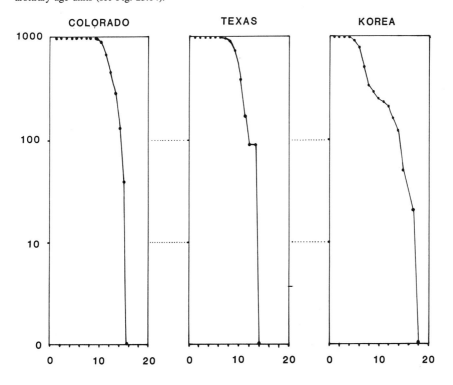

1991). This conclusion suggests the need to integrate track and body fossil data and not base conclusions only on evidence of one type (see Farlow in press for another example of this approach).

Conclusions

1. Some trackway assemblages from particular horizons or ichnofacies reflect the size–frequency distributions of individuals in particular populations and habitats.

2. Trackway size–frequency data is often more abundant and easier to obtain than body fossil data and thus is more useful for population studies.

3. Modern trackway assemblages are reliable indicators of age structure in species (elephants) with known growth rates. Hence, we can conclude that once dinosaur growth rates are more firmly established, tracks will be increasingly useful in studies of population dynamics and age structure.

Figure 23.14. Survivorship curve for elephant tracks, Kenya.

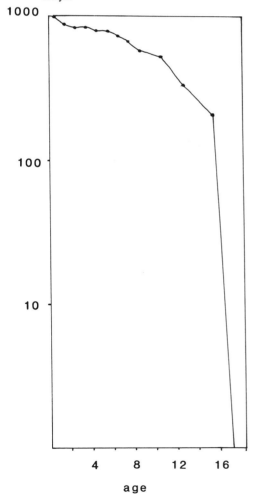

4. Correlation of Cretaceous ornithopod track data with a postulated growth curve for *Maiasaura* indicates a sparse and sketchy record for juveniles, possibly supporting rapid early growth models.

5. Sauropod trackway data show that juveniles and subadults were more common than previously inferred from the body fossil record.

6. A case study of Cretaceous sauropod trackways from South Korea suggests an abundance of juveniles "milling around," possibly close to their place of birth.

Acknowledgments

This work was supported in part by the National Science Foundation. I thank James O. Farlow, Karl Hirsch, and William A. S. Sarjeant for their helpful reviews and comments. The life restoration of *Camarasaurus* in Figure 23.11 was made by David Peters.

References

Alexander, R. McN. 1976. Estimates of the speeds of dinosaurs. *Nature* 261: 129–30.

    1989. *The Dynamics of Dinosaurs and Other Extinct Giants.* (New York: Columbia University Press).

Azuma, Y., & Takeyama, K. 1991. Dinosaur footprints from the Tetori Group, Central Japan, research on dinosaurs from the Tetori Group (4). *Bulletin Fukui Prefectural Museum* 4: 33–51

Bakker, R. T. 1968. The superiority of dinosaurs. *Discovery* 3: 11–22.

    1975. Dinosaur renaissance. *Scientific American* 232: 58–78.

    1986. *The Dinosaur Heresies* (New York: William Morrow and Co).

Bird, R. T. 1944. Did *Brontosaurus* ever walk on land? *Natural History* 53: 61–7.

Boucot, A. J. 1953. Life and death assemblages among fossils. *American Journal of Science* 251: 25–40.

Boughey, A. S. 1971. *Ecology of Populations.* (New York: MacMillan Co., Ltd.).

Britt, B. 1988. A possible "hatchling" *Camarasaurus* from the Upper Jurassic Morrison Formation (Dry Mesa Quarry, Colorado). *Journal of Vertebrate Paleontology* 8 (supplement to No. 3): 9A–10A.

Brown, B., & Schlaikjer, E. M. 1940. The structure and relationships of *Protoceratops*. *Annals of the New York Academy of Sciences* 40: 133–266.

Cadee, G. C. 1988 The use of size-frequency distribution in paleoecology. *Lethaia* 21: 289–90.

Carpenter, K. 1990. Variations in *Tyrannosaurus rex*. *In* K. Carpenter & P. Currie, (eds.), *Dinosaur Systematics.* (New York: Cambridge University Press), pp. 141–5.

    1992. Behavior of hadrosaurs as interpreted from footprints in the "Mesaverde" Group (Campanian) of Colorado, Utah and Wyoming. *University of Wyoming Contributions to Geology* 29: 81–96.

Case, T. J. 1978. Speculations on the growth rate and reproduction of some dinosaurs. *Paleobiology* 4: 320–8.

Chure, D. J., 1992. A baby dinosaur from the land of giants.

Park Science: A Resource Management Bulletin. National Park Service 12: 6.

Cohen, A., Lockley, M.G., Halfpenny J., & Michel, E. 1991. Modern vertebrate track taphonomy at Lake Manyara, Tanzania. *Palaios* 6: 371–89.

Craig, G. Y., & Oertel, G. 1966. Deterministic models of living and fossil populations of animals. *Quarterly Journal of the Geological Society of London* 122: 315–55.

Currie, P. J. 1983. Hadrosaur trackways from the Lower Cretaceous of Canada. *Second Symposium on Mesozoic Terrestrial Ecosystem, Acta Palaeontologica Polonica* 28: 62-74.

Currie, P. J. & Dodson, P. 1984. Mass death of a herd of ceratopsian dinosaurs. *In* W. E. Reif & F. Westphal (eds.), *Third Symposium on Mesozoic Terrestrial Ecosystems* (Tübingen: ATTEMPTO Verlag), pp. 61–5.

Currie, P. J., Nadon, G., & Lockley, M. G. 1991. Dinosaur footprints with skin impressions from the Cretaceous of Alberta and Colorado. *Canadian Journal of Earth Sciences* 28: 102–15.

Currie, P. J., & Sarjeant, W. A. S. 1979. Lower Cretaceous dinosaur footprints from the Peace River Canyon, British Columbia, Canada. *Palaeogeography, Paleoclimatology, Paleoecology* 28: 103–15.

Dodson, P. 1975. Functional and ecological significance of relative growth in *Alligator. Journal of Zoology* 175: 315–55.

1976. Quantitative aspects of relative growth and sexual dimorphism in *Protoceratops. Journal of Paleontology* 50: 929–40.

1990. Sauropod paleoecology. *In* D. Weishampel, P. Dodson, and H. Osmólska (eds), *The Dinosauria.* (Berkeley: University of California Press), pp. 402–7.

Dong Z. 1988. *Dinosaurs From China.* (London: British Museum of Natural History).

1990. On remains of the sauropods from Kelamali Region, Junggar Basin, Xinjiang, China. *Vertebrata PalAsiatica* 28: 43–58.

Dunham, A. E., Overall, K. L., Porter, W. P., & Forster, C. A. 1989. Implications of ecological energetics and biophysical and developmental constraints for life-history variation in dinosaurs. *In* J. O. Farlow, (ed.) *Paleobiology of the Dinosaurs. Geological Society of America Special Paper* 238: 1–20.

Farlow, J. O. 1987. *A Guide to Lower Cretaceous Dinosaur Footprints and Tracksites of the Paluxy River Valley, Somervell Co., Texas.* (Baylor: Baylor University).

1990. Dinosaur energetics and thermal biology. *In* D. B. Weishampel, P. Dodson, & H. Osmólska (eds), *The Dinosauria.* (Berkeley: University of California Press), pp. 43–55.

1991a. *On the Tracks of Dinosaurs: a Study of Dinosaur Footprints* (New York: Franklin Watts).

In press. Sauropod tracks and trackmakers: integrating the ichnological and skeletal records.

Farlow, J. O., Pittman, J. G., & Hawthorne, J. M. 1989. *Brontopodus birdi*, Lower Cretaceous sauropod footprints from the Gulf Coastal Plain. *In* D. D. Gillette & M. G. Lockley (eds.), *Dinosaur Tracks and Traces.* (New York: Cambridge University Press), pp. 371–94.

Horner, J. R. 1987. Ecological and behavioral implications derived from a dinosaur nesting site. *In* S. Czerkas & E. C. Olsen (eds). *Dinosaurs Past and Present*, Vol. II. (Seattle: University of Washington Press), pp. 50–63.

1992. Dinosaur behavior and growth. *In* R. S. Spencer (ed.), *Fifth North American Paleontological Convention, Abstracts and Program. Paleontological Society Special Publication* 6: 135.

Horner, J. R., & Gorman, J. 1988. *Digging for Dinosaurs.* (New York: Workman).

Horner, J. R., & Makela, R. 1979. Nest of juveniles provides evidence of family structure among dinosaurs. *Nature* 282: 296–8.

Horner, J. R., & Weishampel, D. 1988 A comparative embryological study of two ornithischian dinosaurs. *Nature* 332: 256–7.

Jain, S. L. 1989. Recent dinosaur discoveries in India, including eggshells, nests and coprolites. *In* D. D. Gillette & M. G. Lockley (eds.), *Dinosaur Tracks and Traces.* (New York: Cambridge University Press), pp. 99–108.

Kurtén, B. 1953. On the variation and population dynamics of fossil and recent mammal populations. *Acta Zoologica Fennica* 76: 1–122.

Leonardi, G. 1981. Ichnological data on the rarity of young in northeast Brazil dinosaur populations. *Anais da Academia Brasileirade Ciencias* 53: 345–6.

1984. Le impronte fossilidi Dinosauri. *In* J. Bonaparte et al. (eds.), *Sulle Orme dei Dinosauri.* (Venice: Erizzo Editrice), pp. 165–86.

Lim, S.-K. 1991. *Trace Fossils of the Cretaceous Jindong Formation, Koseong, Korea* Unpublished Ph.D. Thesis (Taegu, Korea: Kyungpook National University).

Lockley, M. G. 1985. Vanishing tracks along Alameda Parkway: implications for Cretaceous dinosaurian paleobiology from the Dakota Group Colorado. *In* C. D. Chamberlain et al., (eds.), *A Field Guide to Envionments of Deposition (and Trace Fossils) of Cretaceous Sandstones for the Western Interior.* (Golden, CO: Society of Economic Paleontology and Mineralogy), pp. 131–42.

1986. Dinosaur tracksites. *University of Colorado at Denver, Geology Department Magazine*, Special Issue 1.

1987a. Dinosaur footprints from the Dakota Group of eastern Colorado. *Mountain Geologist* 24: 107–22.

1987b. Dinosaur trackways. *In* S. J. Czerkas & E. C. Olsen, (eds.), *Dinosaur Past and Present*, Vol. II. (Seattle: University of Washington Press), pp. 80–95.

1988. Dinosaurs near Denver. *Colorado School of Mines Professional Contribution* 12: 288–99.

1989a. Tracks and traces: new perspectives on dinosaurian behavior, ecology and biogeography. *In* K. Padian & D. J. Chure (eds.), *The Age of Dinosaurs. Paleontological Society Short courses in Paleontology* 2.

1989b. Summary and prospectus. *In* D. D. Gillette & M. G. Lockley (eds), *Dinosaur Tracks and Traces.* (New York: Cambridge University Press), pp. 441–7.

1990. *A Field Guide to Dinosaur Ridge.* (Denver: Friends of Dinosaur Ridge and University of Colorado at Denver).

1991. *Tracking Dinosaurs: A New Look at an Ancient World.* (New York: Cambridge University Press).

1992. Dinosaur tracks (book review). *Ichnos* 2: 79–83.

Lockley, M. G., & Conrad, K. 1989. The paleoenvironmental context, preservation and paleoecological significance of dinosaur tracksites in the estern USA. In D. D. Gillette, & M. G. Lockley, (eds.), Dinosaur Tracks and Traces (New York: Cambridge University Press), pp. 121–34.

Lockley, M. G., Fleming, R. F., Yang, S.-Y., & Lim, S. K. 1991a. The distribution of dinosaur and bird tracks in the Jindong Formation of South Korea: implications for paleoecology. International Symposium on Origin, Sedimentation and Tectonics of Late Mesozoic to Early Cenozoic Sedimentary Basins at the Eastern Margin of the Asian Continent (Fukuoka, Japan: Kyushu University); p. 61.

Lockley, M. G., & Gillette, D. D., 1989. Dinosaur tracks and traces: an overview. In D. D. Gillette, & M. G. Lockley, (eds.), Dinosaur Tracks and Traces (New York: Cambridge University Press), pp. 3–10.

Lockley, M. G., Houck, K., & N. K. Prince. 1986. North America's largest dinosaur tracksite: implications for Morrison Formation paleoecology. Geological Society of America 97: 1163–76.

Lockley, M. G., Hunt, A., Holbrook, J., Matsukawa, M., & Meyer, C. 1992. The dinosaur freeway: a preliminary report on the Cretaceous megatracksite, Dakota Group, Rocky Mountain Front Range and High Plains, Colorado, Oklahoma and New Mexico. In R. Flores, (ed.), The Mesozoic of the Western Interior. Fieldguide, Society of Economic Mineralogy and Paleontology, Midyear Meeting, Fort Collins, Colorado.

Lockley, M. G., Matsukawa, M., & Obata, I. 1991b. Dinosaurs: Interpretations Based on Footprints. (Tokyo: Maruzen Library).

Lockley, M. G., & Prince, N. K. 1988. The Purgatoire Valley Dinosaur Tracksite Region. Colorado School of Mines Professional Contribution 12: 275–87.

Lockley, M. G., Young, B. H., & Carpenter, K. 1983. Hadrosaur locomotion and herding behavior: evidence from footprints from the Mesa Verde Formation, Grand Mesa Coalfield, Colorado. Mountain Geologist 29: 5–13.

Manabe, M., Hasegawa, Y., & Azuma, Y. 1989. Two new dinosaur footprints from the Early Cretaceous Tetori Group of Japan. In D. D. Gillette, & M. G. Lockley, M. G. (eds.), Dinosaur Tracks and Traces (New York: Cambridge University Press), pp. 309–12.

Mohabey, D. M. 1987. Juvenile sauropod dinosaur from Upper Cretaceous Lameta Formation of Panchmahals District, Gujarat, India. Journal Geological Society of India 30: 210–216.

Morell, V. 1987. Announcing the birth of a heresy. Discover March: 26–50.

Obata, I. and Tomida, Y. 1990. Dinosaur collection of the Museum of the Rockies, Montana State University. (Tokyo: National Science Museum).

Olson, E. C. 1951. Diplocaulus: A study of growth and variation. Fieldiana, Geology 11: 57–154.

Ostrom, J. 1964. A reconsideration of the paleoecology of hadrosaurian dinosaurs. American Journal of Science 262: 975–97.

1969. Terrestrial vertebrates as indicators of Mesozoic climates. North American Paleontological Convention Proceedings Part D: 347–76.

1972. Were some dinosaurs gregarious? Palaeogeography, Palaeoclimatology, Palaeoecology 11: 287–301.

1985. Social and unsocial behavior in dinosaurs. Bulletin Field Museum Natural History 55: 10–21.

Parker, L. R., & Rowley, R. L. 1989. Dinosaur footprints from a coal mine in east-central Utah. In D. D. Gillette, & M. G. Lockley (eds), Dinosaur Tracks and Traces. (New York: Cambridge University Press), pp. 361–6.

Paul, G. 1987. The science and art of restoring the life appearance of dinosaurs and their relatives: a rigorous how-to guide. In S. J. Czerkas, & E. C. Olson, (eds), Dinosaurs Past and Present, Vol. I. (Seattle: University of Washington Press), pp. 4–49.

Paul, G. 1989. The brachiosaur giants of the Morrison and Tendaguru, with a description of a new subgenus Giraffatitan, and a comparison of the world's largest dinosaurs. Hunteria 3: 1–14.

Pittman, J. G. 1989. Stratigraphy, lithology, depositional environment and track type of dinosaur trackbearing beds of the Gulf Coastal Plain. In D. D. Gillette, & M. G. Lockley (eds), Dinosaur Tracks and Traces. (New York: Cambridge University Press), pp. 135–53.

Prince, N. K., & Lockley, M. G. 1989. The sedimentology of the Purgatoire tracksite region, Morrison Formation of southeastern Colorado. In D. D. Gillette, & M. G. Lockley, (eds.), Dinosaur Tracks and Traces (New York: Cambridge University Press).

Raath, M. A. 1990. Morphological variation in small theropods and its meaning in systematics: evidence from Syntarsus rhodesiensis. In K. Carpenter, & P. Currie, (eds.), Dinosaur Systematics (New York: Cambridge University Press), pp.91–106.

Raup, D. M., & Stanley, S. M. 1971. Principles of Paleontology (San Francisco: W. H. Freeman and Co.).

Richards, R. P., & Bambach, R. K. 1975. Population dynamics of some Paleozoic brachiopods and their paleoecological significance. Jornal of Paleontology 49: 775–98.

Richmond, N. D. 1965. Perhaps juvenile dinosaurs were always scarce. Journal of Paleontology 39: 503–505.

Russell, D. A. 1989. The Dinosaurs of North America: An Odyssey in Time. (Minocqua, Wisconsin: Northwood Press).

Sahni, A. 1991. Non-marine Upper Cretaceous biotas of India. Origin, Sedimentation and Tectonics of Late Mesozoic to Early Cenozoic sedimentary basins at the Eastern Margin of the Asian Continent. International Geological Correlation Project-245 International Symposium, Fukuoka, Japan, Abstracts with Program: 72.

Sarjeant, W. A. S. 1981. In the footsteps of the dinosaurs. Explorers Journal 59: 164–71.

Schult, M. & Farlow, J. O. 1992. Vertebrate trace fossils. 34–63, In G. G. Maples, & R. R. West, (eds.), Trace Fossils. Paleontological Society Short Courses in Paleontology No. 5.

Sternberg, C. H. 1932. Dinosaur tracks from the Peace River British Columbia. Bulletin National Science Museum Canada 68: 59–85.

Tandon, S. K., & Sood, A. 1991. Paleoenvironments of Deccan volcanism associated dinosaur bearing Maastrichtian sequences of Jabalpur, Central India. *Origin, Sedimentation and Tectonics of Late Mesozoic to Early Cenozoic Sedimentary Basins at the Eastern Margin of the Asian Continent.* International Geological Correlation Project-245 International Symposium, Fukuoka, Japan. Abstracts with Program: 41.

Thulborn, R. A. 1989. The gaits of dinosaurs. *In* D. D. Gillette, & M. G. Lockley, (eds), *Dinosaur Tracks and Traces* (New York: Cambridge University Press), pp. 39–50.

    1990. *Dinosaur Tracks* (London: Chapman Hall).

Weishampel, D. B., Dodson, P. & Osmólska, H. (eds). 1990.

The Dinosauria. (Berkeley: University of California Press).

Western, D., Moss, C., & Georgiadis, N., 1983. Age estimation and population age strcture of elephants from footprint dimensions. *Journal of Wildlife Management* 47: 1192–97.

Whyte, M. A. 1982. Life and death of the Lower Carboniferous crinoid *Parazeacrinites Konincki* (Bather). *Neues Jahrbuch für Geology und Palaontologie Monatshefte* 1982 :279–296.

Wilson, W. D. 1969 Footprints in the sands of time. *Gems and Minerals* June: 25.

24 Summary and prospectus

KENNETH CARPENTER,
KARL F. HIRSCH, AND
JOHN R. HORNER

The chapters in this book show that large advances have been made in the study of dinosaur eggs, eggshells, and babies. Once a curiosity, dinosaur eggs are fast becoming a recognized subject of scientific study. We now know a considerable amount about the microscopic and macroscopic structure of eggs, who some of the egglayers are, how nests were structured, the spacing of nests in colonial nesting sites, and the behavior of the babies upon hatching. However, the continued discovery of new eggs, new egg types, and new nesting areas reveals that there is still much to be learned.

Eggs and eggshell, due to their nature, their fragile structure, and their biomineralogical composition, are more easily affected by environmental factors and diagenesis than are bones and teeth. For this reason, when studying eggs, eggshells and nests, we must consider the different events to which the specimens may have been subjected prior to burial, during burial, and being exposure to erosion. Some of these factors are presented in Table 24.1.

Studies of eggs, eggshell structure, and nests

Egg studies have come a long way due to advancements in technology. From the original studies of Gervais (see Chapter 2) and van Straelen (1925, 1928) using only normal light and polarized light microscopes (PLM), studies today also use scanning electron microscopy (SEM), computer tomography, and isotope, trace mineral, and amino acid analysis. A new technology showing promise for understanding diagenesis of eggshells is cathodoluminescence. The study of eggs and eggshells is very complex, and as many as possible of these modern techniques should be used because they complement each other. For techniques in preparing eggshell for both light and scanning electron microscopy, see Quinn (in press).

From the various papers in this volume, we may conclude that the ideal study of dinosaur eggs is multifaceted and should include (from the most general to most specific)

1. Nest environment
2. Sedimentological and chemical analyses of the nest and area adjacent to the nest
3. Type, shape, and size of the nest
4. Associated flora and fauna within the nest
5. Arrangement of eggs within the clutch
6. General morphology (macrostructure) of the egg and eggshell, including egg shape and size, shell thickness, ornamentation of shell surface, and pore-opening pattern
7. Egg histostructure (microstructure), which includes shell units, organic core, eisospherite, mammillae, wedges, prisms or continous layer, pore system type, and texture of shell (sequence and composition of horizontal ultrastructures zones), which includes the basal plate groups, aragonite or calcite radial ultrastructure, and tabular or squamatic ultrastructure in the crystallite aggregates
8. Biochemical analysis of the shell, including amino acids and rare elements.

The histostructural studies are best done in radial views of the shells using both polarized light microscopy and scanning electron microscopy. However, a large data base from many specimens is necessary before making conclusions or identifications. A single PLM thin section or SEM view will not necessarily be representative of the eggshell. For example, not every radial view will show the pore canal in its entirety. In addition, the microstructures seen in thin section can change as the shell thickness is reduced by lapping. Ideally, samples from the polar and equatorial regions of an egg should be studied in order to prepare a taxonomic description. In this manner misidentifications can be avoided.

Misidentifications of eggs are common, especially in the older literature. The "oldest" vertebrate egg described by Romer and Price (1939) lacks any descriptions of microstructures of a calcareous egg, and it cannot be a pliable or parchmentlike egg (Hirsch 1979). Erickson's (1978) crocodilian egg from the Upper Cretaceous is in all probability a stomach stone from a deer (Hirsch, 1986).

Eggshell taxonomy

We advocate that a single classification scheme be used for dinosaur eggs. This will make communication about eggs more exact and less confusing. At present, the best system is the parataxonomical classification of Zhao combined with the structural classifi-

Table 24.1. *Factors that influence the preservation of dinosaur eggs*

Preburial period
Egg structure (macro- and microstructures)
Shell pathologies
Amount of calcium absorption by embryo
Condition of egg (fertile, infertile, incubated or hatched)
Nest type (buried in substrate, laid on top of substrate, within vegetation mound)
Nest environment (riverbank, beach, upland, islands)
Biological (e.g., trampling of eggshell, predator–scavenger activity)
Local pH (decay of organic matter in egg, shell membrane, organic matter within crystalline shell layer, or decay of vegetation in nest (Ferguson 1981)
Transport abrasion
Burial period
Matrix (fluvial, pyroclastics, paleosols, etc.)
Rate of sedimentation (slow, paleosols, rapid, flood)
Diagenesis
Pressure (crush or deform eggshell)
Permineralization (include infilling of voids; trace elements from ground water)
Lithification of matrix (steinkerns of eggs with impressions of mammillary layer of shell)
Recrystallization (e.g., loss of shell layer details; herringbone pattern produced)
Replacement (partial to complete loss of structural details)
Postburial period
Weathering
Erosion
Transport
Acidic soil

cation of Mikhailov (see "Introduction" for discussion and references). This system is gaining support, as shown in Chapters 7, 10, 11, and 12. In order to avoid confusion, however, we advise against incorporating a dinosaur name into the parataxonomic classification. We also advise against mixing the unit structure and pore canal systems in a morphotype name.

Baby dinosaurs

We may glean several criteria for recognizing baby dinosaurs from Chapters 14–23. Besides their very small size, they also have a large head relative to body size, an arched braincase, a large orbit relative to skull size, a short muzzle, considerably fewer teeth than the adults, a short tail relative to body length, unfused centra and neural arches, neural canals proportionally larger than in adults, coarsely fibrous periosteal bone, spongy metaphyses, and among some species the apparent lack of ossified body armor.

The future

The washing of sediments and the screening for microfauna, combined with an increased awareness of eggshells, are now producing more and more specimens of thin, tiny eggshell fragments. In addition, some eggshell-like fragments have also been found, confronting us with a new problem: Are these fragments of a new type of shell structure, diagenetically altered eggshell, or some type of naturally occurring inorganic structure?

Because not all dinosaurs were large, we expect that thin dinosaur eggshells will be found (e.g., possibly the eggs described by Buckman 1859, see "Introduction"). It is also possible that evidence will be found for dinosaur eggs with soft or pliable shells. Extant turtles are known to produce eggs that have rigid, pliable and parchmentlike shells; therefore it is not unreasonable that other egg types were produced by dinosaurs. This possibility is based on the observation by one of us (Hirsch) that the French "sauropod" eggs break along the discrete shell-unit boundaries. The "sutures" of adjacent shell units can sometimes be seen. The "loose" suturing of the shell units is similar to that found in extant turtle eggs with a pliable shell. This similarity suggests that these "sauropod" eggs may not have been as rigid during life as they appear to be in their fossilized state. The shell units only abutted together and were bound together by the shell membrane in a manner similar to eggs of the snapping turtle (*Chelydra*). In a dinosaur egg with a pliable shell, the decay of the organic matrix could result in a pile of disaggregated shell units. These shell units might be found in the finer fraction of screened sediments from nesting sites.

Assigning eggs and eggshells or egg morphotypes to specific dinosaur species is still fraught with difficulties. The only positive identifications are based on the presence of embryos or associated hatchlings. As

Figure 24.1. Comparison of baby and adult dinosaur skulls drawn to same skull length. **A, B**. *Coelophysis*; (A) baby (MNA V3318), (B) adult (YPM 41196). **C, D**. *Mussaurus*; (C) nestling (modified from Bonaparte & Vince, 1974), (D) adult (modified from Casamiquela, 1980). **E, F**. *Dryosaurus*; (E) nestling (CM 11340), (F) adult (CM 3392). Scale of A, C = 1 cm; B, E, F = 2 cm: D = 10 cm.

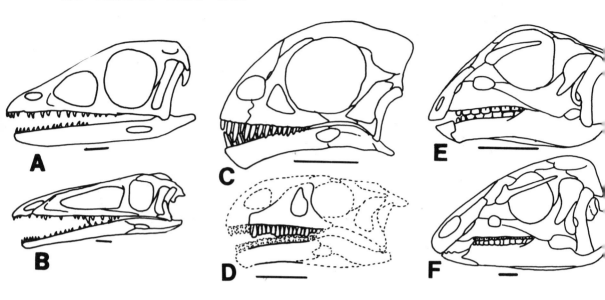

Figure 24.2 Comparison of baby and adult hadrosaur skulls drawn to same skull length. **A–C**. *Maiasaura*; (A) embryo (MOR), (B) nestling (MOR), (C) adult (MOR). **D**. ?*Saurolophus* sp. (reconstructed from Taquet, 1992). **E**. *Saurolophus* (modified from Rozhdestvenky, 1965). **F–H**. *Hypacrosaurus*; (F) embryo (RTMP), (G) nestling (MOR 548), (H) adult (MOR 549). Scale of A, B, D, F, G = 2 cm; C, E, H = 10 cm.

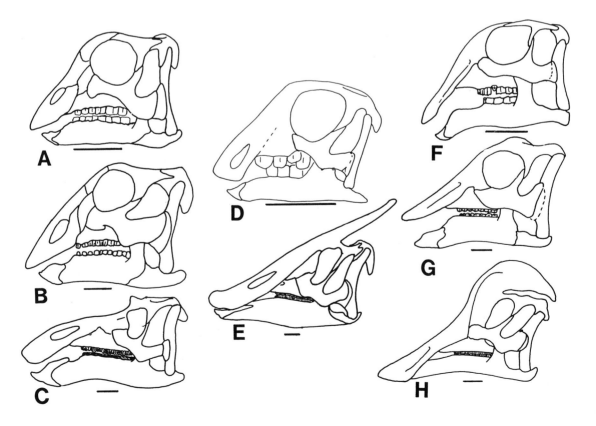

the data base for eggs and babies grows, it will be possible to identify isolated eggs and eggshell pieces more positively. The Romanian eggs described in Chapter 6 are a good example of a difficulty yet to be resolved. Originally the eggs were assigned to the sauropod "*Magyarosaurus*" because of their resemblance to the large "sauropod" eggs from France (Grigorescu et al., 1990). After the discovery of a few embryonic bones near the egg clutches, the eggs were then assigned to the hadrosaur *Telmatosaurus transsylvanicus* (see Grigorescu et al. Chapter 6). However, the presence of nearby embryonic bones may be coincidental, because the shell microstructure most closely resembles that of the "sauropod" shell found in France rather than that of the hadrosaur shell from Montana. This conflict between identity of the embryonic remains and the shell microstructure could be a simple misidentification of the eggs as hadrosaur, or it could reveal that the diversity of eggshell type in the Hadrosauria is much more diverse than hitherto realized.

Diagenesis is an area long neglected. As noted in the "Introduction," trace elements that are added by diagenesis can affect interpretations made on the diet of the parent. In unpublished results one of us (Hirsch) has noted that diagenesis may not only differ between different nesting sites, but also within the same clutch. Perhaps most surprising is the possibility that there may be drastic variations within one shell fragment. Some of

these differences may be linked to events occurring within the nest prior to burial; however this area needs to be explored. Comparisons with nests of extant birds and reptiles will provide many clues.

Parental care will undoubtedly continue to be a subject of discussion, especially concerning which dinosaurs were altricial and which precocial. A new clue might be provided by Bond, Board, and Scott (1988). They note that for altricial birds there is less physical change in the eggs than in the eggs of precocial birds. They suggest that the underdevelopment of altricial hatchling bones would cause less calcium depletion from the egg. The eggshells of *Maiasaura* (presumably altricial) and *Orodromeus* (presumably precocial) could provide a test.

Other areas of future study include the use of shape analysis (Chapman, 1990; Chapman & Brett-Surman, 1990) on the skulls of various babies (Figs. 24.1, 24.2, & 24.3). For some as yet unexplained reason, saurischian babies are less common than ornithischian babies. At present we do not known if there were both precocial and altricial hatchlings among the saurischians. Once this has been been determined, then we may better understand these forms of development among the Dinosauria.

Somewhat related is the study of locomotion in the ontogeny of the dinosaurs. Certainly precocial hatchlings were morphologically better suited for loco-

Figure 24.3 Comparison of baby and adult Ceratopsia skulls drawn to same skull length. **A–C.** *Psittacosaurus*; (A) embryo (modified from Coombs, 1982), (B) nestling (modified from Coombs 1982), (C) adult (modified from Coombs 1982). **D, E.** *Bagaceratops*; (D) nestling (modified from Maryanska & Osmólska, 1975), (E) adult (modified from Maryanska & Osmólska, 1975). **F–H.** *Breviceratops*; (F) embryo (modified from Maryanska & Osmólska, 1975), (G) nestling (modified from Kurzanov, 1990), (H) adult (modified from Kurzanov, 1990). **I, J.** *Protoceratops*; (I) nestling (modified from Taquet, 1992), (J) adult (modified from Brown & Schlaikjer, 1940). Scale of A, B = 1 cm; C–I = 2 cm; J = 10 cm.

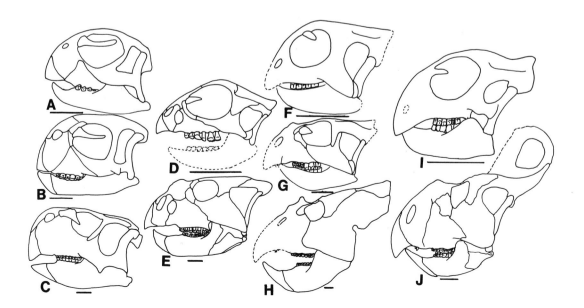

motion (in and out of the nest) than altricial hatchlings, but whether they were as developed as their parents is not yet known.

References

Bonaparte, J. F., & Vince, M. 1979. El Hallazgo del primer nido de dinosaurios Triasicos (Saurischia, Prosauropoda), Triasico Superior de Patagonia, Argentina. *Ameghiniana.* 16: 173–82.

Bond, G., Board, R., & Scott, V. 1988. A comparative study of changes in the fine structure of avian eggshell during incubation. *Zoological Journal of the Linnean Society* 92: 105–13.

Brown, B., & Schlaikjer, E. 1940. The structure and relationships of *Protoceratops. Annals of the New York Academy of Science* 40: 133–266.

Casimiquela, R. 1980. La presencia del genero *Plateosaurus* (Prosauropoda) en el Triasico superior de la Formacion El Tranquilo, Patagonia. *Actas II del Congreso Argentino de Paleontologia y Bioestratigrafia* 1: 143–58.

Chapman, R. 1990. Shape analysis in the study of dinosaur morphology. *In* K. Carpenter & P. Currie (eds.) *Dinosaur Systematics: Approaches and Perspectives.* (New York: Cambridge University Press), pp. 21–42.

Chapman, R., and Brett-Surman, M. 1990. Morphgometric observations on hadrosaurid ornithopods. *In* K. Carpenter & P. Currie (eds.), *Dinosaur Systematics: Approaches and Perspectives.* (New York: Cambridge University Press), pp. 163–77.

Coombs, W. 1982. Juvenile specimens of the ornithischian dinosaur *Psittacosaurus, Palaeontology* 25; 89–107.

Erickson, B. 1978. An ammiote egg from the Upper Cretaceous of Wyoming. *Scientific Publications of the Science Museum of Minnesota* 4: 1–15.

Ferguson, M. 1981. Extrinsic microbial degradation of the Alligator eggshell. *Science* 214: 1135–7.

Grigorescu, D., Seclaman, M., Norman, D. B., & Weishampel, D. B. 1990. Dinosaur eggs from Romania. *Nature* 346: 417.

Hirsch, K. 1979. The oldest vertebrate egg? *Journal of Paleontology* 53: 1068–84.

1986. Not every "egg" is an egg. *Journal of Vertebrate Paleontology* 6: 200–1.

Kurzanov, S. 1990. A new Late Cretaceous protoceratopsid genus from Mongolia. *Paleontological Journal* 4: 85–91.

Maryanska, T., & Osmolska, H. 1975. Protoceratopsidae (Dinosauria) of Asia. *Palaeontologica Polonica* 33: 133–75.

Quinn, E. In press. Preparing fossilized eggshell. *In* P. Leiggi, & P. May, *Vertebrate Paleontological Techniques* (New York: Cambridge University Press).

Romer, A., and Price, L. 1939. The oldert vertebrate egg. *American Journal of Science* 237: 326–9.

Rozhdestvensky, A. 1965. Growth changes in Asian dinosaurs and some problems of their taxonomy. *Palaeontologicheskii Zhurnal* 3: 95–109.

Taquet, P. 1992. *Dinosaures et Mammiferes du Desert de Gobi* (Paris: Muséum National D'Histoire Naturelle).

Straelen, V. Van 1925. The microstructure of the dinosaurian egg-shells from the Cretaceous beds of Mongolia. *American Museum Novitates* 173: 1–4.

1928. Les oeufs de reptiles fossiles. *Palaeontologica* 1: 295–317.

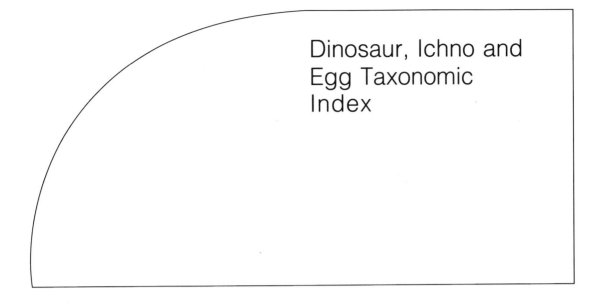

Dinosaur, Ichno and Egg Taxonomic Index